Antimonide-Related Strained-Layer Heterostructures

Optoelectronic Properties of Semiconductors and Superlattices

A series edited by *M. O. Manasreh*, Phillips Laboratory, Kirtland Air Force Base, New Mexico, USA

Volume 1
Long Wavelength Infrared Detectors
Edited by Manijeh Razeghi

Volume 2
GaN and Related Materials
Edited by Stephen J. Pearton

Volume 3
Antimonide-Related Strained-Layer Heterostructures
Edited by M. O. Manasreh

In preparation

Strained-Layer Quantum Wells and Their Applications
Edited by M. O. Manasreh

Structural and Optical Properties of Porous Silicon Nanostructures
Edited by G. Amato, C. Delerue and H.-J. von Bardeleben

The Optics of Semiconductor Quantum Wires and Dots: Fabrication, Characterization, Theory and Application
Edited by Garnett W. Bryant

Vertical Cavity Surface-Emitting Lasers and Their Applications
Edited by Julian Cheng

Characterization of Reduced Dimensional Semiconductor Microstructures
Edited by Fred H. Pollak

Long Wavelength Infrared Emitters Based on Quantum Wells and Superlattices
Edited by Manfred Helm

Semiconductor Quantum Well Intermixing: Materials, Properties and Optoelectronic Applications
Edited by E. Herbert Li

This book is part of a series. The publisher will accept continuation orders which may be cancelled at any time and which provide for automatic billing and shipping of each title in the series upon publication. Please write for details.

Antimonide-Related Strained-Layer Heterostructures

Edited by

M. O. Manasreh

*Phillips Laboratory
Kirtland Air Force Base
New Mexico, USA*

Gordon and Breach Science Publishers
Australia • Canada • China • France • Germany • India • Japan •
Luxembourg • Malaysia • The Netherlands • Russia • Singapore •
Switzerland • Thailand • United Kingdom

Copyright © 1997 OPA (Overseas Publishers Association) Amsterdam B.V. Published in The Netherlands under license by Gordon and Breach Science Publishers.

All rights reserved.

No part of this book may be reproduced or utilized in any form or by any means, electronic or mechanical, including photocopying and recording, or by any information storage or retrieval system, without permission in writing from the publisher. Printed in India.

Amsteldijk 166
1st Floor
1079 LH Amsterdam
The Netherlands

British Library Cataloguing in Publication Data

Antimonide-related strained-layer heterostructures. –
  (Optoelectronic properties of semiconductors and
  superlattices; v. 3)
  1. Heterostructures 2. Antimony compounds
  I. Manasreh, M. O.
  621.3'8152

ISBN 90-5699-544-8

CONTENTS

About the Series   ix

Foreword   xi

1 Introduction   1
   M. O. Manasreh

2 Structural and Electronic Properties of GaSb/InSb and InAs/InSb Superlattices   7
   S. Picozzi, A. Continenza and A. J. Freeman

3 Interfacial Disorder in InAs/GaSb Heterostructures Grown by Molecular Beam Epitaxy   55
   M. E. Twigg, B. R. Bennett, P. M. Thibado, B. V. Shanabrook and L. J. Whitman

4 Type-II (AlGa)Sb/InAs Quantum Well Structures and Superlattices for Opto- and Microelectronics Grown by Molecular Beam Epitaxy   95
   S. V. Ivanov and P. S. Kop'ev

5 Growth and Characterization of InAs/AlSb/GaSb Heterostructures   171
   F. Fuchs, J. Wagner, J. Schmitz, N. Herres and P. Koidl

6 Antimonide-Based Quantum Heterostructure Devices   235
   J. R. Meyer, J. I. Malin, I. Vurgaftman, C. A. Hoffman and L. R. Ram-Mohan

7 Mid-Infrared Strained Diode Lasers   273
   G. G. Zegrya

8 Antimonide-Based Mid-Infrared Quantum-Well Diode Lasers   369
   G. W. Turner and H. K. Choi

9 Mid-Wave Infrared Sources Based on GaInSb/InAs Superlattice Active Layers   433
   R. H. Miles and T. C. Hasenberg

10 InAs/InAs$_x$Sb$_{1-x}$ Type-II Superlattice Midwave
 Infrared Lasers 461
 Yong-Hang Zhang

Subject Index 501

*To
Hannah Grace, Sarah Margaret
and Ann Lee*

ABOUT THE SERIES

The series *Optoelectronic Properties of Semiconductors and Superlattices* provides a forum for the latest research in optoelectrical properties of semiconductor quantum wells, superlattices, and related materials. It features a balance between original theoretical and experimental research in basic physics, device physics, novel materials and quantum structures, processing, and systems—bearing in mind the transformation of research into products and services related to dual-use applications. The following sub-fields, as well as others at the cutting edge of research in this field, will be addressed: long wavelength infrared detectors, photodetectors (MWIR–visible–UV), infrared sources, vertical cavity surface-emitting lasers, wide-band gap materials (including blue-green lasers and LEDs), narrow-band gap materials and structures, low-dimensional systems in semiconductors, strained quantum wells and superlattices, ultrahigh-speed optoelectronics, and novel materials and devices.

The main objective of this book series is to provide readers with a basic understanding of new developments in recent research on optoelectrical properties of semiconductor quantum wells and superlattices. The volumes in this series are written for advanced graduate students majoring in solid state physics, electrical engineering, and materials science and engineering, as well as researchers involved in the field of semiconductor materials, growth, processing, and devices.

FOREWORD

There is increasing interest in antimonide-related heterostructures, due to their unique band structure alignment and for their optoelectronic applications. One of the most salient applications of what are called Sb-related type-II strained-layer superlattices is their use for long wavelength infrared detectors. Recently, there has been growing interest in these heterostructures for their applications as light sources, diode lasers, modulators, filters, switches, nonlinear optics, and field-defect transistors. Several major research groups have been heavily involved in aspects of this class of structures, ranging from theoretical proposals and growth issues to the fabrication of mid-infrared diode lasers. This is the first book to focus on antimonide-related topics, presenting to the reader the latest state-of-the-art results obtained by the best researchers in the field.

The reader will note that the chapters in this volume, while they cover a wide range of topics relating to antimonide-related materials, may be divided into two parts. Chapters 2 through 5 focus on the band structure calculation, growth, and materials characterizations using: (1) structural techniques such as X-ray and transmission electron microscopy; and (2) non-destructive techniques such as optical absorption, photoluminescence, and Raman spectroscopies. In chapters 6 through 10, theoretical proposals for various devices and experimental results of mid-infrared diode lasers are presented. Each chapter discusses unique features; together they form an integrated reference volume that is both accessible to the beginner and rigorous enough for the advanced researcher.

I acknowledge with thanks the efforts of the outstanding scientists and engineers who contributed to this volume: S. Picozzi, A. Continenza, A. J. Freeman, M. E. Twigg, B. R. Bennett, P. M. Thibado, B. V. Shanabrook, L. J. Whitman, S. V. Ivanov, P. S. Kop'ev, F. Fuchs, J. Wagner, J. Schmitz, N. Herres, P. Koidl, J. R. Meyer, J. I. Malin, I. Vurgaftman, C. A. Hoffman, L. R. Ram-Mohan, G. G. Zegrya, G. W. Turner, H. K. Choi, R. H. Miles, T. C. Hasenberg, and Yong-Hang Zhang.

CHAPTER 1

Introduction

M. O. MANASREH

Phillips Laboratory (PL/VTRP), 3550 Aberdeen Ave, SE, Kirtland AFB, NM 87117, USA

The most attractive feature of the antimonide-related materials is their band structure alignment. Due to the development of growth techniques such molecular beam epitaxy (MBE) and metal-organic chemical vapor deposition (MOCVD), one can design and grow antimonide heterostructures with both electrons and holes confined in the same layer (type-I heterostructure), such as AlSb/GaSb, or electron and holes confined in adjacent layers (type-II heterostructure), such as InAs/InGaSb. This class of materials has been the subject of various studies in recent years for many applications ranging from infrared lasers to very long wavelength infrared detectors. In this volume we present several chapters dealing with various aspects of the antimonide-related materials. Our intention in this volume is to provide the readers with the latest research that has been carried out by different groups, keeping in mind a balance between theoretical predictions and experimental measurements.

The construction of the band structure is the most fundamental property of heterostructures and quantum wells. It is the basis for understanding thermal, transport, optical, and electrical properties of these complex materials. Since the band offset of these materials is very important to many applications, several theoretical efforts have focused on the study of this parameter. Other theories are more concerned with the effect of the charge readjustment at the junction and with the role played by the dipole that forms at the interfaces. Chapter two is focused on the theoretical calculations of the band structure based on the density functional method within the local density approximation with the emphasis on the electronic properties of strained-layer GaSb/InSb and InAs/InSb systems. This chapter also points out the lattice mismatch and the growth direction and their influence on all the electronic energy levels and other important properties such as the band gap energy and the band line-up.

The growth of strained-layer antimonide-related heterostructures has been one of the major problems in recent years since the novel properties of these materials depend on the ability to grow thin layers with thickness and composition controlled on an atomic scale. Chapters three, four, and five are dealing in part with the growth issues of the antimonide related heterostructures. Interfacial disorder in InAs/GaSb heterostructures is the subject of chapter three. In forming an interface, the growth temperature needs to be sufficiently high to allow the adatoms arriving on the surfaces to assemble into a flat monolayer before the next monolayer begins. At the same time the temperature cannot be so high as to promote diffusion below the surface and into the bulk. Any deviation from an interface that is perfectly smooth and compositionally abrupt is known as interfacial disorder. All aspects of this type of disorder are undesirable in opto-electronic devices. For examples, interfacial roughness is thought to have a deterious effect on carrier mobilitites and interfacial intermixing can alter the carefully engineered band gaps. This chapter, in particular is dealing with the control of interfacial bonding, disorder, and morphology in InAs/GaSb superlattices using high resolution transmission electron microscopy and other structural characterization techniques.

Type-II (AlGa)Sb/InAs quantum wells and superlattices for optoelectronic and microelectronic applications are the subject of chapter four. In this chapter, detailed discussions of the MBE growth, p- and n-type doping, interface formation, electronic and transport properties of (AlGa)Sb/InAs based epilayers are presented. In addition, devices based on these structures such as mid-wave infrared lasers, resonant tunneling transport, and field-effect transistors are proposed and discussed. Additional results on the growth and characterization of InAs/AlSb/GaSb heterostructures are presented in chapter five. The objective of this chapter, however, is to provide the readers with an analysis of the structural properties and interfacial bonding using a high-resolution X-ray diffraction method and Raman and far-infrared spectroscopies. This chapter provides additional comprehensive characterization of the electronic band- and subband-structures. The interband transitions presenting the fundamental energy gap were studied using infrared photoluminescence technique. This technique has shown that band-to-band transitions can be observed in InAs/AlSb heterostructures. Other techniques to measure a weak absorbance due to type-II superlattices, namely calorimetric absorbance spectroscopy and Fourier-transform photoluminescence excitation, are proven to be powerful techniques in characterizing this class of materials. Other results obtained from techniques such as the optical anisotropy and photocurrent spectroscopy are also presented in this chapter.

The rest of the chapters in this volume are devoted to antimonide-based devices that exhibit advantages over other III–V semiconductor systems. It is shown in chapter six that these advantages become particularly valuable when one moves from the near-infrared to the mid-wave and long-wave infrared spectral regions, where the need for high-performance electro-optic devices has become especially acute for such commercial and military applications as remote chemical sensing, infrared spectroscopy, laser surgery, multi-spectral detection, and future ultra-low-loss fiber communications. This chapter reviews several specific device classes for which the degree of control over the interband and intersubband optical and electronic properties may be used to a particular advantage. These include intersubband-based electro-optical modulators for both normal-incident and waveguide-mode mid-wave and long-wave infrared applications, infrared frequency-conversion devices in which the phase-matching condition can be imposed electrically by varying a bias voltage, and improved mid-wave infrared lasers based on type-II multiple quantum wells as well as a novel type-II interband cascade process. These "wavefunction engineered" devices have been modeled in this chapter using electronic band structures and electro-optical properties derived from 8-band finite element method calculations.

One of the most desired fundamental properties of optoelectronic devices based on interband and intersubband transitions in quantum wells and superlattices is that the photo-generated carriers remain excited until they are collected, or before radiative recombination occurs. However, there are several recombination processes, such as Auger recombination, that affect the photo-excited carriers. The Auger recombination rate is thoroughly studied in chapter seven for mid-infrared strained diode lasers based on InAlAsSb/GaSb structures. Up to now, there has been no lasers of this kind emitting in the wavelength range of $\lambda \geqslant 4\,\mu\mathrm{m}$ at room temperature. In addition to the Auger recombination rate study, this chapter is focused on the study of fundamental physical processes governing the operation of mid-infrared lasers based on strained quantum wells, which include the limits of their operation at room temperature and variety of new effects due to carrier-interface interaction. These and other processes discussed in this chapter are to be taken into account in creating mid-infrared diode lasers. A proposal for a new fundamental approach to developing mid-infrared lasers at room temperature is presented in this chapter as well.

As a practical application of the antimonide-related heterostructures, chapter eight presents experimental results for mid-infrared quantum-well diode lasers. Generally speaking, semiconductor diode lasers emitting in the 2–5 µm spectrum region would be very useful for variety of applications such strong molecular absorption lines and important atmospheric transmission windows. These diode lasers are far preferable as compared to mid-infrared solid state lasers or gas lasers because they are compact, efficient, reliable, and potentially inexpensive. This chapter provides an overview of the current status of antimonide-based mid-infrared quantum well lasers with the emphasis on type-I structures such as GaInAsSb/AlGaAsSb and InAsSb/InAlAsSb quantum wells, which have been grown by the MBE technique. In addition, this chapter describes the properties of antimonide-based materials relevant to the laser performance, discusses the effects of strain on the performance of the mid-infrared lasers, presents the growth and characterization of quantum well structures, describes the performance of the quantum well diode lasers, and gives future directions to improve the performance of this type of lasers.

One of the strained-layer superlattices, namely InAs/InGaSb, has been a subject of numerous studies in recent years. It was studied first as a potential structure for long wavelength infrared detectors in the spectral region of 8–12 µm. The interest in InAs/InGaSb continues as a potential structure for mid-wave infrared sources. Chapter nine briefly reviews the growth of InGaSb/InAs superlattices with AlSb/InAs superlattices employed as clads, and summarizes the intrinsic and extrinsic properties salient to diode lasers. The latest laser results, in addition to practical advantages relative to competing injection devices, are also presented in this chapter.

It is well known that the effective band gap of type-II superlattice can be narrower than any of the constituent materials. In addition, type-II superlattices can be further divided into two categories based on the band gap alignment. The first category is known as a broken-gap alignment, such as InAs/InGaSb superlattices, and the second one is known as staggered alignment, such as InAs/InAsSb superlattices. Chapter ten is focused on InAs/InAsSb staggered type-II superlattices and their application to mid-wave infrared lasers. Owing to a large valence band offset between InSb and InAs, this material system has the following advantages for mid-wave infrared laser applications: (1) Only a small amount of As is needed in the $InAs_xSb_{1-x}$ layers to reach 5 μm. (2) The compressive strain induced in the InAsSb layers and their layer thicknesses can be properly designed to reduce non-radiative Auger recombinations. (3) This structure has an average lattice constant close to that of the InAs substrate, upon which a lattice-matched group-V alloy $AlAs_{0.16}Sb_{0.84}$ can be used as cladding layers, providing both optical and hole confinement. Theoretical consideration and modeling of InAs/InAsSb superlattices, their detailed growth and material characterization, results of cw and quasi-cw optically pumped mid-wave infrared lasers fabricated from this structure are all discussed in chapter ten.

In conclusion, several chapters dealing with various issues related to the antimonide-based heterostructures are presented in this volume. We maintained a balance between theoretical calculations, growth issues, materials characterizations, and device issues and applications. These chapters provide a background on the development of infrared applications of antimonide-based heterostructures and how these structures evolved from the theoretical ideas to the point where the commercial production of mid- and long-wave infrared diode lasers becomes possible.

CHAPTER 2

Structural and Electronic Properties of GaSb/InSb and InAs/InSb Superlattices

S. PICOZZI[1], A. CONTINENZA[1] and A. J. FREEMAN[2]

[1] INFM – Istituto Nazionale di Fisica della Materia, Dipartimento di Fisica, Università degli Studi di L'Aquila, 67010 Coppito (L'Aquila), Italy; [2] Department of Physics and Astronomy and Material Research Center, Northwestern University, Evanston, IL 60208, USA

2.1. Introduction	08
2.2. Survey of Experimental Work	11
2.3. Methodology: Density Functional Theory and its Implementations	14
2.3.1. Fundamentals of density functional theory	14
2.3.1.1. The density functional formalism	15
2.3.1.2. The Kohn–Sham equations	16
2.3.1.3. The exchange-correlation energy and the Local Density Approximation	17
2.3.2. Overview of density functional implementations	18
2.3.3. The full potential augmented plane wave method	20
2.4. Ultrathin Strained-Layer Heterostructures: GaSb/InSb and InAs/InSb Superlattices	23
2.4.1. Bulk binary constituents	23
2.4.1.1. Structural properties	25
2.4.1.2. Electronic properties	26
2.4.2. The 1×1 superlattices	29
2.4.2.1. Structural properties: the importance of growth direction and strain conditions	31
2.4.2.2. Electronic properties	33
2.4.3. The 3×3 superlattices: valence band offsets	41
2.4.3.1. The dependence of the band line-up on growth axis and strain conditions	44
2.4.3.2. The InAs/GaSb interface: validity of the transitivity rule	48
2.5. Conclusions	50
Acknowledgments	51
References	51

ACRONYMS

MBE	Molecular Beam Epitaxy
MOCVD	Metal Organic Chemical Vapor-phase Deposition
FLAPW	Full Potential Linearized Augmented Plane Wave
LDA	Local Density Approximation
VBO	Valence Band Offset
VBM	Valence Band Maximum
CBM	Conduction Band Minimum
CF	Crystal Field
s.o.	spin orbit

2.1 INTRODUCTION

Ternary systems based on III–V semiconductors constitute an important class of new materials of highly relevant scientific interest, since their structural, electronic and transport properties can be opportunely tuned as a function of the constituent materials, composition, strain or doping. This flexibility of their electronic properties has opened new perspectives in the technological field: disordered alloys, superlattices and heterojunctions have recently been proposed as fundamental components in many applications, such as laser diodes, infrared detectors and non-linear optical devices [1].

In the last few years, the development of experimental growth techniques (such as molecular beam epitaxy (MBE) or metal organic chemical vapor phase deposition (MOCVD)) and accurate spectroscopy methods (such as synchrotron radiation photoemission, photoluminescence, photoconductivity) allowed the synthesis and characterization of artificial ultrathin strained-layer superlattices, without misfit and dislocations.

On the other hand, several theoretical approaches have been developed to study real, rather than model, materials and to give precise answers and interpretations to the experimental data. Very often, they have even managed to go further by proposing new structures and predicting new materials with interesting properties. Of course, all this has been possible thanks to the great advances made in the development of supercomputers and fast algorithms that allow the numerical solution of complex equations without idealizing or simplifying any fundamental physical phenomena.

Understanding the electronic properties of a given material is the basis for the full comprehension of its thermal, transport and electric properties; thus, band structure calculations provide a powerful tool to explore real complex materials, giving important information such as density of states, energy level transitions and carrier effective masses. Particular attention has been devoted to the electronic structure of the interface between two semiconductors. The relevant quantity in this case is the relative position of the band gaps (i.e. the difference between the valence topmost level and the lowest conduction state) which determines the potential steps (the so called "valence band offset" (VBO) and "conduction band offset") that the different charge carriers (holes and electrons) have to overcome in order to permit conductivity across the interface. The relative alignment of the potential in the two materials in contact is determined by some intrinsic bulk properties (i.e. the position of the conduction band with respect to the vacuum level) and by the charge readjustment at the interface.

Many theoretical efforts have focused on the study of the heterojunction band offset. In analogy with the Schottky model [2] for the metal–semiconductor interface, some theories make use of the vacuum level as a common reference energy level, to which the energy band position can be related. Within this framework, we mention the original electron affinity rule [3], the Harrison model [4] (which used atomic energy levels as reference for the valence and conduction band edges) and the model solid state approach by Van de Walle and Martin [5], in which the reference level is provided by the electrostatic average potential of the superposition of neutral atomic overlapping charge densities.

Other theories are more concerned with the effect of the charge readjustment at the junction and with the role played by the dipole that forms at the interface. In this class, we can include the so called "charge neutrality point" [6,7], in which Tersoff derives the concept of "midgap energy point", i.e., the energy level in the gap of each semiconductor, for which the character of the metal induced (or semiconductor induced) gap state changes from valence-like to conduction-like. Filling states above or taking out electrons below this level results in a charge imbalance and therefore in a dipole; the relative alignment of these levels in the two semiconductors gives directly the band offset. A similar approach is the one proposed by Lambrecht and Segall [8] which combines first principles calculations for the bulk semiconductors with a study of the interface bonds in terms of a bond-orbital model. Another important theory was proposed by Baroni *et al.* [9]; in this model the actual interface (viewed as a perturbation with respect to a suitably chosen reference periodic system, the so called "virtual crystal") is treated within linear response theory and provides a very accurate description of the electronic structure at the heterojunction.

In our calculations, we use a method which takes into account bulk properties as well as interface properties, and is conceptually very close to the main idea underlying the X-ray photoelectron spectroscopy experimental method to determine the offsets in solids. This procedure, whose details will be given in Section 2.4.3, has been successfully used to predict the valence band line-up in different heterojunctions [10,11]. In the present review, we examine III–V strained layer superlattices (in particular GaSb/InSb and InAs/InSb systems), focusing on what mainly influences the electronic properties of interest and following a theoretical–computational approach based on the widely used density functional theory within the local density approximation (LDA) [14,15]. The computational method used is the full potential linearized augmented plane wave method (FLAPW) [16,17], which is one of the most accurate and precise schemes for total energy estimates and morphological and electronic structure determinations in solids. We use a "super-cell" geometry [13], in which we simulate the superstructure by means of a definite number of alternating layers of different materials with a fixed periodicity in a certain direction.

In the lattice mismatched ternary structures, the strain affects all the electronic energy levels. In particular, it influences important properties, such as the band gap energy and the band line-up: the appreciable difference between the lattice parameters of the two constituents (about 6% in our systems) is thus an additional degree of freedom (compared to lattice matched systems) for band structure tuning. The growth direction is another important parameter to be taken into account, in order to model the electronic energy levels and, hence, the electronic properties of interest in such structures. This is immediately clear if we consider the ternary Brillouin zone as obtained starting from the fcc one (typical of III–V semiconducting constituents) through folding operations: in fact (as it will be largely explained in Section 2.4.2), repulsion effects between electronic levels originate from these operations. Thus, the obvious difference between folding for [001] and [111] directions immediately reflects on the band structure for differently ordered superlattices. We point out that the interest in the [111] ordered structures has been renewed by some recent experimental studies [12], in which a spontaneous (energetically favoured) ordering was observed along the [111] direction, during vapor-phase growth.

The chapter is organized as follows:

- Part 2.2 presents an overview of recent experimental work focused on the III–V semiconducting materials.
- Part 2.3 discusses the theoretical approach followed in our calculations and illustrate density functional theory and the Full Potential Augmented Plane Wave method.
- Part 2.4 shows the results of a study of III–V strained-layer superlattices, grown along the [111] and [001] direction. This part is divided in three sections, regarding the binary constituents, the ultrathin 1×1 superlattices and the GaSb/InSb and InAs/InSb heterojunctions, respectively.
- Part 2.5 summarizes our results and draws some conclusions.

2.2 SURVEY OF EXPERIMENTAL WORK

The spectacular progress made in superlattices and interface physics is due in part to great advances in epitaxial deposition techniques, such as liquid phase epitaxy (LPE) [18], MBE [19] and MOCVD [20]. Thanks to these sophisticated synthesis methods, it is now possible to obtain experimentally the pseudomorphic growth of many artificial structures with a high level of purity and crystallinity. In parallel with the synthesis of these ternary systems, experimental techniques for their structural, electronic and optical characterizations were developed.

Particular attention was devoted to the band line-up measurements. The first investigations, performed with transport techniques (such as the current–voltage (I–V) and capacitance–voltage (C–V) characteristics), presented the intrinsic problem of performing averages in space, thereby producing errors thereby due to the high localization of interface properties [21]. Other methods for the determination of the band line-up were developed, generally based either on photoemission or optical techniques. The practical use and availability of synchrotron radiation facilities and the highly localized character of the band discontinuities explains the frequent and successful use of photoelectron spectroscopy [22]. On the other hand, the study of absorption and emission phenomena associated with quantized levels in quantum well systems [23] gave high accuracy to the evaluation of the valence band offset. Somewhat intermediate in accuracy between the transport and optical techniques is deep-level transient spectroscopy (DLTS) [24].

In order to stress the great attention devoted to artificial systems based on III–V semiconductors, and because of their importance for comparison with results of our calculations, we now briefly discuss some recent experimental studies focused on the materials examined in this chapter (ternary alloys and ordered superlattices based on Ga, In, As and Sb).

In the last decade, strained-layer III–V heterostructures have repeatedly been proposed as good candidates for promising applications, such as lasers [25,26] since the heavy hole effective masses can be reduced by the biaxial strain. As already pointed out, recent advances in the preparation techniques allowed pseudomorphic crystal growth (without misfit dislocations) even in highly lattice-mismatched structures (such as those examined here). It was thus possible to prepare InSb/GaSb quantum well structures by atmospheric pressure metal–organic vapor phase epitaxy (MOVPE) [27]. Their optical and structural characterization through photoluminescence techniques showed sharp emission peaks for well thicknesses varying in the 0.35–0.88 nm range. Only for quantum well thicknesses above 0.9 nm did the broadening of photoluminescence transitions suggest island growth or dislocation formation. This confirms that below a certain critical thickness, "atomically abrupt" interfaces (i.e. the interface region is thin with respect to the carrier diffusion lengths) represent a solid reality and not only a useful approximation in theoretical schemes; furthermore, recent transmission electron micrographs have shown that often the interface region is extremely sharp and can really be made of just two atomic planes.

The $In_xGa_{1-x}Sb$ system has been drawing considerable interest in view of its possible applications for detection devices in the 1.5–6 µm spectral region [28–31]. For example, Qian and Wessels [32] focused on the $In_xGa_{1-x}Sb/GaSb$ (alloy/binary type) system, determined the valence band offset from the energy dependence of photoluminescence emission peaks on well thickness, and used these results as input data for calculations based on a standard finite square-well model (as in Ref. [27]). Following this procedure, they obtained a value for the band line-up, which is consistent with that evaluated through temperature dependent photoluminescence measurements of the activation energies (which depend on the band offset of the quantum wells). In particular, they found a valence band offset of 0.037 eV for the $In_{0.18}Ga_{0.82}Sb/GaSb$ system. Furthermore, the optical properties of strained-layer $In_{0.18}Ga_{0.82}Sb/GaSb$ single and multiple quantum wells have been studied in a recent work [32], focusing on the dependence of the emission energy on well thicknesses. The band-offset was also determined, starting from measurements of the thermal quenching of the luminescence. A similar $In_xGa_{1-x}Sb/GaSb$ system ($0.19 < x < 0.25$), grown by MOCVD, was studied using photoluminescence and photoconductivity techniques [36].

Another alloy/binary type system, the $In_{0.18}Ga_{0.82}Sb/InAs$ superlattice was proposed [34,35] for long-wavelength infrared – 10–12 μm – devices; some important properties of this system (such as the optical absorption coefficient) were found to be as good as in the traditionally used $Hg_xCd_{1-x}Te$ alloy. Within this framework, Miles et al. [33] have succeeded in growing the InAs/InGaSb system by MBE.

Much interest has also been devoted to $InAs_{1-x}Sb_x$ [37] as a new material for infrared applications (such as long-wavelength infrared photodiodes [38]). Osbourn [39] suggested $InAs_{0.4}Sb_{0.6}/InAs_xSb_{1-x}$ ($x \leqslant 0.6$) as a suitable infrared detector material in which strain is used to reduce the band-gap. The spontaneous ordering along the [111] direction of this alloy was experimentally observed [40,41] through transmission electron diffraction [43]; furthermore, the direct band-gap was recently determined through photoluminescence and photoconductivity response measurements on an MBE grown $InAs_{0.62}Sb_{0.38}/InAs_{0.54}Ga_{0.46}$ sample [43]. In this case, theoretical schemes provided a fundamental tool in order to explain experimental results; in fact, the band-gap reduction (with respect to the average of the band gaps of the two binary constituents) observed for this system was interpreted as a consequence of CuPt type ordering theoretically predicted in Ref. [42].

Unfortunately, ultrathin 1×1 and 3×3 superlattices have not yet been realized (at least to our knowledge) and thus our theoretical predictions cannot be directly compared with experimental data regarding their electronic and transport properties. However, in the following Sections, some of our theoretical results will be compared with those experimentally obtained for similar structures (see, for example, Section 2.4.3 for details), providing good tests of our predictions.

Hopefully, the continuous search for new materials, the progress made in crystal growth techniques and the recent theoretical interests focused on Sb-related compounds will, in the near future, encourage experimentalists to grow more and more sophisticated strained layer superlattices, such as those examined in this chapter.

2.3 METHODOLOGY: DENSITY FUNCTIONAL THEORY AND ITS IMPLEMENTATIONS

2.3.1 Fundamentals of Density Functional Theory

The determination of the electronic properties in a solid is an important application of quantum theory to many-body systems. Strictly speaking, one should consider both the electrons and the ions during their motion; however, the electrons velocity (about 10^8 cm/s) is several orders of magnitude greater than that of the ions (about 10^5 cm/s). Thus, the Born–Oppenheimer approximation is generally assumed to be valid and the electrons are therefore considered to instantaneously distribute around the slowly moving nuclei. From a formal point of view, this corresponds to neglecting the ionic kinetic energy term in the Hamiltonian which describes the electronic system, since the nuclei are assumed to be fixed on the lattice sites. As an important consequence, we obtain the total energy and energy eigenvalues dependent on ionic coordinates only as parameters. Even with this approximation, the Schrödinger equation for the N-electron system, depending on the $3N$ electronic spatial coordinates, is still an extremely complicated problem to solve.

In reducing the determination of ground state properties for interacting particle systems to solutions of single particle equations, density functional theory represents a successful scheme to determine the band structure in solids and all related electronic properties. The fundamental points of density functional theory were discussed by Hohenberg and Kohn [14] and Kohn and Sham [15], who derived ground state properties for an N electron interacting system as a function of the ground state charge density, n_{gs}. The total energy, expressed as a unique functional of the density $E[n]$, satisfies a minimum variational principle and can be determined by the solutions of the Kohn–Sham [15] equations (see paragraph 2.3.1). Their solution requires an approximation to compute the exchange-correlation energy $E_{xc}[n]$, for which an exact analytic expression is not known. Kohn and Sham introduced the local density approximation [15], an important simplification in the limit of slow changing density

$$E_{xc}^{LDA} = \int d\mathbf{r}\, n(\mathbf{r})\, \varepsilon_{xc}[n(\mathbf{r})] \tag{1}$$

where $\varepsilon_{xc}[n(\mathbf{r})]$ is the exchange-correlation energy for a homogeneous gas having $[n(\mathbf{r})]$ as charge density.

2.3.1.1 The density functional formalism

We consider N-electrons, occupying positions indicated by $\mathbf{r}_i$ and acted upon by the potential V_{ext} for which their Hamiltonian is

$$H = T + V_{ee} + \sum_{i=1}^{N} V_{ext}(\mathbf{r}_i) \tag{2}$$

where T is the kinetic operator, V_{ee} the non-local electron–electron interaction operator and V_{ext} an external potential (such as the electrostatic interaction between electrons and nuclei). We define the functional

$$F[n] = \min_{|\Psi|^2 \to n} \langle \Psi | T + V_{ee} | \Psi \rangle \tag{3}$$

where the minimum is taken over all the many-body functions $\Psi(\mathbf{r}_1, \mathbf{r}_2, ..., \mathbf{r}_N)$ giving a density

$$n(\mathbf{r}) = \sum_{i=1}^{N} |\Psi_i(\mathbf{r})|^2. \tag{4}$$

We observe that F is a *universal* functional, independent of the specific system and of the external potential V_{ext}.

Denoting E_{gs}, Ψ_{gs} and n_{gs} as the energy, wave function and charge density respectively in the ground state, the most important density functional theory theorems are expressed as

$$E[n] = \int d\mathbf{r}\, V_{ext}(\mathbf{r}) n(\mathbf{r}) + F[n] \geq E_{gs}, \tag{5}$$

$$\int d\mathbf{r}\, V_{ext}(\mathbf{r}) n_{gs}(\mathbf{r}) + F[n_{gs}] = E_{gs}. \tag{6}$$

Let us remark that the $E[n]$ minimization is taken over the density $n(\mathbf{r})$, which depends only on the three spatial coordinates; this represents a significant advantage over the traditional variational principle, which implies minimization over the $3N$ electronic coordinates the energy depends on.

2.3.1.2 The Kohn–Sham equations

The energy functional is expressed as [15]

$$E[n] = T_0[n] + \int d\mathbf{r} \left[V_{ext}(\mathbf{r}) n(\mathbf{r}) + \frac{1}{2} V_H(\mathbf{r}) \right] + E_{xc}[n],$$

$$= T_0[n] + \int d\mathbf{r} \, V_{ext}(\mathbf{r}) n(\mathbf{r}) + \frac{1}{2} \int d\mathbf{r} \, d\mathbf{r}' \frac{n(\mathbf{r}) n(\mathbf{r}')}{|\mathbf{r}-\mathbf{r}'|} + E_{xc}[n] \quad (7)$$

where T_0 is the kinetic term for a non-interacting system having $n(\mathbf{r})$ as charge density, V_{ext} the external potential, $V_H(\mathbf{r})$ the classical Coulomb repulsive potential among electrons and E_{xc} the exchange-correlation term. In Eq. (7) all the terms are exactly evaluable, except for the exchange-correlation. This contribution contains all the (local and non-local) many-body interactions and the difference between the kinetic energy term for the real system and for the fictitious non-interacting one.

The variational principle applied to Eq. (7) leads to

$$\frac{\delta E[n]}{\delta n(\mathbf{r})} = \frac{\delta T_0[n]}{\delta n(\mathbf{r})} + V_{ext}(\mathbf{r}) n(\mathbf{r}) + V_H(\mathbf{r}) + \frac{\delta E_{xc}[n]}{\delta n(\mathbf{r})} = \varepsilon \quad (8)$$

where ε is the Lagrange multiplier for the additional constraint implied by the particle number conservation. This is equivalent to the following relation:

$$\int d\mathbf{r} \, n(\mathbf{r}) = \sum_{i=1}^{N} \int d\mathbf{r} \, \Psi_i^*(\mathbf{r}) \Psi_i(\mathbf{r}) = N. \quad (9)$$

The comparison between Eq. (8) and the corresponding equation for a non-interacting electron system subject to a potential V_{eff}

$$\frac{\delta E[n]}{\delta n(\mathbf{r})} = \frac{\delta T_0[n]}{\delta n(\mathbf{r})} + V_{eff}(\mathbf{r}) n(\mathbf{r}) = \varepsilon, \quad (10)$$

leads to

$$V_{eff}(\mathbf{r}) = V_{ext}(\mathbf{r}) + V_H(\mathbf{r}) + V_{xc}(\mathbf{r}) \quad (11)$$

where $V_{xc}(\mathbf{r}) = (\delta E_{xc}[n]/\delta n(\mathbf{r}))$ (Note that in the above expression the magnitude of the three contributions are in the ratio $100:10:1$).

The variational principle leads to non-linear differential Kohn–Sham equations (equivalent to Schröedinger equations for non-interacting particles)

$$[-\tfrac{1}{2}\nabla^2 + V_{\text{eff}}(\mathbf{r})]\Psi_i(\mathbf{r}) = \varepsilon_i \Psi_i(\mathbf{r}). \tag{12}$$

Since the potential is a functional of the charge density, an iterative solution for Eq. (12) must be required. The solution of the Poisson equation for a trial charge density gives the potential, which is then substituted in Eq. (12) to obtain the $\Psi_i(\mathbf{r})$ orbitals. The Kohn–Sham solutions are then used to construct the new charge density, which will be the input charge density for a new iteration, until good convergence is reached.

All the ground state properties are available through the Kohn–Sham density, as for example the total energy

$$E_{\text{tot}} = \sum_i \varepsilon_i - \frac{1}{2}\iint d\mathbf{r}\,d\mathbf{r}' \frac{n(\mathbf{r})n(\mathbf{r}')}{|\mathbf{r}-\mathbf{r}'|} + E_{\text{xc}}[n] - \int d\mathbf{r}\, V_{\text{xc}}(\mathbf{r})n(\mathbf{r}). \tag{13}$$

We observe that the real eigenvalues ε_i in Eq. (12) are introduced in the density functional formalism as Lagrange parameters and their physical significance are not related to excitation electronic energies – as in the Hartree–Fock approach. A physically important exception is given by the highest (in energy) occupied eigenvalue, ε_N: the formulation of Koopman's theorem [44] in density functional formalism, states that $\varepsilon_N = \mu$, where μ – the chemical potential – is such that the total number of states having Kohn–Sham eigenvalues $\varepsilon_i < \mu$ is equal to the electron number N.

2.3.1.3 The exchange-correlation energy and the local density approximation

As a consequence of the Kohn–Sham theorem, the exchange and correlation energy depends only on the electronic density, $n(\mathbf{r})$. If the exchange correlation effects come predominantly from the immediate vicinity of $\mathbf{r}$ and do not depend strongly on the variations of $n(\mathbf{r})$ near $\mathbf{r}$, the contribution from the volume $d\mathbf{r}$ will be the same as if this volume were surrounded by a constant electronic density of the same value as within $d\mathbf{r}$. Thus, one can assume that the exchange-correlation energy depends only on the local electronic density around each volume $d\mathbf{r}$. This is called the local density approximation

$$E_{\text{xc}}[n] \approx \int n(\mathbf{r})\varepsilon_{\text{xc}}[n(\mathbf{r})]d\mathbf{r}. \tag{14}$$

Note that LDA is a first-principles approach in the sense that the quantum mechanical problem is solved without any arbitrary or system dependent parameters.

Following LDA [15], the total energy (see Eq. (13)) is expressed as

$$E_{tot} \approx \sum_i \varepsilon_i - \frac{1}{2} \iint d\mathbf{r}\, d\mathbf{r}' \frac{n(\mathbf{r})n(\mathbf{r}')}{|\mathbf{r}-\mathbf{r}'|} + \int n(\mathbf{r}) \{\varepsilon_{xc}[n(\mathbf{r})] - \mu_{xc}[n(\mathbf{r})]\}\, d\mathbf{r} \quad (15)$$

where $\mu_{xc} \equiv (d/dn)\{\varepsilon_{xc}[n(\mathbf{r})]n(\mathbf{r})\} = V_{xc}(\mathbf{r})$ is the contribution to the chemical potential for an homogeneous electron gas with constant density $n(\mathbf{r})$.

The evaluation of the exchange-correlation term for a wide range of densities was done according to different approaches, such as many-body perturbation theory [45] and quantum Monte-Carlo methods [46]. In our calculations, we used the Hedin–Lundqvist parametrization [45], which leads to the exchange-correlation energy as obtained by Singwi et al. [47]. The μ_{xc} term is expressed as a product of a factor $\beta[n(r_s)]$ (dependent on the local density and expressed through an analytic function [48]) by the Kohn–Sham–Gàspàr exchange potential:

$$\mu_x(r_s) = -\frac{k_F}{\pi} = -\left(\frac{9}{4\pi^2}\right)^{1/3} r_s^{-1} \quad (16)$$

where k_F is the Fermi k-vector and r_s the Wigner–Seitz radius.

The LDA turns out to be an excellent approximation for determining lattice constants, bulk moduli and Fermi surfaces for a large number of systems. However, weak bonds are often underestimated in LDA and binding energies are found to be larger, compared to experimental data. This has encouraged the development of gradient-corrected density functionals [49–51], where terms depending on the gradient of the electron density (and not only on its value $n(\mathbf{r})$ as usual in LDA) are included.

The LDA underestimation of energy band-gaps (an excited state property) has instead encouraged the use of self-interaction corrections [52] and the so called GW approximation GW [53,54]. This approximation makes use of the LDA Kohn–Sham single-particle wave function to give an approximate solution of the Dyson equation, which is typical of a *quasi-particle* approach to the N-electrons problem.

2.3.2 Overview of Density Functional Implementations

The solution of the Kohn–Sham equations can be obtained through different computational methods, which lead to various formulations of the effective potential. First, we mention the all-electron schemes, in which the Schrödinger equation is solved for all the electrons in the system (core included) and no approximations are made to the shape of the potential. The recently developed full-potential linearized muffin tin orbital [61] and the FLAPW method [16,17] are widely used all-electron mixed-basis methods. As will be extensively discussed in the following regarding the FLAPW method, in both these computational techniques the space is divided into spheres around atoms and in the interstitial region. The two methods treat the sphere region in the same way (i.e., by use of atomic functions), but differ by the treatment of the interstitial space: In the FLAPW method plane waves are used, whereas in the Full Potential Linearized Muffin Tin Orbitals approach localized functions derived from scattering theory (such as Hankel functions) are employed.

In the pseudopotential methods, a different approach is followed [57–60] in both the semi-empirical [55,56] or first-principles (i.e., without any dependence on experimental data, but only on the atomic species involved and on fundamental constants) forms: the core electrons and the atomic nucleus are replaced by a smooth pseudopotential, which becomes the actual potential at a certain cut-off radius away from the atomic site. Within this approximation, all the properties relative to valence electrons are determined in the same way as in all-electron methods, but the computational cost is substantially reduced, due to the smaller number of electrons involved in the calculations. However, the number of plane waves needed to reach good convergence can still be very high in the case of deep pseudopotentials or particular geometries.

If the pseudo-wave function and the all-electron wave function have the same norm (i.e., the integral of the square of the wave function over all space is the same), then the corresponding pseudopotentials are called "norm-conserving". If all the core electrons and the nucleus can be replaced by one pseudopotential which is the same for all the valence electrons, then such a form is called "local pseudopotential". However, it often happens that the valence wave functions with different angular momentum quantum number require different pseudopotentials (see for example the case of oxygen). Thus, the pseudopotential depends not only on the spatial coordinate, but also on the angular momentum and it is then called "non-local pseudopotential". Plane waves are used in the pseudo-wave function expansion, resulting in a high computational efficiency in employing fast Fourier transforms. There are, on the other hand, many situations in which pseudopotential methods fail, such as the determination of those properties that depend critically on the core electrons (for example the hyperfine field or the core level shifts); furthermore, accurate and efficient pseudopotentials for certain materials (such as rare-earth and transition metals) are difficult to construct.

In the above mentioned methods, the computational bottleneck is represented by the conventional diagonalization of the $N \times N$ (N being the total number of basis functions used in the wave function expansion) secular matrix; thus, within density functional theory, other methods (such as Car–Parrinello molecular dynamics [63] and the preconditioned conjugate gradient approach [62]) have been developed to solve the eigenvalue problem, directly minimizing the Kohn–Sham energy functional. Car–Parrinello simulations treat the adiabatic dynamics of a system composed of classical ions and electrons in their fundamental state (represented in terms of the electronic density within the density functional theory formalism). By means of a fictitious dynamics for the electronic degrees of freedom, this method translates the problem of the determination of the minimum for the Kohn–Sham functional during the ions' evolution into an equivalent dynamical classical problem. The electronic wave functions are then updated as the atomic positions are moved, by solving the electronic problem.

2.3.3 The Full Potential Augmented Plane Wave Method

All the results given in this chapter were obtained using a density functional theory implementation (within the local density approximation), namely the FLAPW all-electron *ab initio* self-consistent method. In this highly precise approach, the Bloch function $\Psi_i(\mathbf{k},\mathbf{r})$, for the *i*-th electron with wave vector $\mathbf{k}$ and having lattice periodicity, is usually expanded as

$$\Psi_i(\mathbf{k},\mathbf{r}) = \sum_j C_{ij} \Phi_j(\mathbf{k}+\mathbf{G}_j,\mathbf{r}) \tag{17}$$

where $\mathbf{G}_j$ is a reciprocal lattice vector.

In the FLAPW method the solid is divided in two different regions: muffin tin (MT) spheres S_α (with radius R_α and centered on the atomic sites) and the remaining interstitial zone, I. This choice naturally suggests different basis functions for the wave function expansion, depending on the spatial region in which they are used. The wave function Φ_j is therefore expressed as a Fourier expansion in the interstitial region and through spherical harmonics (evaluated on a discrete logarithmic radial mesh) in the muffin tin region

$$\Phi_j(\mathbf{k}+\mathbf{G}_j,\mathbf{r}) = \begin{cases} (1/\sqrt{\Omega})e^{i\mathbf{K}_j\mathbf{r}} & \mathbf{r} \in I, \\ \sum_{lm}[A_{lm}^{\alpha}(\mathbf{K}_j)u_l(E_l,r_\alpha) \\ \quad + B_{lm}^{\alpha}(\mathbf{K}_j)\dot{u}_l(E_l,r_\alpha)]Y_{lm}(\hat{\mathbf{r}}_\alpha), & \mathbf{r}_\alpha \in S_\alpha \end{cases} \quad (18)$$

where $\mathbf{K}_j = \mathbf{k} + \mathbf{G}_j$, Ω is the unit-cell volume, r_α the coordinate with respect to the α atom position and E_l the energy parameter, usually chosen at the centre of the corresponding band.

The function u_l is the solution of the radial Schrödinger equation

$$\left[-\frac{\partial^2}{\partial r^2} + V_{\text{eff}} - E_l - \frac{l(l+1)}{r^2}\right]u_l(r) = 0 \quad (19)$$

and $\dot{u}_l$ its derivative with respect to the energy.

The coefficients A_{lm} and B_{lm} are determined through continuity conditions at the spherical surface imposed on the function Φ_j and on its spatial derivative.

The variational principle for the Kohn–Sham Eqs. (12) leads to the following secular equation

$$(H_{ij} - \varepsilon_i S_{ij})C_j = 0 \quad (20)$$

where $H_{ij} = \langle \Phi_i|H|\Phi_j\rangle$ is the (ij)-th element of the Hamiltonian matrix and $S_{ij} = \langle \Phi_i|\Phi_j\rangle$ is the overlap integral between the Φ_i and Φ_j orbitals. The wave function linearization, expressed using u_l and $\dot{u}_l$, makes H_{ij} and S_{ij} independent of the energy [64] and this is a remarkable advantage over the original Slater Augmented Plane Wave method [65]. The eigenvalues, ε_i, are thus determined through diagonalization of the matrix $\mathbf{H} - \varepsilon\mathbf{S}$ and then substituted in Eq. (20); this leads to the determination of the coefficients C_{ij} used in Eq. (17).

The charge density is calculated through a summation over the occupied eigenstates

$$n(\mathbf{r}) = \sum_{i=1}^{n_{\text{occ}}} \int_{BZ} d\mathbf{k}\, \Psi_i^*(\mathbf{k},\mathbf{r})\Psi_i(\mathbf{k},\mathbf{r}). \quad (21)$$

In order to reduce the complexity of the Brillouin zone integration, Baldereschi [66] and Chadi and Cohen [67], suggested approximating integrals by a summation over a reduced number of special $\mathbf{k}$ points inside the Brillouin zone, each having a different weight. Following the original Chadi and Cohen approach, one first chooses one (or more) **k**-vector(s), which, through symmetry operations (such as translational invariance, time reversal or point group symmetries) generates the special set. In our work, we followed a similar extended scheme (suggested by Monkhorst and Pack [68]) for the determination of the special **k**-points in the different structures considered.

In a linearized augmented plane wave method, the charge density has the following expansion

$$n(\mathbf{r}) = \begin{cases} \sum_j n_{G_j} e^{i\mathbf{G}_j \mathbf{r}}, & \mathbf{r} \in I, \\ \sum_i n_{\alpha_i}(r_\alpha) K_{\alpha_i}(\hat{\mathbf{r}}_\alpha), & \mathbf{r}_\alpha \in S_\alpha, \end{cases} \qquad (22)$$

where the K_{α_i} functions (called *lattice harmonics* [69]) are linear combinations of spherical harmonics, which fulfill the symmetry requirements for the specific system considered.

Without any shape approximation, we express the potential (as obtained by solving the Poisson equation) as

$$V_{\text{eff}}(\mathbf{r}) = V_{\text{MT}}(\mathbf{r}) + V_{\text{warp}}(\mathbf{r}) + V_{\text{ns}}(\mathbf{r}) \qquad (23)$$

where V_{MT} and V_{ns} indicate the spherical and non-spherical terms, respectively, in the muffin tin region and V_{warp} is the total potential in the interstitial region, including the non-constant terms. The potential expansion is analogous to that for the charge density

$$V(\mathbf{r}) = \begin{cases} \sum_j V_{G_j} e^{i\mathbf{G}_j \mathbf{r}}, & \mathbf{r} \in I, \\ \sum_i V_{\alpha_i}(r_\alpha) K_{\alpha_i}(\hat{\mathbf{r}}_\alpha), & \mathbf{r}_\alpha \in S_\alpha. \end{cases} \qquad (24)$$

In the FLAPW self-consistency cycle, two variations have been implemented: in the first step, the non-spherical ($l \neq 0$) and the spin-orbit terms are neglected in the potential expression. Then, the eigenfunctions obtained from the previous diagonalization are used as basis functions for a so-called "second variation", in which all the terms are included in the potential. The remarkable advantage of this procedure is that the second diagonalization is performed over a reduced set of terms: the number of basis functions used is, in fact, equal to the number of eigenvalues and thus is much smaller (by over a factor of 10) than the total number of basis functions used in the first variation.

The difficulty arising in the solution of the Poisson equation due to the steep and rapid oscillations of the charge density near the nucleus, is side-tracked through a simple observation: the potential in the I region depends only on the multipole moments of the charge density inside the spheres. Therefore, following Weinert [70,16], a pseudo-charge is constructed in the S_α region, having the same multipole moments as the actual charge inside the muffin tin, but with a rapidly convergent Fourier expansion. The Poisson equation is thus exactly solved in the interstitial region and the potential value at the spherical surface constitutes the boundary condition for the Dirichlet problem inside the muffin tin's. In this way, the potential obtained is relative to the real charge density, both in the S_α and the interstitial I regions.

The FLAPW method (see Fig. 1 for the self-consistency cycle) solves the Kohn–Sham equations in a self-consistent way, until good convergence is reached. At the end of each cycle, the output charge density is mixed with the input charge density, following the Broyden scheme [71], in order to accelerate convergence.

As already noted, the FLAPW is an all-electron method, in which the Schröedinger equation is solved for all the electrons of the systems. The core electrons, in fact, are treated fully-relativistically (by solving the Dirac equation for the spherical component of the potential [72]), while the valence electrons are treated scalar-relativistically (including the mass–velocity relativistic terms, but neglecting the spin–orbit coupling term [73]). The small influence of the spin–orbit coupling on the systems considered allows our neglecting this effect during the iterative cycle; however, after self consistency is reached, the spin–orbit interaction is treated in a perturbative second variation approach, starting from the self-consistent solutions.

2.4 ULTRATHIN STRAINED-LAYER HETEROSTRUCTURES: GaSb/InSb AND InAs/InSb SUPERLATTICES

2.4.1 Bulk Binary Constituents

As already pointed out, one of the most important reasons why superlattices are drawing more and more attention is the opportunity to tune their properties, starting from those of their binary constituents. Most of the quantities (T) of interest in a ternary structure ($A_xB_{1-x}C$) can often be derived from the corresponding quantities (B) in the binaries (AC and BC), through linear ($T_{A_xB_{1-x}C} = xB_{AC} + (1-x)B_{BC}$) or quadratic relations ($T_{A_xB_{1-x}C} = xB_{AC} + (1-x)B_{BC} - b_Tx(1-x)$, where b_T is often called the non-linear *bowing parameter* for the property T). Thus a preliminary study of the binary constituents

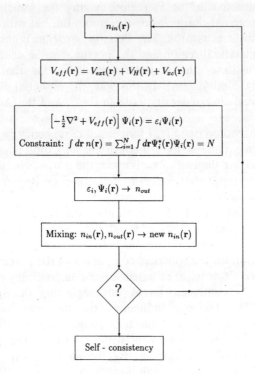

Fig. 1 The self-consistency cycle to solve the Kohn–Sham equations, as implemented in standard electronic structure calculations.

(GaSb, InSb and InAs) is essential for the determination of the properties of interest in the ternary structures.

Our calculations were performed using a cut-off $k_{max} = 3.8$ a.u. for the wave function expansion and a muffin tin radius $R_{MT} = 2.4$ a.u. (equal for all the atomic species involved); these values for the parameters ensure a good convergence ($k_{max} R_{MT} = 9.12$). The maximum value for the angular momentum in the wave function and charge density expansion (inside the muffin tin spheres) is $l_{max} = 8$ and the Brillouin zone sampling is performed using a set of 10 special k-points chosen according to the Monkhorst–Pack scheme [74]. We consider the cations' semi-core d electrons as part of the valence band and the spin–orbit splitting in a perturbative approach, after reaching self-consistency in the calculations.

2.4.1.1 Structural properties

The equilibrium lattice parameter for each semiconductor was determined through a numerical fit of the total energy values obtained for zincblende structures having different lattice constants; the minimum for this curve gives the equilibrium value a_0 for the examined binary. The semi-empirical Murnaghan [75] equation of state was used

$$p = \frac{B_0}{B_0'}\left[\left(\frac{a_0}{a}\right)^{3B_0'} - 1\right] \tag{25}$$

where a is the lattice constant at the pressure p, a_0 its equilibrium value, B_0 the bulk modulus and B_0' its derivative with respect to pressure.

We have also studied the band-gap energy, E_{gap}, as a function of the pressure p, in order to obtain information regarding the valence band maximum (VBM) and the conduction band minimum (CBM) deformation potentials: this allows us to predict how the electronic properties will be modified in the presence of strain. Following the usual experimental approach, we assume E_{gap} to depend linearly [10] on a and quadratically [76] on p

$$E_{gap} = E_0 + 3b\left(\frac{\Delta a}{a_0}\right), \tag{26}$$

$$E_{gap} = E_0 + \alpha p + \beta p^2. \tag{27}$$

Our calculated results are reported in Table I and compared with experiment; the agreement is excellent for the lattice parameters (differing at most by 0.2%) and quite good for the bulk modulus

TABLE I
Calculated and experimental values for the equilibrium lattice constant a, bulk modulus B, its derivative with respect to pressure B' and pressure coefficients: $b = \frac{1}{3}(dE_{gap}/da)$, $\alpha = dE_{gap}/dp$, $\beta = \frac{1}{2}(d^2E_{gap}/dp^2)$.

		a (a.u)	B (Mb)	B'	b (eV)	α (10^{-6}(eV/bar))	β (10^{-12}(eV/bar^2))
GaSb	FLAPW	11.542	0.56	4.9	−6.92	13.0	−37.8
	Expt.	11.520[a]	0.578[a]	—	—	14.8[a]	—
InSb	FLAPW	12.215	0.48	4.5	−5.77	12.3	−41.4
	Expt.	12.243[b]	0.483[b]	—	—	15.0[c]	—
InAs	FLAPW	11.457	0.58	5.1	−5.24	9.2	−28.2
	Expt.	11.448[b]	0.579[b]	—	—	10.0[c]	—

[a]Ref. [77]; [b]Ref. [77], Ref. [79]; [c]Ref. [80].

(maximum error 3.1%). We observe that the direct band gap increases with pressure or, equivalently, decreases as the substrate lattice constant is increased. Note also that the band gap trend as a function of pressure is correctly reproduced by our calculations, despite the well known underestimation of the E_{gap} absolute value, due to the local density approximation.

2.4.1.2 Electronic properties

In Table II, we report the energy eigenvalues (where the energy zero is chosen to be equal to the valence band maximum), calculated at high symmetry points for the zinc blende Brillouin zone and compared with the corresponding experimental values.

We also report the results obtained for the spin–orbit (s.o.) splittings, $\Delta_{s.o.}$, of the valence band maximum (we recall that the conduction band minimum, being the $l = 0$ state, is not affected by the spin–orbit coupling). In fact, this triply degenerate state, having p symmetry, is separated by the relativistic spin–orbit term into a doubly degenerate level (higher in energy and corresponding to $p_{j=3/2}$) and a singly

TABLE II
Calculated and experimental values for valence and conduction bands (in eV) for the binary semiconductors (relative to the valence band maximum). We report in brackets the experimental uncertainty of the different states. Spin–orbit splitting ($\Delta_{s.o.}$) and band-gap (E_{gap}) are also reported.

State	GaSb		InSb		InAs	
	E^{LDA}	Expt.	E^{LDA}	Expt.	E^{LDA}	Expt.
$\Gamma_{1c}(E_{gap})$	−0.23	0.81[a]	−0.42	0.24[b]	−0.51	0.42[b]
Γ_{15v}(VBM)	0	0	0	0	0	0
$\Delta_{s.o.}(\Gamma_{15v})$	0.73	0.75[a]	0.76	0.82[c]	0.36	0.38[c]
Γ_{1v}	−11.66	−11.63[a]	−10.88	−11.45[c]	−11.93	−12.3(4)[c]
Γ_{12v}	−15.19	—	−14.54	−17.29[c]	−14.30	−17.09[c]
Γ_{15v}	−15.24	−19.00[a]	−14.63	−17.31[c]	−14.42	—
X_c	X_{3c} 0.84	1.43[a]	X_{3c} 1.26	—	X_{3c} 1.45	—
X_{5v}	−2.66	−2.72[a]	−2.32	−2.44[c]	−2.35	−2.48[c]
X_{3v}	−6.94	−6.93[a]	−6.05	−6.46[c]	−5.94	−6.3(2)[c]
X_{1v}	−9.21	−9.42[a]	−8.89	−9.26[c]	−9.95	−9.8(3)[c]
L_{1c}	0.25	1.09[a]	0.39	—	0.75	—
L_{3v}	−1.28	−1.32[a]	−1.04	−1.24[c]	−1.01	−0.9(3)[c]
L_{1v}	−6.49	−6.93[a]	−5.74	−6.46[c]	−5.85	−6.3(7)[c]
L_{1v}	−10.03	−10.33[a]	−9.50	−10.53[c]	−10.48	−10.6(3)[c]

[a] Ref. [77]; [b] Ref. [82]; [c] Ref. [81].

degenerate level (corresponding to $p_{j=1/2}$). The calculated energy splittings are in excellent agreement with the experimental [77] values.

Therefore, the resulting band gaps can also be expressed

$$E_{\text{gap}}^{\text{s.o.}} = E_{\text{gap}}^{\text{LDA}} - \tfrac{1}{3}\Delta_{\text{s.o.}} \qquad (28)$$

where the superscripts LDA and s.o. indicate a semi-relativistic and fully-relativistic treatment for the electronic levels at the Γ point. We thus obtain: $E_{\text{gap}}^{\text{s.o.}} = -0.47\,\text{eV}$ for GaSb, $E_{\text{gap}}^{\text{s.o.}} = -0.67\,\text{eV}$ for InSb and $E_{\text{gap}}^{\text{s.o.}} = -0.63\,\text{eV}$ for InSb, and observe that the discrepancy between calculated and experimental values is quite remarkable in that the LDA predicts all the semiconductors examined to show semi-metallic properties. On the other hand, when comparing the energy position of the valence states with the experimental results obtained by photoemission spectroscopy [81], we find a quite good agreement. However, the binding energy of the semicore d states, which represent the lower edge of the valence band, is underestimated by as much as 2.7 eV. This is to be ascribed to core hole relaxation in the ionization processes of deep levels which are neglected in LDA.

The analysis of the charge density in the different states (or in states which are obtained from them through folding relations, see next section) allows one to predict how the electronic structure will be modified by doping, impurities and strain. We therefore report their decomposition inside the muffin-tin spheres in Table III as a function of the angular momentum; the s ($l=0$) character, typical of conduction antibonding states, indicates strong localization on the atomic site (inside the muffin-tin sphere), while the p ($l=1$) character, typical of

TABLE III
Angular decomposition for the charge density of InSb inside the muffin-tin spheres for some of the relevant states.

State	Atom	Q_s	Q_p
	InSb		
Γ_{15v}(VBM)	In	0.000	0.071
	Sb	0.000	0.419
Γ_{1c}(CBM)	In	0.364	0.000
	Sb	0.330	0.000
X_{3c}	In	0.000	0.107
	Sb	0.071	0.000
L_{1c}	In	0.215	0.043
	Sb	0.134	0.055

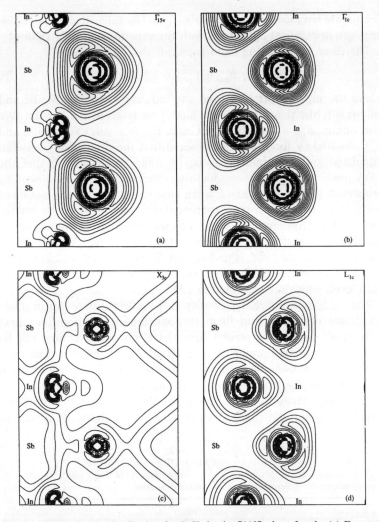

Fig. 2 Charge density distribution for InSb in the [110] plane for the (a) Γ_{15v} state, (b) Γ_{1c} state, (c) X_{3c} state and (d) L_{1c} state. The charge density contour plots are given on a linear scale and separated by $0.5\,e^-$/unit cell.

the valence band maximum state, indicates that the electronic charge is spread along the bond direction.

In Fig. 2, we report the InSb electronic charge density distribution of the Γ_{15v} (VBM) in Fig. 2(a) and of the first conduction band state at the high symmetry points, namely the Γ_{1c} state (CBM) in Fig. 2(b), the

X_{3c} state in Fig. 2(c) and the L_{1c} state in Fig. 2(d). We have chosen InSb as an illustrative example; the charge density distribution and localization are very similar in all of the three binary semiconductors examined. We first observe the bonding (anti-bonding) character of the valence band maximum state (all the conduction states) in heteropolar semiconductors, as illustrated by the appreciable (small) charge density along the bonding direction and confirmed by Table III. We note that the Γ_{15v} electronic charge density is much more concentrated around the anion than around the cation; moreover, the exclusive p character of this state (see the total absence of the s character in Table III) confirms its bonding nature.

The angular decomposition of the conduction state charge density shows strong s character for the Γ_{1c} state (total absence of p character) and also for the L_{1c} state (slight presence of p character which accounts for the small differences between Fig. 2(b) and Fig. 2(d)). The distribution at the X point is significantly different from the other figures, since this state is more delocalized inside the unit cell (as with the group IV semicondutors). These observations confirm that the charge density distribution is strongly dependent on the k point considered (and hence on the symmetry of the wave function in that state), whereas it is much less affected by the atomic species constituting the III–V semiconductor.

2.4.2 The 1×1 Superlattices

In this section we discuss the structural and electronic properties of ultrathin (1×1) GaSb/InSb and InAs/InSb superlattices, focusing our attention on how the quantities of interest (such as electronic energy levels and charge density distributions) can be tuned as the growth direction or the strain conditions are changed.

In fact, we recall that in epitaxial growth processes of strained layer superlattices, the epilayer has a lattice parameter different from that of the substrate. Up to a certain critical thickness, the pseudomorphic growth of a 1×1 superlattice on a given substrate results in alternating single layers of the two constituents in different strain conditions. Moreover, as the substrate is changed, different tensile or compressive stress acts on the monolayers. This results in a modification not only of the structural properties (equilibrium lattice constants and angles between different bond directions after relaxation), but also of the electronic and transport properties of the superstructure.

In analogy with the usual experimental approach, we have considered various growth conditions for strained layer superlattices determined by the different substrates chosen: (i) pseudomorphic growth on one of the binary consituents (in which the lattice parameter parallel to the growth plane is equal to that of the bulk semiconductor composing the substrate); (ii) the "free standing mode" structure in which both the lattice parameters for the two constituents are allowed to relax.

Electronic and structural properties critically depend on the ordering direction: in order to clarify this aspect, we have studied strained structures grown in the [001] and [111] directions, having Cu–Au like and Cu–Pt like structures, respectively. The deposition of alternating monolayers of the two binary constituents along the growth direction leads to a unit cell doubling in real space. As a consequence, the unit cell dimensions are reduced in reciprocal space; in particular, the ternary Brillouin zone can be obtained by folding the fcc Brillouin zone along specified directions (i.e., along the [111] axis for the trigonal unit cell), inducing important modifications in the electronic properties of the ternary systems. In order to illustrate this point, we have reported in Fig. 3 the band structure along the $L - \Gamma$ direction – also known as the Λ direction – for binary InSb in the zincblende (Fig. 3(a)) and trigonal (Fig. 3(b)) structures. We notice that the dashed levels in Fig. 3(b) are obtained starting from those of Fig. 3(a) through reflection with respect to a mirror plane perpendicular to the Λ line, which contains the intermediate point (Z) between L and Γ.

We can thus classify the ternary electronic levels using the labels of the zincblende energy levels from which they are derived through folding operations; in particular, binary states of different symmetry can be folded to superlattice states having the same symmetry. The resulting repulsive interaction effects, which strongly influence the ternary band structure, can be interpreted using a widely adopted scheme [42]. The model considers the superlattice as obtained from an ideal virtual crystal (i.e. a common anion or a common cation system having the non-common atom with intermediate properties between the two different atoms in the binary constituents) to be acted on by a perturbative potential ΔV^{pert} which brings the "virtual" atoms into the real ones. This perturbative term is made of a structural contribution, ΔV^{str} (due to the atomic relaxations consequent to the atomic substitution) and a chemical one ΔV^{chem} (due to electronegativity differences between the atomic species involved)

$$\Delta V^{\text{pert}} = \Delta V^{\text{str}} + \Delta V^{\text{chem}}. \tag{29}$$

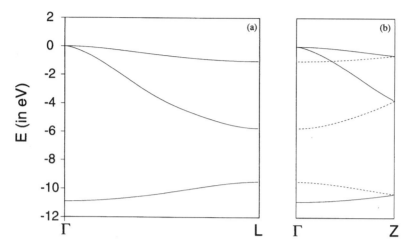

Fig. 3 Band structure for the zincblende binary InSb along the fcc Λ line (a); band folding along the trigonal Λ line in the ternary structures (b). Dashed lines are obtained through reflection on a mirror plane perpendicular to the Γ−L line through the intermediate point (Z).

2.4.2.1 Structural properties: the importance of growth direction and strain conditions

The atomic ordering along the [111] direction gives rise to a trigonal Bravais lattice with the C_{3v}^5 space group (Schöenflies notation). The unit cell (whose particular geometry is illustrated in Ref. [78]) contains four atomic types and its origin is taken on a cation site. Note the different coordination of the common atoms in the trigonal cell; in the GaSb/InSb system, for example, we have the first Sb_{Ga} atom bound with three Ga atoms and one In atom, while the other Sb_{In} atom shows a complementary situation.

In the case of superlattices having the [001] direction as growth axis, we obtain a tetragonal Bravais lattice with D_{2d}^5 space group and a unit cell in real space with four atoms (two of which are equivalent from the coordination point of view); the origin is taken on a cation (anion) site for the GaSb/InSb (InAs/InSb) systems.

The determination of the whole set of free parameters in the unit cell through total energy minimization is an extremely demanding computational problem; this encouraged some simplifications such as considering equal bond lengths between atoms belonging to the same chemical species (i.e., for the GaSb/InSb system, we considered $d(InSb_{Ga}) = d(InSb_{In})$. Within these approximations, we performed total energy minimization of the free parameters for structures grown pseudomorphically on a bulk binary substrate.

We recall that according to the macroscopic theory of elasticity, the structural parameters can be determined by the following relations [83]

$$a_\parallel^{epi} = a^{subs}, \qquad a_\perp^{epi} = a^{epi}[1 - D^{[n_1 n_2 n_3]}(a_\parallel/a^{epi}) - 1]$$

$$D^{[001]} = 2\left(\frac{C_{12}}{C_{11}}\right), \qquad D^{[111]} = 2\left(\frac{C_{11} + 2C_{12} - 2C_{44}}{C_{11} + 2C_{12} + 4C_{44}}\right) \qquad (30)$$

$$\varepsilon_\parallel^{epi} = \frac{a_\parallel}{a^{epi}} - 1, \qquad \varepsilon_\perp^{epi} = \frac{a_\perp^{epi}}{a^{epi}} - 1$$

where the C_{ij} indicate the elastic constants of the bulk epilayer.

We have thus compared the results obtained through the two different procedures (total energy minimization and macroscopic theory of elasticity, using the experimental data [82] for the elastic constants in Eqs. (30)) and found a very good agreement. This confirms the validity of the macroscopic theory of elasticity in predicting the structural properties in strained binaries and justifies its use in the determination of all the unit cell degrees of freedom, particularly for those in the free standing mode structure (where total energy minimization would require great computational effort).

The substrate lattice constants for the free standing mode structures were determined through total energy minimization of the "ideal" AC/BC unrelaxed structures (in which all the atoms are arranged in a zincblende structure with lattice parameter a – the degree of freedom to be optimized – and with bond lengths $d_{AC} = d_{BC} = a(\sqrt{3/4})$). As a result, we obtained an in-plane lattice constant very similar to the average of the bulk constants, as predicted by Vegard's rule; in what follows, the free standing mode structures will be denoted as "Av. subs.".

In Table IV and Table V we report the calculated structural parameters for the ternary common-anion and common-cation systems: the S1 (S2) system is a common-anion superlattice grown on a GaSb (InSb) substrate, while the S3 (S4) system is a common-cation superlattice grown on an InAs (InSb) substrate. We note that GaSb and InAs show similar deformation in equivalent structures, as reasonably expected, due to small differences in their elastic constants [82] and bulk moduli [77]. In addition, we observe that the greater the mismatch between the lattice parameters of the constituents, the larger the strains and percentage deviations from bulk bond lengths: this is the origin of the greater values for Δ and ε in common cation superlattices.

TABLE IV
Bond-lengths (d_{GaSb} and d_{InSb}, in a.u.) and strain parameters (ε^{GaSb} and ε^{InSb}) parallel and perpendicular to the growth plane for $(GaSb)_1/(InSb)_1$ [111] and [001] ordered systems. The quantities denoted by Δ indicate percentage deviations from calculated bulk bond-lengths ($d_{GaSb} = 5.00$ a.u. and $d_{InSb} = 5.29$ a.u.).

	[111]			[001]		
	Av. subs.	GaSb-subs.	InSb-subs.	Av. subs.	GaSb-subs.	InSb-subs.
d_{GaSb}	5.07	5.00	5.15	5.05	5.00	5.11
Δ_{GaSb}	+1.4%	—	+3.0%	+1.0%	—	2.2%
$\varepsilon_{\parallel}^{GaSb}$	+0.029	0	+0.058	+0.029	0	+0.058
$\varepsilon_{\perp}^{GaSb}$	−0.014	0	−0.028	−0.027	0	−0.053
d_{InSb}	5.22	5.16	5.29	5.25	5.24	5.29
Δ_{InSb}	−1.3%	−2.5%	—	−0.8%	−0.9%	—
$\varepsilon_{\parallel}^{InSb}$	−0.028	−0.055	0	−0.028	−0.055	0
$\varepsilon_{\perp}^{InSb}$	+0.016	+0.033	0	+0.030	+0.076	0

TABLE V
Bond-lengths (d_{InAs} and d_{InSb}, in a.u.) and strain parameters (ε^{InAs} and ε^{InSb}) parallel and perpendicular to the growth plane for $(InAs)_1/(InSb)_1$ [111] and [001] ordered systems. The quantities denoted by Δ indicate percentage deviations from calculated bulk bond-lengths ($d_{InAs} = 4.96$ a.u. and $d_{InSb} = 5.29$ a.u.).

	[111]			[111]		
	Av. subs	GaSb-subs	InSb-subs	Av. subs	GaSb-subs	InSb-subs
d_{InAs}	5.04	4.96	5.12	5.01	4.96	5.07
Δ_{InAs}	+1.6%	—	+3.2%	+1.0%	—	2.2%
$\varepsilon_{\parallel}^{InAs}$	+0.033	0	+0.066	+0.033	0	+0.066
$\varepsilon_{\perp}^{InAs}$	−0.019	0	−0.038	−0.036	0	−0.072
d_{InSb}	5.21	5.14	5.29	5.24	5.22	5.29
Δ_{InSb}	−1.5%	−2.8%	—	−0.9%	−1.3%	—
$\varepsilon_{\parallel}^{InSb}$	−0.031	−0.062	0	−0.031	−0.062	0
$\varepsilon_{\perp}^{InSb}$	+0.019	+0.037	0	+0.033	+0.078	0

2.4.2.2 Electronic properties

a. Electronic levels Most of the properties of interest for the ternary structures can be derived from the electronic energy levels, whose determination is one of the main goals of this work. The most important effects acting on the topmost valence levels in the superlattices are due to the non-cubic "crystal field" (CF) and the spin–orbit coupling. The former splits the triply degenerate Γ_{15v} zincblende state into a doubly degenerate $\Gamma_{3v}^{(2)}$ ($\Gamma_{5v}^{(2)}$) state and a singly degenerate Γ_{1v} (Γ_{4v}) state in [111] ([001]) ordered superlattices. The s.o. coupling removes the double degeneracy, yielding three different valence levels, whose energies, according to the "quasi-cubic" model [84] are determined (taking the center of gravity as zero) as follows

$$E_{1,2,3} = \begin{cases} \frac{1}{3}(\Delta_{s.o.} + \Delta_{CF}) \\ -\frac{1}{6}(\Delta_{s.o.} + \Delta_{CF}) \pm \frac{1}{2}\{(\Delta_{s.o.} + \Delta_{CF})^2 - \frac{8}{3}\Delta_{s.o.}\Delta_{CF}\}^{1/2} \end{cases} \quad (31)$$

where Δ_{CF} and $\Delta_{s.o.}$ indicate the CF and s.o. splittings, respectively.

We first discuss the results obtained for the electronic levels at the Brillouin zone center as a function of the strain conditions, neglecting the spin–orbit coupling. The values obtained are reported in Table VI and Table VII for [111] and [001] ordered superlattices respectively, with a numerical uncertainty equal to ± 0.04 eV (the same for all the energies reported in this work, unless otherwise specified); from these values, the crystal field splitting Δ_{CF} is defined as the energy difference

TABLE VI
Calculated electronic energy levels (in eV) with respect to the VBM for the [111] superlattice (neglecting s.o. coupling). The superscripts (1) and (2) indicate the two states involved in the repulsion mechanism. The state multiplicity is given in parentheses.

Superlattice state	Zincblende state	$(GaSb)_1/(InSb)_1$			$(InAs)_1/(InSb)_1$		
		Av. subs	S1	S2	Av. subs	S3	S4
$\Gamma_{1c}^{(2)}$	L_{1c}	0.50	0.70	-0.20	0.79	1.04	0.36
$\Gamma_{1c}^{(1)}$	Γ_{1c}	-0.81	-0.64	-1.47	-1.02	-0.95	-1.42
$\Gamma_{3v}^{(2)}$	Γ_{15v}	0(2)	0(2)	$-0.42(2)$	0(2)	0(2)	$-0.18(2)$
Γ_{1v}	Γ_{15v}	-0.11	-0.59	0	-0.21	-0.77	0
$\Gamma_{3v}^{(1)}$	L_{15v}	$-1.31(2)$	$-1.35(2)$	$-1.68(2)$	$-1.38(2)$	$-1.46(2)$	$-1.50(2)$

TABLE VII
Calculated electronic energy levels (in eV) with respect to the VBM for [001] superlattice (neglecting s.o. coupling). The superscripts (1) and (2) indicates the two states involved in the repulsion mechanism. The state multiplicity is given in parentheses.

Superlattice state	Zincblende state	$(GaSb)_1/(InSb)_1$			$(InAs)_1/(InSb)_1$		
		Av. subs	S1	S2	Av. subs	S3	S4
$\Gamma_{1c}^{(2)}$	X_{1c}	1.12	1.26	-0.78	1.46	1.54	1.01
$\Gamma_{1c}^{(1)}$	Γ_{1c}	-0.49	-0.43	-0.77	-0.68	-0.69	-0.96
$\Gamma_{5v}^{(2)}$	Γ_{15v}	0(2)	0(2)	$-0.20(2)$	0(2)	0(2)	$-0.25(2)$
Γ_{4v}	Γ_{15v}	-0.07	-0.41	0	-0.06	-0.46	0
$\Gamma_{5v}^{(1)}$	X_{5v}	$-2.52(2)$	$-2.52(2)$	$-2.71(2)$	$-2.46(2)$	$-2.46(2)$	$-2.70(2)$

between the states $\Gamma_{3v}^{(2)}$ and Γ_{1v} ($\Gamma_{5v}^{(2)}$ and Γ_{4v}) for the [111] ([001]) ordering, reported in Table VI (Table VII). We notice that Δ_{CF} (taken positive if the doubly degenerate level is higher in energy than the non-degenerate state) is strongly affected by the strain conditions, showing a decreasing trend as the substrate lattice constant is increased, either for Cu–Pt or Cu–Au like structures, in both the common anion and the common cation superlattices. If we substitute in Eqs. (31) the calculated values of the crystal fields and s.o. splittings (the latter is obtained by averaging the $\Delta_{s.o.}$ values in the binary constituents), we find that the self-consistent valence eigenvalues (spin–orbit coupling included) are in good agreement (within 0.05 eV) with the energies determined according to the "quasi-cubic" model. We have thus reported in Fig. 4(a) and Fig. 4(b) (for common anion and common cation, respectively) the relevant electronic levels, labeled according to Eqs. (31) as a function of the substrate lattice parameter. We observe a linear trend for the levels E_1 and E_C (lowest conduction level), while the more complex behaviour for the other levels (E_2 and E_3) is due to the interplay between the CF and the s.o. couplings.

Let us now investigate the effects of the folding operations and of the consequent perturbative potential mentioned in Section 2.4.2. In the second column of Table VI and Table VIII we report the binary zincblende states folded on the ternary Brillouin zone center, which are coupled by the potential. The superscripts (1) and (2) indicate the ternary levels whose energies are shifted downward and upward, respectively.

A first proof of the validity of the band repulsion model is offered by the smaller value of the crystal field splittings in the S4 [111] superlattice compared to the S4 [001] one; this is at variance with the other structures, where $\Delta_{CF}^{[111]}$ is always bigger than $\Delta_{CF}^{[001]}$, in equivalent growth conditions. This apparently strange behaviour is easily understood, if we consider that the interaction (appreciable only for common cation structures) between the $\Gamma_{3v}^{(2)}$ and $\Gamma_{3v}^{(1)}$ states produces an upward shift of the former level. On the other hand, the Γ_{1v} energy level (which is the topmost valence level for S4 systems) is not affected by repulsion effects: the crystal field splitting is thus reduced (as seen in Table VII), compared to the other structures where the valence band maximum is a $\Gamma_{3v}^{(2)}$ state, whose energy upward shift increases the crystal field splitting.

It is important to point out that the band repulsion mechanism is also responsible for the narrowing of the band gap with respect to the

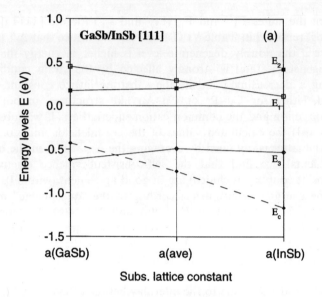
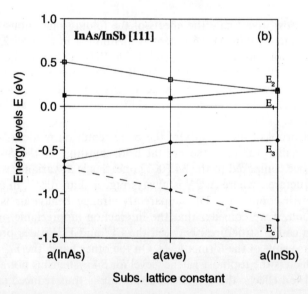

Fig. 4 Calculated highest valence band energy levels (E_1, E_2, and E_3) and lowest conduction state (E_c) at Γ versus substrate lattice constant for the [111] (a) common-anion and (b) common-cation superlattices considered. The center of gravity of the topmost valence bands is taken as zero.

average E_{gap} of the binary constituents. In fact, a similar repulsion mechanism acts on the $\Gamma_{1c}^{(1)}$ and $\Gamma_{1c}^{(2)}$ states resulting in a downward shift of the conduction band minimum; in addition, in all those structures having the topmost valence level with $\Gamma_{3v}^{(2)}$ ([111] growth axis) or $\Gamma_{5v}^{(2)}$ ([001] growth axis) symmetry, the valence band maximum is shifted upward. On the whole, the occurrence of these two effects results in a band gap reduction.

In Table VIII and Table IX we report the energy band gaps as obtained from LDA unperturbed calculations (E_{gap}^{unp}) and after the introduction of the spin–orbit coupling (E_{gap}^{LDA}), for [111] and [001] ordered superlattices, respectively. As mentioned in Section 2.4.1, the band gap is underestimated in the local density approximation and correction algorithms (such as the GW approximation [53,54]) are extremely computationally demanding for complex ternary structures. We have thus corrected our LDA results, using experimental values for the binary constituents and taking into account the effect of strain [85]; the final results (E_{gap}^{emp}) are reported in Table VIII and Table IX

TABLE VIII

Band-gap energies (in eV) for $(GaSb)_1/(InSb)_1$ [111] systems and $(InAs)_1/(InSb)_1$ [111] obtained from unperturbed LDA calculations (E_{gap}^{unp}), with the introduction of spin–orbit coupling (E_{gap}^{LDA}) and corrected using experimental data (E_{gap}^{emp}). We also report the calculated bowing parameter ($b_{gap}^{[111]}$) for the different systems considered.

	$(GaSb)_1/(InSb)_1$			$(InAs)_1/(InSb)_1$		
	ER	S1	S2	ER	S3	S4
E_{gap}^{unp}	−0.81	−0.64	−1.47	−1.02	−0.95	−1.42
E_{gap}^{LDA}	−1.05	−0.90	−1.59	−1.26	−1.20	−1.49
E_{gap}^{emp}	0.05	0.20	−0.47	−0.26	−0.21	−0.51
$b_{gap}^{[111]}$	1.92	1.32	4.08	2.42	2.18	3.34

TABLE IX

Band-gap energies (in eV) for $(GaSb)_1/(InSb)_1$ [001] and $(InAs)_1/(InSb)_1$ [001] symbols as in Table VIII.

	$(GaSb)_1/(InSb)_1$			$(InAs)_1/(InSb)_1$		
	ER	S1	S2	ER	S3	S4
E_{gap}^{unp}	−0.49	−0.43	−0.77	−0.68	−0.69	−0.96
E_{gap}^{LDA}	−0.74	−0.69	−0.96	−0.88	−0.89	−1.07
E_{gap}^{emp}	0.33	0.41	0.13	0.11	0.10	−0.09
$b_{gap}^{[001]}$	0.68	0.48	1.56	0.90	0.94	1.66

(the numerical uncertainty is ± 0.05 eV). We note that the free standing mode structures have a band gap of 0.05 ± 0.05 eV for GaSb/InSb superlattices (semiconducting properties), while for InAs/InSb Sls a band gap of -0.26 ± 0.05 eV (semimetallic properties) is found.

The band-gap trend as a function of the substrate lattice constant and its dependence on the ordering direction are illustrated in Fig. 5, where we report the LDA band-gap (E^{LDA} – solid line) and corrected band-gap (E^{emp} – dashed line) as a function of the substrate lattice constant for $(\text{GaSb})_1/(\text{InSb})_1$ (Fig. 5(a)) and $(\text{InAs})_1/(\text{InSb})_1$ (Fig. 5(b)). A comparison between the superlattice energy band-gaps and the average value (E_{ave}) of the experimental (LDA) band-gap in the pure binaries – indicated by the filled (empty) circles – clearly shows the band-gap narrowing effect.

Some remarkable trends can be noticed for the energy band gaps of the systems studied; these can be summarized in the following predicted [86] relation

$$E_{\text{gap}}^{[111]} < E_{\text{gap}}^{[001]} < E_{\text{gap}}^{\text{ave.}} \tag{32}$$

where $E_{\text{gap}}^{[111]}$ and $E_{\text{gap}}^{[001]}$ are the band gaps in [111] and [001] ordered superlattices (see Section 2.4.1 for the binary band gap energies, which determine $E_{\text{gap}}^{\text{ave.}}$). Both the band gap narrowing – second inequality (as explained above) – and the trend as a function of the growth direction – first inequality – have to be ascribed to the level repulsion mechanism. Within the successful perturbative approach [42] used to explain this effect from a theoretical point of view, the magnitude of the band gaps is determined by the symmetry properties of the matrix element (arising from the perturbational scheme) relative to the states involved. The difference between these quantities in [111] and [001] ordered superlattices is immediately reflected in different energy shifts of the valence band maximum and conduction band minimum levels, thus explaining the dependence of the band gap reduction on the growth axis.

The narrowing is also confirmed by the band gap bowing parameter (related to the difference between the band-gap average of the binaries ($E_{\text{gap}}^{\text{ave}}$) and the superlattice band-gap ($E_{\text{gap}}^{\text{SL}}$) and defined, in analogy with the 50%–50% alloys, as $b_{\text{gap}} = 4(E_{\text{gap}}^{\text{ave}} - E_{\text{gap}}^{\text{SL}})$) reported in Table VIII and Table IX. We note that b_{gap} is larger in [111] than in [001] ordered systems, and that b_{gap} is larger in common cation than in common anion superlattices.

Especially in view of the possible applications of these ternary structures as fundamental components in technological devices, let us

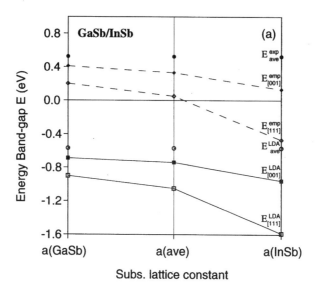

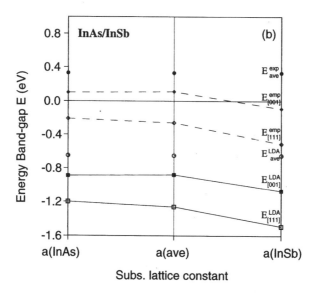

Fig. 5 LDA band-gap (E^{LDA} – solid line) and corrected band-gap (E^{emp} – dashed line) as a function of the substrate lattice constant for the [111] and [001] (a) $(\mathrm{GaSb})_1/(\mathrm{InSb})_1$ superlattices and (b) $(\mathrm{InAs})_1/(\mathrm{InSb})_1$ superlattices. Filled (empty) circles indicate the average value of the experimental (LDA) band-gap in the pure binaries.

now observe that strain conditions offer a good possibility of tuning the superlattice band gap. In fact, we notice that the band gap energy decreases as the substrate lattice constant is increased. Moreover, the range in which E_{gap} varies is as large as 0.7 eV in GaSb/InSb systems and 0.3 eV in InAs/InSb systems,

b. Charge density distribution The chemical term (ΔV^{chem}) of the perturbative potential, mainly determined by the differences in atomic orbital energies, has been shown [87] to affect also the localization of the different states in one of the constituent sublattices of the ternary structures. In order to emphasize this aspect, we report in Table X the angular decomposition for GaSb/InSb systems in the various growth conditions for the relevant states at the Brillouin zone center (we will not discuss the results regarding the charge density decomposition in the common cation superlattices, due to their substantial similarity with the GaSb/InSb ones).

We can see that the charge density distribution for the Γ_{1v} state shows a growing s character on the InSb sublattice and a decreasing p character on the In atom as the substrate lattice constant is increased; at the same time, the s charge density relative to the GaSb monolayer decreases, while the p charge increases on the Ga atom. The charge density distribution of this state is illustrated in Fig. 6(a) for the elastically relaxed GaSb/InSb [111] ordered superlattice. As one can see in Table X, this state derives from p states and from Fig. 6 one can observe that the typical "butterfly" shape along the z axis shows that the state is made of p_z orbitals. A strong bonding character between alternating monolayers and within each of the sublattices is also evident.

TABLE X
Angular decomposition relative to the muffin-tin charge density (for s (Q_s) and p (Q_s) components) of the different states, neglecting s.o. coupling, for $(GaSb)_1/(InSb)_1$ [111].

State		S1				ER				S2			
		Ga	Sb	In	Sb	Ga	Sb	In	Sb	Ga	Sb	In	Sb
Γ_{1v}	Q_s	0.065	0.057	0.006	0.005	0.003	0.001	0.014	0.017	0.007	0.001	0.045	0.023
	Q_p	0.037	0.187	0.060	0.134	0.040	0.218	0.048	0.184	0.074	0.115	0.014	0.241
$\Gamma_{3v}^{(2)}$	Q_s	0.000	0.000	0.000	0.000	0.000	0.000	0.000	0.000	0.000	0.000	0.000	0.000
	Q_p	0.023	0.122	0.061	0.321	0.025	0.117	0.063	0.306	0.026	0.114	0.063	0.294
$\Gamma_{1c}^{(1)}$	Q_s	0.262	0.177	0.074	0.072	0.345	0.232	0.047	0.045	0.349	0.217	0.031	0.030
	Q_p	0.019	0.027	0.002	0.063	0.005	0.019	0.006	0.001	0.011	0.031	0.005	0.001
$\Gamma_{1c}^{(2)}$	Q_s	0.028	0.008	0.188	0.161	0.008	0.000	0.228	0.180	0.000	0.003	0.223	0.180
	Q_p	0.073	0.024	0.000	0.025	0.057	0.006	0.001	0.052	0.001	0.106	0.033	0.004

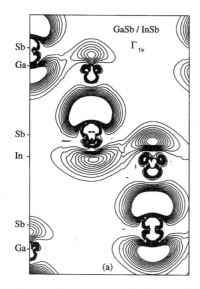 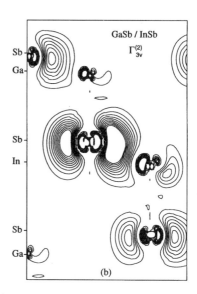

Fig. 6 Charge density distribution (in units of 0.5 e/unit cell) plotted in a plane containing the growth axis for the (a) Γ_{1v} state and (b) the $\Gamma_{3v}^{(2)}$ state in the elastically relaxed [111] ordered common-anion structure.

The $\Gamma_{3v}^{(2)}$ state – whose charge density distribution is drawn in Fig. 6(b) – shows a marked p character on the InSb sublattice, thus affecting the hole carriers spatial localization of the topmost valence level. Furthermore, we point out that the localization of this state is quite independent on strain conditions, as confirmed by Table X.

Let us now discuss the charge density distribution for the two lowest conduction states ($\Gamma_{1c}^{(1)}$ and $\Gamma_{1c}^{(2)}$), illustrated in Fig. 7(a) and (b), respectively. The first conduction state is strongly localized on the GaSb (InAs) sublattice for common anion (common cation) systems; this peculiar feature is strengthened as the substrate lattice constant is increased. A complementary behaviour is shown by the second lowest conduction state, as confirmed by Table X.

On the whole, we obtain a "spatial indirect" band gap: in fact, the valence band maximum is mainly localized on the InSb monolayer, where as the conduction band minimum is localized on the other constituent (GaSb or InAs for the common anion and common cation, respectively).

2.4.3 The 3×3 Superlattices: Valence Band Offsets

The band-offset (i.e., the potential barrier experienced by the charge carriers at the junction of two different semiconductors) is one of the

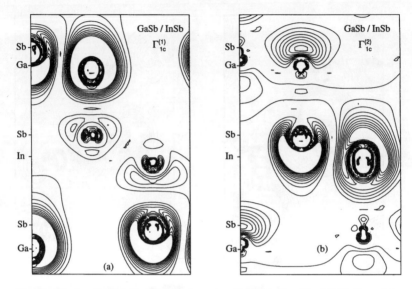

Fig. 7 Charge density distribution (in units of $0.5\,e$/unit cell) plotted in a plane containing the growth axis for the (a) Γ_{1v} (CBM) state and (b) the $\Gamma_{1c}^{(2)}$ state in the elastically relaxed [111] ordered common-anion structure.

most important transport properties, especially in the design of micro- and optoelectronic devices based on quantum wells [88]. In Fig. 8, we illustrate the different valence and conduction band line-ups that can arise at semiconductor interfaces, together with the conventional names used to indicate the relative alignment of the band gaps at the two sides of the heterojunction. The control of the band line-ups at semiconductor heterojunctions has been a great challenge for over 25 years for experimental and theoretical physicists, but the problem of how to influence the alignment when joining two different materials is far from being solved.

In this section, we investigate the effect of growth direction and strain conditions on the band discontinuity at the GaSb/InSb and InAs/InSb interfaces [89], represented by three alternating layers of each binary constituent, leading to 12 atoms in each unit cell (note that the structural parameters are equal to those for the 1×1 superlattices). We should recall that if strain is not taken into account, the valence bands in unstrained GaSb and InSb are predicted to line-up (therefore giving zero valence band offset), whereas the unstrained InSb topmost valence band is expected to be $0.51\,eV$ *higher* in energy than that in unstrained InAs (see Ref. [90] and references therein).

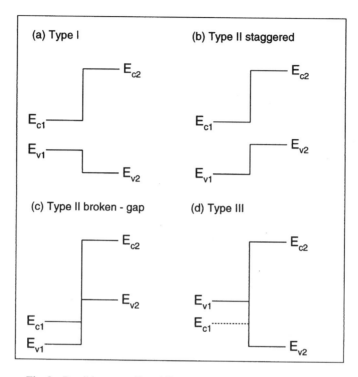

Fig. 8 Possible types of band line-ups at a semiconductor interface.

In our calculations, we have neglected interdiffusion processes which lead to interfacial composition changes, i.e., we consider an atomically abrupt geometry, implying that the interface region is thin with respect to the carrier diffusion length, as confirmed by transmission electron micrographs [27]. Further, we have not taken into account possible relaxations at the interface of the anion–cation distance (bulk bond lengths away from the interface are considered equal to those immediately next to it): this, in fact, is expected [91–93] to introduce little modification of the charge rearrangement at the junction and hence of the VBO.

Following the procedure used in photoemission experiments [94], we have evaluated the valence band offset using core electron binding energies (E_b^c) as reference levels. As illustrated in Fig. 9, the VBO is obtained as the sum of two contributions [10]

$$\Delta E_v = \Delta b + \Delta E_b \tag{33}$$

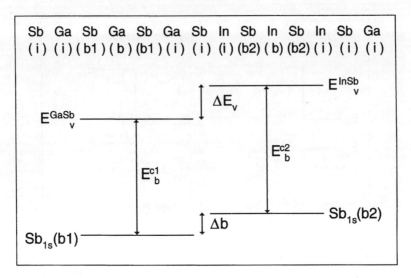

Fig. 9 Schematic diagram of the energy levels in the $(GaSb)_3/(InSb)_3$ superlattices, showing the various quantities involved in the evaluation of the valence band offset.

where Δb is an "interface" term (i.e. the core level alignment of the two atoms belonging to the same chemical species at opposite sides of the interface) and $\Delta E_b = \Delta E_b^{c1} - \Delta E_b^{c2}$ (see Fig. 9) is a "bulk" term (i.e. the binding energy difference, relative to the topmost valence level of the same core levels evaluated in the binary constituents strained as in the superlattice).

In the evaluation of the Δb term, we have considered the s-levels of the common anion in $(GaSb)_3/(InSb)_3$ systems and to the common cation in $(InAs)_3/(InSb)_3$ systems. Other choices of the core levels would lead to VBO values differing from those reported in this work at most by 0.06 eV, which has thus to be considered as our numerical uncertainty.

2.4.3.1 The dependence of the band line-up on growth axis and strain conditions

Our results, reported in Table XI (taking into account the effect of spin–orbit coupling on the topmost valence level), show that the ordering direction seems to influence the valence band line-up, although not very markedly (see the difference in the VBO for [111] and [001] systems in equivalent "average substrate" strain conditions). The dependence of the VBO on the growth axis is not present in lattice matched heterojunctions [95], and thus has to be related only to the appreciable mismatch.

TABLE XI

Interface term (Δb), strained bulk term (ΔE_b) and valence band offset (ΔE_v) for (GaSb)$_3$/(InSb)$_3$ superlattices (taking into account the spin-orbit coupling). Energy differences (in eV) are considered positive if the level relative to the InSb layer is higher in energy with respect to the GaSb (InAs) layer in the common-anion (common-cation) system.

Subs.	(GaSb/InSb) [001]			[111]	InAs/InSb [001]			[111]
	GaSb	Aver.	InSb	Aver.	InAs	Aver.	InSb	Aver.
Δb	+0.20	+0.19	+0.18	+0.29	+0.09	+0.07	+0.05	+0.17
ΔE_b	+0.14	−0.12	−0.34	+0.13	+0.79	+0.47	+0.18	+0.51
ΔE_v	+0.34	+0.07	−0.16	+0.16	+0.88	+0.54	+0.23	+0.68

Let us now discuss the role of strain on the band alignment; we observe that as the substrate lattice constant is varied, the Δb term remains almost constant in all the [001] heterojunctions, suggesting that the charge readjustment is almost uninfluenced by the state of the strain. On the other hand, the bulk contribution is particularly sensitive to the substrate chosen, showing that the core level binding energies are strongly dependent on the pseudomorphic strain conditions. On the whole, we find that the VBO shows a linear decreasing trend as the substrate lattice constant is increased: the smaller the a_{sub}, the more the InSb topmost valence level is higher than in the other superlattice constituent. This trend is illustrated in Figs. 10 and 11 for the common anion and common cation systems respectively. In these figures, our results (filled squares for [001] superlattices and filled circles for [111] superlattices) are compared with other theoretical predictions, obtained from model [83,96], semi-empirical [97] and FLAPW [87] calculations. Note that all the predicted values agree with those of the present work (except those of Ref. [96]), within their uncertainty of a few hundredths of an eV [83] and our error bars, respectively (see Figs. 10 and 11). Incidentally, we observe that a similar disagreement between *ab initio* results and those obtained by Cardona and Christensen [96] was also found in other III–V isovalent heterojunctions, such as GaP/InP [87] and GaAs/InAs [102]. Furthermore, the linear trend of the band offset as a function of the strain found in the present work is in excellent agreement with the predictions of other theoretical work [83,96] and is reasonably expected to reproduce the real situation.

So far, we have completely omitted a discussion of the conduction band offset (ΔE_c), due to well known failures of LDA in predicting the

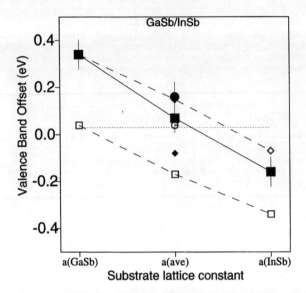

Fig. 10 Valence band offset (in eV) for GaSb/InSb superlattices as a function of the substrate lattice parameter. Our results (together with their error bars) are evidenced by filled squares ([001] superlattices) or filled circles ([111] superlattices) and the solid line. Empty squares, empty diamonds, filled diamonds and empty circles indicate the results of Ref. [95], Ref. [83], Ref. [96] and Ref. [87] respectively. The horizontal dotted line indicates our result for unstrained GaSb and InSb.

correct band gap energies. However, we should now point out that, using an approximate estimate of empirical band gaps (see Ref. [85] for details), we can obtain information on the different kinds of band line-ups as growth conditions are changed. In fact, we find a type I alignment for all the [001] common anion interfaces, while [111] GaSb/InSb shows a type II staggered alignment (partial overlap of the band gaps). On the other hand, for [001] common cation superlattices grown both on an InAs and on an average substrate, we find a type II broken gap line-up, with the InAs conduction band minimum lower in energy than the InSb valence band maximum. Finally, the [001] InAs/InSb grown on InSb and the [111] InAs/InSb heterojunctions contain a semi-metallic compound (InAs), thus leading to a type III alignment.

In order to understand more deeply the effect of strain on the VBO, we have also estimated the band line-up with respect to unstrained binaries for the [001] interfaces. Thus, we substituted in the VBO expression the Δb value obtained for [001] interfaces (which was found to be almost independent of strain effects) and the ΔE_b evaluated

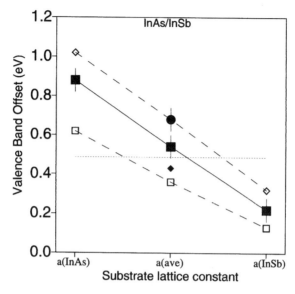

Fig. 11 Valence band offset (in eV) for InAs/InSb superlattices as a function of the substrate lattice parameter. Symbols are the same as those in Fig. 10.

starting from the zincblende bulk unstrained constituents (i.e., disregarding the effect of strain on the binaries valence band maximum). We obtain $\Delta E_v = 0.03$ eV and $\Delta E_v = 0.49$ eV for the GaSb/InSb and InAs/InSb heterojunctions, respectively. These results match perfectly those reported in Ref. [90] and previously mentioned (valence band perfect alignment for the common anion system and $\Delta E_v^{rel} = 0.51$ eV for the common cation case), showing that, if strain is not correctly taken into account, completely different results are found.

It is important to notice that the strain acting on the energy of the valence band maximum level is also responsible for the spatial localization of this state. We find, in fact, that in all of the common anion (common cation) interfaces, the hole carriers are mainly localized on the InSb side of the heterojunction, while in the GaSb/InSb system grown on an InSb substrate we find a complementary situation. This is clear from the decreasing trend of the VBO as the lattice constant is increased and from the sign change in the InSb substrate case (showing that the valence band maximum in GaSb is higher in energy than in InSb). On the whole, our results show that the range of tunability, as a function of growth conditions, is appreciable (about 0.5 eV and 0.7 eV for common anion and common cation respectively), suggesting strain as a valuable tool in designing "*ad hoc*" VBO for technological applications.

Experimentally, many difficulties arise when attempting to pseudomorphically grow the systems in consideration, due to the high mismatch (about 6%) between the lattice constants of the two binary constituents. Recently, however, an InSb quantum well has been realized in GaSb [27]; starting from photoluminescence peak emission energy data and from calculations based on a standard finite square-well model [98,99] (taking strain into account [100,101]), a VBO of 0.16 eV was obtained that is quite different from the one, 0.34 eV, reported in Table XI. Unfortunately, a similar disagreement (which could be due to different strain conditions in the two systems) between theoretical and experimental data obtained using different techniques, was found also for other homopolar isovalent III–V interfaces, such as GaAs/InAs (see Ref. [102] and references therein) and GaP/InP (see Ref. [87] and references therein). Both are photoluminescence systems reasonably close to the common anion studied here. In particular, in a recent experimental work focused on GaAs/InAs superlattices, Ohler *et al.* [103] obtained (from ultraviolet photoelectron spectroscopy measurements of the cation d-core levels) a ΔE_v value in disagreement (by as much as 0.3 eV) with theoretical predictions [83,104,105,102]. Notwithstanding some differences between the experimental and theoretical VBO values, many important observations discussed above are confirmed by the Ohler *et al.* work [103]. For example, the linear trend found [103] for ΔE_v (in GaAs/InAs superlattices) as a function of a_{sub} agrees with other theoretical predictions [102] and with our results. Further, the independence of the interface term on the strain conditions discussed above is experimentally confirmed [103] by the trend of the In $4d_{5/2}$ and Ga $3d_{3/2}$ core-level binding energy difference.

2.4.3.2 The InAs/GaSb interface: validity of the transitivity rule

Recently the InAs/GaSb interface has been the subject of many experimental [106] and theoretical investigations [107,108], due to its peculiar type II band lineup (broken band gap) and the simple growth of this almost lattice matched heterojunction (the two different lattice parameters differ only by 0.6%). According to the transitivity rule (well established to within 0.02 eV for [001] ordered interfaces [93]), we can derive the valence band offset for the InAs/GaSb interface starting from the InAs/InSb/GaSb system, provided that both the InAs/InSb and InSb/GaSb heterojunctions are grown epitaxially on the same substrate (we chose an InAs substrate). We have thus interpolated the calculated common anion VBOs using the theoretical linear trend found as a function of the substrate lattice parameter; we notice, in fact, that this approximation is expected to be reasonably valid, since the substrate lattice constant (a_{InAs}) of this system is very close (within 0.6%) to one of our self-consistent calculated VBO values (for the system grown on a GaSb substrate).

We thus report in Fig. 12 our calculated VBO for the common-cation superlattice on an InAs substrate and the extrapolated value for the ideal common-anion superlattice grown on an InAs substrate. Using the transitivity rule, we obtain:

$$\Delta E_v(InAs/GaSb) = \Delta E_v(InAs/InSb)_{InAs\text{-subs.}} - \Delta E_v(InSb/GaSb)_{InAs\text{-subs.}},$$

so that $\Delta E_v(InAs/GaSb)_{InAs\text{-subs.}} = 0.88 - 0.40 = 0.48 \, eV$. This result is in good agreement with the available experimental values (0.46 eV [109], 0.51 eV [110]) and represents a further proof of the validity of the transitivity rule of for [001] interfaces.

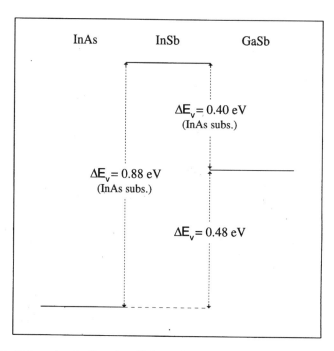

Fig. 12 Valence band offset (in eV) for InAs/GaSb interface, obtained using the transitivity rule.

2.5 CONCLUSIONS

Great interest has been devoted in the past few decades to superlattices and interfaces: the possibility of tuning "*ad hoc*" electronic and transport properties in such artificial structures as a function of different parameters (i.e., constituent materials, compositon, layer thickness etc.), has encouraged the development of a new and promising field in materials science for optoelectronic applications, the so called "band engineering" [111].

Strain and ordering direction constitute valid tools in order to appreciably modify the electronic properties of interest (such as energy band gap and band alignment at a semiconductor heterojunction) in superlattices. In this work, we have focused our attention on Sb-related structures, in particular on ultrathin [001] and [111] common anion GaSb/InSb and common cation InAs/InSb superlattices in different strain conditions. Density functional theory and its powerful implementations, such as the *all-electron* FLAPW method, have been used to predict the structural and electronic parameters in such structures.

The relevant results for the band gaps can be summarized as follows:

1. Due to repulsion mechanisms involved in band folding [42], the band gap in the superlattices is smaller than the band gap average of the binary constituents ($E_{\text{gap}}^{\text{ave}}$), showing a dependence on the ordering direction expressed by the relation $E_{\text{gap}}^{[111]} < E_{\text{gap}}^{[001]} < E_{\text{gap}}^{\text{ave}}$;
2. The range in which the gap varies as a function of pseudomorphic growth conditions is as large as 0.7 eV (0.3 eV) in [111] GaSb/InSb (InAs/InSb), showing good opportunites for band gap tuning;
3. A decreasing trend is observed in the band gap energy as the substrate lattice parameter is increased.

In view of their possible applications as fundamental components in micro- and optoelectronic devices, particular attention was devoted to the band line-up at the GaSb/InSb and InAs/InSb interfaces. Our results seem to indicate that in the same strain conditions, the growth axis slightly affects the band offset, probably because of the different structural relaxation of the interface atoms at the [001] and [111] heterojunctions. On the other hand, a much more important effect is due to pseudomorphic growth on different substrates: the appreciable tunability of the VBO (about 0.5 eV and 0.7 eV for common-anion and for common-cation superlattices, respectively) is evidenced by its linear decreasing trend as the substrate lattice constant is increased, mainly due to the bulk contribution to the band line-up.

Finally, the transitivity rule has been used to determine the [001] InAs/GaSb type II valence band offset, and good agreement between our theoretical predictions and available experimental data is obtained.

In conclusion, strained layer superlattices represent a powerful resource in new materials science, due to the additional flexibility – arising from various strain conditions and different growth directions in tuning the electronic properties with respect to traditional lattice matched [001] ordered systems. It is clear that the promising field of artificial structures based on semiconductor constituents is likely to develop further in the next few years, both from the experimental and theoretical points of view. On one hand, solid state physicists will look for new tools (such as intralayer deposition, doping, etc.) in order to obtain further degrees of freedom to tune the electronic properties of technological interest. On the other hand, they will also investigate the microscopic effects which determine the resulting electronic properties, in order to reach complete understanding and control of the band structure in these systems.

Acknowledgments

Work at Northwestern University supported by the MRL Program of the National Science Foundation, at the Materials Research Center of Northwestern University, under Award No. DMR-9120521, and by a grant of computer time at Pittsburgh Supercomputing Center. A partial support by a supercomputing grant at Cineca (Bologna, Italy) through the Consiglio Nazionale delle Ricerche (CNR) is also acknowledged. We thank B. W. Wessels for helpful discussions.

References

1. See, for example, "*Materials for* Infrared Detectors and Sources", R. Farrow, J. F. Schetzina and J. J. Cheung (Materials Research Society, Pittsburg, 1987), Vol. 90.
2. W. Schottky, Z. Physik **113**, 367 (1962).
3. R. L. Anderson, *Solid State Electron.* **5**, 341 (1962).
4. W. Harrison, *J. Vac. Sci. Technol.* **16**, 1492 (1979).
5. C. G. Van de Walle and R. Martin, *Phys. Rev. B* **37**, 4801 (1988) and references therein.
6. J. Tersoff, *Phys. Rev. Lett.* **52**, 465 (1984).
7. J. Tersoff and W. A. Harrison, *J. Vac. Sci. Technol. B* **5**, 1221 (1987).
8. W. R. Lambrecht and B. Segall, *Phys. Rev. B* **41**, 2832 (1990).
9. S. Baroni, R. Resta, A. Baldereschi and M. Peressi, in "*Spectroscopy of Semiconductor Microstructures*", Vol. 206 of NATO Advanced Study Institute, Series B: Physics,

edited by G. Fasol, A. Fasolino and P. Lugli (Plenum, New York, 1989); A. Baldereschi, S. Baroni and R. Resta, *Phys. Rev. Lett.* **61**, 734 (1988).
10. S. Massidda, B. I. Min and A. J. Freeman, *Phys. Rev. B* **35**, 9871 (1987).
11. S. H. Wei and A. Zunger, *Phys. Rev. Lett.* **59**, 144 (1987).
12. For a review see A. Zunger and S. Mahajan in *"Handbook of semiconductors"*, 2nd ed., edited by S. Mahajan (Elsevier, Amsterdam, in press), Vol. 3 (and references therein).
13. D. L. Smith and C. Mailhot, *Reviews of Modern Physics* **62**(1), 173 (1990).
14. P. Hohenberg and W. Kohn, *Phys. Rev.* **136**, B864 (1964).
15. W. Kohn and L. J. Sham, *Phys. Rev.* **140**, A1113 (1965); W. Kohn and L. J. Sham, *Phys. Rev.* **145**, 561 (1966)
16. H. J. F. Jansen and A. J. Freeman, *Phys. Rev. B* **30**, 561 (1984); M. Weinert, H. Krakauer, E. Wimmer and A. J. Freeman, *Phys. Rev. B* **24**, 864 (1981) and references therein.
17. M. Weinert, E. Wimmer and A. J. Freeman, *Phys. Rev. B* **26**, 4571 (1982).
18. M. B. Small and I. Crossley, *J. of Crystal Growth* **27**, 35 (1974).
19. M. A. Herman and H. Sitter, *"Molecular Beam Epitaxy – Fundamentals and current status"* Springer Ser. Mater. Sci. Vol. 7 (Springer, Berlin, 1989).
20. G. B. Stringfellow, *"Organo-Metallic Vapour-Phase Epitaxy – Theory and Practice"* (Academic, NY, 1990).
21. H. Kroemer, *Surface Sci.* **132**, 543 (1983).
22. See the pioneering articles: P. Perfetti *et al.*, *Appl. Phys. Lett.* **33**, 667 (1978); R. S. Bauer and J. C. McMenamin, *J. Vac. Sci. Technol.* **15**, 1444 (1978); R. W. Grant, J. R. Waldrop and E. A. Kraut, *Phys. Rev. Lett.* **40**, 656 (1978).
23. R. Dingle, W. Weigman and C. H. Henry, *Phys. Rev. Lett.* **33**, 827 (1974).
24. D. V. Lang, A. M. Sergent, M. B. Panish and H. Temkin, *Appl. Phys. Lett.* **33**, 667 (1978).
25. J. E. Schriber, I. J. Fritz, and L. R. Dawson, *Appl. Phys. Lett.* **46**, 187 (1985).
26. A. R. Adams, *Electron Lett.* **22**, 249 (1986).
27. L. Q. Qian and B. W. Wessels, *Appl. Phys. Lett.* **63** (5), 628 (1993).
28. G. Bougnot *et al.*, *J. Cryst. Growth* **107**, 502 (1991).
29. I. H. Campbell *et al.*, *Appl. Phys. Lett.* **59**, 846 (1991).
30. Y. K. Su, F. S. Juang and C. H. Su, *J. Appl. Phys.* **71**, 1368 (1992).
31. R. J. Warburton *et al.*, *Surf. Science* **288**, 270 (1990).
32. L. Q. Qian and B. W. Wessels, *J. Vac. Sci. Technol. B* **11**(4), 1652 (1993).
33. R. H. Miles, D. H. Chow, J. N. Schuman and T. C. McGill, *Appl. Phy. Lett.* **57**, 801 (1990).
34. C. Mailhot and D. Smith, *J. Vac. Sci. Technol. B* **5**, 1268 (1987).
35. C. Mailhot and D. Smith, *J. Vac. Sci. Technol. A* **7**, 445 (1989).
36. S. M. Chen, Y. K. Su and Y. T. Lu, *J. Appl. Phys.* **74**, 7288 (1993).
37. S. Y. Lin *et al.*, *Appl. Phys. Lett.* **57**, 1015 (1990).
38. S. R. Kurtz *et al.*, *Appl. Phys. Lett.* **52**, 831 (1958).
39. G. C. Osbourn, *J. Vac. Sci. Technol. B* **2**, 176 (19 4).
40. J. R. Yen, Y. Ma and G. B. Stringfellow, *Appl. Phys. Lett.* **54**, 1154 (1989).
41. Y. E. Ihm *et al.*, *Appl. Phys. Lett.* **51**, 2013 (1987).
42. S. H. Wei and A. Zunger, *Phys. Rev. B* **39**, 3279 (1989).
43. S. R. Kurtz *et al.*, *Phys. Rev. B* **46**, 1909 (1992).
44. T. C. Koopman, *Physica* **1**, 104 (1933).
45. L. Hedin and B. I. Lundqvist, *J. Phys. C.* **4**, 2064 (1971).
46. D. Ceperley and B. Alder, *Phys. Rev. Lett.* **45**, 566 (1980).
47. K. S. Singwi, A. Sjölander, M. P. Tosi and R. H. Land, *Phys. Rev. B* **1**, 1044 (1970).
48. O. Gunnarsson and B. J. Lundqvist, *Phys. Rev. B* **13**, 4274 (1976).
49. J. P. Perdew, *Phys. Rev. B* **33**, 8822 (1986).
50. A. D. Becke, *Phys. Rev. A* **38**, 3098 (1988).

51. J. P. Perdew et al., *Phys. Rev. B* **46**, 6671 (1992).
52. J. P. Perdew and A. Zunger, *Phys. Rev. B* **23**, 5048 (1981).
53. M. S. Hybertsen and S. G. Louie, *Phys. Rev. B* **34**, 5390 (1986).
54. R. W. Godby, M. Schlüter and L. J. Sham, *Phys. Rev. Lett.* **56**, 2415 (1986).
55. J. C. Phillips, *Phys. Rev.* **112**, 685 (1958).
56. M. L. Cohen and V. Heine, *Solid State Physics*, **24**, 37 (1970).
57. D. R. Hamann, M. Schlüter and C. Chiang, *Phys. Rev. Lett.* **43**, 1494 (1979).
58. A. Zunger and M. L. Cohen, *Phys. Rev. B* **20**, 4082 (1979); A. Zunger and M. L. Cohen, *Phys. Rev. Lett.* **56**, 2656 (1979).
59. G. Kerker, *J. Phys. C* **13**, L189 (1980).
60. G. B. Bachelet, D. R. Hamann and M. Schlüter, *Phys. Rev. B* **26**, 4199 (1982).
61. M. Methfessel, C. O. Rodriquez and O. K. Andersen, *Phys. Rev. B* **40**, 2009 (1989).
62. M. C. Payne et al., *Phys. Rev. Lett.* **56**, 2656 (1986).
63. R. Car and M. Parrinello, *Phys. Rev. Lett.* **55**, 2471 (1985).
64. D. D. Koelling and G.O. Arbman, *J. Phys. F: Metal Phys,* **5**, 2041 (1975); O.K. Andersen, **12**, 3060 (1975).
65. J. C. Slater, *Phys. Rev.* **51**, 846 (1937).
66. A. Baldereschi, *Phys. Rev. B* **7**, 5212 (1973).
67. D. J. Chadi and M. L. Cohen, *Phys. Rev. B* **8**, 5747 (1973).
68. H. J. Monkhorst and J. D. Pack, *Phys. Rev. B* **13**, 5188 (1976).
69. F. C. Von der Lage and H. A. Bethe. *Phys. Rev.* **71**, 612 (1947).
70. M. Weinert, *J. Math. Phys.* **22**, 2433 (1981).
71. C. G. Broyden, *Math. Comp.* **19**, 577 (1965).
72. J. D. Bjorken and S. D. Drell in *Relativistic Quantum Mechanics*, McGraw-Hill, New York (1964).
73. D. D. Koelling and B. N. Harmon, *J. Phys. C.* **10**, 3107 (1977).
74. H. J. Monkhorst and J. D. Pack, *Phys. Rev. B* **13**, 5188 (1976).
75. F. D. Murnaghan, *Proc. Natl. Acad. Sci. USA.* **30**, 244 (1944).
76. H. Müller, R. Trommer, M. Cardona and P. Vogl, *Phys. Rev. B* **21**, 4879 (1980).
77. *Landolt-Börnstein: Numerical Data and Functional Relationships in Science and Technology*, edited by O. Madelung, M. Schulz and H. Weiss (Springer-Verlag, Berlin, 1982), Vol. 17a.
78. R. Magri and C. Calandra, *Phys. Rev. B* **40**, 3896 (1989).
79. *Handbook of Chemistry and Physics*, 69th ed., edited by R. W. Weast (Chemical Rubber Co., Boca Raton, FL, 1989).
80. R. Zallen and W. Paul, *Phys. Rev.* **155**, 703 (1967).
81. L. Ley, R. A. Pollak, F. R. McFeely, S. P. Kowallczyk and D. A. Shirley, *Phys. Rev. B* **9**, 600 (1974).
82. W. A. Harrison "*Electronic Structure and the Properties of Solids*" (Freeman, San Francisco, 1980).
83. C. G. Van de Walle, *Phys. Rev. B* **39**, 1871 (1989).
84. J. J. Hopfield, *J. Phys. Chem. Solids* **15**, 97 (1960).
85. S. Picozzi, A. Continenza and A. J. Freeman, *Phys. Rev. B* **52**, 5247 (1995).
86. S. H. Wei and A. Zunger, *Appl. Phys. Lett.* **58**, 2685 (1991).
87. A. Franceschetti, S. H. Wei and A. Zunger, *Phys. Rev. B* **50**, 8094 (1994).
88. For a review on semiconductor heterojunctions see G. Margaritondo, "*Electronic structure of semiconductor heterojunctions*", (Kluwer Academic Publishers – Jaca Book, Milan, 1988).
89. S. Picozzi, A. Continenza and A. J. Freeman, to appear in *Phys. Rev. B*.
90. D. L. Smith and C. Mailhot, *J. Appl. Phys.* **62**(6), 25545 (1987).
91. A. Continenza, S. Massidda and A. J. Freeman, *Phys. Rev. B* **42**, 3469 (1990).
92. R. G. Dandrea, C. B. Duke and A. Zunger, *J. Vac. Sci. Technol. B* **10**(4), 1744 (1992).
93. Y. Foulon and C. Priester, *Phys. Rev. B* **45**, 6259 (1992).
94. A. D. Katnani and R. S. Bauer, *Phys. Rev. B* **33**, 1106 (1986).

95. S. Baroni, M. Peressi, R. Resta and A. Baldereschi, *Theory of Band Offsets at Semiconductor Heterojunctions*, in *Proceedings of the 21th International Conference on the Physics of Semiconductors*, edited by Ping Jiang and Hou-Zhi Zheng (World Scientific, Singapore, 1993), p. 689.
96. M. Cardona and N. Christensen, *Phys. Rev. B* **35**, 6182 (1987).
97. A. Ichii, Y. Tsou and E. Garmine, *J. Appl. Phys.* **74**(3), 2112 (1993).
98. G. Bastard, *Phys. Rev. B* **24**, 5693 (1981).
99. KlAlavi et al., *Electron. Lett.* **19**, 227 (1983).
100. H. Asai and K. Oe, *J. Appl. Phys.* **54**, 2052 (1983).
101. G. Ji et al., *J. Appl. Phys.* **62**, 3366 (1987).
102. N. Tit, M. Peressi and S. Baroni, *Phys. Rev. B* **48**, 17607 (1993).
103. C. Ohler et al., *Phys. Rev. B* **50**, 7833 (1994).
104. C. Priester, G. Allan and M. Lannoo, *Phys. Rev. B* **38**, 9870 (1988).
105. A. Taguchi and T. Ohno, *Phys. Rev. B* **39**, 7803 (1989).
106. K. F. Longebach, L. F. Luo and W. I. Wang, *Appl. Phys. Lett.* **57**(15), 1554 (1990).
107. Jun Shen, Shang Yuan Ren and John D. Dow, *Phys. Rev. B* **46**, 6938 (1992).
108. L. A. Hemstreet, C. Y. Fong and J. S. Nelson, *J. Vac. Sci. Technol. B* **11**(4), 1693 (1993).
109. M. Jaros, *Phys. Rev. B* **37**, 7112 (1988).
110. H. Kroemer, *J. Vac. Sci. Technol. B* **2**, 433 (1984).
111. F. Capasso, *Ann. Rev. Mat. Sci.* **16**, 263 (1986).

CHAPTER 3

Interfacial Disorder in InAs/GaSb Heterostructures Grown by Molecular Beam Epitaxy

M. E. TWIGG[1], B. R. BENNETT[1], P. M. THIBADO[2],
B. V. SHANABROOK[1] and L. J. WHITMAN[2]

[1]*Electronics Science and Technology Division;* [2]*Chemistry Division, Naval Research Laboratory, Washington, DC 20375-5347, USA*

3.1.	Introduction	55
3.2.	Control of Interfacial Bonding	62
3.3.	Using High-Resolution Transmission Electron Microscopy to Study Interfacial Disorder	70
	3.3.1. HRTEM contrast of InAs/GaSb superlattices	71
	3.3.2. Quantitative image processing of HRTEM data	77
	3.3.3. The Assessment of interfacial disorder via HRTEM	79
3.4.	Interfacial Morphology	84
3.5.	Conclusions	89
	Acknowledgments	91
	References	91

3.1 INTRODUCTION

Heterostructures of InAs/GaSb have served as the archetypal system for the study of the type II semiconductor band alignment [1]. Similar to other type II heterostructures, the valence and conduction band offsets have the same sign so that one type of layer confines electrons (InAs), while its companion (GaSb) provides potential wells for holes. However, the InAs/GaSb system is unique in that the InAs band gap is smaller than the valence band offset between GaSb and InAs. As a result, the top of the GaSb valence band is located in energy above the bottom of the InAs conduction band.

By forming an InAs/GaSb superlattice (SL), the effective band gap can be tuned within a well-defined range. Solving the Schrodinger

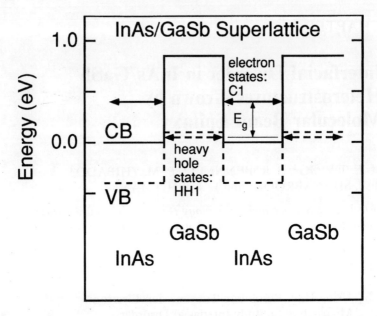

Fig. 1 Band diagram of InAs/GaSb superlattice. C1 is the lowest lying electron state; HH1 is the highest valence band state. Eg, the difference between the energies of C1 and HH1, is the gap for the superlattice sub-band.

equation for the periodic potentials defined by the SL, as shown in Fig. 1 (for a SL grown on a (100) plane) one arrives at SL states for the conduction band and valence band. The lowest lying conduction band state is designated as C1; the highest valence band state, HH1, is a heavy hole state. As the SL period increases, the bottom of the SL conduction sub-band moves toward the bottom of the InAs conduction band while the SL heavy hole sub-band approaches the top of the GaSb valence band (as shown in Fig. 2(a)). Because the GaSb valence band is above the InAs conduction band, the rise in the energy of HH1 sub-band edge and the fall in the energy of C1 sub-band edge eventually leads to a crossing of these two bands at a superlattice period of 15 nm. Therefore, the effective bandgap of a InAs/GaSb superlattice can be made to assume almost any value less than 0.5 eV – a most intriguing attribute [2].

Because of the ability of InAs/GaSb SLs to assume a band gap less than 0.5 eV, it would seem that this material could function as an infrared detector for any wavelength greater than 2.5 μm. By growing a SL with a period of 10 nm, an effective bandgap of 0.12 eV could be

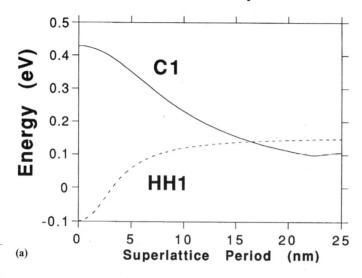

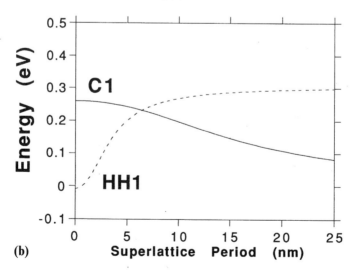

Fig. 2 Positions of C1 conduction sub-band and HH1 heavy hole valence sub-band edges as a function of SL period. The thickness of InAs and GaSb layers are set to be equal. (a) For InAs/GaSb; (b) For InAs/Ga$_{0.6}$In$_{0.4}$Sb.

achieved, corresponding to the technologically important 10 μm wavelength regime. Unfortunately, due to the real space separation between the electrons and holes, as the superlattice period increases, the wavefunction overlap decreases rapidly. Because of this small overlap, the optical absorption coefficient of InAs/GaSb SLs is below that of the principal material currently used in infrared detectors: $Hg_{1-x}Cd_xTe$ [3].

Because of the large SL period required to achieve the desired bandgap of 0.12 eV, InAs/GaSb is not a suitable material for infrared detectors. A shorter SL period can be obtained through the substitution of indium for a significant atomic fraction of the gallium in the SL. $Ga_{1-x}In_xSb$ has a larger lattice parameter than does GaSb, leading to larger coherent strains in the SL layers. Because biaxial tension lowers the InAs conduction band and biaxial compression raises the $Ga_{1-x}In_xSb$ heavy hole band, the band gap for a given superlattice period increases with the indium atomic fraction [4,5]. As shown in Fig. 2(b), for an $InAs/Ga_{0.6}In_{0.4}Sb$ SL strained to the in-plane GaSb (100) lattice parameter (as would be the case for such a SL grown on GaSb (100)), the effective bandgap falls to 0.12 eV at a SL period of less than 5 nm. Because a narrow band gap is achieved at such a small value of the SL period, wavefunction overlap is greater. As a result, an $InAs/Ga_{1-x}In_xSb$ SL can have an absorption capability on par with bulk $Hg_{1-x}Cd_xTe$ over the technologically important 8–14 μm range. In addition, $InAs/Ga_{1-x}In_xSb$ SLs have lower leakage currents, greater uniformity, and the ability to withstand higher processing temperatures, as compared to bulk $Hg_{1-x}Cd_xTe$ [6].

The novel properties of a small-period semiconductor SL, however, depend on the ability to grow thin layers with thickness and composition controlled on an atomic scale. This degree of control can be achieved by molecular beam epitaxy (MBE), a form of vapor phase growth in which molecular or atomic beams deposit a thin film onto a substrate surface which is maintained at an elevated temperature in ultra-high vacuum [7,8]. The challenge, however, is in controlling the structure of a strained SL given the large volume fraction consisting of interfaces. In forming an interface, the growth temperature needs to be sufficiently high (and the growth rate sufficiently low) to allow the adatoms arriving on the surface to assemble into a flat monolayer before the next monolayer begins; at the same time the temperature cannot be so high as to promote diffusion below the surface and into the bulk. At each thermal extreme there is a potential complication. At too low a temperature, the deposited monolayer may not have the opportunity to smoothly distribute itself (and thereby reduce surface roughness). In growing at too high a temperature, intermixing may occur at the interface. All aspects of interfacial disorder are, of course, undesirable in opto-electronic devices. Interfacial roughness, the result of too low a growth temperature, is thought to have a deleterious effect on carrier mobilities [9]. Interfacial intermixing, due to too high a growth temperature, can alter the carefully engineered bandgap [10].

By the term *interfacial disorder*, we mean any deviation from an interface that is perfectly smooth and compositionally abrupt. In Fig. 3 we show the two possible configurations for a perfectly ordered InAs/GaSb interface: one consisting of InSb-like bonds and the other consisting of GaAs-like bonds. In large part, there are two major components to interfacial disorder (as shown in Fig. 4). The first component, interfacial roughness, is due to the formation of steps and

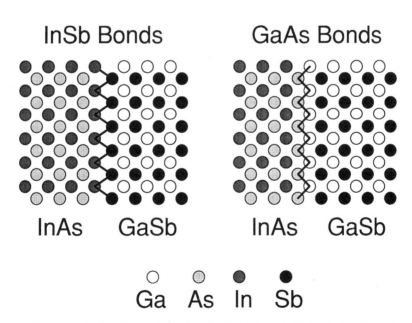

Fig. 3 Interfacial bonding configuration for InSb-like and GaAs-like bonds at the InAs/GaSb interface as viewed along the [100] direction.

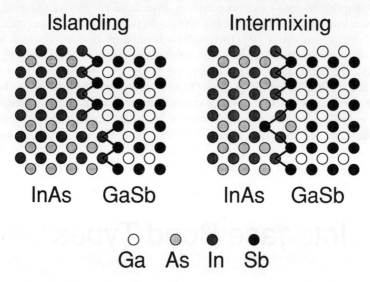

Fig. 4 Interfacial bonding configuration for InSb-like interfaces perturbed by roughness and intermixing as viewed along the [100] direction.

islands which in turn lead to uneven coverage of the heteroepitaxial surface [11–14]. The second component, interfacial diffuseness, is due to mixing driven by both stochastic processes (simple diffusion) and differences in bonding energies (exchange reactions) [15]. Strain may also play a role in diffusion kinetics. Because the chemical potential increases with strain, and atomic diffusion follows the gradient of the chemical potential, strain at the interface may undermine interfacial stability and promote interfacial diffuseness [16–18].

The precarious balance that must be maintained in MBE growth may also be undermined by the complexities of heteroepitaxy. Because heteroepitaxy inevitably leads to growing a monolayer of one lattice parameter and binding energy upon a surface with a different lattice parameter and binding energy, one is faced with the problem of maintaining at least two different kinds of surfaces in equilibrium while growing the metastable film. The first constituent may be easily deposited at a given temperature and growth rate, with the surface in equilibrium over the course of the growth and the bulk kinetics held in abeyance. The second constituent, may, however, be more strongly bonded than the first constituent, so that reduced surface diffusion occurs and therefore the surface is not able to maintain equilibrium during growth [19]. Differences in lattice parameter and the strain that accompanies this difference may also prevent the formation of large flat terraces on the surface.

The existence of a wide range of binding energies among the adatoms forming a heterostructural interface enhances the likelihood of exchange reactions and contributes to interfacial diffuseness. In Fig. 5, we show the range of binding energies and lattice parameters for the constituents of the InAs/GaSb heteroepitaxial growth system [20]. This wide range in binding energies and lattice parameters is likely to frustrate any attempt to achieve the desired layer by layer growth that is necessary for perfect epitaxy [21]. One might expect for GaAs (or InAs) on GaSb to present more of a problem than for InSb (or GaSb) on InAs, since the antimonides have lower binding energies than the arsenides. The energy of a given surface might then be reduced by the exchange of a surface As atom with a sub-surface Sb

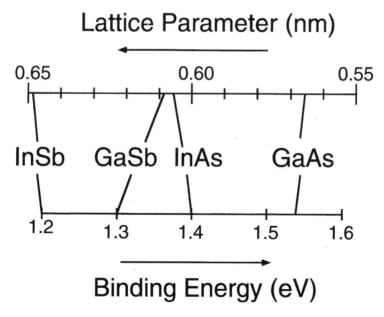

Fig. 5 Range of lattice parameters and binding energies for InSb, GaSb, InAs, and GaAs, the four possible cation–anion bonding configurations of the InAs/GaSb system.

atom. Therefore exchange reactions would tend to occur when arsenides are grown on antimonides.

The nature of the growth surface itself may contribute to the difficulties in establishing a smooth and abrupt heteroepitaxial interface. Even when the first monolayer can be grown under equilibrium conditions and thereby establish 2-dimensional (2-D) growth (due to the low surface energy of the deposited layer relative to the substrate), we are still faced by the problem of 3-dimensional (3-D) growth occurring as the second or third ML goes down. This transition from 2-D to 3-D growth in heteroepitaxy, known as Stranski–Krastinov growth, is predicted by equilibrium thermodynamics [22,23]. The formation of 3-D islands, of course, is a source of interfacial roughness. It is also important to consider whether the interface is grown under tension or compression. There is evidence that a surface or interface grown under compression is rougher than when grown under tension [24,25]. In InAs/GaSb SLs, GaSb layers are under compression when grown on an InAs buffer layer, whereas InAs layers are under tension when grown on a GaSb buffer layer. Therefore, one might expect an InAs/GaSb SL grown on a GaSb buffer layer to have less interface roughness than an InAs/GaSb SL grown on an InAs buffer layer.

Coming to a proper understanding of the configuration of a heteroepitaxial interface needs to begin with the morphology of the growth surface. Recent observations using *in situ* scanning tunneling microscopy (STM) suggest aspects of the growth surface morphology that had not been anticipated by earlier models. In some cases, heteroepitaxial strains may need to be relieved by the formation of vacancy lines between adjacent 2-D islands [26,27]. There is also the problem of growing on top of a reconstructed surface. In the case of GaAs [28] and InAs [29], an anion-terminated (100) surface is capable of reconstructing in such a way that the arsenic density is only half that of the bulk crystal. On the other hand, GaSb (100) surfaces are capable of reconstructing such that the surface layer consists of as much as 1.7 atomic layers of antimony [30]. In either case, the question arises as to how a proper interface would be formed upon a reconstructed surface, given a surplus or deficit of anions covering the surface.

3.2 CONTROL OF INTERFACIAL BONDING

Given the requirement that the arriving adatoms must evenly distribute themselves across the surface, one would do well to consider a growth procedure which allows the surface atoms to assemble into an equilibrium configuration. One possible growth procedure for achieving this goal is migration enhanced epitaxy (MEE), where one alternates between the deposition of cation and anion monolayers [31]. By depositing the cations as a separate monolayer, they are granted the opportunity to atomically diffuse over an anion-terminated surface before the next layer of anions bonds to the cations and reduces the diffusion rate. Furthermore, this growth procedure may also reduce the intermixing because it allows the formation of a complete cation monolayer prior to the deposition of the anion overlayer. This cation monolayer is thought to serve as a buffer between the the subsurface anion monolayer and the subsequent anion overlayer. Because such a cation monolayer would prevent exchange reactions involving the anion layers it separates, we speak of this cation monolayer as a "cation firewall" [25].

In this section, we will present X-ray diffraction [32] and Raman spectroscopy data [33,34] for our InAs/GaSb SLs, and review the cross-sectional scanning tunneling microscopy (XSTM) results of Feenstra et al. [35,36]. Using MEE at the interfaces, we grew 40-period SLs with nominal structures of 8 monolayers (MLs) InAs and 12 MLs GaSb. These structures were grown on a 1 μm buffer of either GaSb or InAs on a semi-insulating GaAs (100) substrate. The growth temperature, measured by infrared transmission thermometry [37], was 400°C. Other growth details are given in Bennett et al. [32].

In order to achieve a better understanding of local interface abruptness, InAs/GaSb SLs were also grown on a 0.1 μm GaSb buffer layer on a lattice-matched GaSb (100) substrate. For these samples, the GaSb buffer was grown at 500°C (a suitable temperature for achieving a smooth GaSb surface) with frequent growth interrupts. During the growth interrupts, we observed an increase in the specular RHEED intensity, indicating the formation of large, atomically flat, terraces on the surface. Following the completion of the buffer layer, the temperature was reduced to 400°C and the SL was grown. (The formation of the large terrace structures during the growth interrupt is also verified by the observation of strong RHEED oscillations when GaSb was grown on GaSb at 400°C.)

One measure of the success of MEE is provided by X-ray diffraction (XRD) results [32]. The position of XRD peaks is related to the average lattice constant of the SL via Bragg's law. When the average lattice constant of the SL differs from that of the GaSb buffer, tetragonal distortion alters the spacing between atomic planes parallel to the growth surface (e.g. (400)). The (400) lattice spacing (which can be calculated from the the lattice parameters and elastic constants using elasticity theory) is proportional to the average lattice parameter of the SL measured parallel to the growth direction. Because the InSb lattice parameter is larger than that of GaAs, we expect for the (400) spacing to be larger for a SL with InSb bonds (e.g. 8-8-InSb) than for the same SL with GaAs bonds (e.g. 8-8-GaAs). Because the Bragg angle is inversely proportional to the lattice spacing, we expect a given XRD peak for a SL with GaAs-like interfaces to be at a larger Bragg angle than for a sample with InSb-like interfaces, as is indeed the case. For a SL with alternating GaAs-like and InSb-like interfaces, a given XRD peak assumes an intermediate value.

In Fig. 6, we plot the lattice constants parallel to the growth direction for a number of SLs (derived from the XRD measurements

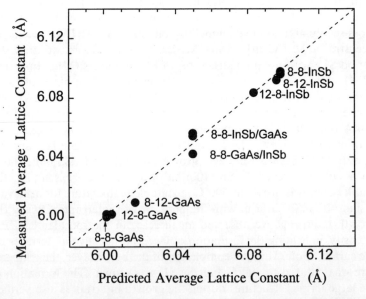

Fig. 6 X-ray lattice parameter measurements (for the lattice parameter parallel to the growth direction) corresponding to a number of different InAs/GaSb SLs plotted as a function of the calculated lattice parameter assuming the nominal layer thicknesses and pseudomorphic growth. The notation "i_{InAs}-j_{GaSb}-k" used in labeling these structures reserves the first index "i_{InAs}" for the thickness of the InAs layer in MLs, the second index "j_{GaSb}" for the thickness of the GaSb layer in MLs, and the third index "k" for the type of interface. For the case where two alternating interfaces are grown, the notation i_{InAs}-j_{GaSb}-k_{InAs}/k_{GaSb} is used, where k_{InAs} is the type of interface grown on InAs and k_{GaSb} is the type of interface grown on GaSb.

of the (400) Bragg angle) as a function of the average vertical lattice constant expected from simple tetragonal distortion. The lattice constant differences due to interface bonds are largest for the 8-8-X structures because the interfaces are a larger fraction of the total structure. In all cases, the expected lattice parameter and that derived from XRD experiments agree to within 7×10^{-4} nm.

Shanabrook and coworkers [33,34] identified the bonding at the interface through the use of Raman spectroscopy. Planar vibrational modes (PVM's), associated with the GaAs-like interfaces occurring in InAs/GaSb superlattices, were shown to be highly localized and therefore sensitive to the structure of the interface. In Fig. 7(a), we see the Raman spectra from several interfaces with varying values of interfacial composition. Raman spectra for GaAs-like interfaces are distinctly different from the spectra of InSb-like interfaces. In all but the purely InSb-like interface (A), there is evidence of the PVM associated with the GaAs-like interface. Furthermore, a systematic shift in wavenumber is observed as the interface bond type is altered

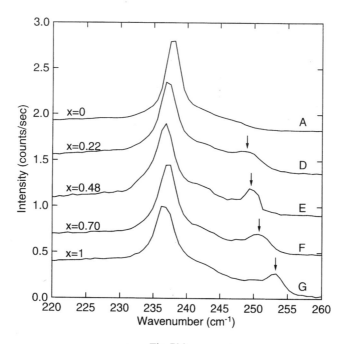

Fig. 7(a)

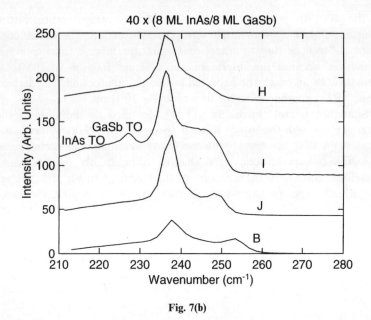

Fig. 7(b)

Fig. 7 Raman spectra showing displacement of planar vibrational mode (PVM) peak for InAs/GaSb SLs. (a) Raman spectra displayed as a function of the composition of the As_xSb_{1-x} interface. Because the PVM peak is easily discerned for an arsenic fraction of 0.22, we suspect that the nominally InSb-like interface have an arsenic volume fraction of less than 0.1. (b) Raman spectra obtained from sample subjected to various growth conditions: (H) MEE, InSb-like bonds; (B) MEE, GaAs-like bonds; (I) Sb_4 soaks at both interfaces; (J) no MEE or anion soaks (see text). Note that the wavenumber of the PVM for I and J corresponds to that expected for a mixed (both InSb-like and GaAs-like) interface.

from completely InSb-like (A) to completely GaAs-like (G). This shift is also in accord with the model of the vibrational properties of the As_xSb_{1-x} interface [33]. One strength of the Raman technique in this case is that the GaAs-like interface gives rise to a phonon mode that is easily observed in vibrational spectra. Because this feature is absent from the Raman spectra for SLs with only InSb-like interfaces, we conclude that this growth procedure provides a high degree of order in InSb-like interfaces of these SLs. Unfortunately, there is not a similar strong InSb-like peak in Raman spectra, making it more difficult to judge the degree of order for GaAs-like interfaces.

Two examples of the consequences (SLs J and I) of growing without the MEE technique are shown in the Raman spectra shown in Fig. 7(b). The other two spectra (H and B) correspond to MEE SLs grown with

InSb-like and GaAs-like interfaces respectively. Superlattice J was grown with no attempt to control the interface stoichiometry. At the interfaces, the gallium and antimony shutters were closed at the same time that the indium and arsenic shutters were opened and vice-versa. Analysis of the XRD data reveals an average lattice spacing between that of a purely GaAs bonded SL and a purely InSb bonded SL, suggesting a mixture of interfacial bonds for sample J. This result is confirmed by the Raman measurement which reveals a GaAs interfacial mode which is shifted toward lower energies. The energy of the Raman peak (248 cm^{-1}) is similar to that of sample E (Fig. 7(a)), suggesting an interfacial composition near $As_{0.48}Sb_{0.52}$. For sample I, the interfaces were grown using only an Sb_4 soak, that is, without the separate cation monolayer. We see that the energy of the PVM indicates a mixed interface (i.e. an interface of both InSb-like and GaAs-like character with approximately 15% GaAs-like bonds) rather than a purely InSb-like interface. The presence of nominally forbidden TO (transverse optical) phonons also attests to the defective nature of this interface [38].

We expect that the mixed character of the interfaces grown in sample I can be traced to anion exchange [39]. There is evidence that for the case where an InAs layer is grown on a GaSb layer, the Sb exchanges with the As. This exchange is expected because antimonide bonds tend to be weaker than arsenide bonds. Since antimonide bonds are weaker, antimonide surface energies are expected to be smaller than arsenide surface energies. By exchanging As with Sb, the surface achieves a lower energy. One of the consequences of such an exchange, of course, is interfacial disorder.

The question of exchange reactions and the structure of InAs/GaSb superlattices has also been addressed by Feenstra *et al.* [35,36] using cross-sectional scanning tunneling microscopy (XSTM). In Figs. 8(a) and (b), we show XSTM images obtained from an MBE-grown 15 ML InAs/8 ML GaSb SL grown at a temperature of 380°C and using Sb_2 soaks at the interfaces. A gray-scale filled-state STM image is shown in Fig. 8(a), and a derivative image is shown in Fig. 8(b). The growth direction is from right to left. The GaSb layers are bright and the InAs layers are dark. Atomic planes, with spacings of 0.6 nm (2 MLs) are clearly visible in the image. Interfacial disorder between InAs and GaSb layers is apparent. In addition, there are numerous bright spots within the InAs layers. Such bright spots appear most clearly in filled state images, indicating that they arise from anion replacement. Thus, the bright spots are associated with Sb atoms incorporated

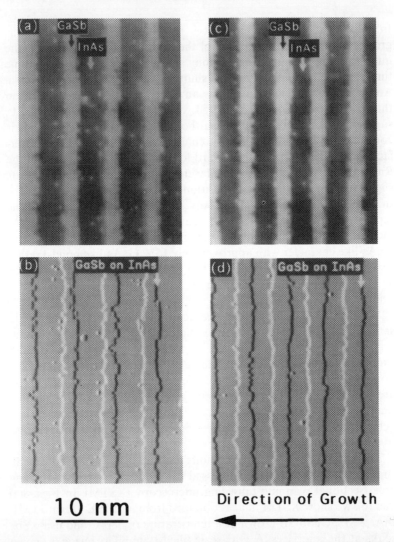

Fig. 8 Cross-sectional STM images of InAs/GaSb superlattices. Growth direction is from the right to left, as indicated. Images were acquired with sample voltage of -2.0 V. Sb incorporation in the InAs gives rise to the bright spots in the InAs layers. In constant-current topographs, the gray-scale range is 0.2 nm. (a) Constant-current topograph from SL grown by MBE at 380°C. (b) Derivative image, computed from (a). (c) Constant-current topograph from SL grown by MBE at 460°C (d) Derivative image, computed from (a). (From R. M. Feenstra, D. A. Collins and T. C. McGill, *Superlattices and Microstructures* **15**, 215 (1994).)

substitutionally in the InAs. This interpretation is consistent with the view that Sb floats on top of the InAs layers during growth.

In Figs. 8(c) and (d) are STM images obtained from an InAs/GaSb SL grown by MBE at an elevated growth temperature of 460°C. A gray-scale STM image is shown in Fig. 8(c), and a derivative image is shown in Fig. 8(d). For the GaSb on InAs InSb-like interfaces, the results of Fig. 8(d) are similar to those of Fig. 8(b), with the interfaces containing many flat sections of length 5–10 nm. For InAs on GaSb, however, careful inspection of Fig. 8(d) compared with Fig. 8(b) (as well as detailed Fourier analysis) reveals that these interfaces are somewhat smoother in Fig. 8(d), with the interfaces now containing some flat sections of length 5 nm. It is also apparent in Fig. 8(d) that the Sb incorporation in the InAs layers is greatly reduced (by about a factor of 4) compared to that of Fig. 8(b).

An analysis of these observations indicates that the InAs-on-GaSb interface has significantly more intermixing than the GaSb-on-InAs interface. This was found to be the case for both Sb_2 and As_2 soaks. This observation is rationalized in terms of Sb having a lower surface free energy than As, producing exchange of Sb and As when InAs is grown on GaSb, but not for growth of GaSb on InAs. It is also apparent that the degree of interfacial roughness in these interfaces is less for the higher growth temperature (460°C). Evidently, a higher growth temperature allows surface diffusion to better smooth the surfaces that serve as templates for the interfaces. The degree of Sb incorporation in the InAs layers is also less at the higher temperature, possibly due to enhanced Sb desorption. There is also the possibility that the smoother surfaces formed at high temperatures present fewer surface steps that may serve as sites for anion exchange reactions. This tendency for an enhanced exchange of Sb and As at the InAs-on-GaSb interface (as compared to the GaSb-on-InAs interface), however, does not seem to be present in the MEE-grown SLs grown by Bennett et al. [32].

It should be noted that a limitation of XSTM arises from the atomic arrangement of the (110) surface imaged. Due to the zinc blende structure of a (110) surface, only every-other (001) bilayer plane is exposed on the (110) surface. Therefore, the cleavage surface may reveal the two layers above or below an interface rather than the interface itself. By implication, XSTM is unable to resolve the difference between 1 ML and 2 ML of interface roughness. This morphological artifact may explain why Feenstra et al. [36] do not report any difference in interface roughness between GaAs and InSb-like interfaces in InAs/GaSb SLs.

3.3 USING HIGH-RESOLUTION TRANSMISSION ELECTRON MICROSCOPY TO STUDY INTERFACIAL DISORDER

One of the best-known consequences of the translational symmetry of a crystal lattice is Bloch's Theorem, which states, in part, that the electron density in such a solid must have the same periodicity as the lattice [40]. As is obvious to anyone acquainted with the physics of semiconductor devices, Bloch's Theorem is useful in understanding electron transport in crystalline solids. Bloch's Theorem is also important in understanding the transport of high energy electrons (accelerated by several hundred thousand volts) that pass through a thin (5–1000 nm) film sample imaged in a transmission electron microscope (TEM) [41]. The TEM imaging technique projects a highly-magnified image of the high-energy electron wavefunction (which consists of a series of Bloch waves) onto a phosphor screen or recording medium, and in so doing reveals the structure (including lattice periodicity) of the crystal under study.

The electron wavefunction q($\underline{r}$) describing the electrons, as they emerge from the exit surface of a thin TEM sample, is modified by the lens system of the electron microscope, yielding the wavefunction $\Psi(\underline{r})$. The product of this modified wavefunction and its complex conjugate (i.e. $\Psi(\underline{r})\Psi(\underline{r})^*$) gives the intensity of the electrons falling on an electron-sensitive film or a detector. A map of these intensities corresponds to the image of the electron-transparent sample. If a number of stringent conditions are met in the TEM imaging experiment, the resulting image should reflect many of the features present in the original Bloch wavefunction, including the lattice periodicities present in the sample. These stringent experimental conditions include the use of magnetic lenses with minimal spherical and chromatic aberration, and a TEM sample that is extremely thin (a few tens of nms or less). These conditions must be met in order to achieve the imaging mode known as high-resolution transmission electron microscopy (HRTEM) [42,43].

By careful control of such experimental parameters as defocus, sample thickness, and sample tilt, the microscopist can hope to record an image containing atomic scale information. The interpretation of such an image, however, requires an understanding of how to solve the Schrödinger equation for high energy electrons in a crystal. This is no trivial feat, and a number of people have dedicated their careers to this problem [44]. Thanks to their efforts, however, we now have a number of commercially available software packages for calculating HRTEM images via the so-called "multi-slice" approach [45]. In multi-slice theory, the sample is modeled as a stack of thin kinematically-diffracting slices (each slice usually consists of only one or two atomic layers). The effect of each successive slice on the incident electron plane wave is calculated using a physical analogy to a series of thin lens elements in optics. The intelligent application of these programs can be used, for example, to distinguish a column of Ga atoms from a column of Sb atoms in GaSb in a properly recorded HRTEM image.

3.3.1 HRTEM Contrast of InAs/GaSb Superlattices

In order to use HRTEM in measuring changes in composition, such as those occurring at an interface, one needs to be able to image along a zone axis that includes reflections which are particularly sensitive to changes in atomic number. From a simple geometrical point of view, imaging with the electron beam direction parallel to the [001] zone axis allows one to view the cations and anions in separate columns (as shown in Fig. 3), thereby allowing one atomic species to be distinguished from another. The {200} reflections of the [001] zone axis are particularly useful because the strength of each {200} reflection is proportional to the difference between the cation and anion scattering factors [46,47]. Therefore, a {200} reflection is strong for zinc blende materials with a large difference in atomic number between cation and anion, such as InAs or GaSb, the components of our SL layers. Similarly, {200} reflections would be expected to be weak for zinc blende materials with a small difference between cation and anion atomic number, such as GaAs or InSb, which are responsible for the bonding at InAs/GaSb interfaces [48–52].

An example of this imaging approach is shown in Fig. 9. This image was recorded at 300 kV on a Hitachi H-9000 UHR HRTEM instrument with a top entry goniometer and a Scherzer resolution of 0.19 nm [53]. It should be noted that imaging InAs/GaSb heterostructures at 400 kV promotes electron beam damage, whereas imaging at 300 kV appears free from this drawback [50]. The sample used in making this specimen was grown on a GaSb (100) substrate, cleaved along two intersecting $\langle 110 \rangle$ directions, and then mounted in the [100] cross-sectional TEM (XTEM) configuration [54]. It should be noted that InAs/GaSb heterostructures grown on GaAs (100) substrates do not appear to survive the cleaving process. An inspection of such samples in the TEM reveals that the SL shears off the top of the

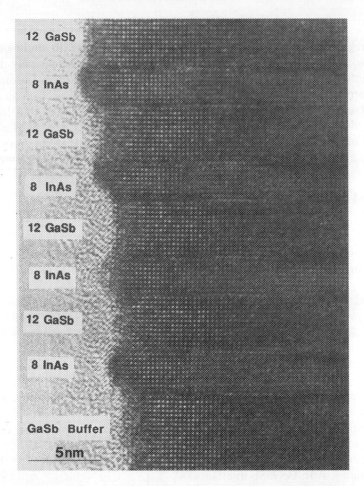

Fig. 9 HRTEM image of cleaved sample taken from an InAs/GaSb SL grown on (100) GaSb. The image was recorded at a defocus of 60 nm.

sample when cleaved. In general, it seems that the cleaving process only works well when the epitaxial layer is similar to the substrate. For this reason, samples consisting of InAs/GaSb heterostructures grown on GaAs (100) were ion-milled at liquid nitrogen temperature.

Using the focusing action of the objective lens, the HRTEM can be tuned to accent the contribution of the {200} reflections of the [001] zone axis. Applying non-linear imaging theory, we used multislice simulations to calculate the {200} contribution to the intensity at the cation site for both GaSb and InAs [54,55]. As shown in Fig. 10(a), for

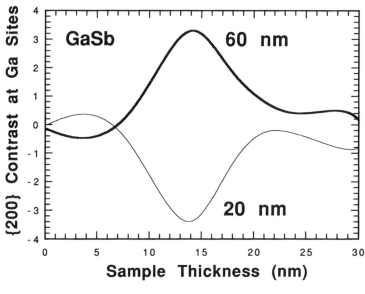

Fig. 10(a)

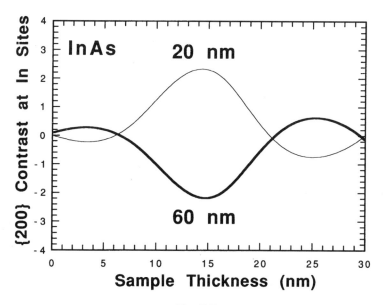

Fig. 10(b)

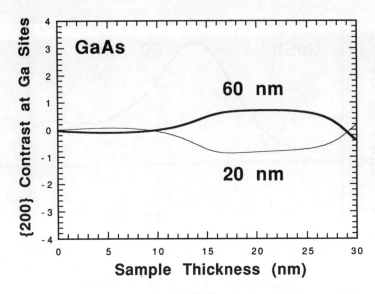

Fig. 10(c)

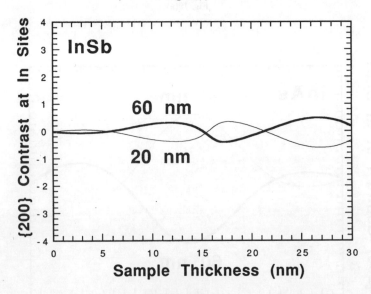

Fig. 10(d)

Fig. 10 Intensity of (200) spatial frequency as a function of sample thickness for two different defocus conditions (20 nm for the fine line; 60 nm for the bold line). (a) GaSb. (b) InAs. (c) GaAs. (d) InSb.

GaSb with a sample thickness in the vicinity of 15 nm, the {200} contribution to the image intensity at the gallium site is strongly negative at a defocus (i.e. underfocus) value of 20 nm, and strongly positive at a defocus value of 60 nm. For a 15 nm thick InAs sample (as shown in Fig. 10(b)), the {200} contribution to the intensity at the indium site is strongly positive at a defocus of 20 nm while strongly negative at a defocus of 60 nm. For both GaSb and InAs, the {200} contrast is greatest for samples of thickness in the neighborhood of 15 nm. For GaAs and InSb, however, the {200} contribution to the intensity at the cation sites is small, as shown in Figs. 10(c) and (d), respectively. This small contribution is in accord with the small {200} scattering factors for GaAs and InSb, where the difference between the group III and group V scattering factors (and atomic numbers) is small. Similarly, the contribution of {200} anion contrast can also be calculated from non-linear imaging theory. It should be noted that at a defocus of 60 nm, the interpretation of the image is in accord with simple intuition: atomic columns corresponding to small atomic numbers (i.e. Ga and As) appear bright; atomic columns corresponding to large atomic numbers (i.e. In and Sb) appear dark. At a defocus of 20 nm, we have the opposite effect: atomic columns corresponding to small atomic numbers (Ga and As) appear dark; atomic columns corresponding to large atomic numbers (In and Sb) appear bright, as shown in the simulation in Fig. 11. This interpretation of the 60 nm

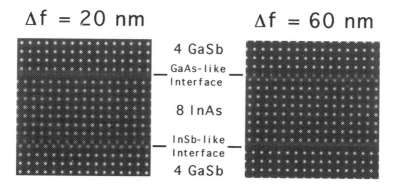

Fig. 11 Simulation of HRTEM image of abrupt InSb-like and GaAs-like interfaces for the InAs/GaSb system for two different defocus conditions ($\Delta f = 20$ nm and $\Delta f = 60$ nm).

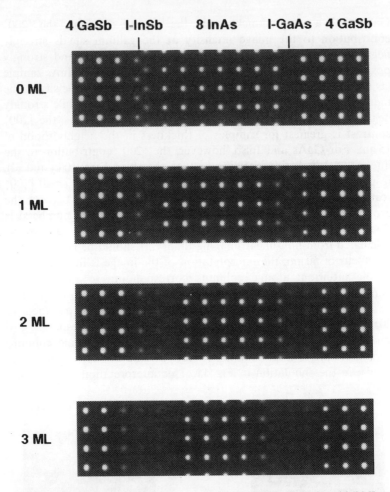

Fig. 12 HRTEM simulations of [100]-oriented InAs/GaSb superlattice with 8 MLs of InAs separated from 8 MLs of GaSb by alternating GaAs and InSb interfaces. Simulations correspond to a defocus of 20 nm, a spherical aberration coefficient of 0.9 mm, and an accelerating voltage of 300 kV. Interface widths in the simulations range from 0 MLs (those completely devoid of roughness) to 1, 2, or 3 MLs.

defocus condition is analogous to that proposed by Ourmazd et al. in their study of $Al_xGa_{1-x}As$/GaAs [46].

The results of multislice simulations of HRTEM images corresponding to the 20 nm defocus condition are shown in Fig. 12. We simulated SL images with abrupt GaAs-like and InSb-like interfaces as well as those in which the interfaces were replaced by 1–3 MLs of the composition $In_{0.5}Ga_{0.5}As_{0.5}Sb_{0.5}$ in order to mimic the contrast effects of 1 or 2 MLs of interface disorder. By comparing these simulated images with experimental results, the degree of interfacial disorder can be estimated.

3.3.2 Quantitative Image Processing of HRTEM Data

In order to make a quantitative assessment of the images recorded by HRTEM, we digitized portions of the negatives as 256×256 8-bit arrays using a CCD (Charge Coupled Device) camera. An example of such a digitized image, taken from an ion milled specimen recorded at a defocus of 20 nm (InSb-bonded InAs/GaSb superlattice grown on 1 μm GaSb on GaAs (100)) is shown in Fig. 13(a). The digital filtering process begins with taking the 2-dimensional Fast Fourier Transform (FFT) of the recorded image, as shown in Fig. 13(b) [56]. In order to digitize the image in such a way as to avoid the undersampling that leads to aliasing, we first need to know what sampling interval to use (i.e. the number of pixels per unit distance). According to the Sampling Theorem, imaging data would need to be sampled at twice the highest spatial frequency in order to prevent aliasing [57]. From the Fourier spectrum, it appears that highest spatial frequency is {420}. For InSb, the largest unit cell of the four III–V compounds involved in our study (lattice parameter 0.64794 nm) d_{420} is 0.14488 nm. For GaAs, with the smallest unit cell in our study (lattice parameter 0.56533 nm) d_{420} is 0.12641 nm. Even if we were to consider satellites of {420} at {520}, we would have a value of $d_{520} = 0.104$ for GaAs. If we were to sample at twice the spatial frequency of GaAs {520}, we would find ourselves sampling at a rate of once every 0.052 nm (i.e. about twice every Angstrom). Because our CCD camera is sampling once every 0.025 nm (i.e four times every Angstrom), there is no danger of undersampling.

From non-linear imaging theory for the [001] zone axis, we know that only the {200} and {420} spatial frequencies contribute to the intensity of the {200} reflections [55]. Therefore, we multiply the FFT by a masking function (shown in Fig. 13(c)) so that only the spatial frequencies near the {200} and {420} positions contribute to the filtered image (shown in Fig. 13(d)). It should be noted that the vertical (i.e. perpendicular to a horizontal interface) extent of the spatial frequency bandwidth allowed by the mask is fairly large. We found that a large vertical value for the spatial frequency bandwidth was necessary in order to include sufficient Fourier components to allow the interface in the processed image to be abrupt. However, we limited

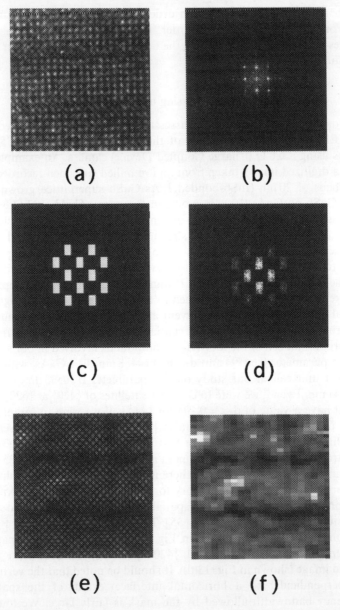

Fig. 13 Image processing of HRTEM images of InAs/GaSb interfaces. (a) Recorded image; (b) Amplitude of FFT of image; (c) The mask for {200} and {420} spatial frequencies; (d) Masked FFT; (e) Filtered-image obtained by inverse FFT of masked FFT; (f) Resampling the filtered image yields the final processed HRTEM image.

the vertical bandwidth so as to minimize the contribution of satellites of the {220} spatial frequencies, which are not sensitive to the differences in the cation/anion scattering factors [46,47].

In the filtered image, both cation and anion columns are present. This situation is, of course, different from the original image (recorded at a defocus of 20 nm or 60 nm) in which only one (either cation or anion) type of column is present. By removing the contribution of the {220} reflections from the image, the spatial frequencies that allow cation and anion columns to be distinguished from one another are eliminated. Even though the filtered image in Fig. 13(e) is more sensitive to differences between anion and cation scattering factors than the recorded image shown in Fig. 13(a), additional processing is still required. A plot of locally averaged {200} intensity as a function of distance from the interface would be more useful in characterizing the extent of interfacial disorder. If we were to plot the intensity variations away from one interface (as in Fig. 13(e)) we would obtain intensity oscillations due to the atomic columns. The problem of atomic-scale oscillations can be avoided by simply resampling the image so that we are only reading intensities from areas corresponding to a pair of anion/cation columns. For the 256×256 pixel image in Fig. 13(f), the resampling area is 6×12 pixels, which corresponds to 0.15×0.3 nm (i.e. one atomic column vertically by two atomic columns horizontally – essentially one cation and one anion column side by side). We chose to orient this sampling area laterally so that we would have one atomic layer resolution perpendicular to the interface. In this way we arrived at a processed image that is an intensity map for the compositionally sensitive {200} reflections. Because the InSb and GaAs interfaces give rise to much smaller {200} diffracted intensities than for those which are excited in the InAs and GaSb layers in the SL, the processed image in Fig. 13(f) shows much better definition of the interface width than the original recorded image in Fig. 13(a).

3.3.3 The Assessment of Interfacial Disorder via HRTEM

The difference between InSb-like and GaAs-like interfaces in SLs grown on a 1 μm GaSb buffer layer on GaAs (100) is shown in Figs. 14(a) and (b). In Figs. 14(c) and (d), we show processed images of SLs grown on a 1 μm InAs buffer layer on GaAs (100). The quantification of simulated images that have undergone image processing is shown in Fig. 15. The results of averaging several of these images are shown in Fig. 16. Our interpretation of these images should, however,

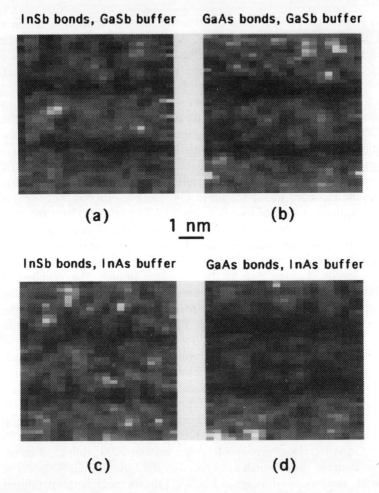

Fig. 14 Processed HRTEM images of 8 MLs InAs/12 MLs GaSb SLs grown on GaSb buffer layers on GaAs (100) substrates. The InAs layer is in the middle with a GaSb layer on top and bottom, and the growth direction is from bottom to top. (a) InSb-like interfaces for a SL grown on GaSb have a roughness of 1 ML; (b) GaAs-like interfaces for a SL grown on GaSb have roughness of 2 MLs; (c) InSb-like interfaces for a SL grown on InAs have a roughness of 2 MLs; (d) GaAs-like interfaces for a SL grown on InAs have a roughness of 3 MLs.

be viewed in the context of Raman spectroscopy results (that is, that the interfaces are locally abrupt) [32–34]. Therefore, we consider any measurement of the degree of interfacial disorder in these SLs to be due to interfacial roughness [48].

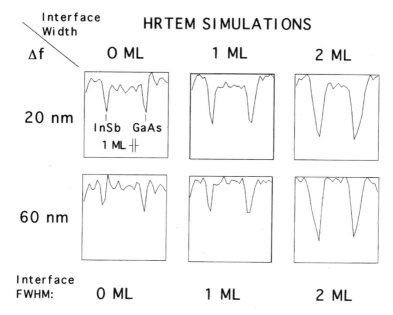

Fig. 15 Intensity profiles derived from image processing of multislice simulations for two defocus values (20 nm and 60 nm) and three values of interface width (0 ML, 1 ML, and 2 MLs). Note that the full width at half maximum (FWHM) of the profiles increases with interface width. In the InAs/GaSb superlattices grown in this study, the interface width is attributed to interface roughness.

A comparison of the recorded images with the simulated images suggests that the interfacial disorder (which we interpret as interfacial roughness) for SLs grown on a GaSb buffer layer is on the order of 1 ML for the InSb-like interface and 2 MLs for the GaAs-like interface. The interface roughness for SLs grown on an InAs buffer is on the order of 2 MLs for the InSb-like interface and 3 MLs for the GaAs-like interface. From these results, it seems that not only is interface roughness greater for GaAs-like interfaces than for InSb-like interfaces, but the roughness is greater for SLs grown on an InAs buffer layer than for SLs grown on a GaSb buffer layer. For the Ge_xSi_{1-x} system, a rougher surface occurs for films under compression than under tension [24]. Such may also be true in our system, since the GaSb SL layers are under compression when grown on an InAs buffer layer, whereas the InAs SL layers are under tension when grown on a GaSb buffer layer [25]. These values of interface roughness are similar to those measured for InAs/GaSb superlattices by De Cooman *et al.* using dark-field TEM [58].

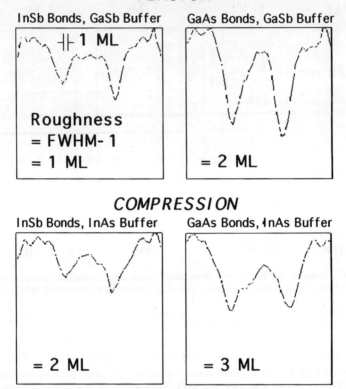

Fig. 16 Intensity profiles derived from image processing of several images for each combination of interface type and buffer layer shown in Fig. 14. Quantification scheme suggested in Fig. 15 was used to determine the degree of interfacial roughness.

The large lattice mismatch of both GaSb and InAs with respect to a (100) GaAs substrate, must, of course lead to threading dislocations in the GaSb or InAs buffer layer [59]. Although the dislocation density in the SLs grown on a 1 µm GaSb or InAs buffer layer is less than $10^7/cm^2$, there is still the possibility of dislocations affecting the surface and interface morphologies [60]. In order to rule out the possible effects of these dislocations on interfacial disorder, we also grew the SL on (100) GaSb substrates. Growing the SL on a GaSb substrate also allows us to prepare the HRTEM sample by cleaving, a process that results in a sample that is generally cleaner and better

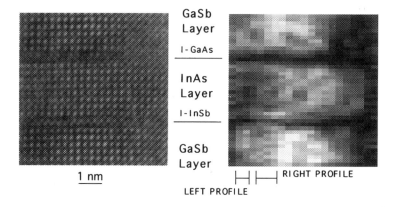

Fig. 17 To the left is a HRTEM image of cleaved specimen taken from an InAs/GaSb SL grown on a GaSb buffer layer and (100) GaSb substrate. Defocus value is 60 nm. The processed image on the right, taken from the HRTEM image to the left, maps intensities associated with {200} reflections. Note that the left portion of the InSb interface (left profile) is more abrupt than right portion (right profile) of the InSb interface.

defined than a sample thinned by ion milling. In Fig. 17 we show an HRTEM image recorded from a cleaved specimen at a defocus of 60 nm. The thickness of the sample increases from right to left in Fig. 17. There are two types of interfaces in this sample. The bottom interface is InSb-like; the top interface is GaAs-like. As was the case for InAs/GaSb SLs grown on a GaSb buffer layer on GaAs (100), the interface roughness was on the order of 1 ML for InSb-like interfaces and 2 MLs for GaAs-like interfaces. We estimate that the thinnest part of the sample shown in Fig. 17 is 10 nm thick and the thickest part of the sample is 20 nm thick. Careful inspection of the processed image in Fig. 17 reveals that the portion of the InSb-like interface defined by the left bracket appears narrower than the portion of the interface defined by the right bracket. We make a quantitative comparison of these two regions in Fig. 18, where we plot the integrated intensities of the processed image (i.e. the intensity of the {200} reflections) as a function of the distance from the interfaces for these two regions. The left profile of the InSb-like interface in Fig. 18 closely resembles the simulated profile for an abrupt (0 ML wide) InSb interface recorded at a defocus of 60 nm (as shown in Fig. 15), indicating that this portion of the InSb-like interface is indeed abrupt. The right profile of the InSb-like interface in Fig. 18 resembles a rougher (1 ML wide) InSb-like interface recorded at a defocus of 60 nm (as also shown in Fig. 18).

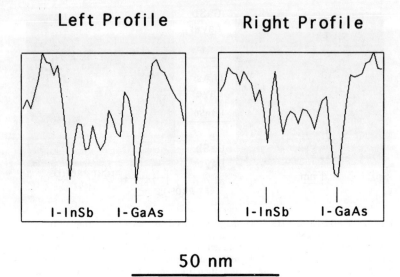

Fig. 18 The {200} intensity profiles from the processed image in Fig. 17. The left profile is taken from an average of the two columns allotted to this profile in Fig. 17. The right profile is taken from an average of the three columns allotted to this profile in Fig. 17. Note that in each case the InSb-like interface is more abrupt than the GaAs-like interface. Also, the InSb-like interface in the left profile is more abrupt than the InSb-like interface in the right profile.

The abrupt InSb-like interface in the image corresponds to a region of the sample where the imaging electrons of the HRTEM traverse a 15 nm path over a portion of the specimen devoid of steps or terraces. The existence of such an abrupt InSb-like interfacial region would not be possible if significant interdiffusion had occurred. The rougher portion of the InSb-like interface corresponds to a region where steps and terraces are present. There is also evidence for the existence of a locally abrupt GaAs-like interface in the left profile of the processed image in Fig. 18 (as attested by the left profile in Fig. 18).

3.4 INTERFACIAL MORPHOLOGY

In order to further characterize interfacial disorder in InAs/GaSb heterostructures, we performed a complimentary study utilizing *in situ* scanning tunneling microscopy (STM) on a number of interfacial surfaces in plan-view. With plan-view STM, both the three-dimensional morphology and the atomic-scale structure of the evolving epitaxial interfaces can be observed directly. The growth program used for the cleaved HRTEM samples (grown on a GaSb (100) substrate) described above was used for these samples as well. However, after the samples were grown up to the point of interest, an appropriate anion-soak was used to terminate the growth. The sample was then quenched to room temperature in the absence of any flux and transferred to the surface analysis chamber. STM images were acquired with a constant current of 0.1 nA and sample bias between -1.8 and -2.2 V. In order to minimize any possible effects of surface contamination on the superlattice morphology, after examination in the STM, each interface was buried under at least 0.5 μm of GaSb and the superlattice was regrown up to the next interface of interest [61].

An STM image of the surface of a typical GaSb buffer layer grown on a GaSb (100) substrate is shown in Fig. 19(a). The surface is Sb-terminated and consists of large, atomically-smooth terraces (~ 50 nm wide) separated by monolayer-height (0.3 nm) steps, with very few adatom or vacancy islands. This surface appears to be close to thermodynamic equilibrium, so that the average terrace width is determined by the misorientation of the sample with respect to (100). (Some regions of the sample may have larger terraces than others due to local variation in the orientation due to polishing.) Atomically-resolved images on each terrace (not shown) reveal the Sb-terminated 1×3 surface reconstruction [30], consistent with RHEED. The stability of the [01$\bar{1}$]-oriented rows of Sb dimers inherent to this reconstruction give the terrace edges their characteristically straight [01$\bar{1}$]-oriented and jagged [011]-oriented edges [28]. Thin GaSb films (8 ML) grown at 400°C within the superlattice have a very similar surface structure to the GaSb buffer (atomically-smooth terraces with few vacancy or adatom islands), but with more rounded terrace edges. Our *in situ* STM characterization of the GaSb/InAs interfaces will focus on the disorder due to roughness, defined as the number of additional monolayers present on each terrace at the completion of interface growth. It can be characterized on any length scale, but we will focus on two: (1) the total roughness on each substrate terrace – a good indication of the overall roughness associated with the growth; and (2) the roughness within a typical 20 nm-long line oriented in the $\langle 100 \rangle$ direction – a sampling comparable to that viewed by HRTEM. The roughness on the clean GaSb (100) surfaces, as defined here, is 0 ML on both length scales.

The addition of an interface layer to a GaSb (100) 1×3 surface causes significant changes to the surface morphology. Following the

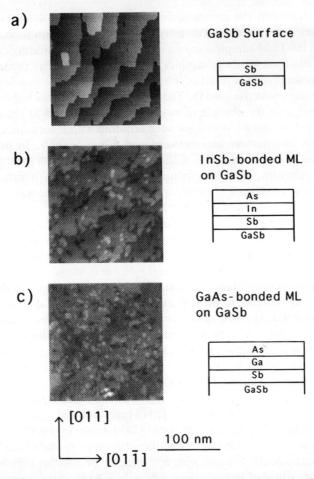

Fig. 19 STM images of surfaces corresponding to various internal interfaces occurring in the first period of an InAs/GaSb SL grown on a GaSb (100) buffer layer and substrate. The images were acquired with a constant current of 0.1 nA and sample biases between −1.8 and −2.2 V. (a) GaSb (100) buffer layer; (b) InSb-like interface on GaSb; (c) GaAs-like interface on GaSb. The top-most layers at each surface are indicated in a schematic diagram to the right of each image.

growth of an As-terminated InSb interface (Fig. 19(b)), small 1 ML-deep vacancy islands and 1 ML high adatom islands (10 nm diameter) are observed on each terrace, giving the terraces a roughness of 2 ML. Due to the low density of these features, the typical roughness on the length scale sampled in HRTEM (20 nm) is only 1 ML (i.e. along this sampling length either a vacancy or adatom island would typically be encountered).

Following the deposition of an As-terminated GaAs surface on GaSb (100) (corresponding to a GaAs-like interface in the SL) (Fig. 19(c)) a greater degree of roughening is observed. Small vacancy and adatom islands are now observed on each terrace, similar to the InSb-like interface, but with approximately equal areas and twice the density; this surface has a terrace roughness of 2 ML. The surface roughness averaged over 20 nm along a $\langle 001 \rangle$ direction, as occurs in HRTEM imaging, is also 2 ML. The GaAs surface is also noticeably rougher on the atomic scale than the InSb interface.

The As-terminated InAs starting surface consists of 8 ML of InAs grown on a (100) GaSb buffer layer. (Since the InAs layer is well under the critical layer thickness, it is coherently strained.) As shown in Fig. 20(a), this surface consists of large terraces with very few islands or pits, similar to the GaSb (100) starting surface (terrace roughness averaged over 20 nm is 0 ML), but with terrace fingers elongated along the $[0\bar{1}1]$ direction; there is also much more atomic-scale disorder. In contrast to the clean GaSb surfaces, the clean InAs surfaces do not appear to be as close to thermodynamic equilibrium. Although the atomic-scale structure is not well-ordered, the finger-like shape of the terrace edges indicates that there is some reconstruction-related local order (consistent with RHEED): the $[01\bar{1}]$-oriented row-like structure of As-terminated InAs (100) 2×4 promotes growth along this direction [29].

The addition of an Sb-terminated InSb layer (corresponding to a InSb bonded interface in the SL) to the strained InAs film further roughens the surface (Fig. 20(b)), with many large (30–50 nm), elongated adatom islands appearing along with some generally smaller elongated vacancy islands (giving a terrace roughness of 2 ML). The terrace edges are also more jagged, which can be attributed to the growth mode whereby elongated islands are incorporated incompletely into the terrace edges. The asymmetric nature of the surface features is further indication of a strong directional anisotropy in the growth of indium on the InAs surface. Although this surface appears rougher than the InSb/GaSb interface surface, the roughness averaged over 20 nm is also approximately 1 ML (due to the larger island size).

The roughest surface examined was the Sb-terminated GaAs-like interface on InAs (Fig. 20(c)), a surface with a very high density of interconnected islands. The islands are elongated in the $[01\bar{1}]$ direction as in the InSb/InAs case, but with noticeably rounder edges. Note that many islands have clearly become attached to the terrace edges,

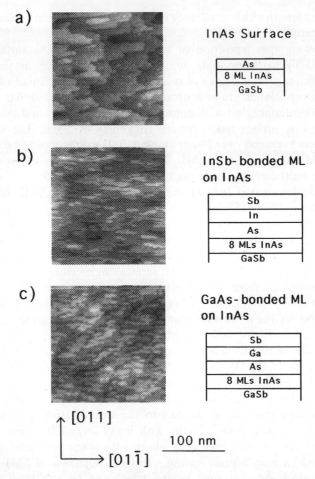

Fig. 20 STM images of surface corresponding to various internal interfaces occurring on an eight monolayer-thick strained InAs (100) epilayer on GaSb. (a) As-terminated InAs layer; (b) InSb-like interface on InAs; (c) GaAs-like interface on InAs. The topmost layers at each surface are indicated in a schematic diagram to the right of each image.

making it difficult to discern the underlying substrate terraces. However, based on the typical terrace width on the InAs surface, we estimate the roughness per terrace to be 3 ML (four layers are present, but the fourth layer is sparse). We find that a 20 nm-long line along $\langle 001 \rangle$ would typically encompass three layers on this surface, corresponding to a roughness of 2 ML on this length scale.

Under our growth conditions, we find that the interfacial surfaces on GaSb are smoother than those on strained InAs, and that the InSb-like interfaces are generally smoother than GaAs-like interfaces for both surfaces (a roughness averaged over 20 nm of 1 ML vs. 2 ML). These observations are consistent with the widths of identically grown interfaces measured with HRTEM. Most significantly, the roughness we observe on the 20 nm length scale, associated with the vacancy and adatom island shapes and size distributions, completely accounts for the interface widths observed via HRTEM, as shown in Fig. 16.

3.5 CONCLUSIONS

We have used a number of experimental techniques to determine the nature of interfacial disorder in InAs/GaSb SLs. From X-ray diffraction, we know that our overall control of composition at the interfaces is good [32]. From Raman spectroscopy we have found that the nominally InSb-like interfaces have an As_xSb_{1-x} anion composition where the fraction of As is less than 0.1 [33]. The reproducibility of the Raman spectra for SLs grown with GaAs-like interfaces suggests that GaAs-like interfaces are also well-controlled. These results also suggest that these heteroepitaxial interfaces are locally abrupt. The MBE technique used in growing these interfaces, MEE, deposits alternating cation and anion layers, with the less mobile cation monolayer isolating the adjacent anion layers so as to minimize intermixing or exchange reactions. The importance of the MEE procedure in growing InAs/GaSb SLs is evident from attempts to establish InSb interfaces using Sb_4 soaks. It should also be noted that Feenstra et al. studied material which was grown using an Sb_2 soak to establish the InSb-like interface; it was clear that the InAs on GaSb interface suffered from intermixing of the anion species [35,36].

HRTEM imaging experiments found that there was a degree of disorder in InAs/GaSb interfaces grown on GaSb buffer layers [48]. This degree of disorder was found to be on the order of 2 MLs for GaAs-like interfaces and 1 ML for InSb-like interfaces. The Raman spectroscopy results indicate that intermixing occurs for less than 10% of the atoms forming an InSb-like interface, a result that is incompatible with a full 1 ML of interfacial disorder, unless most of this disorder is ascribed to a cause other than intermixing. The HRTEM and Raman results are, however, consistent with the contention that InSb-like interfaces are locally abrupt, suggesting that interfacial disorder assumes the guise of islanding (resulting in interfacial roughness) rather than interdiffusion or exchange reactions (resulting in diffuse interfaces). Indeed, small regions of atomically abrupt interfaces have been observed in our HRTEM study of InAs/GaSb SLs with InSb-like and GaAs-like interfaces [25]. We have also studied the interfaces for InAs/GaSb SLs grown on InAs buffers. These interfaces were found to exhibit a greater degree of roughness than those grown on GaSb buffers. Again, GaAs-like interfaces proved rougher than the InSb-like interfaces. From the results of HRTEM imaging experiments, it appears that the interfaces may be less rough when grown on a GaSb buffer layer (as compared to an InAs buffer layer) because the SL is grown under tension instead of compression.

It is interesting to note that XSTM observations of Feenstra et al. show that there is no significant difference in the degree of disorder in GaAs-like and InSb-like interfaces [36]. It is possible that this apparent similarity between GaAs-like interfaces and InSb-like interfaces may be due to the fact that cross-sectional STM is limited to a resolution of 2 MLs in the growth direction, while the difference in disorder between these two interfaces is on the order of 1 ML. It is also possible, of course, that the degree of disorder for the InSb-like and GaAs-like interfaces is similar in the SLs studied by Feenstra et al. Because the material studied by Feenstra et al. [35,36] was grown by another MBE process (i.e. establishing interfaces through anion soaks rather than via MEE, and using an Sb_2 source rather than an Sb_4 source), it may well differ significantly from the material studied by Twigg et al. [25,48,49] and Thibado et al. [61].

In order to come to a more detailed understanding of the contribution of surface morphology to interface roughness, we conducted *in situ* STM experiments imaging interfacial surfaces of thin GaSb or InAs layers grown on a GaSb (100) substrate. The degree of surface roughness indicated by *in situ* STM experiments agreed with the degree of interface roughness determined by HRTEM imaging experiments. That is, for both interfaces and interfacial surfaces, InSb-like surfaces and interfaces were smoother than GaAs-like surfaces and interfaces.

These differences in morphology may be ascribed to either differences in the kinetics of forming each interface, or to differences in the equilibrium configuration of each interface. The GaAs bond is significantly stronger than the InSb bond. From the standpoint of kinetics, one would then expect for diffusion to proceed more quickly over an InSb-bonded surface than for a GaAs-bonded surface, so that the InSb-like surface would be smoother than the GaAs-like surface. It should be noted, however, that even in equilibrium, an InSb-bonded surface might be smoother than a GaAs-bonded surface. The presence of a surface or interface with a large energy per unit area may not be stable. Equilibrium thermodynamics might favor the formation of GaAs islands to lower the contribution of the GaAs-like interface to the system.

Despite our progress in achieving a better understanding of interfacial disorder in InAs/GaSb heterosructures, a host of unanswered questions remain. How much smoother can the interfacial surface become if the newly-arrived adatoms have sufficient time to diffuse? Is there a danger of intermixing during surface diffusion? How is the surplus or deficit of anions terminating reconstructed surfaces compensated during epitaxial growth? What is the role of steps in anion exchange reactions? It is conceivable that these questions can eventually be answered, especially by a systematic *in situ* STM, cross-sectional STM, and HRTEM study of InAs/GaSb interfaces and surfaces prepared under a wide range of growth conditions.

Acknowledgments

This work was sponsored by the Office of Naval Research and the NRL/NRC Post-doctoral Fellowship Program (P.M.T.). We thank Larry Ardis for expert technical assistance. We also thank James Waterman, Robert Wagner, and Ming-Jey Yang for their insights regarding narrow band gap materials and heterostructures.

References

1. G. Bastard, *Wave Mechanics Applied to Semiconductor Heterostructures* (John Wiley and Sons, New York, 1988), p. 98.
2. L. Esaki in *Narrow Gap Semiconductors: Physics and Applications*, Lecture Notes in Vol. 133, Ed. by W. Zawadzki (Springer-Verlag, New York, 1980), pp. 302–323.
3. D. K. Arch, G. Wicks, T. Tonaue and J. -L. Staudenmann, *J. Appl. Phys.* **58**, 3933 (1985).
4. D. L. Smith and C. Mailhiot, *J. Appl. Phys.* **62**, 2547 (1987).
5. C. Mailhiot and D. L. Smith, *J. Vac. Sci. Technol. A* **7**, 445 (1989).
6. R. H. Miles, D. H. Chow, J. N. Schulman and T. C. McGill, *Appl. Phys. Lett.* **57**, 801 (1990).
7. M. A. Herman and H. Sitter, *Molecular Beam Epitaxy* (Springer-Verlag, New York, (1989).
8. J. Y. Tsao, *Materials Fundamentals of Molecular Beam Epitaxy* (Academic Press, Boston, 1993).
9. C. A. Hoffman, J. R. Meyer, F. J. Bartoli and W. I. Wang, *Phys. Rev. B* **48**, 1959 (1993).

10. D. H. Chow, R. H. Miles and A. T. Hunter, *J. Vac. Sci. Technol. B* **10**, 888 (1992).
11. A. Ourmazd, D. W. Taylor, J. Cunningham and C. W. Tu, *Phys. Rev. Lett.* **62**, 933 (1989).
12. C. A. Warwick, W. Y. Jan and A. Ourmazd, *Appl. Phys. Lett.* **56**, 2666 (1990).
13. D. Gammon, B. V. Shanabrook and D. S. Katzer, *Phys. Rev. Lett.* **67**, 1547 (1991).
14. N. Grigorieff, D. Cherns, M. J. Yates, M. Hockly, S. D. Perrin and M. R. Aylett, *Philos. Mag.* **68**, 121 (1993).
15. J. Schmitz, J. Wagner, F. Fuchs, N. Herres, P. Koidl and J. D. Ralston, *J. Crystal Growth* **150**, 858 (1995).
16. J. M. Moison, C. Guille, F. Houzay, F. Barthe and M. Van Rompay, *Phys. Rev. B* **40**, 6149 (1989).
17. J. M. Moison, F. Houzay, F. Barthe, J. M. Gerard, B. Jusserand, J. Massies and F. S. Turco-Sandroff, *J. Crystal Growth* **111**, 141 (1991).
18. B. R. Bennett, B. V. Shanabrook and R. Magno, *Appl. Phys. Lett.* **68**, 958 (1996).
19. M. Copel, M. C. Reuter, E. Kaxiras and R. M. Tromp, *Phys. Rev. Lett.* **63**, 632 (1989).
20. M. Yano, H. Yokose, Y. Iwai and M. Inoue, *J. Crystal Growth* **111**, 609 (1991).
21. M. W. Wang, D. A. Collins, T. C. McGill, R. W. Grant and R. M. Feenstra, *J. Vac. Sci. Technol. B* **13**, 1689 (1995).
22. E. Bauer, Z. Kristallogr., **110**, 372 (1958).
23. C. W. Snyder, B. G. Orr and H. Munekata, *Appl. Phys. Lett.* **62**, 46 (1993).
24. Y. H. Xie, G. H. Gilmer, C. Roland, P. J. Silverman, S. K. Buratto, J. Y. Cheng, E. A. Fitzgerald, A. R. Kortan, S. Schuppler, M. A. Marcus and P. H. Citrin, *Phys. Rev. Lett.* **73**, 3006 (1994).
25. M. E. Twigg, B. R. Bennett and B. V. Shanabrook, *Appl. Phys. Lett.* **67**, 1609 (1995).
26. X. Chen, F. Wu, Z. Zhang and M. G. Lagally, *Phys. Rev. Lett.* **73**, 850 (1994).
27. C. Priester and M. Lannoo, *Phys. Rev. Lett.* **75**, 93 (1995).
28. E. J. Heller, Z. Y. Zhang and M. G. Lagally, *Phys. Rev. Lett.* **71**, 743 (1993).
29. V. Bressler-Hill, A. Lorke, S. Varma, P. M. Petroff, K. Pond and W. H. Weinberg, *Phys. Rev. B* **50**, 8479 (1994).
30. G. E. Franklin, D. H. Rich, A. Samasvar, E. S. Hirshorn, F. M. Leibsle, T. Miller and T. -C. Chiang, *Phys. Rev. B* **41**, 12619 (1990).
31. Y. Horikoshi, M. Kawashima and H. Yamaguchi, *Jpn. J. Appl. Phys.* **25**, L868 (1986).
32. B. R. Bennett, B. V. Shanabrook, R. J. Wagner, J. L. Davis and J. R. Waterman, *Appl. Phys. Lett.* **63**, 949 (1993).
33. B. V. Shanabrook, B. R. Bennett and R. J. Wagner, *Phys. Rev. B* **48**, 17172 (1993).
34. B. V. Shanabrook and B. R. Bennett, *Phys. Rev. B* **50**, 1695 (1994) and references therein.
35. R. M. Feenstra, D. A. Collins, D. Z. -Y. Ting, M. W. Wang and T. C. McGill, *Phys. Rev. Lett.* **72**, 2749 (1994).
36. R. M. Feenstra, D. A. Collins and T. C. McGill, *Superlattices and Microstructures* **15**, 215 (1994).
37. B. V. Shanabrook, J. R. Waterman, J. L. Davis and R. J. Wagner, *Appl. Phys. Lett.* **61**, 2338 (1992).
38. For the sake of completeness, it should be noted that the question of interfacial integrity in InAs/GaSb SLs grown by metalorganic vapor phase epitaxy (MOVPE) has been addressed by the group at Oxford: G. R. Booker, P. C. Klipstein, M. Lakrimi, S. Lyapin, N. J. Mason, I. J. Murgatroyd, R. J. Nicholas, T. -Y. Seong, D. M. Symons and P. Walker, *J. Crystal Growth* **145**, 778 (1995); G. R. Booker, P. C. Klipstein, M. Lakrimi, S. Lyapin, N. J. Mason, I. J. Murgatroyd, R. J. Nicholas, T. -Y. Seong, D. M. Symons and P. Walker, *J. Crystal Growth* **146**, 495 (1995).
39. M. W. Wang, D. A. Collins, T. C. McGill and R. W. Grant, *J. Vac. Sci. Technol. B* **11**, 1418 (1993).

40. J. M. Ziman, *Principles of the Theory of Solids* (Cambridge Universty Press, Cambridge, 1972) pp. 15–19.
41. P. Hirsh, A. Howie, R. B. Nicholson, D. W. Pashley and M. J. Whelan, *Electron Microscopy of Thin Crystals* (Krieger, Huntington, New York, 1977).
42. J. C. H. Spence, *Experimental High-resolution Electron Microscopy* (Clarendon, Oxford, 1981).
43. P. Buseck, J. Cowley and L. Eyring, *High-Resolution Transmission Electron Microscopy and Associated Techniques* (Oxford University Press, New York, 1988).
44. J. M. Cowley, *Diffraction Physics* (North Holland, New York, 1981).
45. P. Stadelmann, *Ultramicrosc.* **21**, 131 (1987).
46. A. Ourmazd, F. H. Baumann, M. Bode and Y. Kim, *Ultramicrosc.* **34**, 237 (1990).
47. B. D. Cullity, *Elements of X-ray* Diffraction (Addison-Wesley, Reading Mass., 1967) pp. 134–135.
48. M. E. Twigg, B. R. Bennett, B. V. Shanabrook, J. R. Waterman, J. L. Davis and R. J. Wagner, *Appl. Phys. Lett.* **64**, 3476 (1994).
49. M. E. Twigg, B. R. Bennett and B. V. Shanabrook, in *Microscopy of Semiconducting Materials, Inst. Phys. Conf. Ser. No. 146* (Institute of Physics, Philadelphia, 1995) pp. 349–352.
50. I. J. Murgatroid, N. J. Mason, P. J. Walker and G. R. Booker, *ibid.*, pp. 353–356.
51. K. Scheerschmidt, S. Ruvimov and D. Timpel, *ibid.*, pp. 39–42.
52. K. Scheerschmidt, S. Ruvimov, P. Werner, A. Hopner and J. Heydenreich, *J. Microsc.* **179**, 214 (1995).
53. For a proper discussion and definition of the Scherzer resolution as applied to HRTEM, refer to Buseck, Cowley and Eyring (ref. 43 of this Chapter) pp. 19–24.
54. S. Thoma and H. Cerva, *Ultramicrosc.* **35**, 77 (1991).
55. S. Thoma and H. Cerva, *Ultramicrosc.* **38**, 265 (1991).
56. The Fourier transform of the HRTEM image is not that same as the diffraction pattern. The diffraction pattern is the Fourier transform of the wave function multiplied by the complex conjugate of its Fourier transform ($F[\Psi(\underline{r})]\ F[\Psi(\underline{r})]^*$). The Fourier transform of the HRTEM image is the Fourier transform of the product of the wave function and its complex conjugate ($F[\Psi(\underline{r})\ \Psi(\underline{r})^*]$).
57. R. N. Bracewell, *The Fourier Transform and its Applications*, 2nd Ed., (McGraw-Hill, New York, 1978), pp. 197–198.
58. B. C. De Cooman, C. B. Carter, G. W. Wicks, T. Tanoue and L. F. Eastman, *Thin Solid Films* **170**, 49 (1989).
59. J. M. Kang, M. Nouaoura, L. Lassabatere and A. Rocher, *J. Crystal Growth*, **143**, 115 (1994).
60. P. M. Thibado, B. R. Bennett, M. E. Twigg, B. V. Shanabrook and L. J. Whitman, *J. Vac. Sci. Technol. A* **14**, 885 (1996).
61. P. M Thibado, B. R. Bennett, M. E. Twigg, B. V. Shanabrook and L. J. Whitman, *Appl. Phys. Lett.* **67**, 3578 (1995).

CHAPTER 4

Type-II (AlGa)Sb/InAs Quantum Well Structures and Superlattices for Opto- and Microelectronics Grown by Molecular Beam Epitaxy

S. V. IVANOV and P. S. KOP'EV

A. F. Ioffe Physico-Technical Institute of RAS, 26, Politekhnicheskaya Str., St. Petersburg 194021, Russia

4.1.	Introduction	96
4.2.	MBE Growth and Doping of Type-II (AlGa)Sb/InAs Strained Heterostructures	99
	4.2.1. MBE growth and characterization of undoped (AlGa)Sb/InAs-based epilayers and heterostructures	99
	4.2.2. P- and N-type doping of (AlGa)Sb and InAs in MBE	112
	4.2.3. (AlGa)Sb/InAs interface formation in QWs and SLSs	115
4.3.	Electronic Properties and Transport Characteristics of Type-II (AlGa)Sb/InAs QW Heterostructures	125
	4.3.1. Electronic properties of type-II (AlGa)Sb/InAs heterostructures	125
	4.3.2. Carrier mobility and concentration in unintentionally doped AlSb/InAs/AlSb single QW	133
4.4.	Devices Based on Type-II (InGa)(AsSb)/(AlGa)(SbAs) QW Structures	140
	4.4.1. Middle-wave IR lasers	141
	4.4.1.1. Type-I $In_xGa_{1-x}As_ySb_{1-y}/Al_xGa_{1-x}As_ySb_{1-y}$ heterostructure lasers	142
	4.4.1.2. Type-II heterostructure lasers	146
	4.4.2. Resonant tunneling transport	152
	4.4.3. (AlGa)Sb/In(AsSb) based transistors	158
4.5.	Conclusion	165

| Acknowledgments | 165 |
| References | 166 |

4.1 INTRODUCTION

Since late 1980s, quantum well (QW) heterostructures and superlattices (SL) based on an (AlGa)Sb/InAs narrow bandgap material system attracts considerable interest of the experts working in the fields of middle-wave infrared (2–5 µm range) optoelectronics and ultra-high-speed electron devices.

Table I presents the most important transport and optical characteristics of bulk III–V compounds. Bandgaps of the III–V compounds and alloys are shown in Fig. 1 as a function of a lattice parameter. As can be seen from Fig. 1, besides the intensively employed GaAs/AlAs lattice-matched material system ($E_g = 1.42$–2.14 eV), the only InAs/GaSb/AlSb combination among other III–V compounds provides wide range of the energy gap variation ($E_g = 0.36$–1.63 eV) at the relatively small lattice mismatch between the constituent binary compounds ($\Delta a/a = 0.6$–1.3%). The latter is of particular importance for device applications, since the lower is the lattice constant difference between the adjacent layers in heterostructure, the less severe are the thickness restrictions for obtaining epilayers of high structural quality.

The most attractive feature, peculiar to these materials as compared to other III–V's, is the unique conduction and valence band alignment (Fig. 2), allowing to design practically any required type of a heterojunction, including Type-I (AlSb/GaSb), Type-II staggered (AlSb/InAs), and Type-II "broken gap" (GaSb/InAs). The energy of spatially-indirect optical transition at the Type-II staggered heterointerface is less than the bandgaps of the adjacent materials and can be reduced to zero, for instance, by changing the AlSb content in the (AlGa)Sb alloy from 100 to 30%. The system flexibility can be further enhanced by introducing the small amount of As into AlGaSb or Sb into InAs, as well as by using the QW thickness, in QW structures or SLs, as an additional parameter in variation of carrier energy spectrum.

The superior room temperature transport characteristics demonstrated by InAs (see Table I) as compared to that of GaAs or InGaAs and the largest InAs/AlSb conduction band offset among all III–V compounds (Fig. 2) lead to extremely high conductivity of two-dimensional electron gas (2DEG) in AlSb/InAs/AlSb QWs, giving a

TABLE I
Transport and optical parameters of bulk III–V semiconductor materials.

Material	E_g, (eV) (300/77 K)	m_e	m_{hh}	m_{lh}	μ_e, (cm²/(Vs)) (300 K)	μ_h, (cm²/(Vs)) (300 K)	$\varepsilon(0)$	n	a, (Å)
GaAs	1.42/1.52	0.0665	0.45	0.087	~8500	~400	12.5	~3.3	5.65325
InAs	0.36/0.42	0.023	0.41	0.025	~33000	~460	14.5	~3.5	6.0583
InSb	0.18/0.25	0.0139	0.45	0.021	~78000	~750	16.8	~4.0	6.47937
AlSb	1.63/1.69	0.33	0.4	0.12	~200	~420	11.5	—	6.1355
GaSb	0.70/0.81	0.042	0.33	—	~4000	~1400	15.0	—	6.0959

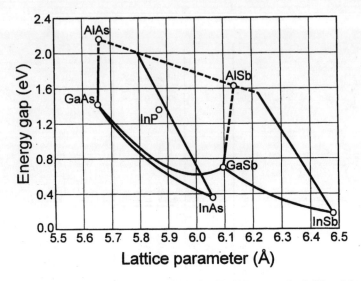

Fig. 1 Energy gaps versus lattice parameters for III–V compounds. Solid and dashed lines connecting the binary compounds show direct-gap and indirect-gap ternary alloys, respectively.

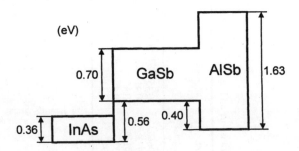

Fig. 2 Band alignment of binary compounds in the InAs/GaSb/AlSb material system.

chance to a new generation of ultra-low-noise and ultra-high-speed electron devices.

This chapter considers a wide spectrum of problems related to technology, physics, and device applications of the Type-II (AlGa)Sb/InAs QW structures. Section 4.2 is concerned with theoretical and experimental aspects of molecular beam epitaxial (MBE) growth and doping of the individual layers and heterostructures as well as with the problem of interface formation, when both cation and anion elements are changed across the interface. Electronic properties of the AlGaSb/InAs QWs and strained-layer superlattices (SLS) are discussed in Section 4.3, along with the description of experimental studies of the effect of various InAs/AlGaSb QW structure parameters on its transport and optical characteristics. The final Section 4.4 includes three subsections, each representing the different type of opto- and microelectronic devices fabricated from these QW structures and SLSs. Brief reviews of the most important achievements are given for: (1) middle-wave infrared (MWIR) (2–5 µm) injection lasers, (2) resonant-tunneling diodes, and (3) heterostructure field effect transistors (HFETs).

4.2 MBE GROWTH AND DOPING OF TYPE-II (AlGa)Sb/InAs STRAINED HETEROSTRUCTURES

4.2.1 MBE Growth and Characterization of Undoped (AlGa)Sb/InAs-based Epilayers and Heterostructures

To obtain high quality Type-II InAs/(AlGa)Sb heterostructures, optimization of MBE growth parameters for individual constituent layers, as wells as for the structure as a whole, is extremely important. This section summarizes the results of various experimental and theoretical studies related to the effect of basic MBE growth parameters, such as substrate temperature (T_S) and V/III flux ratio, on optical and electrical properties and on the structural quality of undoped GaSb, (AlGa)Sb, and InAs epilayers and related heterostructures, grown on lattice-matched (GaSb) and strongly lattice-mismatched GaAs ($\Delta a/a \sim 7\%$) substrates. The GaAs substrates are used for MBE of these compounds both to study their electrical properties and to fabricate InAs-channel electronic devices with a carrier transport, normal to growth direction, since GaSb semi-insulating substrates are not available.

The quality of substrate surface preparation prior to MBE growth plays a decisive role in obtaining the growing films of good structural quality. The preparation process usually involves two stages: (1) *ex situ* chemical treatment and (2) high temperature annealing in ultra-high vacuum. According to a standard chemical procedure, both GaAs and GaSb substrates are first consequently degreased in boiling solvents: trichloroetilene, acetone, and methanol, to eliminate residual organic contamination [1,2]. Thereafter, GaAs substrates are etched in the (4–8):1:1 $H_2SO_4:H_2O_2:H_2O$ solution [3–5] for several minutes. The

0.01% bromine (or 0.5% Br_2) solution in methanol is used as the etchant for GaSb substrates [3,5]. Finally, substrates of both types are rinsed in deionized water to obtain a thin protective oxide layer. Additionally, a $1:20:30:250$ $HNO_3:HBr:HCl:CH_3COOH$ solution has been proposed as the finishing etchant for GaSb [1]. As soon as blown dry with filtered nitrogen, GaAs or GaSb substrate is mounted on molybdenum blocks using In solder, and oxide is desorbed by heating in an ultra-high vacuum chamber under As_4 or Sb_4 molecular fluxes, respectively. The degassing temperature within the 600–640°C range is used for GaAs substrates resulting in the normal (2 × 4) reconstructed surface, as measured *in situ* by reflection high energy electron diffraction (RHEED). GaSb substrates are usually heated to 580–590°C and exhibit (1 × 3) RHEED surface reconstruction prior to growth.

In the case of the GaSb/GaAs growth, a quite thin ($\sim 0.2\,\mu m$) GaAs buffer layer is normally initially grown at $T_S \geqslant 580-600°C$ to improve the surface quality. It is followed by decreasing T_S to 500–550°C, where the GaSb growth is started. Due to the large lattice mismatch between the GaSb film and the substrate ($\Delta a/a \sim 7\%$), the elastic strain in the film relaxes after growing 1 (or 2) monolayers by formation of high misfit dislocation density ($\sim 10^9\,cm^{-2}$) [6,7]. However, it does not cause a long stage of the three-dimensional (3D) growth, reflected by the spotty RHEED pattern. Usually, after the growth of next several monolayers, the streaky clear (1 × 3) [3,5] (sometimes (2 × 3) [7]) antimony stabilized surface reconstruction appears, corresponding to a smooth atomically stepped growth. After this point, no significant difference in the growth mode between hetero- (GaSb/GaAs) and homoepitaxial (GaSb/GaSb) MBE growth can be observed. Changing the GaSb growth parameters from Sb-rich to Ga-rich, until the surface is covered with gallium droplets, does not result in variation of the surface reconstruction [3,5]. Actually, the (1 × 3) RHEED pattern becomes dimmer, gradually disappearing. At $T_S < 400°C$, under Sb_4 flux without growth, the (1 × 3) reconstruction transforms into (1 × 5) [3,5] or (2 × 5) [8]. Yano *et al.* [4] reported polycrystalline structure of GaSb films grown at $T_S < 450°C$. Nevertheless, the special study of the (AlGa)Sb surface structure has revealed apparent transition of the RHEED surface reconstruction from c(2 × 6)Sb-rich to c(8 × 2)Ga-rich at $T_S > 540°C$ [9].

Lee *et al.* [3] have presented experimental data on the Sb_4/Ga beam equivalent pressure (BEP) ratio versus T_S for a large number of epilayers grown under both Sb- and Ga-stabilized conditions within the T_S range of 500–600°C. The activation energy of Sb desorption has been found from the boundary condition line to be approximately 1.2 eV. On the other hand, no temperature dependence of the Sb incorporation rate was observed at $T_S = 350-530°C$ [8]. Finally, it has been found [7], that the increase in T_S from 500 to 530°C at the initial growth stage of GaSb on GaAs requires the nearly order of magnitude enhancement of the Sb_4/Ga flux ratio relative to $Sb_4/Ga = 1:1$ at 500°C, to avoid the surface morphology degradation either by the Ga droplets formation or by the dramatic increase in the surface defect density (5,000–10,000 cm^{-2}) [10]. Similar effect of hindered Sb incorporation at the initial stage of the GaSb/GaAs heteroepitaxial growth was also observed in [11]. Unlike this phenomenum, an AlSb nucleation buffer layer started on GaAs at $T_S = 550°C$ showed mirror-like surface morphology with low defect density, despite the 3D nucleation RHEED pattern transforms to streaky (1 × 3) one still after the 10–15 nm of AlSb growth [10]. The (1 × 3)Sb-rich [8–10] and (4 × 2)Al-rich [10] reconstructions are usually observed in AlSb MBE growth. The RHEED study of Sb incorporation in AlSb has revealed [8] no dependence on T_S in the 420–610°C range.

To find correlation between the experimental data on hetero- and homoepitaxial GaSb growth, Ivanov et al. [7] have proposed a theoretical description for the lattice-matched (GaSb/GaSb) and lattice-mismatched (GaSb/GaAs) MBE growth, which is based on the thermodynamic model, developed by Kop'ev and Ledentsov [12] and Ivanov et al. [13] for MBE of unstrained III–V compounds and alloys. The main equilibrium parameters of the model are substrate temperature and partial pressures corresponding to atomic and molecular fluxes evaporated from the substrate surface.

Thermodynamic analysis of the antimony tetramer dissociation in a gaseous phase over a surface with the temperature $T = T_S$

$$Sb_4(g) = 2Sb_2(g) \tag{1}$$

employed the entropies and formation heats summarized in Table II [14–17]. It has been shown that the gaseous phase at more than 90% consists of Sb_4 molecules at T in the range of 450–530°C and the total antimony pressures $P_{tot} = 10^{-7} - 5 \times 10^{-6}$ Torr, typical for the GaSb MBE. This conclusion is in agreement with experimental data of Yata [18], who found that the mass-spectrum of Sb molecules evaporated from a Si surface at T_S up to 820 K ($\sim 550°C$) contains only the Sb_4 line, provided that the Sb cell produces the Sb_4 incident flux. It was shown that, taking account of a GaSb(s) surface does not significantly change the relationship between the Sb_4 and Sb_2 pressures in the

TABLE II
Material parameters used in thermodynamic calculation.

Components	ΔH^0_{298}, (kcal/mole)	S^0_{298}, (cal/mole)	Refs.
$Sb_4(g)$	46.2	80.0	[14,15,22]
$Sb_2(g)$	59.8	60.5	[14,15,22]
$Ga(g)$	65.0	39.1	[16,22]
$GaSb(s)$	−10.55	18.18	[24]
$Sb(s)$	0	11.4	[17]

temperature range of interest. Thus, for typical GaSb growth regimes, Sb_4 tetramers are the predominant antimony molecules in the evaporating flux, which determine the antimony equilibrium pressure over the GaSb surface.

Hence, in the case of unstrained GaSb/GaSb MBE growth, the reaction describing the GaSb formation from Ga(g) and Sb_4(g) molecules is

$$Ga(g) + 1/4\ Sb_4(g) = GaSb(s), \quad (K_{GaSb}), \qquad (2)$$

where K_{GaSb} is the reaction equilibrium constant. Using the values of the relevant constants for the components of Eq. (2) from Table II, the action mass equation for Eq. (2) can be written as follows

$$P_{Ga} P_{Sb_4}^{1/4} = K_{GaSb}^{-1} = 8.66 \times 10^8 \exp(-3.78\,(eV)/kT), \qquad (3)$$

where P_{Ga}, P_{Sb_4} are the partial equilibrium pressures of Ga and Sb_4 molecules evaporating from the GaSb surface, kT is in eV.

The minimal antimony pressure in the incident flux, which corresponds to the transition from Sb- to Ga-rich conditions at a given growth rate (P^0_{Ga}) and T_S, i.e. the pressure allowing the GaSb layer to grow without formation of Ga droplets on the surface, can be written as follows (a weak square root temperature dependence is neglected here)

$$P^0_{Sb_4,\min} = 1/4\sqrt{m_{Sb_4}/m_{Ga}}(P^0_{Ga} - P^{Ga-L}_{Ga}) + P^{Ga-L}_{Sb_4}, \qquad (4)$$

where m_{Sb_4}, m_{Ga} are the molar masses, P^{Ga-L}_{Ga} and $P^{Ga-L}_{Sb_4}$ are the Ga and Sb_4 equilibrium partial pressures over the Ga–GaSb liquidus of the GaSb phase diagram. It should be mentioned that this equation may give a good quantitative evaluation of the minimum Sb_4 beam equivalent pressure (BEP), taking into account the experimental facts that four Sb atoms are incorporated for each Sb_4 molecule impinging on the surface of GaSb [8] and AlSb [8,10] for T_S ranging from 350 to 530°C and from 420 to 610°C, respectively.

Next, the expression for P_{Ga}^{Ga-L} is as follows

$$P_{Ga}^{Ga-L}(GaSb) = P_{Ga}^L[Ga_L]\gamma_{Ga}; \quad [Ga_L] + [Sb_L] = 1, \quad (5)$$

where γ_{Ga} and $[Ga_L]$ are Ga activity coefficient and concentration (in mole %) in a liquid phase, respectively, and P_{Ga}^L is the gallium equilibrium pressure over the pure Ga melt. The experimental expression for Sb concentration in the liquid phase [19] is

$$[Sb_L] = 2.67 \times 10^3 \exp(-0.754(eV)/kT), \quad (6)$$

and $\gamma_{Ga} = 0.998 - 1.004 \cong 1$ within the 430–590°C temperature range. Thus, from Eq. (5) we obtain

$$P_{Ga}^{Ga-L} = 2.88 \times 10^5 \exp(-2.74(eV)/kT)(1-[Sb_L]), \quad (7)$$

and from Eqs. (3) and (7) it follows that

$$P_{Sb_4}^{Ga-L} = 8.1 \times 10^{13} \exp(-4.16(eV)/kT)(1-[Sb_L])^{-4}. \quad (8)$$

To estimate the maximum possible excess Sb_4 pressure, at which Sb(s) precipitates begin to appear on the GaSb growth surface, i.e. the Sb equilibrium pressure over the GaSb-Sb solidus of the GaSb phase diagram ($P_{Sb_4}^{Sb-S}$), one can use the following equation

$$1/4\, Sb_4(g) = Sb(s). \quad (9)$$

Since the chemical potentials of the components in Eq. (9) are equal, one can obtain from Eq. (9) and Table II

$$P_{Sb_4}^{Sb-S} = 3.26 \times 10^7 \exp(-2.0(eV)/kT). \quad (10)$$

It follows from the estimation using Eq. (10) in the T_S range of interest that this antimony pressure cannot be achieved under the common MBE conditions.

For the lattice-mismatched GaSb/GaAs growth, one should take into consideration the additive strain-induced Gibbs free energy (ΔG_{str}) [20], which is maximal during the growth of the first several monolayers before relaxation of the GaSb film, producing high misfit dislocation density. The expression for strain-induced variation of the formation heat is as follows

$$\Delta H_{str} = 2G\left[(1+v)/(1-v)\right]V_m\left[(a-a_0)/a_0\right]^2, \quad (11)$$

where $v = C_{12}/(C_{11} + C_{12})$ is the Poisson ratio, V_m is the substrate molar volume (28.6 cal^3/mole [21]), $G = 1/2C_{44}$ (for the (100) plane), C_{11}, C_{12}, C_{44} are the elastic coefficients (2.11 × 10^4, 0.96 × 10^4, 1.03 × 10^4 cal/cm^3, respectively [22]) and a is the layer lattice constant (6.0959 × 10^{-8} cm). The calculation according to Eq. (11) with the above values of the parameters gives $\Delta H_{str,max} = 0.15$ eV ($\Delta a/a_0 = 7.5\%$). P_{Ga}^{Ga-L}, being the pressure over the Ga–GaSb liquidus, is not affected by the presence of strain, and, therefore, $P_{Sb_4}^{Ga-L}$ is the only variable parameter in Eq. (3) due to the ΔG_{str} change. Since during the pseudomorphic growth of the first GaSb monolayers on the GaAs substrate the crystalline order does not change, the reaction entropy remains the same, i.e., $\Delta S_{str} = 0$. Hence, substituting the additive strain-induced Gibbs free energy $\Delta G_{str} = 0.15$ eV results in the following equation for the Sb$_4$ equilibrium pressure over the Ga–GaSb liquidus

$$(P_{Sb_4}^{Ga-L})_{str} = 8.1 \times 10^{13} \exp(-3.56(eV)/kT)(1-[Sb_L])^{-4}. \quad (12)$$

Fig. 3 presents the calculated temperature dependencies of the gallium and antimony equilibrium partial pressures over the Ga–GaSb liquidus: for the lattice-matched GaSb/GaSb growth according to Eqs. (7) and (8) (solid lines), for the strained layer GaSb/GaAs growth according to Eq. (12) (broken line). The dotted line shows the minimal antimony incident BEP [see Eq. (4)] for the unstrained and strained growth at a given growth rate [7].

Analysis of the dependencies from Fig. 3, related to the unstrained GaSb/GaSb MBE growth, permits the conclusion to be made that the GaSb growth rate is independent of the substrate temperature up to $T_S \sim 630 \sim 650°$C, and is only determined by the incident Ga flux, because the Ga re-evaporation from the substrate surface (P_{Ga}^{Ga-L}) is negligible within the T_S range used. On the other hand, GaSb thermal dissociation ($\Delta E_d = 4.16$ eV), giving rise to the increase in the minimal antimony incident pressure required for keeping the Sb-stabilized surface conditions, can be experimentally observed for T_S above 600°C. Hence, the activation energy of antimony desorption ($\Delta E_d \sim 1.2$ eV) determined for T_S in the 500–600°C range [3] reflects the temperature dependence of the Sb$_4$ incorporation coefficient caused by the kinetic limitations for the Sb$_4$ incorporation into a GaSb film. According to the Sb incorporation behavior at $T_S < 530°$C [8], the Sb$_4$ incorporation coefficient saturates at T_S lower than 500–530°C.

In the case of strained heteroepitaxial growth of first few GaSb monolayers on GaAs before the film relaxation, the dramatic strain-induced

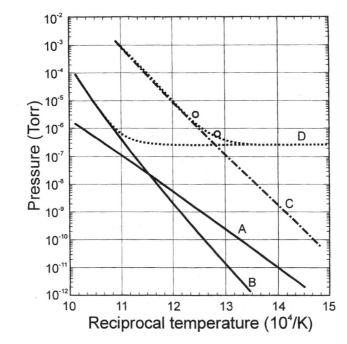

Fig. 3 The calculated temperature dependences of (A) gallium P_{Ga}^{Ga-L} and (B) antimony $P_{Sb_4}^{Ga-L}$ equilibrium partial pressures along the Ga–GaSb liquidus according to Eqs. (7) and (8) for lattice-matched GaSb/GaSb growth; (C) maximal antimony equilibrium partial pressure along the Ga–GaSb liquidus $(P_{Sb_4}^{Ga-L})_{str}$ according to Eq. (12) for strained layer GaSb/GaAs growth; (D) the minimum possible Sb$_4$ incident BEP $P_{Sb_4,min}^0$ according to Eq. (4).

increase in antimony evaporation flux in the 500–550°C T_S range $((P_{Sb_4}^{Ga-L})_{str})$ requires approximately order of magnitude enhancement of the minimal incident $P_{Sb_4,min}^0$ pressure at $T_S \sim 530$°C as compared to the unstrained growth, to compensate the Sb loss. This conclusion is in good agreement with the respective experimental values $P_{Sb_4,min}^0$ [7], shown in Fig. 3 by open circles at 500 and 530°C. The different degree of the antimony surface depletion at the initial growth stage may result either in the complete surface morphology degradation [7] or in the appearance of the high oval-shaped surface defect density [10].

To avoid this undesirable effect of strain, two-stage substrate temperature growth regime has been proposed [7], which included the growth of 0.2–0.5 μm GaSb at $T_S \sim 500$°C or even lower, where the effect of strain is not significant, and the growth of the rest of structure at higher temperature. Similar growth regime has been successfully used by Bennet *et al.* [23].

The growth of an AlSb nucleation layer on GaAs substrate even at higher temperature [10] appears to be the other way to solve the problem of the dramatic Sb loss. Since the Al–Sb binding energy is higher, than Ga–Sb one [24], it results in the much lower Sb_4 equilibrium pressure over the Al–AlSb liquidus of the AlSb phase diagram even under the strain conditions in the T_s range of interest.

The epilayers, lattice-mismatched to a substrate, whose thickness exceeds the critical value (1–2 ml in the (AlGa)Sb/GaAs case), relax with the formation of high misfit dislocation density ($\sim 10^{10}\,cm^{-2}$) in the interface. The majority of misfit dislocations occurring in the interface plane have been found [7] to be pure edge dislocations. They are parallel to both [110] and [1$\bar{1}$0] directions and form a square network with characteristic spacing of about 50 Å in the interface plane (see the transmission electron microscopy (TEM) image in Fig. 4). Fig. 5 shows high resolution TEM (HREM) image of the GaSb/GaAs interface cross-section (the (100) plane image). The misfit dislocation cores are marked by arrows. One can see that these dislocations are generally of the Lomer type with the Burgers vector $b = a/2 \langle 1\bar{1}0 \rangle$ lying in the interface. The average distance between the

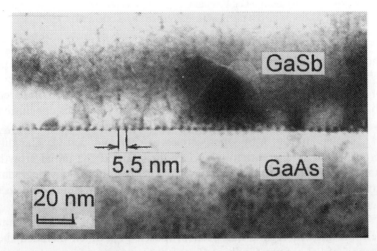

Fig. 4 Cross-section TEM image of the GaSb/GaAs interface. The equally spaced dark spots are the misfit dislocations.

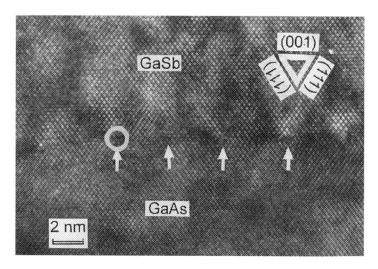

Fig. 5 HREM image of the GaSb/GaAs interface. Arrows indicate the misfit dislocation cores.

dislocations, shown in Fig. 4, (~ 55 Å) is in good agreement with the theoretical value (L)

$$L = b/f \approx 55 \text{ Å}, \tag{13}$$

where the Burgers vector $b = 4.1$ Å and the lattice mismatch $f = 7.5 \times 10^{-2}$. This result means that the mismatch strain between the substrate and the epilayer should be nearly completely compensated by dislocations at room temperature.

It has been shown [7] by the comprehensive TEM and X-ray diffraction (XRD) studies, that the density of threading dislocations, originating at the GaSb/GaAs interface, decreases significantly from $\sim 10^{10}$ cm^{-2}, within the first (0.2–0.4) μm of GaSb after the GaSb/GaAs interface, to $(2-3) \times 10^8$ cm^{-2} near the structure surface. The plane view electron micrographs of these regions are given in Figs. 6(a) and (b), respectively. The propagation of threading dislocations and their density are strongly affected by the growth regime [7] and introduction of a strained-layer SL (SLS) [25]. Rossi et al. [26] inserted a (GaAs$_1$/GaSb$_1$) 10-period SLS at the GaSb/GaAs interface to relieve strain and reduce the dislocation density. Incorporation of a (50 Å-AlSb/50 Å-GaSb)$_{10}$ SLS at ~ 0.5 μm from the GaSb/GaAs interface has been found [7] to decrease the concentration of non-radiative recombination and scattering centers in the top GaSb layer, thus increasing the integral PL intensity and reducing the degree of compensation. However, the SLS did not prevent the propagation of threading dislocations through the structure.

The non-uniform defect distribution in the epilayer has been demonstrated [4] to affect significantly the electrical characteristics of the 1.3 μm thick GaSb layer measured at various distances from the interface. An order of magnitude increase in the hole concentration and respective two-times reduction of the carrier mobility from the surface to the interface have been observed. Homoepitaxial GaSb epilayers usually exhibit more than order of magnitude higher integral photoluminescence (PL) intensity as compared to those grown on

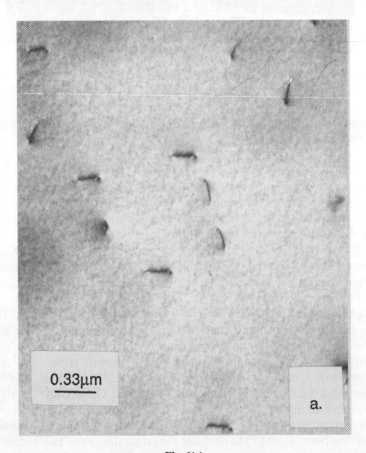

Fig. 6(a)

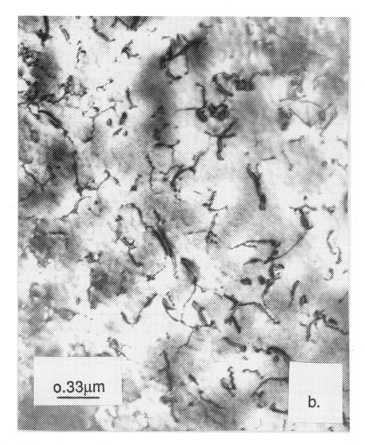

Fig. 6 Plan-view TEM images of GaSb layers (a) near the surface ($N_d \approx 2 \times 10^8 \, \text{cm}^{-2}$) and (b) near the GaSb/GaAs interface ($N_d \approx 10^{10} \, \text{cm}^{-2}$).

GaAs substrates under nominally the same growth conditions [3]. Nevertheless, the substrate temperature and the Sb_4/Ga flux intensity ratio predominantly affect both electrical and optical characteristics of the GaSb epilayers. The T_S range of 500–550°C has been found [4,27] to be preferential for obtaining the high quality layer with lower residual acceptor concentration and higher mobility (Fig. 7), while at $T_S < 400°C$ the conductivity tended to be even n-type, probably due to the enhancement of residual donor concentration at low temperatures.

A strong correlation has been found between the GaSb layer quality and the Sb_4/Ga atomic flux ratio [3,7,27]. The best optical and electrical

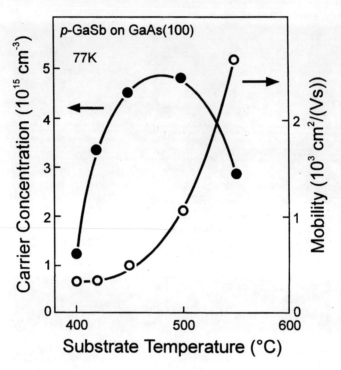

Fig. 7 Mobilities, ○, and carrier concentrations, ●, of undoped GaSb epilayers as a function of the growth temperature [27].

properties were obtained using the minimum excess antimony pressure providing Sb-stable growth at given T_S. The lowest reported hole concentration was 7.8×10^{15} cm^{-3} with the respective 300 K mobility of 950 cm^2/(Vs) [3]. Less than two-times decrease in the Sb$_4$/Ga flux ratio at $T_S = 530°C$ led to the PL intensity increase of approximately two orders of magnitude. The similar MBE growth regime turned out to be also optimal for obtaining the best structural and optical characteristics of AlSb/GaSb short period SLS [28]. The low Sb$_4$/Ga ratio at $T_S = 530°C$, combined with the two-T_S stage growth regime, led to the observation of the free exciton PL emission (809 meV) in the 2 μm thick GaSb layer grown on GaAs substrate with the AlSb/GaSb SL buffer structure [7]. The PL spectrum of the GaSb/GaAs epilayer, presented in Fig. 8, contains also the luminescence lines identified as the donor-shallow acceptor (~793 meV) and donor-deep native acceptor (~775 meV) related transitions, as well as four lines attributed to the

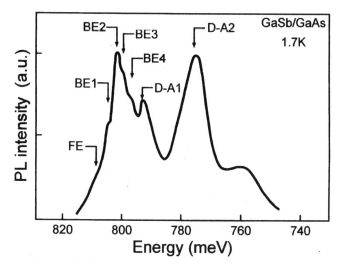

Fig. 8 Photoluminescence spectra at 1.7 K and $I_{ex} = 5$–$10\,\text{W/cm}^2$ of GaSb/GaAs (100) layer grown at low Sb/Ga flux ratio. The arrows show the PL lines observed in the LPE GaSb/GaSb layer: FE – free exciton, BE1–BE4 – bound excitons, D-A2 – deep native acceptor related donor–acceptor-pair recombination, D-A1 recombination radiation line was observed only for the MBE epilayer.

bound exciton (796–804 meV) recombination transitions. This PL spectrum is characterized by the considerably enhanced bound-exciton/deep-acceptor peak intensity ratio, demonstrating the improved optical quality of the GaSb epilayer.

InAs layers, grown on GaSb substrates, exhibited the behavior of electrical properties versus the MBE growth parameters quite similar to that of GaSb [27]. Undoped InAs layer usually show n-type conductivity. It has been found that high electron mobilities could be reproducibly obtained under the InAs surface stoichiometry conditions intermediate between the (2×4) As-stabilized and the (4×2) In-stabilized surfaces. As can be seen from Fig. 9, the relatively high electron mobilities of $(3\text{–}4) \times 10^4\,\text{cm}^2/(\text{Vs})$ and the low residual electron concentrations of $(0.6\text{–}1.2) \times 10^{16}\,\text{cm}^{-3}$ can be obtained in a wide range of the growth temperatures of 400–500°C. The increase in T_s higher than 500°C caused the dramatic enhancement of the arsenic re-evaporation from the surface, which cannot be compensated by the V/III flow ratio increase [29]. It resulted in the degradation of the epilayer quality and in the inferior electrical characteristics.

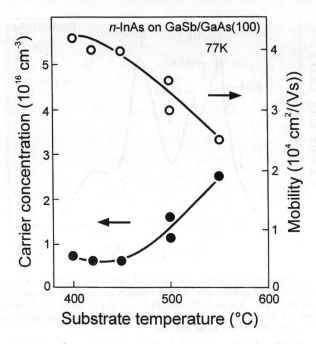

Fig. 9 Mobilities, O, and carrier concentrations, ●, of undoped InAs on GaSb epilayers as a function of the growth temperature [27].

4.2.2 P- and N-type Doping of (AlGa)Sb and InAs in MBE

As has been shown in the previous section, undoped GaSb layers grown by MBE on GaAs substrates show p-type conductivity with the 300 K free hole concentration in the $(0.8–8.0) \times 10^{16}$ cm^{-3} range and mobility of 400–950 cm^2/(Vs), depending on growth conditions and the layer thickness. Residual deep acceptor, responsible for the PL line at 773–776.5 meV, has been considered to originate from the native Ga_{Sb}-V_{Ga} defect complex [30,31].

To achieve p-type doping level in MBE grown GaSb and AlGaSb layers as high as 10^{19} cm^{-3} Be dopant, common for GaAs MBE, was initially employed [32], displaying lower diffusion ability at the doping level of 10^{19} cm^{-3} and substrate temperature of 530°C [33], than it could be expected from a comparison of the GaAs and GaSb lattice parameters. First attempts to substitute Be by more pure and less toxic Si, which was commonly used as a p-type dopant in melt-grown GaSb, resulted in a closely autocompensated material [34]. The first low compensated p-type GaSb MBE layers doped with Si have been reported by Rossi et al. [26]. They obtained in GaSb:Si, grown at $T_S = 460°C$, the room temperature free-hole concentration monotonically increasing from $4.0 \times 10^{15} cm^{-3}$ to $4.3 \times 10^{18} cm^{-3}$ with the increase in the silicon cell temperature up to 1250°C. Comparing these results with the experiments on the n-type Si doping of MBE GaAs, with the same silicon source temperatures used, revealed nearly equal Si doping efficiency in both cases. PL spectra of GaSb:Si layers, characterized by a single near band-edge emission line broadened with increasing the doping level, demonstrated dramatic suppression of the native defect-like acceptor formation.

In contrast to p-type, the n-type doping behavior of MBE grown (AlGa)Sb is quite different from that of (AlGa)As compounds. Since silicon and tin, the two main donors in (AlGa)As MBE, exhibited the p-type and amphoteric properties, respectively, in molecular beam epitaxial GaSb, most of the efforts to obtain the highly reproducible n-type doping level in group-III antimonides have been focused on employing group-VI chalcogenide elements: tellurium [4,10,32,33], selenium [35], and sulphur [5,36]. The principal problem, why these elements were not intensively used in MBE of (AlGa)As compounds, is the dramatic decrease in dopant incorporation with increasing growth temperature. The lower growth temperatures used in (AlGa)Sb MBE were expected to make this effect less severe.

The first successful n-type doping of GaSb using a conventional Te cell has been reported by Yano et al. [4]. They achieved satisfactory doping behavior in the $5 \times 10^{17} - 1 \times 10^{18} cm^{-3}$ concentration range at the Te cell temperatures of 200–220°C, while the maximum mobilities were still about $2000 cm^2/(Vs)$. However, even this less volatile group-VI element has sufficiently high vapor pressure at typical MBE setup bakeout temperatures (200–250°C), which may result in both undesirable cross contamination of other sources and high Te background level in a growth chamber. Moreover, Te can react with Ga at the surface forming Ga_2Te_3 or Ga_2Te, which have been shown to deteriorate the growth [37]. The next attempt to increase the operation temperature of Te-containing cell to at least 350°C, using Ga_2Te_3 as a Te "captive source" instead of elemental Te, has been made by Ohmori et al. [32], who obtained GaSb:Te with $n = 1 \times 10^{18} cm^{-3}$ and N-$Al_{0.2}Ga_{0.8}Sb$ doped with Te up to $n = 2 \times 10^{18} cm^{-3}$.

To use the more volatile S dopant, two different approaches have been proposed. In the first one [36] a H_2S_3 gas was taken as the S source. However, Gotoh et al. [36] failed to reach the free carrier concentration in GaSb greater than $4 \times 10^{16}\,cm^{-3}$ under practical MBE growth conditions. The highest room temperature mobility achieved for the low doped samples was about $2000\,cm^2/(Vs)$. In addition, sulphur was found to be a deep donor with an activation energy of 75 meV, which, in turn, also contributed to the reduced doping efficiency.

In the second approach, Pool et al. [5] used the AgS_2 electrochemical source earlier developed by Davies et al. [38]. The flux from this cell is an exponential function of applied voltage, allowing the rapid and accurate control. On the other hand, there is no significant evaporation of compound at typical system bakeout temperatures. They achieved the maximum sulphur doping level in GaSb in excess of $1 \times 10^{18}\,cm^{-3}$, as measured by secondary ion mass-spectroscopy (SIMS). The reported capacitance–voltage measurements gave the ionized donor concentration of about $1.5 \times 10^{17}\,cm^{-3}$ for the sample with practically the same S doping level of $1.7 \times 10^{17}\,cm^{-3}$ (SIMS), demonstrating the nearly 100% electrical activity. However, the sulphur incorporation exhibited the strong dependence on the substrate temperature, varying from $1.5 \times 10^{18}\,cm^{-3}$ at T_S below 435°C to $2.0 \times 10^{16}\,cm^{-3}$ at $T_S = 525°C$. Furthermore, increasing the antimony-to-gallium flux ratio by a factor of 4 at T_S in the 490–550°C range led to approximately the same increase in S incorporation before the saturation. This flux ratio dependence, inverse to that expected for the S_{Sb} substitutional donor, was qualitatively explained by reduction of the surface concentration of free gallium atoms at high Sb/Ga flow ratios. This, in turn, decreases significantly a probability of volatile Ga_2S forming, which has been shown for the case of GaAs:S [39] to be an effective channel for the S loss. As a result, the optimal S doping conditions ($T_S \leqslant 450°C$, $Sb/Ga \geqslant 4:1$) turned out to be too far from the MBE growth conditions required for high quality GaSb epilayer ($T_S = 530$–$540°C$, $Sb/Ga \approx 1:1$) [3,7], where the sulphur incorporation factor was estimated to be as low as $\sim 0.4\%$.

The same but much steeper Sb/Ga flux ratio dependence has been obtained by McLean et al. [35], who used PbSe as Se "captive" source. To reach the maximal reported free electron concentration in GaSb:Se of $5 \times 10^{17}\,cm^{-3}$, they had to use the Sb/Ga atomic flux ratio as high as 40:1. Even these high flux ratios could not prevent dramatic dopant loss from the GaSb surface at $T_S > 530°C$. Finally, selenium, similar to sulphur, appeared to be undesirable deep donor in GaSb [40].

The most successful results on n-type doping of GaSb and AlSb with Te have been achieved by Subbanna et al. [10] by using Pb-enriched

PbTe source, which previously was used to dope GaAs up to 2×10^{19} cm^{-3} at the substrate temperature of 530°C [41]. The source evaporates mainly nondissociated PbTe molecules. The molecules are decomposed on the substrate surface, leading to incorporation of Te and evaporation of lead. The PbTe source temperature was varied in the 360–540°C range providing the carrier concentration control in GaSb from 1.2×10^{16} to 1.6×10^{18} cm^{-3}. High 300 K Hall mobilities of 4200 cm^2/(Vs) were measured for the lightly doped GaSb layers. The maximum electron concentration reported for AlSb was $(6-7) \times 10^{17}$ cm^{-3} in the 550–650°C T_S range. Using SIMS profiling, tellurium incorporation has been found by to be a strong function of MBE growth parameters. However, significant decrease in Te incorporation in GaSb was observed only at T_S above 500°C, and no distinct reduction of Te doping level in AlSb was registered up to at least 650°C. In contrast to other chalcogenide dopants, an increase in the Sb/Ga flux ratio was found to reduce the Te incorporation. Under the MBE growth conditions of interest, the Te doping efficiency in GaSb and AlSb was approximately 50% of that in GaAs. The symmetric shape of SIMS profiles obtained at very thin Te doping spikes indicated no significant Te surface segregation. Among the disadvantages of the PbTe doping source, the high background Pb doping level ($\sim 2.6 \times 10^{18}$ cm^{-3}) is worth to be noted. A GaTe doping source with typical operating temperatures around 500–600°C, proposed by Furukawa and Mizuta [33], seems to eliminate this high Pb background problem, showing nearly the same Te doping level in GaSb and $Al_{0.5}Ga_{0.5}Sb$ MBE epilayers.

For MBE doping of InAs, Si and Be have been commonly used as donor and acceptor dopants, respectively, similar to the case of GaAs with nearly the same doping efficiency [42]. They demonstrated high doping levels, close to those in GaAs, low diffusion coefficients and practically unity incorporation coefficients.

4.2.3 (AlGa)Sb/InAs Interface Formation in QWs and SLSs

Atomic scale interface (IF) structure of (AlGa)Sb/InAs QWs and SLSs strongly affects their structural, optical, and electronic properties. Because both the anion and cation are different on either side of the interface, two possible IF atom arrangements may be realized. An "InSb-like" interface is formed when the InAs layer is terminated by the In atomic plane, while the Al(Ga)Sb layer starts from the Sb plane. In the other case, the As atoms from the InAs side and the Al or Ga atoms from the Al(Ga)Sb side form a "Al(Ga)As-like" IF structure with the Al–As or Ga–As bonds across the interface [43]. Both types of the interface are schematically shown in Fig. 10. The large difference in the lattice parameters between the InSb or Al(Ga)As IF monolayers and the InAs or Al(Ga)Sb constituent layers (see Fig. 1) may result in the marked additional tensile or compressive strain, respectively, in the Type-II QW or SLS structures.

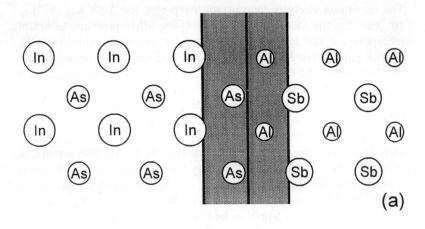

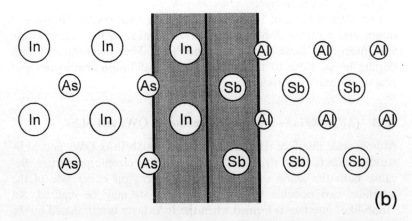

Fig. 10 Interface structures for the (a) AlAs-like interface and (b) InSb-like interface.

Special technological efforts should be taken to obtain the well-defined type of the IF bond. Possible shutter operation sequence for supplying the alternating molecular fluxes onto the growth surface and allowing to control the Al(Ga)Sb/InAs interface planes layer by layer has been suggested by Tuttle *et al.* [43]. In this method, the shutter of the selected terminal element of the bottom layer, for instance Sb of AlSb, was left open for some "soak" time to saturate all the Al–Sb bonds on the surface. The indium shutter was opened for the time necessary to deposit exactly one monolayer of InSb simultaneously with the interruption of the Sb flux. The growth of the InAs QW commenced by opening the As shutter. Then the In flux was left open to recover the surface with the one monolayer In plane, which was followed by consequent deposition of Sb to soak the surface and to form the top InSb IF layer. The growth of AlSb was initiated by opening the Al shutter. The shutter sequences used to obtain InAs/AlSb QW with either InSb-like or AlAs-like interfaces on bottom and top are shown in Fig. 11. InAs/AlSb QWs and SLSs, fabricated in this manner, have demonstrated the essentially different transport [43] and optical [44,45] characteristics, with the type of the bottom IF bond playing a crucial role. These phenomena are considered below in a separate section.

To separate the effect of the IF composition and of the quality of the constituent well and barrier layers on the QW and SLS properties, and to optimize the growth parameters and conditions for formation of the required IF bond configuration in Al(Ga)Sb/InAs QW, it is necessary to provide direct control of the IF composition. Raman scattering by IF vibrational modes has been found to be a most powerful and sensitive technique to study the IF composition [23,45–49] and sharpness [47], internal strain in SLS [23,45,46,48,49] and even SQWs [47,50,51], because those IF mechanical modes are strongly localized in the vicinity of the interface [52]. The penetration depth of the laser excitation beam of commonly used photon energy ($E_{exc} = 2.41$ eV) is as low as several hundred angstroms, which strictly limits the distance between the surface and an interface under study. The typical Raman scattering spectrum for InAs/AlSb QW with both InSb-like interfaces [47] is shown in Fig. 12. Table III outlines the room temperature IF mode frequencies obtained in several studies, together with those of longitudinal-optical (LO) and transverse-optical (TO) phonon modes of the adjacent layers. The LO modes of GaAs and AlAs are shown for comparison with the respective frequencies assigned to their IF modes. Note quite similar difference of 40–50 cm^{-1} for both GaAs and AlAs.

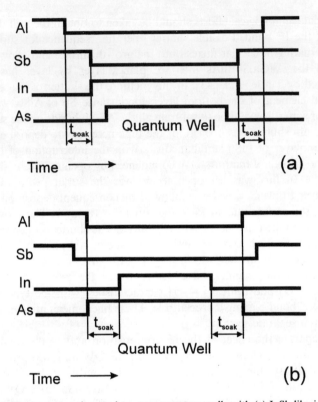

Fig. 11 Shutter sequencing used to grow quantum wells with (a) InSb-like interfaces on top and bottom and (b) AlAs-like interfaces [43].

X-ray diffraction (XRD) can be used only as the additional indirect method of the IF composition characterization in SLS [23,44], because the angle position of a zero-order SLS peak is also affected by both the unintentional cross contamination between the constituent layers, resulting in the formation of alloys instead of binaries [53], and the variation of well to barrier thickness ratio in SL.

Numerous Raman scattering and XRD studies of InAs/Al(Ga)Sb QWs and SLSs grown by MBE have shown that the IF composition and sharpness on an atomic scale are the result of interplay between different growth parameters and conditions. At first, the simultaneous use of two volatile group-V elements with high V/III ratios for both may cause the unintentional cross contamination in the Al(Ga)Sb and InAs layers, including the interfaces. For instance, unintentional

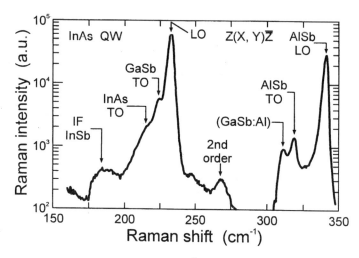

Fig. 12 Raman scattering spectrum for 150 Å InAs/AlSb quantum well with InSb-like interfaces grown at 515°C. Data were collected in the $Z(X,Y)\bar{Z}$ configuration [47].

TABLE III
Room temperature Raman scattering experimental data for the relevant materials and interfaces.

Mode	Observation range (cm^{-1})	Refs.
LO (GaSb/InAs)	234–240	[23,45–47,49,50]
TO GaSb	224	[47]
TO InAs	218	[47]
LO AlSb	338–349	[45,47,50,51]
TO AlSb	319	[47]
LO GaSb$_{0.99}$As$_{0.01}$	240–245	[47]
LO AlSbAs	342	[51]
IF GaAs	250–253	[23,46,49]
LO GaAs	292	[47]
TO GaAs	269	[47]
IF InSb	185–190	[45,47,49–51]
IF AlAs	104, 352	[45]
LO AlAs	~404	[54]

incorporation of As in GaSb can result in $GaAs_xSb_{1-x}$ with x as high as 0.07–0.30 at high growth temperatures [53–55]. On the other hand, the temperature decrease gave rise to the markedly enhanced Sb incorporation in MBE grown GaAsSb [53,56] and AlAsSb [57]. Possible way to solve this problem, reported by Bennett et al. [23], is to reduce by factor 4–5 the InAs growth rate relative to that of GaSb and to provide both the Sb/Ga and As/In flow ratios as low as possible. Using this minimization technique they have reached about 1%-As/0.3%-Sb residual contamination in GaSb/InAs SLS. In addition, the As and Sb flux transients ($\sim 30s$ for As [23]) can change dramatically the IF composition through the substitution reactions during the IF formation. The 10s GaAs surface exposure to the Sb_4 flux at the substrate temperature $T_S = 470°C$ has been found [58] to result in the formation of one GaSb surface monolayer. The similar anion exchange reactions with characteristic time of $\sim 20s$ have been observed on the InAs surface under the Sb flux by RHEED and X-ray photoelectron spectroscopy (XPS) [59].

The effect of substrate temperature on the composition and quality of the InAs/Al(Ga)Sb interface was usually related to the unfavorable intermixing reaction between the adjacent InAs and Al(Ga)Sb layers [60] at elevated T_S, and to the improvement of the top AlSb/InAs IF sharpness if the InAs QW was grown at lower T_S [47]. However, in terms of thermodynamics, formation of the IF monolayer of any type (InSb or Al(Ga)As) in this material system is equivalent to the initial stage of pseudomorphic growth on strongly lattice-mismatched substrate, considered above in detail [7] (see Section 4.2.1). Therefore, the strain-induced giant enhancement of the group-V equilibrium pressure over this monolayer, being a steep function of T_S, may exceed the group-V molecule pressure in the incident flux, preventing the proper formation of this monolayer. This effect seems to be most important for the InSb-like interface, having the lowest temperature of the congruent evaporation [61]. Despite the highest congruent evaporation temperature among other compounds in this system, AlAs has the lattice mismatch to InAs and AlSb as high as 7.0–8.4%, which may significantly decrease its dissociation temperature. Thus, there should exist maximal possible growth temperature at a certain incident group-V BEP for each interface and IF bond combination in this material system.

Contrary to this, the buffer layer should be grown at the highest possible T_S to provide flat growth surface prior to the QW or SLS deposition [43,47]. The low temperature AlSb growth may result in inferior surface morphology which does not allow to obtain nominally pure InSb or AlAs interface due to bondings between different atomic steps.

For InAs/GaSb SLS grown at rather low $T_s \sim 400°C$ on a GaSb buffer layer, it has been shown by Raman scattering, that both types of possible interface bonds (InSb- or GaAs-like) can be prepared selectively by applying an appropriate quite simple shutter sequence during the growth [23,46,49,62]. However, special cares were taken to obtain the pure InSb (free from the GaAs IF mode) interface [23].

Unlike InAs/GaSb, the InAs/AlSb QWs grown on the AlSb buffer layer usually demonstrated the predominant formation of the InSb IF bonds, independent of the beam alternating sequence [47], and even at three AlAs monolayers deposited at each QW interface [51]. On the other hand, the detection of a pure AlAs-like interface mode in the Raman scattering spectra turned out to be very difficult problem, which is probably due to different technological reasons. Among them are (1) AlSb buffer layer growth mode, resulting in more than one monolayer height steps on its surface; (2) inappropriate shutter sequence; (3) relatively high temperature of IF formation stimulating the As-Sb exchange reactions at the InAs/AlSb interface. For instance, Wagner et al. [51] assigned the intensive line, $\sim 4\,cm^{-1}$-shifted to higher frequencies from the AlSb LO buffer layer mode, to the $AlSb_{1-x}As_x$ LO mode associated to the As contaminated barrier layer.

The observation of the pure AlAs IF Raman mode at $325\,cm^{-1}$ in the 10 period $(InAs)_{10}(AlSb)_{10}$ SLS with nominally AlAs-like IF has been reported only in Ref. [45], where the low growth temperature of 350°C, low InAs and AlSb growth rates (0.25 ml/s), and the special in situ RHEED controlled beam alternating sequence were used (see Fig. 13). The growth mode combined conventional MBE growth of AlSb or InAs layers with the two-cycle migration-enhanced epitaxy (MEE) termination of each layer before starting the IF formation. In particular, they observed a (1×1) RHEED pattern, attributed to the residual Sb surface contamination, after the first 1 ml Al pulse, whereas the second 1 ml Al pulse transformed the surface to (4×1) Al-rich [Fig. 13(c)]. This residual Sb contamination might be the origin of the predominant formation of the InSb bond, when only one MEE shutter cycle [43] was used. Typical Raman spectra obtained by Yano et al. [45] for InAs/AlSb SLS grown on a AlSb buffer layer with both IF types are shown in Fig. 14. Careful analysis of the strain-induced energy shift of the InAs and AlSb LO lines relative to their strain-free position revealed the strong influence of the both types of IF bonds

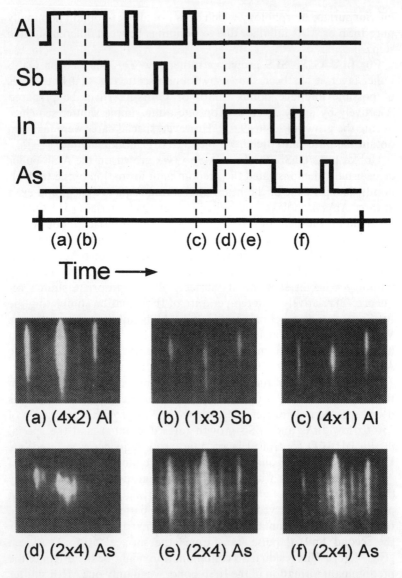

Fig. 13 Typical one period of beam supply sequence used for the growth of SLSs with AlAs interface [45]. One or two monomolecular planes of InAs and AlSb near the heterointerface were deposited layer by layer, although a major part of the SLSs was grown by a conventional MBE process. The RHEED patterns (a)–(f) were observed at respective time of (a)–(f) in the given beam supply sequence.

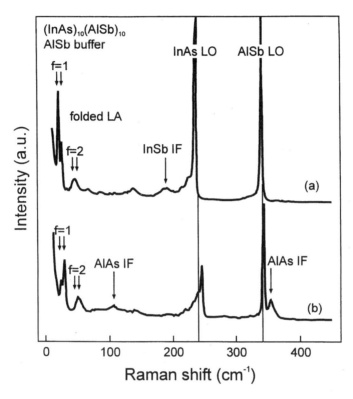

Fig. 14 Typical Raman spectra from $(InAs)_{10}(AlSb)_{10}$ SLSs on AlSb buffer layers [45]. The spectra (a) and (b) were observed for SLSs with InSb and AlAs interface bonds, respectively. f is the folding order of zone-folded LA phonon modes. The straight lines at 240 and 341 cm^{-1} show the strain-free LO phonon frequencies of InAs and AlSb, respectively.

and the buffer layer material on strain relaxation in InAs/AlSb SLSs grown on the AlSb and InAs buffer layers [45]. Schematic cross sections of four SLS types with the profiles of lattice parameter variation, expected from the Raman and PL analysis (dashed line), are shown in Fig. 15. It is clearly seen that, except the case of AlAs-like IF on the AlSb buffer, where the constituent layers of SLS relax separately at each interface, all the other combinations of buffer material and IF bond type form nearly pseudomorphic SLSs. This apparent peculiarity of the "AlAs on AlSb" case, which has been later confirmed by direct TEM observation of the severely disordered crystalline structure of this type SLS with defect generation originating at the buffer – SLS

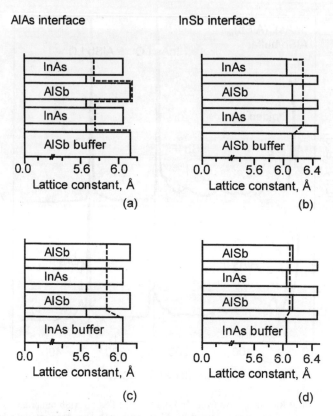

Fig. 15 Schematic cross sections of four types of InAs/AlSb SLS: (a) AlAs interface bond on AlSb buffer; (b) InSb interface bond on AlSb buffer; (c) AlAs interface bond on InAs buffer; and (d) InSb interface bond on InAs buffer [45]. The solid lines are the relative strain-free lattice constants of the constituent layers. The dashed lines show the strained SLS lattice constants that are expected from the Raman and PL analysis.

interface [44], may be understood from the comparison of the lattice mismatch between the IF bonds and the buffer layers [45]. InSb and AlAs have 6.5% and 7.0% lattice mismatches to InAs, and 5.3% and 8.4% to AlSb, respectively. Thus, due to the 8.4% lattice mismatch the critical thickness of the AlAs pseudomorphic growth on AlSb can be less than 1 ml, resulting in its instant relaxation with the formation of "free stranding" monomolecular plane. This conclusion, as is considered below, is in very good agreement with the interface-bond-related behavior of the transport [43] and optical [44,45] properties of InAs/AlSb QWs and SLSs.

4.3 ELECTRONIC PROPERTIES AND TRANSPORT CHARACTERISTICS OF TYPE-II (AlGa)Sb/InAs QW HETEROSTRUCTURES

4.3.1 Electronic Properties of Type-II (AlGa)Sb/InAs Heterostructures

As illustrated in Fig. 2, the InAs/GaSb/AlSb system provides great possibilities for design of different types of heterojunctions, including Type-I (GaSb/AlSb), staggered Type-II (InAs/AlSb), and staggered Type-II with broken gaps (InAs/GaSb). This section is concerned with electronic properties of the Type-II superlattices and QWs, which are characterized by strong interaction between the conduction and valence bands. The "Type-II broken gap" band alignment is a unique feature of the InAs/GaSb heterojunction, providing the unusual case of band separation, with the valence bandedge of GaSb lying above the conduction bandedge of InAs. This kind of heterostructures was first proposed by Sai-Halasz et al. [63] in 1977. Early investigations of these structures were reviewed by Chang and Esaki [64]. Therefore, only the key feature of InAs-GaSb superlattices and QWs, namely, a semiconductor-to-semimetal transition is discussed in this section, leaving a space for more recent investigations of InAs/AlGaSb and InAs/AlSb heterostructures.

First band calculations of InAs/GaSb superlattices were performed in the framework of Kane's two-band model [63] and with the LCAO method [65], ignoring the charge redistribution at the interface. The bandedge energy separation $\Delta = E_{c,InAs} - E_{v,GaSb} = -150\,meV$ was taken from the absorption measurements [66]. Note that this value remains unchanged up to now. Fig. 16 shows the calculated energy subband positions of electrons as well as heavy and light holes for the superlattices with equal well and barrier thicknesses. As the superlattice period is increased, the band gap decreases and becomes eventually negative at a critical value of 170 Å. This corresponds to the condition of a semiconductor-to-semimetal transition. Experimentally, the superlattice semimetallic state was observed by far-infrared magnetoabsorption [67] and magnetotransport [68] measurements.

In the semimetallic superlattice, electrons transfer from the GaSb valence band to the InAs conduction band with the formation of the strongly localized dipole layer, which requires a self-consistent calculation. Comprehensive self-consistent calculations of electronic structure in the envelope-function approximation with the three-band **k p** formalism have been performed for InAs/GaSb superlattices [69]. The

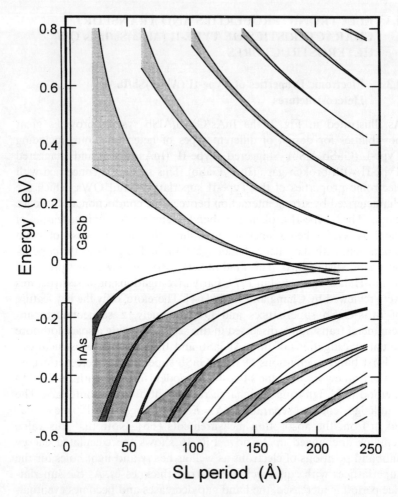

Fig. 16 Calculated subband energies and widths for electrons and heavy and light holes as a function of the period of InAs/GaSb superlattice [64]. The cross-over of the electron and heavy hole subbands at 170 Å indicates semiconductor–semimetal transition.

intrinsic sample was shown to exhibit no truly semimetallic band structure, but rather a zero- or very small bandgap ($E_g \leqslant 10\,\text{meV}$) semiconductor one. Therefore, Altarelli [69] argued that the observed semimetallic behavior is dominated by extrinsic reasons like doping and temperature effects.

Self-consistent calculations were also performed for the GaSb/InAs/GaSb QWs [70], showing the existence of semiconductor-to-semimetal transition when the InAs thickness exceeds the critical value of about 100 Å. Fig. 17 shows the schematic energy diagram of the InAs/GaSb QW with the thickness above the critical value. GaSb valence electrons are transferred to the adjacent InAs layers, generating an equilibrium concentration of spatially separated electron- and hole-like two-dimensional gases. Fig. 18 demonstrates the schematic subband dispersion for such QW. The lower part of the electron subband is broadened into a resonance and the Fermi surface is ring shaped, with inner radius k_{Fh} and outer radius k_{Fe}. The hole-like states near k_{Fh} are localized in GaSb and electron-like states near k_{Fe} in InAs.

In an ideal QW, the electron concentration in the InAs layer is equal to the sum of the hole concentrations in the two GaSb barriers. However, the results of magnetotransport measurements [71–73] in the MBE grown samples contradict to this simple picture, showing a large excess concentration of electrons with respect to holes. This observation ruled out the transfer from the GaSb valence band as the only source of electrons. The imbalance of electron and hole concentrations was explained by Altarelly et al. [74] by the surface acting as the additional source of electrons. The suggested model accounted also for the experimentally observed appearance of quantum-Hall plateaus [71].

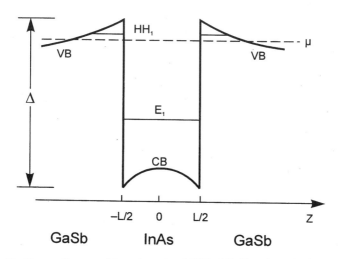

Fig. 17 Energy diagram of the valence band (VB) of GaSb and conduction band (CB) of InAs for a GaSb/InAs/GaSb heterostructure [70]. Band bending occurs as a result of electron transfer from GaSb to InAs. E_1 and HH_1 are the electron and heavy hole energy levels, μ is the Fermi level position.

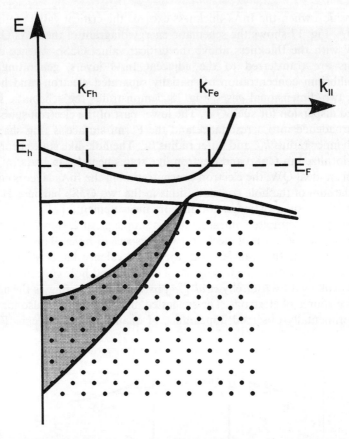

Fig. 18 Schematic subband k_II-dispersion for the InAs/GaSb quantum well [74]. The dotted region denotes the GaSb valence band continuum. The broadening of the E_1 electron subband into a resonance is indicated by the shaded zone. E_h denotes the $k=0$ position of the hole subband.

Luo et al. [75] proposed an alternative explanation of the carriers imbalance. Magnetotransport measurements demonstrated a large concentration of electrons in the GaSb/InAs/GaSb QW with the width of 75 Å, which is below the characteristic value for the semiconductor–semimetallic transition. Positively charged centers located in the vicinity of the interfaces were believed to act as the electron sources. Furthermore, beating pattern in the Shubnikov-de Haas oscillations was observed, indicating the lack of inversion symmetry [76]. Luo et al. [75] believed that this asymmetry indicates the presence of the built-in electric field arising from the asymmetrical distribution of the centers between the two interfaces. An appearance of significant photo-galvanic current, observed experimentally by Dmitriev *et al.* [77] in the InAs/GaSb single QW structure, was also attributed to the effect of the interface-related electric field.

The problem of the extrinsic electron sources is particularly important for the AlSb/InAs/AlSb and AlGaSb/InAs/AlGaSb QW heterostructures. As can be seen from Fig. 2, the InAs/AlSb interface represents the staggered Type-II heterojunction. Slightly staggered valence band offset $\Delta E_v[\text{InAs} \rightarrow \text{AlSb}] = 0.10–0.15\,\text{eV}$, deduced from the known valence band offsets $\Delta E_v[\text{InAs} \rightarrow \text{GaSb}] = 0.5\,\text{eV}$ [64] and $\Delta E_v[\text{AlSb} \rightarrow \text{GaSb}] = 0.35–0.40\,\text{eV}$ [78] using the transitivity postulate, gave the conduction band offset of about 1.30–1.35 eV, which is well consistent with that of $1.35 \pm 0.05\,\text{eV}$ obtained from electrical C–V measurements [79]. Magnetotransport measurements [80] reflect the semimetallic behavior in $\text{Ga}_{1-y}\text{Al}_y\text{Sb}/\text{InAs}/\text{Ga}_{1-y}\text{Al}_y\text{Sb}$ with y less than 0.3. By increasing the alloy composition, the conductivity nature changes from semimetallic-like to semiconductor-like. So, the solid alloy containing 30% of AlSb corresponds to the transition from the "broken gap" band pattern to the simple staggered valence band offset. However, regardless of the alloy composition, large two-dimensional electron gas (2DEG) concentration has been observed in the InAs QW confined by the $\text{Ga}_{1-y}\text{Al}_y\text{Sb}$ barriers [81–83] confirming the presence of additional electron sources. Cyclotron resonance (CR) measurements were performed on 2DEG in AlSb/InAs/AlSb quantum wells [84]. Spin-splitting of CR peaks was obtained, resulting from the lack of inversion symmetry in the quantum wells with the InAs/AlSb interfaces. This asymmetry was believed to originate from the difference in the interface state concentrations, similar to the case of InAs/GaSb QWs [75]. The problem of electron sources in $\text{Ga}_{1-y}\text{Al}_y\text{Sb}/\text{InAs}/\text{Ga}_{1-y}\text{Al}_y\text{Sb}$ is considered in more detail in the separate section related to the transport properties (see Section 4.3.2).

The electronic properties were also studied as a function of the interface composition of InAs/AlSb heterojunctions grown by MBE. As discussed above (Section 4.2.3), the ideal abrupt interface between two binary semiconductors, differing in both cation and anion elements, can be formed with two possible compositions. Depending on the atomic layer sequences, either AlAs-like interface or InSb-like interface can occur with strongly different interplanar spacing across the interfaces, which necessarily results in a local interface strain. In addition to the Raman scattering studies considered in Section 4.2.3, the interface-induced strains have been observed experimentally in InAs/GaSb superlattices by ion channeling [85] and more recently by studying the far-infrared reflectivity spectra [86].

Generally, this strain can influence such important heterostructure parameters, as band offsets. Waldrop et al. [87] used X-ray photoemission spectroscopy to measure the dependence of the valence band offset ΔE_v on the interface composition (AlAs-like versus InSb-like) for strained InAs epilayers grown on AlSb by MBE with accurate interface composition control, as in Ref. [43]. Valence band offsets of 0.14, 0.16 and 0.17 eV were found for three heterojunctions nominally grown to have an InSb interface, and the offsets of 0.15 and 0.22 eV for two heterojunctions with the AlAs interface composition. Large discrepancy between the two values measured for the AlAs-like interface was attributed to the technological difficulties in obtaining this type of heterojunctions. Based on empirical tight-binding calculations, the value of 0.22 eV was interpreted as characteristic for the AlAs interface, thus resulting in the 60 meV variation in band offset versus the interface composition.

Dandrea and Duke [88] performed a comprehensive calculation of band offsets within the first-principles local density functional theory for strained InAs grown on AlSb with different interface types. Calculation gave the valence band offset of 0.19 eV, in good agreement with the 0.18 ± 0.04 eV, revealed from the photoemission experiments [87]. However, the calculated valence band offset was found to be independent of the interfacial bonding type and, therefore, Dandrea and Duke [88] suggested another interpretation of the experimental data of Waldrop et al. [87]. Namely, they reasoned that the attempts to grow AlAs interface on AlSb layer could lead to more radical variations in interfacial atomic structure than the considered simple isovalently bound chemically abrupt structure. This suggestion is well consistent with the peculiarity of "AlAs-on-AlSb" case [45] considered in Section 4.2.3. The calculations accounted also for the strain reducing the InAs band gap. The resultant band offsets are given in Fig. 19. The most remarkable feature of the band lineup is the nearly vanishing the overall heterojunction band gap, in contrast to the band lineups of unstrained layers (for comparison see Fig. 2).

Dandrea and Duke [88] also analyzed electronic properties of short-period superlattices (SPSL) with the identical InAs and AlSb layer thicknesses but different interfacial bond type. The quantum confined states were found to be very sensitive to the interfacial nature, even though the band offsets are independent of the interface type. As is shown in Fig. 19, the band gap of the $(InAs)_5(AlSb)_4$ superlattice with both the AlAs interfaces is by 0.24 eV larger than that with the InSb interfaces, which was explained in terms of the superlattice wave functions. Similar results have been obtained by Shaw et al. [89], predicting the change in the band gap of SPSLs with different interface configurations from *ab initio* pseudopotential calculations. The band gap difference was shown to be due to localized states at perfect InSb interfaces in the AlSb/InAs superlattices.

Experimentally, the band gap of InAs/AlSb SPSLs with different types of interfaces was studied using photoluminescence (PL) spectroscopy [44,45,90]. The pronounced 96 meV difference between the

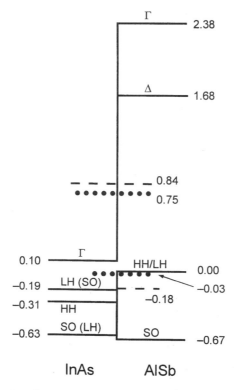

Fig. 19 Band offsets for InAs grown on AlSb (001) [88]. The valence band offset of 0.19 eV is calculated: the experimental $T = 0$ band gaps (corrected for strain in the case of InAs) are added to these calculated levels. LH(SO) and SO(LH) are the edges of the mixed light-hole and spin-orbital bands in InAs. Also shown are the band edge quantum well states for two short-period $(InAs)_5/(AlSb)_4$ (001) superlattices, one with InSb interface (dotted lines), and another with an AlAs interface (dashed lines).

radiative transition energies was detected, with the higher-energy peak observed for AlAs-like interfaces [44,90]. Spitzer *et al.* [90] assumed that observation of the energy difference smaller than that predicted by Dandrea and Duke [88], might have been due to slightly larger SPSL period (the $(InAs)_6(AlSb)_6$ experimental SPSLs versus the $(InAs)_5(AlSb)_4$ superlattice used in the calculations) and other effects, like strain relaxation in some samples and the presence of Al_{Sb} and As_{Al} antisite defects. The better agreement has been reached between the experimental results from Refs. [44,90] and calculations namely for the $(InAs)_6(AlSb)_6$ superlattice [89]. Opposite experimental results were obtained by Yano *et al.* [45], who demonstrated the lower energy PL emission in the case of InAs/AlSb superlattices and QWs with the AlAs interfaces, whereas the higher energy PL emission of SPSLs and QWs with the InSb interfaces showed more than order of magnitude higher intensity. This contradiction apparently stimulates further efforts towards reproducible growth of the structures with the specific electronic properties.

Brar *et al.* [91] have reported on PL study of the InAs/AlSb heterostructures containing five narrow QWs of different thickness. Fig. 20 shows the low-temperature PL spectrum containing emission lines related to different QWs. Significant discrepancy was found

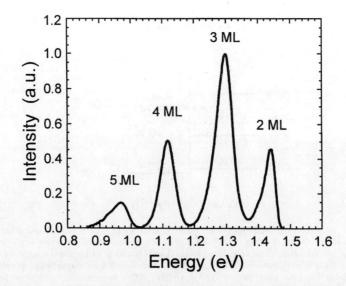

Fig. 20 Low-temperature PL spectra from InAs/AlSb multiquantum wells [91].

between the measured PL transition energies for the wells less than 5 ml thick and the values calculated using the envelope-function model including non-parabolicity corrections (Fig. 21). Different possible explanations for this disagreement were discussed, including the effect of interface roughness, the validity of the envelope-function approximation itself, possible interaction between Γ-valley states in the InAs QW and X-valley states in the AlSb barriers, and the possibility for electrons in the InAs well to be poorly confined by the AlSb Γ-valley barrier. Conduction-band states in such superlattices were studied by Boykin [92] using the ten-band tight-binding model. Finally, the envelope-function approximation has been shown to be inadequate for very narrow InAs/AlSb QWs. The calculation also suggested the importance of the interaction between Γ-valley well states and X-valley barrier states in decreasing the transition energies.

4.3.2 Carrier Mobility and Concentration in Unintentionally Doped AlSb/InAs/AlSb Single QW

InAs/(AlGa)Sb system is currently considered to be very promising for high-speed electronic applications. Two most attractive features of these Type-II QW heterostructures are high low-field mobility of

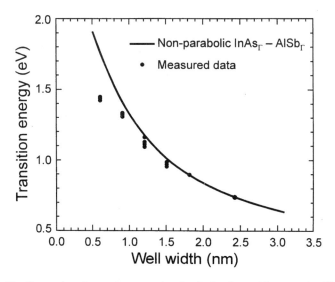

Fig. 21 Comparison between measured and calculated transition energies. The calculation assumes a spatially indirect transition [91].

electrons in InAs and very large conduction-band offset of 1.35 eV between InAs and AlSb. As compared to GaSb barriers or $Al_yGa_{1-y}Sb$ barriers with low Al content, the AlSb barriers have the advantage of eliminating the complications of broken-gap band structure at the interface. Therefore, most of papers, which are not focused especially on the semimetallic properties, have centered on the InAs/AlSb heterostructures, despite their larger lattice mismatch and the requirement for more mature MBE technology. First Ga-free InAs/AlSb QWs and superlattices were studied by Chang et al. [82] in 1984. Although the samples suffered from relatively low mobilities [$\sim 15,000\, cm^2/(Vs)$], the two-dimensional electron behavior was demonstrated, including first observation of the quantum Hall effect in such structures. Since that time, great improvements have been made in the technology, resulting in the mobilities as high as $850,000\, cm^2/(Vs)$ at 4.2 K [93]. High resolution TEM image of a 220 Å-thick AlSb/InAs/AlSb QW structure with low-temperature mobility in excess of $200,000\, cm^2/(Vs)$ [50] is shown in Fig. 22. It should be noted that this QW structure with the InAs QW width exceeding the critical thickness exhibited the relatively high mobility despite the onset of threading dislocations directly observed at the bottom InAs/AlSb interface.

Low-temperature magnetoresistance and Hall resistance were also measured with the magnetic field normal and parallel to the 2D electron layer [94]. These data give evidence for the high quality 2DEG channel in the InAs/AlSb QW, indicating practically no parallel conducting paths in these samples, grown on nearly semi-insulating AlSb buffer layer. However, the low-temperature mobilities are still lower than the best values achieved in GaAs quantum wells, which has stimulated investigations of various mechanisms restricting realization of the maximum possible mobilities. Along with the problem of electron sources in unintentionally doped structures, these studies are considered in this section.

It has long been understood [95] that high electron sheet concentrations of the order $10^{12}\, cm^{-2}$ with typical low-temperature mobilities above $300,000\, cm^2/(Vs)$ can be obtained in the unintentionally doped AlSb/InAs/AlSb QWs. The high sheet electron density could not be due to any background donor in the InAs itself, because the concentration was at least two decades higher than the known background. Furthermore, the high mobilities observed are incompatible with the amount of impurity scattering such a high donor concentration would have introduced. The valence band of AlSb was

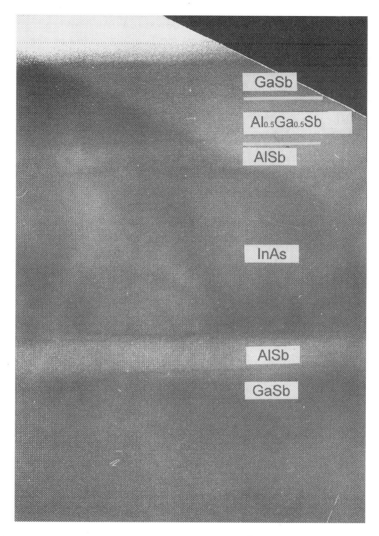

Fig. 22 High resolution TEM image of single 220 Å-thick AlSb/InAs/AlSb QW structure.

also ruled out as a possible source of electrons (in contrast to the broken-gap InAs/GaSb structures) because the estimated position of the Fermi level was well above the AlSb valence bandedge. Therefore, Tuttle et al. [95] concluded that the high electron concentration indicated the presence of a deep donor in the AlSb barriers. Subsequently, the source of electrons in unintentionally doped InAs/AlGaSb heterostructures has been attributed to various mechanisms, including the trapping levels at InAs/AlGaSb heterointerfaces of intrinsic [43] or extrinsic [96] nature, surface levels [97–99] and donor levels in the bulk barrier material [100,101].

Tuttle et al. [43] studied the electron concentration and mobility in the MBE grown QW structures as dependent on the type of both InAs/AlSb heterointerfaces. The samples with different interfaces (either InSb-like or AlAs-like, as illustrated in Fig. 10) were obtained by using proper alternating sequence of the Al, Sb, As, and In shutters during the interface formation (Fig. 11). Electron mobility and concentration were found to depend strongly on the manner in which the quantum well's interfaces were grown. The wells having AlAs bottom interfaces exhibited much higher electron concentrations and much lower mobilities than the wells with InSb bottom interface. The effect of the top interface type was found to be negligible. This behavior was explained by allowing for the presence of a high concentration of As_{Al} antisite defects which arose during forming the AlAs interface by the exposure of Al-rich AlSb surface to the As soak flux.

Nguyen et al. [97,98] studied electron concentration in the InAs/AlSb heterostructures with InSb-like interfaces as a function of the top barrier thickness in single-well structures and the internal barrier thickness in multiwell structures. The samples were capped with 10 nm of GaSb to protect the AlSb surface from oxidation. Strong and systematic decrease in the sheet concentration was observed with increasing the thickness of the top barrier, showing that a large fraction of the high electron concentrations, usually found in the wells with thin top barriers, comes from the surface donors. The Fermi level has been shown to be pinned at 850 ± 50 meV below the conduction bandedge of the AlSb top barrier. With increasing the top barrier thickness up to 0.5 µm, the sheet concentrations eventually leveled out, indicating the low-temperature limit of about 3×10^{11} cm^{-2}. The remaining electrons were attributed to internal electron sources, like bulk donors in the AlSb barriers and interface donors at the InAs/AlSb interfaces. Furukawa [99] studied the role the material of thin surface layer plays in electron accumulation in InAs/AlSb QWs. The electron density transferred from the surface into the well was found to be much smaller, when the GaSb cap layer was changed by the InAs one. This result was explained by the difference in the Fermi level surface pinning position between InAs and GaSb (Fig. 23). Furukawa [99] has reported the surface pinning position in InAs to be 0.2 eV lower than that in GaSb.

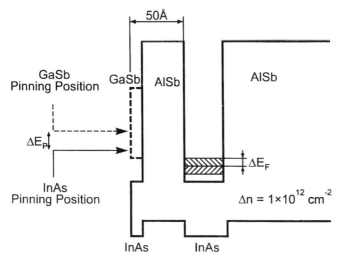

Fig. 23 Band structure of AlSb/InAs QW [99]. The surface is on the left-hand side. The solid line and broken line indicate surface InAs and GaSb band lineup, respectively.

Ideshita et al. [100] attempted to identify the dominant contributions to sheet electron density among the interface and bulk levels. For this purpose they studied two series of InAs/AlSb QW heterostructures shown schematically in Fig. 24. All samples had the top AlSb layer thick enough to exclude the effect of surface levels. In the first series, the period of the AlSb/InAs superlattice was varied at both sides of the 15 nm InAs well. Thus, the number of interfaces could be changed without changing the effective barrier width. The electron concentration was found to be almost independent of the number of interfaces [see Fig. 24(a)], showing that the interface levels are not the dominant electron sources in the QW system. In the other series, the total AlSb barrier thickness was varied, as shown in Fig. 24(b). The electron concentration increased linearly with the increase in the AlSb thickness, indicating that a bulk donor level exists in AlSb and supplies electrons into the InAs wells.

Chadi [101] examined the problem of electron accumulation in undoped InAs/AlSb QWs theoretically, focusing on the deep levels in bulk AlSb. Both the deep-donor and the compensating deep-acceptor antisite defects were considered and were found to have energy levels, which are able to supply electrons into the quantum well. However, a comparison of the predicted temperature dependence of the electron density with the available experimental data demonstrated that the

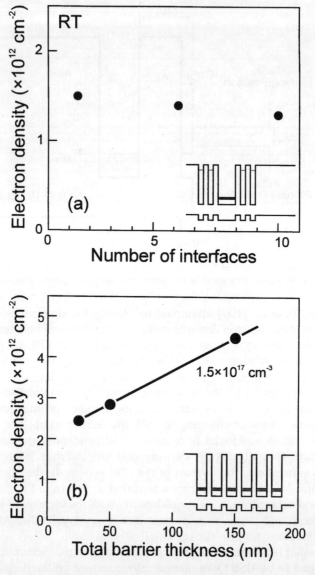

Fig. 24 (a) Room temperature dependence of the accumulated electron concentration on the number of InAs/AlSb interfaces. The inset is the sample structure with several periods of AlSb(2 nm)/InAs(2 nm) SL on both sides of the 15 nm InAs well. The thickness of AlSb cap layer is about 50 nm. (b) Dependence of the accumulated electron concentration on the AlSb barrier thickness. The inset is the sample structure with six 15 nm thick InAs wells. The thickness of AlSb cap layer is about 50 nm [100].

deep-donor Sb_{Al} antisite defects in AlSb are the most important source of electrons for at least some samples. The dependence of the sheet electron density in the well on the concentration and binding energy of the deep-donor centers in AlSb barriers has been derived.

In parallel with sheet electron concentration, low temperature mobility was studied as a function of the structure design, including the type of interface [43], the thickness of the top AlSb layer [97] and the width of the InAs QW [102,103]. High mobilities in AlSb/InAs/AlSb QWs were observed only if the bottom QW interface was InSb-like [43]. Tuttle et al. [43] attributed this effect to enhanced concentration of As_{Al} antisite interface donors, which could occur at the AlAs-like interface. These donors introduce strong ionized-impurity scattering into the well, resulting in the drastic mobility reduction. The other explanation, proposed by Yano et al. [45], is based on the experimental observation of dramatically disturbed AlAs interface structure, when it is formed on an AlSb buffer layer. It was suggested that due to the extremely large lattice mismatch ($\sim 8.4\%$) between AlAs and AlSb, the critical thickness of a pseudomorphic AlAs growth on AlSb is less than 1 ml, which results in the monolayer degradation by forming a large number of point and extended interface defects. These defects may serve as the scattering centers and supply high electron concentration into the InAs/AlSb QW. The model of the structurally disordered AlAs-on-AlSb interface also accounts for the specific role of the bottom IF type.

The mobility has also demonstrated drastic dependence on the thickness of the top barrier [97]. Both low-temperature mobility and electron concentration increased with decreasing the top layer thickness, except for the well with the highest electron concentration, i.e. closest to the surface. Bolognesi et al. [102,103] studied the electron transport as a function of the well width. Both room-temperature and low-temperature mobilities were significantly reduced in narrow wells due to the dominance of interface roughness scattering. Using Gold's [104] theory of interface roughness scattering and comparing the computed and measured mobilities, Bolognesi et al. [103] have found that correlation length of the interface roughness is $\Lambda \leqslant 60$ Å. They assumed that this roughness scale is characteristic for the bottom (InAs/AlSb) interface. The mobility peaks for well widths around 125 Å, and then decreases, most probably due to the onset of scattering by misfit dislocations originating at the bottom interface, because the width of the quantum well exceeds the critical thickness at given lattice mismatch between the well and barrier layers.

Another interesting feature of InAs/AlSb QWs is low-temperature persistent photoconductivity. Depending on the wavelength of the illuminating light, either the persistent reduction or the persistent increase of the carrier density in the well was observed, with the negative persistent photoconductivity characteristic for the incident light energy larger than the AlSb bandgap. At temperatures below 100 K, the changes persisted for hours after the light was turned off, but they disappeared if the sample was heated. Positive persistent photoconductivity, observed for light energies smaller than about 1.55 eV, might be explained by photoionisation of deep donors, whose nature and ionization energies are generally unknown [105]. The effect of negative persistent photoconductivity was studied in more detail both experimentally [94,95,105] and theoretically [101]. Tuttle et al. [95] suggested the straightforward explanation of the effect: the illumination by the light with the energy larger than the AlSb band gap results in the creation of electron-hole pairs. The holes are driven by the band bending towards the well, where they can recombine with the electrons, depleting the QW. The persistence of the effect at sufficiently low temperatures was believed to be due to the remaining repulsive band bending barrier preventing the well from being refilled. Gauer et al. [105] refined this picture, showing that the negative persistent photoeffect in those structures was determined by the holes photogenerated in the GaSb cap layer. The photogenerated electrons in this model do not contribute to the photoconductivity, neutralizing the surface donor. Chadi [101] proposed an alternative microscopic explanation for the negative persistent photoconductivity, based on the antisite defect model. In this model, the light-induced transition from stable- to metastable-state occurs for the neutral state of the antisite, shifting the donor levels of the antisite defect in the AlSb barriers below the valence-band maximum. The antisite in the metastable state cannot supply any electrons to the QW, which leads to the negative-persistent-photoconductivity effect. The effect was expected to be observed only at low temperatures preventing the metastable- to stable-state retransformation of the donor defect, which is well consistent with the experimental data.

4.4 DEVICES BASED ON TYPE-II (InGa)(AsSb)/(AlGa)(SbAs) QW STRUCTURES

Great interest in antimonide-based Type-II heterostructures, with GaSb/InAs as the simplest one, has been initially caused by a possibility to create the highest conductivity 2D electron channel in the InAs QW, nearly lattice-matched to GaSb barriers ($\Delta a/a \approx 0.6\%$). In this broken-gap heterostructure, very promising for high speed and low noise electronic device applications, the 2D electrons were supposed to be supplied into the InAs QW only by the interband transition from the GaSb valence band, contrary to the intentional doping which could give additional source of noise at high frequencies. However, technological problems, related mainly to the lack of semi-insulating GaSb substrates and poor quality of Schottky gate, as well as a strong competition with well developed InGaAs/InAlAs pseudomorphic heterostructures on InP substrates [106] have resulted in shifting the main direction of research activity, especially from the early nineties, to the middle-wave infrared (MWIR) (2–5 μm) light emitter applications.

This section represents different types of opto- and microelectronic devices fabricated from the narrow bandgap (InGa)(AsSb)/(AlGa)(SbAs) heterostructures with a main attention paid to those including Type-II heterojunctions. First, the results on fabrication and study of the 2–5 μm range laser structures are considered. Then, the brief reviews of resonant-tunneling diodes and heterostructure field effect transistors (HFETs) are presented.

4.4.1 Middle-Wave IR Lasers

Semiconductor lasers in the 2–5 μm range are of increasing importance for ecology, chemical and military industries, as well as for optical communications. Many important atmospheric molecules, such as HCl, CO_2, CH_4, CO, N_2O, have their most intensive absorption lines within this wavelength range. It contains also the regions of low atmospheric absorption and high fluoride fiber optic transmission. Room temperature operation of these mid-IR lasers is a key point for most of the applications. However, fabrication of high-performance continuous wave (CW) lasers using the narrow bandgap III–V material system meets severe problems related to (1) dramatic enhancement of Auger recombination of non-equilibrium carriers, (2) increased light absorption by free carriers, (3) weak overlap of electron and hole wave function, and (4) poor optical mode localization due to a small difference in refractive index between the respective narrow bandgap materials. Nevertheless, during the last few years main achievements in this field were attributed to the (InGa)(AsSb)/(AlGa)(SbAs) heterostructures grown on GaSb substrates. The emission wavelength of these lasers varied within 1–4 μm.

4.4.1.1 Type-I $In_xGa_{1-x}As_ySb_{1-y}/Al_xGa_{1-x}As_ySb_{1-y}$ heterostructure lasers

Since 1980s the main research activity has been focused on Type-I $In_xGa_{1-x}As_ySb_{1-y}/Al_xGa_{1-x}As_ySb_{1-y}$ laser heterostructures lattice-matched to GaSb, which are briefly considered here in their progress to the longer operating wavelengths achieved at higher temperatures.

The first pulse operated InGaAsSb/GaSb single heterostructure laser diode with $\lambda \approx 1.9\,\mu m$ at 90 K was fabricated in 1978 by Dolginov et al. [107], using a liquid phase epitaxy (LPE) technique. Further improvements in the optical and carrier confinement by using the $In_{0.05}Ga_{0.95}As_{0.04}Sb_{0.96}/Al_{0.2}Ga_{0.8}As_{0.02}Sb_{0.98}$ double heterostructure (DH) [108] has resulted in the room temperature laser operation with a threshold current density of $5\,kA/cm^2$ and a characteristic temperature $T_0 = 112\,K\,(77–300\,K)$. Temperature dependence of the lasing wavelength is shown in Fig. 25. Caneau et al. [109] have succeeded in reducing the 300 K threshold current density down to $I_{th} = 1.7\,kA/cm^2$ at $\lambda = 2.2\,\mu m$ by the increase in aluminum content in AlGaAsSb cladding layers to provide a better optical confinement. To release the internal strain in the $In_{0.16}Ga_{0.84}As_{0.15}Sb_{0.85}/Al_{0.4}Ga_{0.6}As_{0.04}Sb_{0.96}$ double

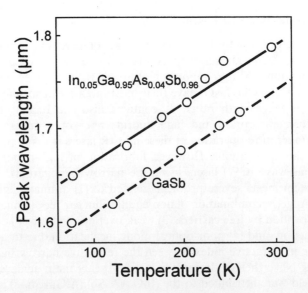

Fig. 25 Temperature dependence of the lasing wavelengths for $In_{0.05}Ga_{0.95}As_{0.04}Sb_{0.96}/Al_{0.2}Ga_{0.8}As_{0.02}Sb_{0.98}$ and $GaSb/Al_{0.2}Ga_{0.8}As_{0.02}Sb_{0.98}$ DH lasers [108].

heterostructure, they introduced 0.2 μm thick layers with lower Al content at both sides of the active region. The emission wavelength as high as 2.4 μm at 300 K in LPE grown InGaAsSb/AlGaAsSb laser diodes has been reported by Bochkarev et al. [110].

The first MBE grown $In_{0.16}Ga_{0.84}As_{0.15}Sb_{0.85}/Al_{0.35}Ga_{0.65}Sb$ DH laser diode has been reported by Chiu et al. [111] to exhibit the RT pulse operation at $\lambda \approx 2.2$ μm with $I_{th} = 4.2$ kA/cm^2 and characteristic temperature as low as $T_0 = 26$ K. The subsequent increase in the Al content in the $Al_xGa_{1-x}As_ySb_{1-y}$ claddings up to $x = 0.5$ [112] and $x = 0.75$ [113] was accompanied by both the decrease in threshold current density lower than 1 kA/cm^2 at the cavity length of $L = 1000$ μm and the increase in T_0 up to 50 K (Fig. 26). Note that the latter characteristics were achieved at $\lambda = 2.2$ μm under the CW operation conditions.

Choi and Eglash [114] introduced a multiple-quantum-well (MQW) active region, consisting of five 100 Å $In_{0.16}Ga_{0.84}As_{0.14}Sb_{0.86}$ QW's separated by the 200 Å $Al_{0.2}Ga_{0.8}As_{0.02}Sb_{0.98}$ barriers (Al content in cladding layers was as high as 0.9), which allowed them to obtain the 2.1 μm lasers with $I_{th} = 260$ A/cm^2 (at $L = 2$ mm). This threshold current density turned out to be more than 3 times lower than that for DH

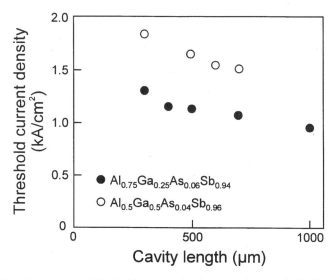

Fig. 26 Dependence of threshold current density on cavity length for broad-stripe GaInAsSb/AlGaAsSb lasers with $Ga_{0.84}In_{0.16}As_{0.14}Sb_{0.86}$ active layer and two different cladding layer compositions [113].

lasers [113]. Fig. 27 represents the 300 K pulsed threshold current density of this MQW laser as a function of the reverse cavity length. The introduction of the strained quantum wells led to much lower probability of Auger recombination, which, in its turn, caused the dramatic T_0 enhancement up to 113 K (near 300 K), as compared to DH structure [113]. The differential quantum efficiency of the MQW laser measured at 300 μm stripe diode was 70% at the internal loss of about 10 cm^{-1}. Optimization of MBE growth allowed one to reach the high performance 2 μm RT CW laser with the lowest threshold current density $I_{th} = 143$ A/cm^2 and the highest at that time output power of 1.3 W [115].

This material system has also been used to demonstrate a $In_{0.24}Ga_{0.76}As_{0.16}Sb_{0.84}/Al_{0.25}Ga_{0.75}As_{0.02}Sb_{0.98}$ MQW pulsed laser emitting at 2.78 μm at 15°C, which is the longest emission wavelength achieved with a RT III–V laser to date [116]. The emission spectrum of the pulse laser is shown in Fig. 28. This laser diode demonstrated $I_{th} = 10$ kA/cm^2, $T_0 = 58$ K, the 9% quantum efficiency, and the 30 mW maximal output power at 15°C. CW single mode operation of the similar MQW heterostructure laser at $\lambda = 2.7$ μm and maximal operating temperature as high as 234 K was reported by Garbuzov et al. [117].

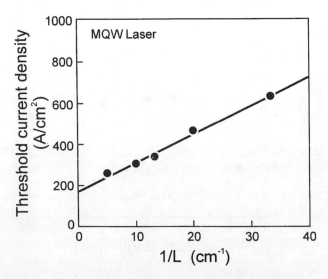

Fig. 27 Room-temperature pulse threshold current density vs inverse cavity length for 100 μm-wide GaInAsSb/AlGaAsSb multiple-quantum-well lasers.

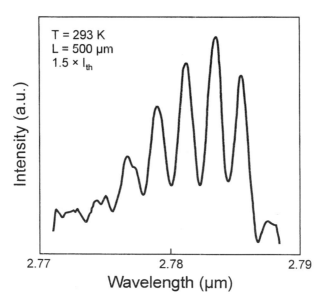

Fig. 28 Spectral output for a 2.78 μm pulse laser operating at 293 K. Operating conditions were 1.4 A peak current, 0.4 μs pulse duration and a 660 Hz pulse repetition rate [116].

High performance lasers, emitting at 3 μm and above, have proven to be more serious problem. The only way to increase the lasing wavelength using the Type-I $In_xGa_{1-x}As_ySb_{1-y}$/AlGaAsSb heterostructures lattice-matched to GaSb is to increase the In content in the $In_xGa_{1-x}As_xSb_{1-x}$ active region. However, increasing the In content up to 1 results in nearly zero valence band offset (ΔE_v) between $InAs_{0.91}Sb_{0.09}$ and $AlAs_{0.08}Sb_{0.92}$, which is the best for this system lattice-matched cladding layer in terms of the refractive index step and maximal possible ΔE_c. Since decreasing x in $In_xGa_{1-x}As_ySb_{1-y}$ from 1 to 0.6 rises ΔE_v up to 0.13 eV, while the InGaAsSb energy gap is nearly constant [118] at 0.28 eV (the respective wavelength is 4.2 μm), the use of the $In_xGa_{1-x}As_ySb_{1-y}$ active layer with $x \sim 0.6$ is more preferable. The refractive index difference between the latter and AlAsSb also does not change significantly. The LPE growth of the InGaAsSb quaternaries is limited to $x < 0.22$ or $x > 0.84$ [119] due to a miscibility gap, which limits the generation wavelength at $\lambda = 2.4$ μm for this technique. On the other hand, high quality InGaAsSb metastable alloys within this composition range have been grown by MBE [120].

$In_{0.54}Ga_{0.46}As_{0.48}Sb_{0.52}/Al_{0.9}Ga_{0.1}As_{0.08}Sb_{0.92}$ diode lasers, emitting at 3 μm with maximal operating temperature as high as 255 K, have been fabricated by MBE [121]. The threshold current density varied from 9 A/cm^2 at 40 K to 13.4 kA/cm^2 at 255 K demonstrating the characteristic temperature of about 30 K. These diodes exhibited a maximal output power of 45 mW (at $I = 850$ mA) and differential quantum efficiency $\sim 75\%$ at 100 K. To date, these laser diodes are the best among those emitting in the 3–3.5 μm range [122,123].

First laser diodes operating at 4 μm were fabricated by Eglash and Choi [124] from $InAs_{0.91}Sb_{0.09}/AlAs_{0.08}Sb_{0.92}$ DH. Although ΔE_v at this heterointerface is practically equal to zero, the large conduction band offset ($\Delta E_c \sim 1.3$ eV) forms a high potential barrier for holes at the InAsSb/n-AlAsSb interface [125]. The maximal pulse operating temperature obtained was 155 K with $T_0 = 17$ K. This low T_0 value seems to be related to the enhanced probability of Auger recombination for the narrower bandgap materials. Employing the $InAs_{0.85}Sb_{0.15}/In_{0.9}Al_{0.1}As_{0.9}Sb_{0.1}$ MQW active region allowed them to reduce further I_{th} and to demonstrate 3.9 μm CW laser operation at 123 K [126].

4.4.1.2 Type-II heterostructure lasers

All the Type-I laser heterostructures mentioned above use ternary or quaternary alloys composed of GaSb, AlSb, InAs, and InSb in the active region and cladding layers. The longer operation wavelength requires variation of composition of the active region, which causes the respective variation of its lattice parameter and refractive index, especially at $\lambda = 3$ μm. The need for both good optical confinement and lattice-matching significantly restricts the set of suitable cladding alloys. It may result in serious difficulties in increasing the maximal operating temperature of long wavelength lasers up to 300 K. Among the other possible disadvantages of Type-I heterostructure lasers is a multi-mode generation due to slow diffusion of the non-equilibrium carriers into the active region, as compared to the radiative recombination rate.

The above problems may be solved by using a Type-II heterostructure active region. The main disadvantage of the Type-II heterojunction is that it cannot be used in the injection laser, for the carrier diffusion through it is improbable. These problems have been overcome by Baranov et al. [127], who have realized for the first time a non-injection semiconductor laser on the basis of a single Type-II n-$In_{0.1}Ga_{0.9}Sb_{0.91}As_{0.09}$/p-GaSb heterojunction. Epitaxial structure of the device with the schematic band diagram at zero bias are shown in Figs. 29(c) and (a), respectively. When the direct bias higher than the contact potential difference is applied [Fig. 29(b)], two selfconsistent potential wells for the electrons (in n-InGaSbAs) and for holes (in p-GaSb) are formed at the interface. In this structure the radiative recombination occurs between holes tunneling through the heterointerface into n-region and electron tunneling into p-region, which is improbable in terms of classical mechanics. The energy gap between electron and hole energy levels in the quantum wells, which is equal

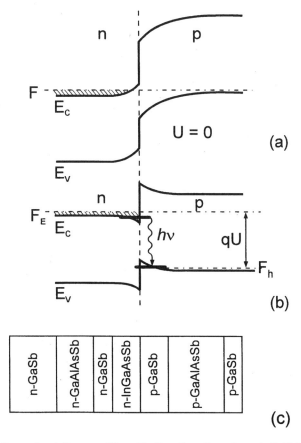

Fig. 29 Energy band diagram of Type-II n-$In_{0.1}Ga_{0.9}Sb_{0.91}As_{0.09}$/p-GaSb heterojunction laser (a) in thermodynamic equilibrium (zero bias), F is the Fermi level, and (b) under the forward bias applied, F_e and F_h are quasi-Fermi levels for electrons and holes, respectively; (c) schematic layer sequence of the structure [127].

to the emitting wavelength, must be lower than the narrowest bandgap of materials forming a heteropair. The increase in InAs mole fraction in the InGaSbAs alloy from 0 to ~ 0.7 results in the variation of energy gap between the alloy conduction bandedge and the GaSb valence bandedge from 0.7 eV to zero, and hence, in the respective variation of the emitting wavelength. One should note, that this variation causes no noticeable change either in lattice parameter or in refractive index of the active region. The large steps of refractive index at the GaSb/AlGaSbAs and InGaSbAs/AlGaSbAs interfaces, and the relatively small step ($\Delta n = 0.03$) at the p-n interface provided good confinement of the optical mode in the active region with its maximum near the p-n junction. Moreover, the recombination current losses were significantly reduced in this structure due to the elimination of the carrier injection through the interface. This structure showed room temperature CW operation at 2 μm with $I_{th} = 7.6$ kA/cm^2 and the lowest I_{th} at 100 K at that time.

The buried stripe laser diodes fabricated using this structure demonstrated $I_{th} = 4.8$ kA/cm^2 (300 K) and $T_0 = 89$ K [128]. It is worth noting that the operating wavelength is longer than that corresponding to the energy gap of n-InGaSbAs.

A limitation of the above non-injection laser is a fairly small overlap of the electron and hole wave functions at the interface. Another problem is related to the depth of the selfconsistent potential wells, which is of the order of kT. It dramatically increases the probability for the carries to leave the quantum well at high operating temperature.

To avoid these disadvantages the 2–5 μm noninjection semiconductor laser based on a thin Type-II QW heterostructure has been suggested in Ref. [129]. Schematic band diagram of the active region under thermodynamic equilibrium (at zero bias), is illustrated in Fig. 30(a). A thin InAs layer sandwiched between n- and p-type GaSb layers forms a quantum well for electrons with the parameters controlled by the variation of its thickness and by the doping level of the n-GaSb region. The barriers for current are formed by charged donors and acceptors in the n- and p-regions, respectively. Due to very small thickness of the InAs well (about 10–20 Å) its charge is negligible, and the electronic energy level lies above the Fermi level. When forward bias is applied to the structure, the quasi Fermi level for electrons rises over the energy level in the InAs QW [Fig. 30(b)]. The QW is filled with electrons which decrease the potential barrier for holes in p-GaSb, resulting in formation of the hole accumulation layer at the interface. It should be noted, that if the InAs layer is thinner than

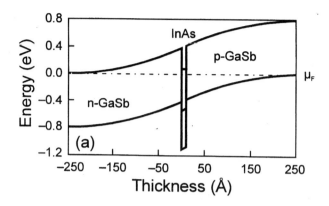

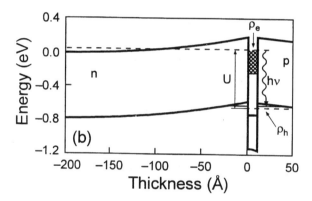

Fig. 30 Schematic band diagram of n-GaSb/InAs/p-GaSb QW heterostructure (a) in thermodynamic equilibrium (zero bias) and (b) with applied bias.

~ 20 Å, the typical expansion of an electron wave function exceeds the QW thickness, giving rise to a marked enhancement of the overlap of electron and hole wave functions. The numerical estimations give the maximal threshold current density at room temperature as low as $1\,\text{kA/cm}^2$ even at $\lambda = 4\,\mu\text{m}$ [130]. To prevent the carrier injection from n- to p-region, which contributes to the threshold current, and to reduce thermal losses of the QW electrons at room temperature, it is necessary to provide a 5–10 kT potential barrier for electrons, which is mainly determined by the doping level of the n-GaSb region nearest to the well. On the other hand, the sheet carrier density of electrons localized in the InAs QW should not exceed $\sim 5 \times 10^{12}\,\mathrm{cm}^{-2}$. Otherwise, it increases the threshold current and the probability of the device electric breakdown. Fig. 31(a) represents photoluminescence spectra at 77 K of Type-II InAs/GaSb MQW structures, which have

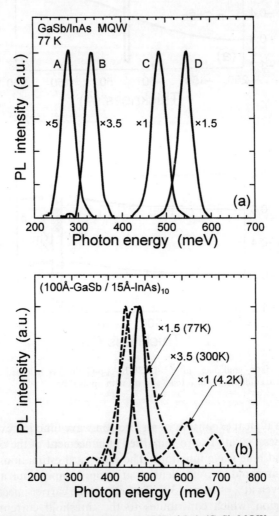

Fig. 31 (a) Photoluminescence spectra at 77 K of InAs/GaSb MQW structures with different InAs quantum well thickness: (A) – 30 Å, (B) – 20 Å, (C) – 15 Å, and (D) – 10 Å. The low excitation density of 1 W/cm² was used. Note the practically equal FWHM values for all lines; (b) Photoluminescence spectrum of InAs/GaSb MQW structure with 15 Å thick InAs quantum wells at different observation temperatures.

been proposed as an active region for such laser structure. The InAs QW thickness was varied from 10 Å to 30 Å, while that of GaSb barriers was kept constant at 100 Å [131]. All PL spectra were dominated by highly efficient emission line, demonstrated the strong red shift from 2.3 to ~ 4.5 μm with the increase in InAs QW thickness. The anomalous temperature behavior of the PL spectra, as shown in Fig. 31(b) for the structure with 15 Å thick QWs, may, presumably, be attributed to the spatially indirect radiative transitions between the electron states localized in the InAs layers and the extended hole states in the MQW valence band.

Similar idea has been realized in the first optically pumped laser structure with (77 Å-InAs/23 Å-InAs$_x$Sb$_{1-x}$)$_{30}$ Type-II SL, emitting at 3.3–3.4 μm [132]. However, contrary to the previous structure, here the holes are the localized carriers (Fig. 32). The maximal CW lasing

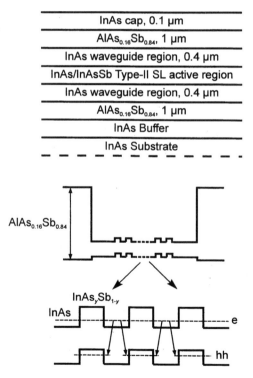

Fig. 32 Schematic sample layer structure and active region bandedge diagram of the typical InAs/InAs$_x$Sb$_{1-x}$ laser structure [132].

TABLE IV
Parameters of MWIR lasers developed during last decade.

Laser structure	λ (μm)	I_{th} (A/cm²) {regime}	T_{max} (K)	T_0 (K) {regime}	W (mW) {regime}	η (%) {regime}	Operation mode	Growth method	Refs.
DH InGaAsSb/AlGaAsSb	1.8	5000 {300 K}	300	112 {77–300 K}	—	100 {77 K} 10 {300 K}	pulse	LPE	[108]
DH InGaAsSb/AlGaAsSb	2.2	1700 {300 K}	290	60 {80–300 K}	—	8 {80 K} 16 {290 K}	pulse	LPE	[109]
DH InGaAsSb/AlGaAsSb	2.2	4200 {300 K}	300	26 {10–30°C}	—	—	pulse	MBE	[111]
DH InGaAsSb/AlGaAsSb	2.2	940 {p*, 300 K}	320	49 {0–30°C}	10.5 {300 K}	—	cw & pulse	MBE	[113]
MQW InGaAsSb/AlGaAsSb	2.1	260 {p*, 300 K}	420	113 {0–50°C}	190 {cw}	70 {p*, 300 K}	cw & pulse	MBE	[114]
MQW InGaAsSb/AlGaAsSb	2.0	143 {p*, 300 K}	300	—	1300 {cw, 300 K}	47 {cw, 300 K}	cw	MBE	[115]

TABLE IV (Continued).

Laser structure	λ (μm)	I_{th} (A/cm^2) {regime}	T_{max} (K)	T_0 (K) {regime}	W (mW) {regime}	η (%) {regime}	Operation mode	Growth method	Refs.
MQW InGaAsSb/AlGaAsSb	2.78	10,000	330	58 {0–40°C}	30	9	pulse	MBE	[116]
MQW InGaAsSb/AlGaAsSb	2.7	—	234{cw}	62{cw} 110{p*}	~0.1 {cw, 234 K}	41{170 K} 0.6{234 K}	cw & pulse	MBE	[117]
DH InGaAsSb/AlGaAsSb	3.0	9{40 K} 13400 {255 K}	255{p*} 170{cw}	35 {40–120 K}	45 {cw, 100 K}	75 {cw, 100 K}	cw & pulse	MBE	[121]
DH InGaAsSb/AlGaAsSb	4.0	33{50 K}	155{p*} {80 K}	17 {50–155 K}	—	—	cw & pulse	MBE	[124]
HJ p-GaSb/n-GaInAsSb	2.0	120{4.2 K} 280{78 K} 4800{300 K}	320	89	—	—	cw	LPE	[128]
MQW InGaAsSb/AlGaAsSb	3.5	406{80 K}	160	—	—	—	pulse	MBE	[133]

* p – denotes pulse operation.

temperature was 95 K with the equivalent threshold current density of ~ 56 A/cm^2.

Finally, the first pulsed 3.5 µm diode laser with the MQW active region based on a Type-II (35 Å-Ga$_{1-x}$In$_x$Sb/17 Å-InAs)$_{4.5}$ SL quantum wells separated by 400 Å Ga$_{0.75}$In$_{0.25}$As$_{0.22}$Sb$_{0.78}$ barriers, has been demonstrated by Hasenberg et al. [133]. It showed the maximal operating temperature of 160 K and the threshold current density of 406 A/cm^2 at 80 K.

The main parameters of the IR lasers described are summarized in Table IV. Combining the wide variation of emission wavelength characteristic for the Type-II heterostructures and the advantages of the Type-I heterostructure injection lasers seems to be most promising and fruitful approach to fabricating room temperature CW middle-wavelength IR laser diodes.

4.4.2 Resonant Tunneling Transport

InAs/GaSb/AlSb material system is also attractive due to specific resonant tunneling transport through the structures, formed by thin layers with different types of band alignment. Two features peculiar to this system have been of particular importance. First, among all nearly lattice-matched III–V binary combinations, the AlSb/InAs heteropair provides the largest barrier for electrons, increasing such important characteristic of resonant tunneling devices as peak-to-valley current ratio (PVR) due to suppression of the inelastic carrier transport. Second, narrow bandgap results in additional effects governed by the interaction between respective valence and conduction bands.

Sai-Halasz et al. [63] were the first, who investigated theoretically the tunneling probability in an InGaAs/GaSbAs single barrier structure, establishing a negative differential resistance (NDR) region in the current–voltage characteristic. NDR in single barrier tunneling originates mainly from the narrow energy gap between the conduction band edge in emitters and the valence-band edge in barrier. The predicted NDR was experimentally observed up to room temperature in InAs/AlGaSb/InAs single-barrier heterostructures [134,135].

Sweeny and Xu [136] have developed the concept of resonant interband tunneling (RIT) devices in more detail. The basic idea was that the interband tunneling and the resonant tunneling in QW were utilized in device operation to combine both high peak-to-valley current ratios of conventional tunnel diodes and exceptional high-frequency properties of resonant tunneling devices. Fig. 33 shows schematic band

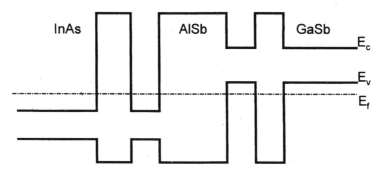

Fig. 33 Schematic band diagram of a polytype heterostructure resonant interband tunnel diode [136].

diagram of the RIT heterostructure proposed by Sweeny and Xu [136] Doping of the constituent layers is determined by the structure of the quantum wells, which allows the interband tunneling without the requirement of degenerate doping, necessary for conventional tunneling diodes. This is particularly important for high-speed operation due to the reduction of the capacitance. On the other hand, the device uses an advantage of the bandgap blocking mechanism, which is very effective in suppressing both inelastic and elastic tunneling processes.

RIT effects were subsequently studied for different InAs/GaSb/AlSb two-terminal tunneling devices, both experimentally [137–147] and theoretically [148–155]. The main types of the device structures are illustrated in Fig. 34. The InAs/AlSb device shown in Fig. 34(a) is conceptually close to a conventional resonant tunneling double-barrier heterostructure. However, to calculate accurately the tunneling properties, the staggered band offset at the InAs/AlSb heterojunction requires the valence band to be taken into account [145]. The diodes of this type demonstrated room-temperature peak current density of 3.7×10^5 A/cm^2 and PVR of 3.2. These resonant tunneling diodes oscillated up to 712 GHz [146]. The devices in Figs. 34(b) and (c) are the examples of RIT devices, containing GaSb (InAs) QW, sandwiched between two thin AlSb barriers and two n-type InAs (p-type GaSb) electrodes. The energy-band diagram of the electron version of this device is shown in Fig. 35. Resonant transport in these structures occurs due to tunneling through the quasi-bound valence (conduction-) band state. Devices with PVR as high as 20:1 at room temperature and 88:1 at 77 K have been realized [138]. Another class of RIT devices consisting of a thin GaSb layer sandwiched between two n-type InAs electrodes is

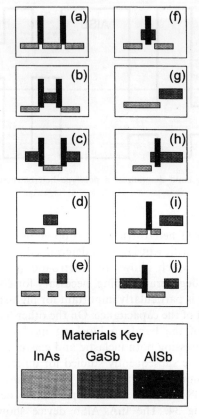

Fig. 34 Schematic energy band diagrams of ten different two-terminal tunnel devices that have been fabricated in the InAs/GaSb/AlSb system.

presented in Fig. 34(d). Fig. 36 shows the corresponding energy band diagram. Resonant tunneling in this structure also involves quasi-bound states in the GaSb valence band, but in contrast to the states found in the double barrier structures mentioned above, which originate from the presence of classically forbidden barrier regions, the quasi-bound states in the InAs/GaSb/InAs structure are due to the partial reflection of the carriers at the InAs/GaSb interfaces. Resonant tunneling without classically forbidden barriers exhibits characteristic broad transmission resonances, which result in NDR with peak current densities exceeding $10^4 \, A/cm^2$ and ultra-short tunneling times theoretically estimated as $\sim 25 \, fs$.

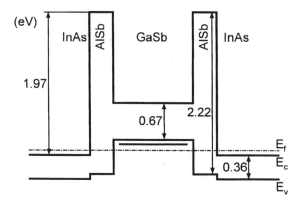

Fig. 35 Band-edge diagram (Γ-point) for the RIT structure at room temperature. Experimental band offset values have been reported for the InAs/GaSb interface ($\Delta E_v = 0.51$ eV) and the AlSb/GaSb interface ($\Delta E_v = 0.40$ eV) only. The valence bandedge of the GaSb and the quantized hole state of the well are both above the InAs conduction bandedge at zero bias. Band bending has been neglected here.

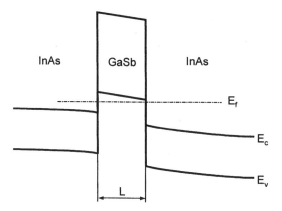

Fig. 36 Energy band diagram of an InAs/GaSb/InAs broken-gap interband tunneling device. Note that an electron can tunnel through the GaSb layer via the GaSb valence bands, as indicated by the dot-dash line.

Fig. 34(g–i) present the next class of RIT structures, using the interband tunneling through the InAs(n)/GaSb(p) interface [143,144]. In these structures the negative differential resistance is due to the GaSb bandgap blocking the current. The formation of quantum well between the InAs/GaSb interface and the AlSb barrier in the structures represented in Figs. 34(h) and (i) resulted in resonant enhancement of the peak current density as compared to the simple InAs(n)/GaSb(p) interface structure. Peak current densities above 10^5 A/cm^2 have been observed for the structures shown in Figs. 34(h) and (i).

Transmission coefficients for the InAs/GaSb/AlSb structures have been computed using both the simple two-band models, which included only the interaction between conduction-band and light-hole states, and the multiple band models. It was found that while the main interband transport mechanism arises from coupling between the InAs conduction-band states and the GaSb light-hole states, the presence of heavy-hole states gives rise to additional heavy-hole resonances and introduces substantial hole mixing effects in the structures containing GaSb QW [150,151]. The calculated in-plane dispersion relations [155] demonstrate the coupling between the light- and heavy-hole bands. These calculations show that the heavy-hole contribution to the tunneling current increases with increasing the in-plane momentum.

Besides the high peak-to-valley ratios, the main goal in the development of RIT devices for digital applications is the achievement of low valley current densities (VCD). The reduction of VCD has been reported for InAs/AlSb/GaSb/AlSb/InAs RIT structures grown with monolayers of AlAs deposited at the InAs/AlSb and both GaSb interfaces [156]. It has been speculated that these AlSb(As) layers act as "hole tunneling barriers", suppressing elastic hole tunneling. Raman spectroscopy was used to study compositional and structural properties of AlSb(As) tunneling barriers in such structures. It has been demonstrated that the tunneling barriers in the structure with the AlAs monolayers at the interfaces consist of pseudoternary Al(SbAs) rather than of binary AlAs and AlSb. Moreover, the structural quality of the Al(SbAs) barriers was found to be inferior to the quality of binary AlSb barriers. These results obviously stimulate further investigations of the resonant tunneling characteristics as a function of structural and compositional properties of layers and interfaces.

4.4.3 (AlGa)Sb/In(AsSb) Based Transistors

Among various microelectronic applications of Type-II (AlGa)Sb/InAs heterostructures, InAs channel field-effect transistors (FET) were intensively studied during the last decade due to their potential superiority to both GaAs and InGaAs channel devices. InAs shows the maximal low-field mobility among III–V's, except only InSb, but unlike to the latter, it has a series of nearly lattice-matched wider bandgap (AlGaIn)(AsSb) compounds with any required band alignment. InAs also has large direct-indirect minima separation in conduction band, which should yield higher carrier transient and overshoot velocities than in InP and GaAs [157,158]. Moreover, InAs/AlSb electron quantum well is the deepest among III–V's.

On the other hand, the narrow bandgap of InAs (0.36 eV) results in undesirable breakdown phenomena due to impact ionization at moderate electric fields, which has been estimated to be approximately 6 kV/cm for bulk n-type InAs [157,159]. If applied directly to 1 μm InAs FETs, this electric field would limit operation to drain-source voltages of ~ 1 V.

The first AlSb/InAs quantum well FET operated at room temperature has been fabricated by Tuttle and Kroemer [160]. It demonstrated a current gain of about 10. However, this attempt and other initial efforts to fabricate InAs channel heterojunction FETs [161,162] did not reach the predicted performance characteristics, because of (1) high dislocation density (10^7 cm^{-2}), resulting in significant gate leakage current, (2) low Schotky barrier height of (AlGa)Sb, and, probably, (3) lack of mature technology for these materials.

The first demonstration of high performance InAs/(AlGa)Sb heterojunction FET (HFET) operated at room temperature under conditions very close to enhancement-mode has been reported by Yoh et al. [163]. The energy band diagram and the schematic diagram of the InAs HFET are shown in Figs. 37(a) and (b) respectively. This energy band diagram is typical for InAs channel HFETs.

The main reason for using GaAs substrate, which is strongly lattice-mismatched to the InAs/(AlGa)Sb system by more than 7%, is the lack of GaSb (or InAs) semi-insulating substrates. Usually a thick (2–3 μm) buffer layer (AlSb or AlGaSb) is required to improve the crystal quality of the InAs channel layer. The undoped (AlGa)Sb buffer layer, which is generally semi-insulating, electrically isolates the channel from the substrate and unintentionally p-doped GaSb nucleation layer (if any). The buffer structure is usually completed by thin undoped $Al_{0.5}Ga_{0.5}Sb$ layer to passivate the highly hygroscopic AlSb buffer layer after the device mesa etching. These buffer layers are followed by the 100–150 Å InAs channel sandwiched between two 50–60 Å AlSb spacer layers (in the structure shown in Fig. 37(a) the top spacer was not used). On the one hand, using the AlSb spacers is believed to increase carrier concentration in the channel owing to the unintentional donors in AlSb [95] or at the InAs/AlSb interface [43]. On the other hand, potential barriers for electrons in the AlSb/InAs/AlSb QW are the largest in the InAs/(AlGa)Sb system. Special growth procedure

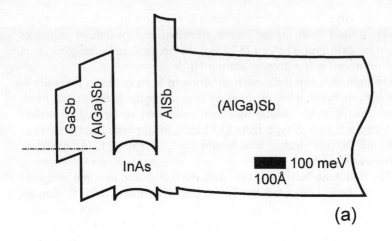

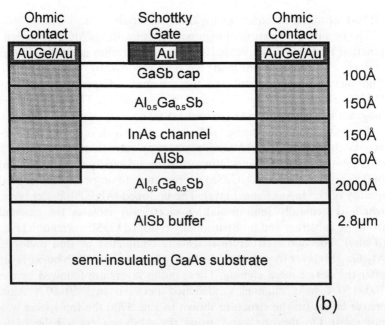

Fig. 37 (a) Typical energy band diagram of InAs/(AlGa)Sb HFET, (b) typical layout of the heterojunction FET [163].

is commonly used to form an "InSb-like" interface at the AlSb/InAs boundary, which has been shown [43] to enhance dramatically the low-field electron mobility. The top layer sequence contains the 100–400 Å AlGaSb (AlSb) layer for effective barrier for holes (0.7 eV for Au/p-AlSb Schottky diodes [164]), and the 50 Å GaSb cap layer to protect the underlying Al-containing layer. The overall distance of the InAs channel from the surface determines the operation mode of the FET device, either depletion (thicker) or enhancement one. The depletion-mode operation needs higher gate voltage, which may lead to undesirable increase in output conductance due to the onset of carrier multiplication within the channel. The structure presented in Fig. 37(a) has shown the low-field mobility of 15,100 and 32,700 cm^2/(Vs) at 300 and 77 K, respectively. The respective sheet carrier densities were 1.9×10^{12} and 1.4×10^{12} cm^{-2}. Such low mobilities, as compared to the maximal values, reported for the InAs/AlSb 2DEG heterostructures [93], were probably due to the thick AlGaSb buffer layer, used instead of GaSb, significantly deteriorating the structure morphology and mobility.

The 1.7 μm gate-length HFET fabricated using this structure (Fig. 37(b)) demonstrated the highest so far room temperature extrinsic transconductance of 460 mS/mm (at the drain-source voltage $V_{ds} = 0.5$ V) and of 509 mS/mm at $V_{ds} = 1.0$ V, and good pinch-off characteristics (Fig. 38). However, the dramatic increase in output conductance up to 200 mS/mm at the V_{ds} values in excess of 0.8 V indicates the onset of carrier multiplication by impact ionization due to the narrow bandgap of InAs.

Similar HFET structure containing a 3 μm thick AlSb binary buffer layer, thinner InAs channel of 125 Å, and two Te δ-doping sheets in the upper 400 Å thick AlSb barrier has been reported [165] to demonstrate the markedly enhanced room temperature mobility of 21,000 cm^2/(Vs) and the sheet electron density of 3.8×10^{12} cm^{-2}. The first high-performance depletion-mode HFET devices with 2 μm gate-length, demonstrating the better pinch-off characteristics at a gate bias of about -1.0 V, have been fabricated from this structure. Within the useful V_{ds} operating range between 0.3 and 0.7 V they exhibited high peak extrinsic transconductance of 473 mS/mm (at 300 K) and the lowest output conductance of 20 mS/mm. The doubled, with respect to the previous structure, sheet electron density in the channel resulted in the 1.5 times higher operating current density, which was 400 mA/mm. The onset of carrier multiplication within the channel, resulting in noticeable increase in the output conductance, has been observed in these HFETs at slightly higher V_{ds}. It is worth noting, that a remarkable gate leakage

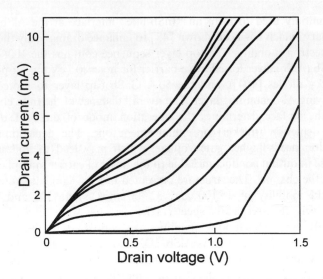

Fig. 38 Measured drain current characteristics of the 1.7 μm (gate length) InAs FET at 300 K. The gate voltages are taken from −0.1 to 0.25 V in 50 mV steps [163].

current of 50–70 μA in the operating region ($V_{ds} = 0.3$–0.7 V) was demonstrated by all the devices.

Significant improvement in InAs/AlSb HFET's performance has been achieved by Li et al. [166], who introduced a 250 Å $AlSb_{0.9}As_{0.1}$ upper barrier layer, with the higher potential barrier for holes, and reduced the InAs channel thickness to 100 Å. The room temperature drain-to-source current–voltage characteristics of the 1 μm gate-length InAs/AlSbAs HFET, given in Fig. 39, showed no impact ionization within the whole range of V_{ds} up to 2.2 V. Estimation of the respective electric field in the channel gave the value as high as 20 kV/cm, which is about 3 times higher than the predicted threshold for impact ionization in a bulk InAs [157]. In addition, these devices exhibited the lowest reported output conductance at $V_{ds} > 1.0$ V for InAs channel devices, the voltage gains of the order of 10, and the maximum extrinsic transconductance of 414 mS/mm at $V_{ds} = 1.0$ V. The drain current was intentionally limited at the level of 450 mA/mm, while the room temperature gate leakage current was less than 20 μA for $V_{gs} = -0.9$ V and $V_{ds} = 0-2.2$ V. The former was due to the enhanced electron sheet density in the InAs channel, and the latter was most probably due to the higher valence band offset at the GaSb/AlAsSb interface as compared to the case of GaSb/AlSb. The estimated cutoff frequency

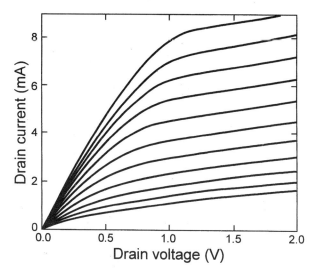

Fig. 39 Room-temperature drain-to-source current–voltage characteristics for a 1 μm × 20 μm gate AlSbAs/InAs HFET [166]. V_{gs} is varied from 0.1 V (top curve) to −0.9 V in 0.1 V increments.

of 39 GHz was more than twice as high as that for 1 μm gate-length GaAs-based FETs.

Among the reasons proposed for such two- or three-fold increase in breakdown field over all previous results we would note two points, which appear to be consistent with the results described above [163,165]. At first, the increase in effective bandgap due to the energy quantization in the InAs channel rises the threshold for breakdown. The latest structure [166] contained the narrowest (100 Å) InAs quantum well. Secondly, the decrease in gate-length of the FET's, fabricated using the InAs channel with high carrier mobility and, hence, long carrier mean-free path, may reduce significantly the probability for carriers to contribute into impact ionization collisions.

Finally, these results have shown that InAs-channel HFET's may operate at the supply voltages, characteristic for practical use. The device performance is expected to be improved significantly as the device dimensions are reduced and the technology matures.

Another quite interesting and promising field, where the InAs/(AlGa)Sb material system may be successfully used due to its great flexibility in choosing band alignment, is related to a fabrication of a quantum-size-effect transistor. The main goal is to employ very high speed of quantum phenomena in fabricating extremely fast electronic circuits and high frequency switches. One of such devices is a Stark effect transistor (SET) intended to operate by means of current modulation by the Stark shift of energy levels in a quantum well. In the SET concept, suggested in Refs. [167,168], the lowest base layer is spatially and electrically separated from the upper emitter and collector by the thick potential barrier layer, which suppresses dramatically the base-collector current as compared to the emitter–collector one. Applying bias across the thick barrier to the collector quantum well, would electrostatically modulate the emitter–collector current. The first room-temperature SET demonstrating a current gain as high as 50 has been fabricated using p-GaSb/InAs/AlSb/p-GaSb heterostructure by Collins et al. [169]. Schematic bandedge diagram of the transistor is presented in Fig. 40 together with the device layout. It should be noted

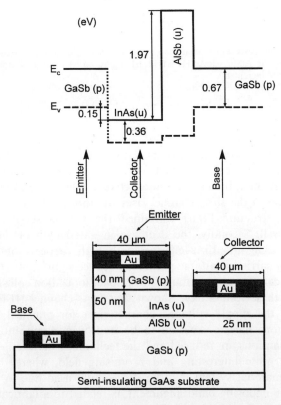

Fig. 40 (a) Schematic bandedge diagram (neglecting band bending) of Stark effect transistor (b) Layout of the device after processing [169].

that even though there is no quantum barrier between the InAs collector and GaSb emitter, the InAs/GaSb interface partially reflects incident electrons, leading to the formation of resonant level in the thin InAs collector layer sandwiched between GaSb on one side and AlSb on the other. This transistor structure is potentially useful in the fabrication of ultra-high-speed electronic circuits, because the emitter–collector current conduction in this structure is due to the resonant transmission, controlled by the base biases at a very small base current.

4.5 CONCLUSION

In conclusion, the narrow bandgap InAs/(Al, Ga)Sb material system is one of most intenisvely developed today. Its main attractive features are:

(1) relatively small lattice mismatches between the respective binary compounds;
(2) very specific controllable InAs/(Al, Ga)Sb interface structure when both anion and cation are changed across the interface;
(3) unique band alignment allowing to obtain heterostructures with both type-I and type-II alternating heterojunctions;
(4) wide effective bandgap variation from zero (for type-II InAs/(Al, Ga)Sb heterostructure) to 1.63 eV (for AlSb);
(5) Superior transport characteristics of InAs with respect to GaAs and InGaAs.

The powerful set of characteristics provide various possibilities in designing new types of heterostructures both for fundamental studies and for device applications among which the mid-wave infrared room temperature lasers, resonant-interband tunneling diodes, and high speed heterostructure field effect transistors are of primary importance. Finally, the advantages of this material system could be practically realized owing to the progress in the MBE technique, which providing a monolayer and even a sub-monolayer accuracy of the epitaxial growth control.

Acknowledgments

The authors are grateful to Prof. Zh. Alferov for permanent encouragement of this study, to Drs. A. Toropov, V. Ustinov, B. Meltser, and A. Tsatsul'nikov for useful and fruitful discussions, and to P. Necliudov and S. Shaposhnikov for valuable assistance in preparing the chapter.

References

1. F. W. Dasilva et al., *Semicond. Sci. Technol.* **4**, 7 (1989).
2. Y. K. Su, S. C. Chen, and F. S. Juang, *Solid-State Electron.* **32**, 733 (1989).
3. M. E. Lee, D. J. Nicholas, K. E. Singer, and B. Hamilton, *J. Appl. Phys.* **59**, 2895 (1986).
4. M. Yano et al., *Jpn. J. Appl. Phys.* **17**, 2091 (1978).
5. I. Poole et al., *J. Appl. Phys.* **63**, 395 (1988).
6. G. R. Jonson et al., *Semicond. Sci. Technol.* **3**, 1157 (1988).
7. S. V. Ivanov et al., *Semicond. Sci. Technol.* **8**, 347 (1993).
8. J. R. Waterman, B. V. Shanabrook, and R. J. Wagner, *J. Vac. Sci. Technol. B* **10**, 895 (1992).
9. J. Riao, R. Beresford, and W. I. Wang, *J. Vac. Sci. Technol. B* **8**, 276 (1990).
10. S. Subbanna, G. Tuttle, and H. Kroemer, *J. Electron. Materials* **17**, 297 (1988).
11. L. Topfer, O. Brandt, E. Tournie, and K. Ploog, in *Proceedings of the VII International Conference on Molecular Beam Epitaxy, Schwabisch Gmünd*, 1992, p. Tu 1.13.
12. P. S. Kop'ev and N. N. Ledentsov, *Sov. Phys. Semicond.* **22**, 1093 (1988).
13. S. V. Ivanov, P. S. Kop'ev, and N. N. Ledentsov, *J. Cryst. Growth* **104**, 345 (1990).
14. W. Ruhle et al., *Phys. Status Solidi B* **73**, 255 (1976).
15. O. Kubaschewski, E. L. Evans, and C. B. Alcock, *Metallurgical Thermochemistry*, 4th edn. (Pergamon, London, 1967).
16. R. E. Honig, *RCA Rev.* **23** (1962).
17. J. Y. Tsao, *J. Cryst. Growth* **110**, 595 (1991).
18. M. Yata, *Thin Solid Films* **137**, 79 (1986).
19. M. B. Panish and M. Ilegems, *Prog. Solid State Chem.* **7**, 39 (1972).
20. G. B. Stringfellow, *J. Appl. Phys.* **43**, 3455 (1972).
21. R. E. Nahory, M. A. Pollack, E. D. Beebe, and J. C. DeWinter, *J. Electrochem. Soc.* **125**, 1053 (1978).
22. B. A. Pyabin, M. A. Ostroumov, and T. F. Swit, *Handbook on thermodynamic properties of substances* (Chemistry, Leningrad, 1977).
23. B. R. Bennett et al., *Appl. Phys. Lett.* **67**, 949 (1993).
24. A. V. Novoselova et al., *Handbook on Physical and Chemical Properties of Semiconductors* (Science, Moscow, 1978).
25. P. L. Gourley, R. M. Biefeld, and L. R. Dawson, *Appl. Phys. Lett.* **47**, 482 (1985).
26. T. M. Rossi, D. A. Collins, D. H. Chow, and T. C. McGill, *Appl. Phys. Lett.* **57**, 2256 (1990).
27. H. Munekata, T. P. Smith III, and L. L. Chang, *J. Cryst. Growth* **95**, 235 (1989).
28. Y. Suzuki, Y. Ohmori, and H. Okamoto, *J. Appl. Phys.* **59**, 3760 (1986).
29. J.-Y. Shen and C. Chatillon, *J. Cryst. Growth* **106**, 543 (1990).
30. Y. J. Van der Meulen, *J. Phys. Chem. Solids* **28**, 25 (1967).
31. J. Allegre and M. Aveerous, *Inst. Phys. Conf. Ser.* **46**, 379 (1979).
32. Y. Ohmori, S. Tarucha, Y. Horikoshi, and H. Okamoto, *Jpn. J. Appl. Phys.* **23**, L94 (1984).
33. A. Furukawa and M. Mizuta, *Electron. Lett.* **24**, 1378 (1988).
34. T. D. McLean et al., *Inst. Phys. Conf. Ser.* **74**, 145 (1984).
35. T. D. McLean et al., *J. Vac. Sci. Technol. B* **4**, 601 (1986).
36. H. Gotoh et al., *Jpn. J. Appl. Phys.* **20**, L893 (1981).
37. A. Y. Cho and J. R. Arthur, *Prog. Solid State Chem.* **10**, 157 (1975).
38. G. J. Davies, D. A. Andrews, and R. Heckingbottom, *J. Appl. Phys.* **52**, 7214 (1981).
39. D. A. Andrews, R. Heckingbottom, and G.J. Davies, *J. Appl. Phys.* **54**, 4421 (1983).
40. N. B. Brandt et al., *Sov. Phys.-JETP* **59**, 847 (1984).
41. J.-D. Sheng, Y. Makita, K. Ploog, and H. J. Queisser, *J. Appl. Phys.* **53**, 999 (1982).

42. G. H. Döhler, K. Ploog, *Periodic doping structures in GaAs*. -in *Progress in Crystal Growth and Characterization*, Ed. B. R. Pamplin (Pergamon Press, Oxford, 1981).
43. G. Tuttle, H. Kroemer, and J. H. English, *J. Appl. Phys.* **67**, 3032 (1990).
44. B. Brar, J. Ibbetson, H. Kroemer, and J. H. English, *Appl. Phys. Lett.* **64**, 3392 (1994).
45. M. Yano, M. Okuizumi, Y. Iwai, and M. Inoue, *J. Appl. Phys.* **74**, 7472 (1993).
46. C. Lopez *et al.*, *Surf. Sci.* **267**, 176 (1992).
47. I. Sela, C. R. Bolognesi, L. A. Samoska, and H. Kroemer, *Appl. Phys. Lett.* **60**, 3283 (1992).
48. J. Spitzer *et al.*, *Appl. Phys. Lett.* **62**, 2274 (1993).
49. M. Inoue *et al.*, *Semicond. Sci. Technol.* **8**, S121 (1993).
50. P. D. Wang *et al.*, *Solid State Commun.* **91**, 361 (1994).
51. J. Wagner, J. Schmitz, D. Behr, J. D. Ralston, and P. Koidl, *Appl. Phys. Lett.* **65**, 1293 (1994).
52. A. Fasolino, E. Molinari, and J. C. Maan, *Phys. Rev. B* **33**, 8889 (1986).
53. D. H. Chow, R. H. Miles, J. R. Soderstrom, and T. C. McGill, *J. Vac. Sci. Technol. B* **8**, 710 (1990).
54. I. Sela, C. R. Bolognesi, and H. Kroemer, *Phys. Rev. B* **44**, 16142 (1993).
55. J. T. Zborowski *et al.*, *J. Appl. Phys.* **71**, 5908 (1992).
56. H. C. Chen *et al.*, *J. Vac. Sci. Technol. B* **13**, 706 (1995).
57. Y.-H. Zhang and D. H. Chow, *Appl. Phys. Lett.* **65**, 3239 (1994).
58. F. Hatami *et al.*, *Appl. Phys. Lett.* **67**, 656 (1995).
59. D. A. Collins, M. W. Wang, R. W. Grant, and T. C. McGill, *J. Appl. Phys.* **75**, 259 (1994).
60. M. Yano, H. Yokose, Y. Iwai, and M. Inoue, *J. Cryst. Growth* **111**, 609 (1991).
61. H. Seki and A. Koukitu, *J. Cryst. Growth* **78**, 342 (1986).
62. J. R. Waterman *et al.*, *Semicond. Sci. Technol.* **8**, S106 (1993).
63. G. A. Sai-Halasz, R. Tsu, and L. Esaki, *Appl. Phys. Lett.* **30**, 651 (1977).
64. L. I. Chang and L. Esaki, *Surf. Sci.* **98**, 70 (1980).
65. G. A. Sai-Halasz, L. Esaki, and W. A. Harrison, *Phys. Rev. B* **18**, 2812 (1978).
66. G. A. Sai-Halasz *et al.*, *Solid State Commun.* **27**, 935 (1978).
67. Y. Guldner *et al.*, *Phys. Rev. Lett.* **45**, 1719 (1980).
68. L. L. Chang *et al.*, *Appl. Phys. Lett.* **35**, 939 (1979).
69. M. Altarelly, *Phys. Rev. B* **28**, 842 (1983).
70. G. Bastard, E. E. Mendez, L. L. Chang, and Esaki, *J. Vac. Sci. Technol.* **21**, 531 (1982).
71. E. E. Mendez *et al.*, *Surf. Sci.* **142**, 215 (1984).
72. S. Washburn *et al.*, *Phys. Rev. B* **31**, 1198 (1985).
73. H. Munekata, E. E. Mendez, Y. Iye, and L. Esaki, *Surf. Sci.* **174**, 449 (1986).
74. M. Altarelly, J. C. Maan, L. L. Chang, and L. Esaki, *Phys. Rev. B* **35**, 9867 (1987).
75. J. Luo, H. Munekata, F. F. Fang, and P. J. Stiles, *Phys. Rev. B* **38**, 10142 (1988).
76. Yu. L. Ivanov, P. S. Kop'ev, S. D. Suchalkin, and V. M. Ustinov, *Pis'ma Zh. Eksp. Teor. Fiz.* **53**, 470 (1991) [*JETP Lett.* **53**, 493 (1991)].
77. A. P. Dmitriev, S. A. Emel'yanov, S. V. Ivanov, and Ya. V. Terent'ev, *JETP Lett.* **62**, 633 (1995).
78. U. Cebulla *et al.*, *Phys. Rev. B* **37**, 6278 (1988).
79. A. Nakagava, H. Kroemer, and J. H. English, *Appl. Phys. Lett.* **54**, 1893 (1989).
80. H. Munekata *et al.*, *J. Phys.* (Paris) **48**, C5–151 (1987).
81. C.-A. Chang, E. E. Mendez, L. L. Chang, and L. Esaki, *Surf. Sci.* **142**, 598 (1984).
82. C.-A. Chang *et al.*, *J. Vac. Sci. Technol. B* **2**, 214 (1984).
83. H. Munekata, L. Esaki, and L. L. Chang, *J. Vac. Sci. Technol. B* **5**, 809 (1987).
84. S. D.Suchalkin *et al.*, in *Proceedings of the 22nd International Conference on the Physics of Semiconductors, Vancouver, 1994*, p. 735.
85. W. K. Chu *et al.*, *Phys. Rev. B* **26**, 1999 (1982).

86. C. Gadaleta, G. Scamarcio, F. Fuchs, and J. Schmitz, *J. Appl. Phys.* **78**, 5642 (1995).
87. J. R. Waldrop *et al.*, *J. Vac. Sci. Technol. B* **10**, 1773 (1992).
88. R.G. Dandrea and C.B. Duke, *Appl. Phys. Lett.* **63**, 1795 (1993).
89. M. J. Shaw, P. R. Briddon, and M. Jaros, *Phys. Rev. B* **52**, 16341 (1995).
90. J. Spitzer *et al.*, *J. Appl. Phys.* **77**, 811 (1995).
91. B. Brar, H. Kroemer, J. Ibbetson, and J. H. English, *Appl. Phys. Lett.* **62**, 3303 (1993).
92. T.B. Boykin, *Appl. Phys. Lett.* **64**, 1529 (1994).
93. A. Nakagawa, H. Kroemer, and J. H. English, *Appl. Phys. Lett.* **54**, 1893 (1989).
94. P. F. Hopkins *et al.*, *Appl. Phys. Lett.* **58**, 1428 (1991).
95. G. Tuttle, H. Kroemer, and J. H. English, *J. Appl. Phys.* **65**, 5239 (1989).
96. P. S. Kop'ev *et al.*, *Sov. Phys.-Semicond.* **24**, 450 (1990).
97. C. Nguyen, B. Brar, H. Kroemer, and J. H. English, *J. Vac. Sci. Thechnol. B* **10**, 898 (1991).
98. C. Nguyen, B. Brar, H. Kroemer, and J. H. English, *Appl. Phys. Lett.* **60**, 1854 (1992).
99. A. Furukawa, *Appl. Phys. Lett.* **62**, 3150 (1993).
100. S. Ideshita, A. Furukawa, Y. Mochizuki, and M. Mizuta, *Appl. Phys. Lett.* **60**, 2549 (1992).
101. D. J. Chadi, *Phys. Rev. B* **47**, 13478 (1993).
102. C. R. Bolognesi, H. Kroemer, and J. H. English, *J. Vac. Sci. Technol. B* **10**, 877 (1992).
103. C. R. Bolognesi, H. Kroemer, and J. H. English, *Appl. Phys. Lett.* **61**, 213 (1992).
104. A. Gold, *Phys. Rev. B* **35**, 723 (1987).
105. Ch. Gauer *et al.*, *Semicond. Sci. Technol.* **8**, S137 (1993).
106. L. D. Nguyen, A. S. Brown, M. A. Tompson, and L. M. Jelloian, *Microwave Journal*, June-1993, p. 96.
107. L. M. Dolginov *et al.*, *Sov. J. Quantum Electron.* **5**, 703 (1978).
108. N. Kobayasi, Y. Horikoshi, and C. Uemura, *Jpn. J. Appl. Phys.* **19**, L30 (1980).
109. C. Caneau *et al.*, *Appl. Phys. Lett.* **51**, 764 (1987).
110. A. E. Bochkarev *et al.*, *Sov. J. Quantum Electron.* **12**, 869 (1985).
111. T. H. Chiu, W. T. Tsang, J. A. Ditzenberger, and J. P. van der Ziel, *Appl. Phys. Lett.* **49**, 1051 (1986).
112. H. K. Choi and S. J. Eglash, in *Proceedings of the 12 IEEE International Semiconductor Laser Conference, Davos*, 1990, p. PD-9.
113. H. K. Choi and S. J. Eglash, *Appl. Phys. Lett.* **59**, 1165 (1991).
114. H. K. Choi and S. J. Eglash, *Appl. Phys. Lett.* **61**, 1154 (1992).
115. G. W.Turner *et al.*, *J. Vac. Sci. Technol.* **12**, 1266 (1994).
116. H. Lee *et al.*, *Appl. Phys. Lett.* **66**, 1942 (1995).
117. D. Z. Garbuzov *et al.*, *Appl. Phys. Lett.* **67**, 1346 (1995).
118. J. C. DeWinter, M. A. Pollack, A. K. Srivastava, and J. L. Zyskind, *J. Electron. Mater.* **14**, 729 (1985).
119. G. B. Stringfellow, *J. Cryst. Growth* **58**, 194 (1982).
120. T. H. Chiu *et al.*, *Appl. Phys. Lett.* **46**, 408 (1985).
121. H. K. Choi, S. J. Eglash, and G. W. Turner, *Appl. Phys. Lett.* **64**, 2474 (1994).
122. M. Aidaraliev *et al.*, *Sov. Tech. Phys. Lett.* **15**, 600 (1990).
123. A. Ishida, K. Muramatsu, H. Takashida, and H. Fujiyashu, *Appl. Phys. Lett.* **55**, 430 (1989).
124. S. J. Eglash and H. K. Choi, *Appl. Phys. Lett.* **64**, 833 (1994).
125. Y. Tsou, A. Ichii, and E. Garmire, *IEEE J. Quantum Electron.* **28**, 1261 (1992).
126. H. K. Choi and G. W. Turner, *Appl. Phys. Lett.* **67**, 332 (1995).
127. A. N. Baranov *et al.*, *Sov. Phys.–Semicond.* **20**, 2217 (1986).
128. A. N. Baranov *et al.*, *Sov. Tech. Phys. Lett.* **14**, 1671 (1988).

129. P. S. Kop'ev, N. N. Ledentsov, A. M. Monakhov, and A. A. Rogachev, Patent of Russian Federation 2019895 "Non-injection light emitting diode", 1991.
130. A. S. Filipchenko *et al.*, in *Proceedings of International Symposium on Nanostructures: Physics and Technology*, St. Petersburg, 1993, p. 44.
131. P. S. Kop'ev *et al.*, in *Proceedings of the VIII International Conference on Molecular Beam Epitaxy*, Osaka, 1994, p. 542.
132. Y.-H. Zhang, *Appl. Phys. Lett.* **66**, 118 (1995).
133. T. S. Hasenberg *et al.*, *Electron. Lett.* **31**, 275 (1995).
134. H. Munekata, T. P. Smith, and L. L. Chang, *J. Vac. Sci. Technol. B* **7**, 324 (1989).
135. R. Beresford, L. F. Luo, and W. I. Wang, *Appl. Phys. Lett.* **54**, 1899 (1989).
136. M. Sweeny and J. Xu, *Appl. Phys. Lett.* **54**, 546 (1989).
137. L. F. Luo, R. Beresford, and W. I. Wang, *Appl. Phys. Lett.* **53**, 2320 (1988).
138. J. R. Söderström, D. H. Chow, and T. C. McGill, *Appl. Phys. Lett.* **55**, 1094 (1989).
139. L. F. Luo, R. Beresford, and W. I. Wang, *Appl. Phys. Lett.* **55**, 2023 (1989).
140. R. Beresford, L. F. Luo, K. F. Longenbach, and W. I. Wang, *Appl. Phys. Lett.* **56**, 2023 (1990).
141. K. F. Longenbach, L. F. Luo, and W. I. Wang, *Appl. Phys. Lett.* **57**, 1554 (1990).
142. E. T. Yu *et al.*, *Appl. Phys. Lett.* **57**, 2675 (1990).
143. D. A. Collins *et al.*, *Appl. Phys. Lett.* **57**, 683 (1990).
144. D. Z.-Y. Ting *et al.*, *Appl. Phys. Lett.* **57**, 1257 (1990).
145. J. R. Söderström *et al.*, *Appl. Phys. Lett.* **58**, 275 (1991).
146. E. R. Brown *et al.*, *Appl. Phys. Lett.* **58**, 2291 (1991).
147. J. F. Chen, L. Yang, and Y. Cho, *IEEE Electron Device Lett.* **11**, 532 (1990).
148. D. Z.-Y. Ting *et al.*, *J. Vac. Sci. Technol. B* **8**, 810 (1990).
149. J. R. Söderström *et al.*, *J. Appl. Phys.* **68**, 1372 (1990).
150. D. Z.-Y. Ting, E. T. Yu, and T. C. McGill, *Appl. Phys. Lett.* **58**, 292 (1991).
151. D. Z.-Y. Ting, E. T. Yu, and T. C. McGill, *Phys. Rev. B* **45**, 3583 (1992).
152. T. B. Boykin, R. E. Carnahan, and R. J. Higgins, *Phys. Rev. B* **48**, 14232 (1993).
153. J.-C. Chiang, *Appl. Phys. Lett.* **64**, 1956 (1994).
154. A. Sigurdardottir, V. Krozer, and H. L. Hartnagel, *Appl. Phys. Lett.* **67**, 3313 (1995).
155. J. Genoe, K. Fobelets, C. Van Hoof, and G. Borghs, *Phys. Rev. B* **52**, 14025 (1995).
156. J. Wagner, J. Schmitz, H. Obloh, and P. Koidl, *Appl. Phys. Lett.* **67**, 2963 (1995).
157. K. Brennan and K. Hess, *Solid State Electron.* **27**, 347 (1984).
158. A. Cappy and B. Carney. *IEEE Trans. Electron. Devices* **ED-27**, 2158 (1980).
159. G. Bauer and F. Kuchar, *Phys. Lett. A* **30**, 399 (1969).
160. G. Tuttle and H. Kroemer, *IEEE Trans. Electron. Devices* **ED-34**, 2358 (1987).
161. L. F. Luo, R. Beresford, W. I. Wang, and H. Munekata, *Appl. Phys. Lett.* **55**, 789 (1989).
162. J. Werking *et al.*, *Appl. Phys. Lett.* **57**, 905 (1990).
163. K. Yoh, T. Moriuchi, and M. Inoue, *IEEE Electron. Device Lett.* **11**, 526 (1990).
164. S. Sadiq and A. Joullie, *J. Appl. Phys.* **65**, 15 (1989).
165. J. D. Werking *et al.*, *IEEE Electron. Device Lett.* **13**, 164 (1992).
166. X. Li, K. F. Longenbach, Y. Wang, and W. I. Wang, *IEEE Electron. Device Lett.* **13**, 192 (1992).
167. A. R. Bonnefoi, D. H. Chow, and T. G. McGill, *Appl. Phys. Lett.* **47**, 888 (1985).
168. F. Beltram *et al.*, *Appl. Phys. Lett.* **53**, 219 (1988).
169. D. A. Collins, D. H. Chow, and T. C. McGill, *Appl. Phys. Lett.* **58**, 1673 (1991).

CHAPTER 5

Growth and Characterization of InAs/AlSb/GaSb Heterostructures

F. FUCHS, J. WAGNER, J. SCHMITZ, N. HERRES and P. KOIDL
*Fraunhofer-Institut für Angewandte Festkörperphysik,
Tullastrasse 72, D-79108 Freiburg, Germany*

5.1.	Introduction	171
5.2.	Molecular-Beam Epitaxial Growth	173
5.3.	Analysis of Structural Properties and Interfacial Bonding	181
	5.3.1. High-resolution X-ray diffraction	181
	5.3.2. Raman spectroscopy	187
	5.3.3. Far-infrared spectroscopy	194
5.4.	Electronic Band- and Subband-Structure	201
	5.4.1. Interband spectroscopy of the fundamental energy gap	204
	5.4.1.1. Infrared photoluminescence of InAs/AlSb heterostructures	204
	5.4.1.2. Calorimetric absorbance	208
	5.4.1.3. Optical anistropy	210
	5.4.1.4. Photocurrent spectroscopy on InAs/GaSb superlattices	214
	5.4.2. Intersubband spectroscopy by inelastic light scattering	216
	5.4.3. Ellipsometric spectroscopy of higher lying band gaps	222
5.5.	Results on Devices	225
5.6.	Summary	229
	References	229

5.1 INTRODUCTION

There is current interest in InAs/AlSb/GaSb heterostructures for fundamental studies and device applications. The band overlap between InAs and GaSb and the large conduction band offset between

InAs and AlSb result in unique physical properties of such heterostructures and allows the realization of a variety of devices. Recent advances in molecular-beam epitaxial growth made it possible to grow III–V antimonide heterostructures with a high precision including the controlled formation of the heterointerfaces.

InAs/AlSb quantum-wells (QW) offer the advantage of a low effective mass for electrons in the InAs QW in combination with a large conduction-band offset of 1.35 eV between InAs and AlSb [1]. Such deep QW's for high-mobility electrons are attractive for the fabrication of, e.g., high-speed QW field-effect transistors [2]. InAs/AlSb QW's, in combination with superconducting Nb electrodes, have recently made possible the investigation of electronic interactions at superconductor–semiconductor interfaces [3].

The band overlap between InAs and GaSb allows the realization of resonant interband-tunneling diodes with high peak current densities and peak-to-valley ratios for potential applications in high-speed analog devices as well as in multiple-valued logic circuits [4,5]. In InAs/GaSb superlattices (SL's) with sufficiently thin InAs and/or GaSb layers the sum of the quantisation energies of electrons and holes exceeds the band overlap between the InAs conduction band and the GaSb valence band. As a consequence, a spatially indirect SL band gap opens between the topmost quantized hole state in the GaSb layers and the lowest confined electron level in the InAs layers. The corresponding band gap energy, and thus the optical properties of InAs/GaSb SL's, can be tuned over a wide range covering the far- and mid-infrared spectral range by varying the individual layer thicknesses [6]. Therefore InAs/GaSb SL's can be used for the fabrication of IR detectors [7] and IR lasers [8].

In InAs/GaSb (InAs/AlSb) heterostructures, where both the group III and the group V atoms change at the heterointerface, either In–Sb or Ga–As (Al–As) bonds can be formed across the interface (IF), depending on the growth procedure [9–11]. The formation of these IF bonds, not present in either of the bulk materials, leads to the appearance of InSb-like or GaAs-like (AlAs-like) IF modes [12–16]. These modes are localized at the IF regions and are expected at energies either above the optical modes (GaAs-like or AlAs like IF) or in the gap between the optical and acoustic modes (InSb-like IF). It has been shown that the band overlap at the InAs/GaSb heterojunction [17–19] as well as the IR absorption cutoff wavelength of InAs/GaSb SL's [20] depend on the type of IF bond.

In this chapter we shall focus in Section 5.2 on the molecular-beam epitaxial growth of InAs/AlSb/GaSb heterostructures. In Section 5.3

the issue of structural characterization of InAs/GaSb SL's and the analysis of interfacial bonding at InAs/AlSb and InAs/GaSb interfaces (IF's) will be discussed. Section 5.4 deals with the experimental determination of the electronic band and subband structure of InAs/GaSb SL's and InAs/AlSb QW's. In Section 5.5 experimental results on InAs/GaSb p-i-n SL photodiodes and InAs/AlSb/GaSb resonant interband-tunneling diodes will be presented as examples for devices based on well characterized and optimized epitaxial layer structures. A brief summary will be given in Section 5.6.

5.2 MOLECULAR-BEAM EPITAXIAL GROWTH

Solid-source molecular-beam epitaxy (MBE) is a widely used technique for the growth of InAs/AlSb/GaSb heterostructures [21]. These heterostructures can be grown almost lattice matched, with differences in the lattice parameters of < 1%, on GaSb or InAs substrates [22,23] or, at the expense of a large lattice mismatch of 7.8%, heteroepitaxially on GaAs substrates [24] (see Fig. 1). For the growth on GaAs substrates thick strain-relaxed GaSb, AlSb, or InAs buffer layers are placed between the substrate and the actual heterostructure to accommodate the lattice mismatch [25].

An important issue for the growth of group III arsenide–antimonide heterostructures is the unintentional incorporation of As into the antimony-containing layers from the As background in the MBE

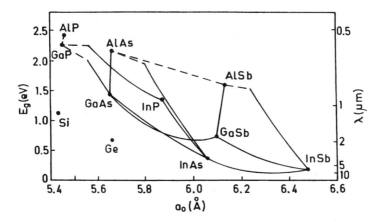

Fig. 1 Band gap energy E_g vs. lattice parameter a_0 for various III–V semiconductors.

growth chamber. It has been shown that using dimeric Sb (Sb_2) and tetrameric As (As_4) from a conventional mechanically shuttered arsenic effusion cell, As concentrations as large as 10–20% may be incorporated into nominally binary GaSb and AlSb layers, both during growth and on stationary surfaces [26,27]. The dependence of the unintentional incorporation of arsenic on the As background in the growth chamber is illustrated in Fig. 2. Secondary ion mass spectrometry (SIMS) data are shown of GaSb-capped InAs/AlSb QW's grown under different As background conditions [26,27]. The 15 nm-wide InAs QW sandwiched between AlSb barriers is placed 20 nm below the surface. Sample A was grown using As_4 from a mechanically shuttered As effusion cell producing, an As background pressure (BGP) of 2×10^{-8} Torr when the mechanical shutter is closed. Samples B and C were grown using As_2 from a valved cracker effusion cell. The As BGP with the mechanical shutter closed was either 4×10^{-9} Torr when the valve was open or 5×10^{-10} Torr when both the shutter and the valve were closed. Fig. 2

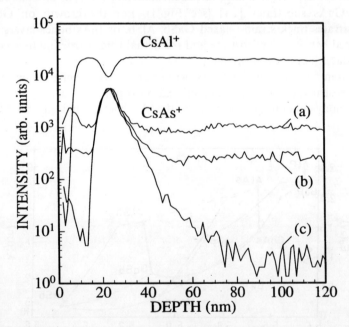

Fig. 2 SIMS depth profiles of Al ($CsAl^+$) and of As ($CsAs^+$) for GaSb capped InAs/AlSb QW structures; (a) sample A, (b) sample B, and (c) sample C. The As background pressure in the MBE growth chamber decreased from 2×10^{-8} Torr for sample A to 5×10^{-10} Torr for sample C (Ref. [26]).

shows SIMS depth profiles of CsAs$^+$ and, for reference purposes, also of CsAl$^+$. For all three samples, there is a maximum in the CsAs$^+$ signal at the expected depth of the InAs QW, accompanied by a minimum in the CsAl$^+$ signal. The strength of the CsAs$^+$ signal observed in the regions above and below the InAs QW, which consist of nominally binary AlSb, depends critically on the As BGP. There is a decrease by almost three orders of magnitude in the As background concentration when going from sample A grown with the highest As BGP to sample C grown with the lowest As BGP. In the latter case the CsAs$^+$ signal is close to or even below the detection limit of the SIMS apparatus.

An independent confirmation of the above finding was obtained from Raman spectroscopy. Fig. 3 shows Raman spectra from samples A, B, and C, which were recorded for excitation in resonance with the $E_1 + \Delta_1$ band gap of GaSb to enhance the Raman scattering signal from the GaSb capping layer over that of the InAs QW [26,27]. With increasing As BGP the AlSb LO phonon line shifts to higher frequencies. This behavior was explained by the formation of pseudo-ternary $AlSb_{1-x}As_x$ barrier layers rather than binary AlSb barriers due to the unintentional incorporation of increasing amounts of As into the AlSb [26,27]. The GaSb LO phonon line splits for samples A and B into two modes. The splitting into a GaAs-like mode ("GaAs") and a GaSb-like mode ("GaSb") indicates the formation of a $GaSb_{1-x}As_x$ capping layer. $GaSb_{1-x}As_x$ is known to show a two-mode behavior [28] rather than a single-mode behavior observed for $AlSb_{1-x}As_x$ [29]. A quantitative analysis of the mode shifts and splittings indicates an As concentration of 8% and 18% in the $AlSb_{1-x}As_x$ layers of samples B and A, respectively [29]. The corresponding concentrations of As in the $GaSb_{1-x}As_x$ capping layers are 7% and 19%, respectively [28]. The As concentrations deduced from the Raman measurements are consistent with the SIMS data (Fig. 2) which show a CsAs$^+$ signal by a factor of three lower for sample B than for sample A. The above data clearly demonstrate that a close control of the As BGP, achieved e.g. by using a valved cracker As effusion cell, is essential for the growth of purely binary antimonide layers in group III arsenide–antimonide heterostructures.

An important issue related to the heteroepitaxial growth of InAs/AlSb/GaSb heterostructures on GaAs substrates using buffer layers composed of materials with lower band gaps than the GaAs substrate material, such as GaSb and InAs, is the control of the substrate temperature during buffer layer growth [30]. For In-free mounted radiatively heated substrates, *in situ* optical transmission experiments

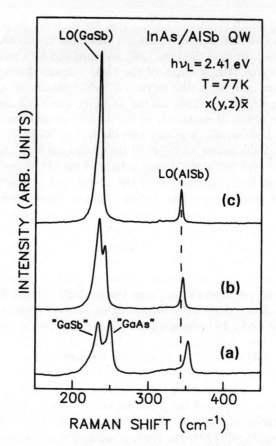

Fig. 3 Depolarized $[x(y,z)\bar{x}]$ low-temperature (77 K) Raman spectra of GaSb capped InAs/AlSb QW structures excited at 2.41 eV. The spectra were recorded (a) from sample A, (b) from sample B, and (c) from sample C with As background pressure in the MBE growth chamber decreasing from 2×10^{-8} Torr for sample A to 5×10^{-10} Torr for sample C (Ref. [27]).

showed a strong decrease of the GaAs band gap energy, reflecting an increase in the substrate temperature by up to 150°C, when layers of GaSb or InAs with thicknesses ranging from 0.1 to 1.0 µm were deposited [30]. This change in substrate temperature was not observed for homoepitaxial growth of GaAs or when growing materials with a larger band gap energy than GaAs, such as AlSb [30]. The observed increase in the substrate temperature for the growth of materials with lower band gap energies than the substrate material was explained by an increase in the fraction of the emitted heater power absorbed in the epitaxial layer [30].

An alternative approach to measure the change in substrate temperature occurring for the growth of GaSb or InAs on GaAs, is to record the change in the thermocouple reading at which a certain change in surface reconstruction, occurring at a specific substrate temperature, is observed by *in situ* reflection high-energy electron diffraction (RHEED) [31]. Fig. 4 shows the change in the nominal substrate temperature as measured by the thermocouple, required to maintain a constant temperature of the growth surface as monitored by RHEED, versus the thickness of the deposited GaSb or InAs layer. The temperature of the GaSb surface was monitored by the change in the GaSb surface reconstruction from (1×3) to (1×5) under an incident Sb_2 flux [22]. For InAs the temperature was monitored using the change in surface reconstruction from As-stabilized (2×4) to In-stabilized (4×2) without any incident flux at the surface [32]. For the growth of InAs on GaAs, a decrease in the thermocouple reading by 150°C, corresponding to a increase of the substrate temperature by

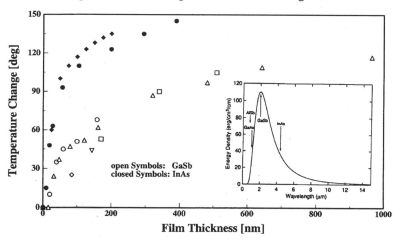

Fig. 4 Change of substrate temperature vs. epitaxial layer thickness for growth of GaSb (open symbols) and InAs (filled symbols) on semi-insulating GaAs substrates (Ref. [31]). The different symbols indicate different growth runs. The filled diamonds represent data for InAs taken from Ref. [30]. The inset shows the spectral distribution of power emitted by a black body with a temperature of 1150°C together with the band gap energies of GaAs, AlSb, GaSb, and InAs at 350°C (Ref. [30]).

roughly the same amount when keeping the heater power constant, is observed for an InAs layer thicknesses of 0.4 μm. Growing GaSb on GaAs this increase in temperature is somewhat smaller saturating at a value of about 110°C for GaSb layer thicknesses approaching 1 μm. The inset of Fig. 4 illustrates the physical reason for the increase in absorbed heater power for the growth of GaSb and InAs on GaAs. There the band gap energies of the various materials of interest at 350°C are shown together with the spectral distribution of the power emitted by a black body with a temperature of 1150°C, which represents the emission characteristic of the substrate heater [30].

Both the optical transmission [30] and the RHEED measurements [31] show a significant increase in the substrate temperature for the growth of GaSb and InAs buffer layers on GaAs substrates. There is also agreement regarding the magnitude of this effect. For the growth of InAs/AlSb/GaSb heterostructures on GaAs using such buffer layers, this increase in the absorbed fraction of the power emitted by the heater has to be compensated by a reduction of the heater power in order to control the substrate temperature accurately. This control is necessary for the reproducible growth of high-quality InAs/AlSb/GaSb heterostructures.

As both the group III and the group V atoms change across the InAs/GaSb (InAs/AlSb) heterointerface, in principle two types of chemical bonds, either InSb-like or GaAs-like (AlAs-like) bonds, can be formed across the IF [12–16]. This is shown schematically in Fig. 5 for an InAs/GaSb heterostructure grown along a ⟨100⟩ crystallographic direction. Using solid-source MBE the formation of these IF bonds can be controlled by using the technique of monolayer epitaxy for the growth of the IF region [9–11]. As illustrated in Fig. 6 for the growth of InAs on GaSb with InSb-like IF bonds the deposition of GaSb is terminated by closing first the Ga-shutter and then, after the time interval necessary for the deposition of one atomic layer of Sb, closing the Sb-shutter. The deposition of InAs is started by first opening the In-shutter for the deposition of one atomic layer of In followed by the opening of the As-shutter for the growth of the InAs layer. For the intended growth of an GaAs-like IF the deposition of GaSb is terminated by first closing the Sb-shutter while maintaining the Ga flux for the time necessary to deposit one atomic layer of Ga. The growth of InAs is then initiated by first opening the As shutter with a delayed opening of the In shutter.

It has been shown by several groups that in InAs/GaSb heterostructures both InSb-like and GaAs-like IF bonds can be realized using

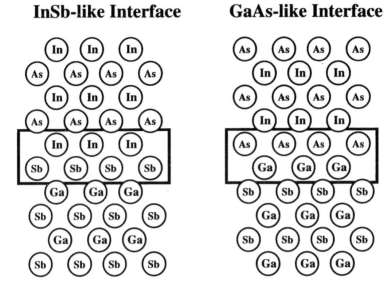

Fig. 5 Ideal structure of the interface region of an InAs/GaSb heterostructure grown along a ⟨100⟩ crystallographic direction. The formation of InSb-like and GaAs-like interface bonds is illustrated in the left and the right part of the figure, respectively.

monolayer epitaxy, as verified by Raman spectroscopy [10,11,33]. At the InAs/AlSb heterointerface a strong tendency is observed towards the formation of InSb-like IF bonds [34] and there is a controversy whether AlAs-like IF bonds can be formed at all at this IF [35–37]. A more complete discussion of the experimental verification of the different types of interfacial bonding in InAs/AlSb/GaSb heterostructures will be given in Section 5.3.

The roughness and intermixing at the heterointerfaces in InAs/GaSb SL's have been studied by both high-resolution transmission electron microscopy [38,39] (HRTEM) and cross-sectional tunneling microscopy [40,41] (STM). The IF's in InAs/AlSb SL's have been analyzed so far only by HRTEM [42]. In detailed HRTEM studies of InAs/GaSb SL's grown on both GaSb and InAs buffer layers by using the technique of monolayer epitaxy for the deposition of the either InSb-like or GaAs-like IF's, it has been found that the width of the InSb-like IF is always smaller than that of the GaAs-like IF [38,39]. An IF width of 1 monolayer (ML) and 2 ML for SL's with InSb-like IF's grown on GaSb and InAs buffer layers, respectively, and 2 ML and

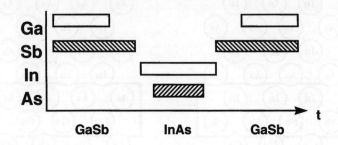

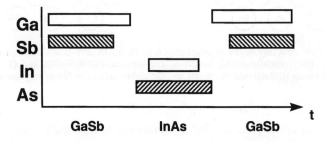

Fig. 6 Shutter sequence for the MBE growth of an InAs/GaSb heterostructure (top) with GaAs-like interface bonds and (bottom) with InSb-like interface bonds.

3 ML for SL's with GaAs-like IF's on GaSb and InAs buffer layers, respectively, has been observed.

It was concluded that interface roughness rather than intermixing is causing these IF widths [39]. This conclusion was confirmed by an *in situ* STM study of the interface of InAs/GaSb SL's, which revealed smoother interfaces on unstrained GaSb than on strained InAs and less roughness for InSb-like than for GaAs-like IF's [43]. Cross-sectional STM results on InAs/GaSb SL's, where the interfacial bonding was controlled by applying either Sb_2 or As_2 soaks to the stagnant growth surface, indicated more intermixing at the InAs-on-GaSb IF than at the GaSb-on-InAs IF independent of the IF type [40]. Similar observations were made for InAs/$Ga_{1-x}In_xSb$ SL's grown with nominally InSb-like IF's by applying a Sb_2 soak [44]. In another cross-sectional STM study InAs/GaSb SL's with IF's prepared either by the Sb_2/As_2 soak technique or by the monolayer epitaxy technique were compared and, for a given growth temperature, a reduced IF width was found for SL's grown using the latter technique [41].

5.3 ANALYSIS OF STRUCTURAL PROPERTIES AND INTERFACIAL BONDING

5.3.1 High-Resolution X-ray Diffraction

High-resolution X-ray diffraction (HRXRD) including the recently developed technique of HRXRD space mapping [45–47] are valuable tools for the structural characterization of semiconductor heterostructures, including in particular SL's. Structural parameters like the period and the lattice parameter of the SL and, for heteroepitaxial growth, also the strain situation in the buffer layer and in the SL stack, can be evaluated when combining information extracted from symmetric and asymmetric reflection profiles. In the following, the application of HRXRD to the structural analysis of InAs/GaSb SL's will be discussed in detail.

The InAs/GaSb SL's under study were grown by solid-source MBE on (100) GaAs substrates. The buffer layer sequence consists of a 100-nm-thick AlSb nucleation layer followed by a 10-period GaSb/AlSb smoothing SL and a 1.1-μm-thick strain-relaxed GaSb buffer layer. On top of this buffer layer 100 period InAs/GaSb SL's were grown with a fixed GaSb layer thickness of 10 ML and InAs layer thicknesses varying from 4 to 14 ML. The shutter sequence for the growth of the InAs/GaSb IF's was chosen either for the deposition of 1 ML of InSb, promoting the formation of InSb-like IF's, or for the deposition of 1 ML of GaAs to induce GaAs-like IF bonds [10,11,33]. The complete layer sequence of the InAs/GaSb SL samples is displayed in Fig. 7.

HRXRD profiles covering the symmetric 004 reflection range and the asymmetric 115(−) reflection range are shown in Figs. 8 and 9 for a pair of 10 ML InAs/10 ML GaSb SL's with InSb-like and GaAs-like IF's [48]. The measured profiles show diffraction peaks originating from the GaAs substrate, the AlSb nucleation layer, and the strain-relaxed GaSb buffer layer. In addition well-resolved SL diffraction peaks up to the fourth order are observed indicated by "0", "± 1", etc. The widths of these peaks are significantly narrower for the SL with InSb-like IF's (Fig. 8) than for the SL with GaAs-like IF's (Fig. 9). This observation holds for all samples studied irrespective of the SL period [48]. Also shown in Figs. 8 and 9 are simulated diffraction profiles

100 periods undoped SL: N ML InAs 10 ML GaSb	N=4,6,8,10,14	410 °C
undoped GaSb buffer	11000 Å	500 °C
(25 Å GaSb/ 25 Å AlSb)x10 SL	500 Å	530 °C
AlSb nucleation layer	1000 Å	570 °C
4 ML AlAs (2ML GaAs-4 ML AlAs)x2	100 Å	580 °C
GaAs layer	1000 Å	580 °C
semi insulating GaAs substrate	500 µm	

Fig. 7 Layer sequence of InAs/GaSb SL samples grown with either InSb-like or GaAs-like IF's on (100) GaAs substrates using a thick strain-relaxed GaSb buffer layer (Ref. [48]).

which reproduce well the peak positions and relative peak widths [48]. To model the peak widths observed for the SL's with GaAs-like IF's, an increasing strain relaxation across the SL stack had to be assumed with the first few periods closest to the GaSb buffer layer being coherently strained to the in-plane lattice parameter of the buffer layer (see below) [48].

Results of more detailed HRXRD measurements on a 4 ML InAs/ 10 ML GaSb SL sample with GaAs-like IF's are shown in Fig. 10. In the top panel diffraction profiles covering the 004 reflection range are plotted, which are recorded either with a fully open detector (upper curve) or with a narrow slit placed in front of the detector (lower curve). The use of a slit in front of the detector results in considerably reduced peak widths and thus in a significant improvement in the resolution of the various diffraction peaks [48]. The lower panel in Fig. 10 shows a two-dimensional angular diffraction space map obtained by measuring several $\omega/2\Theta$-scans with different ω offsets. Here, ω denotes the angle between the sample surface and the incident beam and 2Θ denotes the angle between the incident and the diffracted

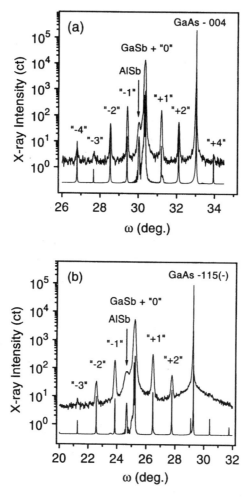

Fig. 8 Measured (upper trace) and simulated (lower trace) HRXRD profiles of a 10 ML InAs/10 ML GaSb SL with InSb-like IF's covering (a) the 004 reflection range and (b) the 115(−) reflection range. The simulated profiles have been smoothed and shifted vertically (Ref. [48]).

beam. The intensity along the median line ($\omega = 0$) corresponds to the better resolved diffraction profile shown in the top panel of Fig. 10. In the diffraction space map the SL diffraction peaks show a considerable broadening perpendicular to the median line, whereas diffraction peaks from the buffer layers and from the GaAs substrate exhibit a nearly circular shape in the diffraction space map.

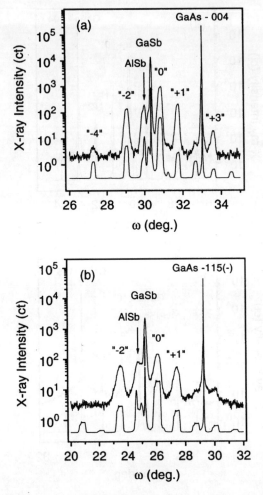

Fig. 9 Measured (upper trace) and simulated (lower trace) HRXRD profiles of a 10 ML InAs/10 ML GaSb SL with GaAs-like IF's covering (a) the 004 reflection range and (b) the 115(−) reflection range. The simulated profiles have been smoothed and shifted vertically (Ref. [48]).

The largest broadening of the SL peaks in the diffraction space map is found at the low-angle side of each peak. At the high-angle side these peaks are more closely confined to the median line. This behavior has been explained by an inhomogeneous strain distribution across the SL stack [48]. Some periods of the SL, most likely those closest to the GaSb buffer, are strained to the in-plane lattice parameter of that

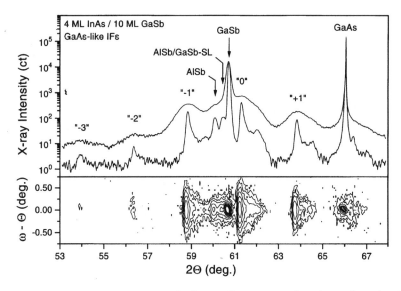

Fig. 10 One-dimensional X-ray reflection profiles (top panel) and two-dimensional angular diffraction space map (bottom panel) of a 4 ML InAs/10 ML GaSb SL with GaAs-like IF's taken near the 004 reflections of the GaAs substrate and the AlSb and GaSb buffer layers (Ref. [48]).

buffer and give rise to the diffraction signal at the high-angle side of the SL peaks. With increasing distance from the SL-buffer IF the SL shows an increasing degree of strain relaxation with the topmost portion of the SL stack being fully relaxed. Those parts of the SL contribute to the broadened diffraction signal observed at the low-angle side of the SL peaks [48]. Mosaicity tilts constitute the major contribution to the broadening [48].

Fig. 11 shows the average lattice parameter of the SL stack parallel [$a_\parallel$, Fig. 11(a)] and perpendicular to the growth surface [$a_\perp$, Fig. 11(b)], as deduced from HRXRD measurements, plotted versus the nominal width of the InAs layers. For the SL's with InSb-like IF's the lattice parameter $a_\parallel$ is close to the lattice constant of the strain-relaxed GaSb buffer layer, which indicates a lattice matched growth of the SL on the GaSb buffer. For InAs layer thicknesses exceeding 6 ML also the lattice parameter $a_\perp$ is close to that of GaSb, which indicates a nearly cubic metric for the SL or, i.e., very little residual strain. For the SL's with GaAs-like IF's and InAs layer thicknesses $\geqslant 6$ ML the lattice parameter $a_\perp$ is significantly smaller than that of the GaSb buffer layer,

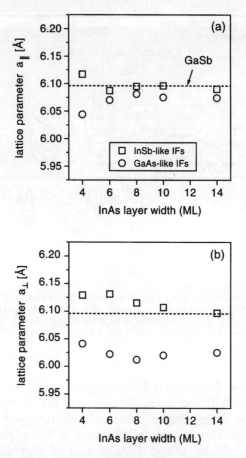

Fig. 11 Average lattice parameter (a) parallel to the growth surface and (b) perpendicular to the growth surface of N ML InAs/10 ML GaSb SL's with either InSb-like IF's (squares) or GaAs-like IF's (circles) plotted vs. the nominal InAs layer width (Ref. [48]).

whereas $a_\parallel$ is still close to the lattice parameter of the buffer layer. This observation indicates that the SL's with GaAs-like IF's are subjected to a biaxial tensile strain [48]. The 4 ML InAs/10 ML GaSb SL with GaAs-like IF's shows the same lattice parameter parallel and perpendicular to the growth surface and both values are smaller than that for the GaSb buffer layer, indicating a complete strain relaxation within the SL stack.

To summarize, the HRXRD analysis revealed that InAs/GaSb SL's grown on a GaSb buffer with InSb-like IF's are essentially unstrained, because the larger lattice parameter of the InSb IF layers at least partially compensates the smaller lattice parameter of the InAs layers. SL's with GaAs-like IF's, in contrast, experience a significant tensile strain. This strain is caused by the smaller lattice parameters of both the InAs layers and the GaAs IF layers, which add up to an average SL lattice parameter smaller than that of the GaSb buffer.

A similar study of the structural properties of InAs/AlSb SL's has been carried out, using a combination of HRXRD and HRTEM [42]. SL's grown with a shutter sequence promoting the formation of AlAs-like IF's showed a lower structural quality than SL's grown with InSb-like IF's because of inhomogeneous strain relaxation and intermixing of As [42].

5.3.2 Raman Spectroscopy

Raman spectroscopy has found widespread use for the study of vibrational modes in semiconductor QW's and SL's [49]. For InAs and GaSb the bulk phonon dispersion curves overlap because the atomic weight per unit cell for InAs and GaSb is roughly the same. Therefore, InAs/GaSb SL's exhibit a spectrum of quasiconfined optical SL phonons. This spectrum is composed of alternating extended and confined modes, the latter being localized in either the InAs or GaSb layers [12–14]. The formation of InSb-like or GaAs-like IF bonds at the InAs/GaSb heterointerface leads to the appearance of "mechanical" interface modes not present in either of the two bulk materials. These modes are sharply localized at the interface regions and their frequencies lie either above the optical phonon modes (GaAs-like IF) or in the gap between the optical and acoustic modes (InSb-like IF) of both InAs and GaSb. Raman spectroscopy is a convenient technique to detect these "mechanical" IF modes. It allows therefore a direct assessment of the type of IF bonds formed [10,11]. The backfolding of the acoustic phonon dispersion branches onto the reduced Brillouin zone given by the SL period leads to the appearance of doublets of folded longitudinal acoustic (LA) phonon modes in the Raman spectrum [49]. The center frequency of these doublets is determined by the average sound velocity and by the SL period [49].

Fig. 12 shows Raman spectra of a pair of 4 ML InAs/10 ML GaSb SL's with either InSb-like or GaAs-like IF's [48,50]. Besides the longitudinal optical SL phonons (LO(SL)) and the InSb-like and GaAs-like IF modes at frequencies above and below the LO(SL) modes, well-resolved and intense zone-folded LA phonon doublets are observed

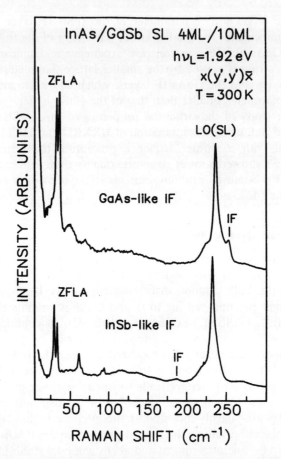

Fig. 12 Room-temperature Raman spectra of a pair of 4 ML InAs/10 ML GaSb SL's with (bottom) InSb-like and (top) GaAs-like IF's. Optical excitation was at 1.92 eV. Polarization of the incident and scattered light was parallel to the same $\langle 110 \rangle$ crystallographic direction $[x(y',y')\bar{x}]$. ZFLA denotes zone-folded longitudinal acoustic phonon modes (Ref. [48]).

for both types of IF bonds. The spectral range of the LO(SL) phonon and IF modes is shown more closely in Fig. 13. For the intended growth of GaAs-like IF's the GaAs-like IF mode is a very prominent feature in the low-temperature Raman spectrum. From the IF mode frequency of 256 cm^{-1} it can be concluded that essentially pure GaAs-like IF bonds were formed with only a negligible admixture of Sb to the Ga–As IF layer [51]. The growth of the IF regions with a shutter sequence promoting the formation of In–Sb IF bonds results in a complete

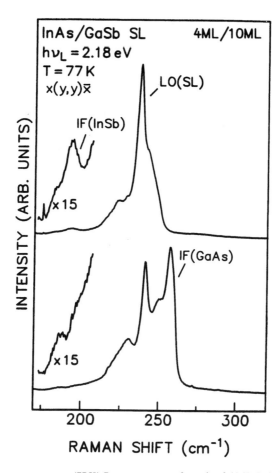

Fig. 13 Low-temperature (77 K) Raman spectra of a pair of 4 ML InAs/10 ML GaSb SL's with (top) InSb-like and (bottom) GaAs-like IF's. Optical excitation was at 2.18 eV. Polarization of the incident and scattered light was parallel to the same $\langle 100 \rangle$ crystallographic direction $[x(y,y)\bar{x}]$ (Ref. [48]).

disappearance of the GaAs-like IF mode and instead in the appearance of an InSb-like IF mode at 190 cm^{-1}. This InSb-like IF mode is, in contrast, not detected for the SL grown with nominally GaAs-like IF's. The present Raman spectra clearly demonstrate that the growth of InAs/GaSb SL's with essentially pure GaAs-like or InSb-like IF's can be achieved and verified by Raman spectroscopy.

Zone-folded LA phonon spectra are shown in Fig. 14 on an expanded frequency scale for three N ML InAs/10 ML GaSb SL's with

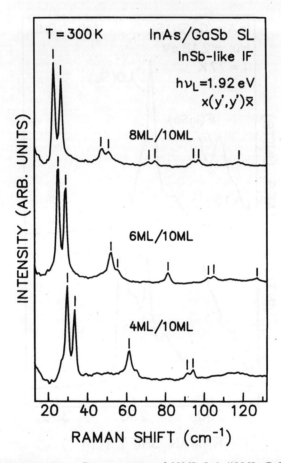

Fig. 14 Room-temperature Raman spectra of N ML InAs/10 ML GaSb SL's with InSb-like IF's and different InAs layer widths N. Zone-folded LA phonon doublets are marked by vertical lines. Optical excitation was at 1.92 eV. Polarization of the incident and scattered light was parallel to the same ⟨110⟩ crystallographic direction $[x(y',y')\bar{x}]$ (Ref. [50]).

InSb-like IF's and InAs layer widths N from 4 to 8 ML, corresponding to SL periods ranging from 14 to 18 ML [50]. Optical excitation was at 1.92 eV to provide maximum resonance enhancement [50]. Well resolved doublets up to the fifth order are observed, which shift to higher frequencies with decreasing SL period. The presence of a large number of zone-folded LA phonon modes indicates a high perfection of the SL periodicity [49]. This behavior is analogous to the known dependence of the number and intensities of SL diffraction satellites observed in HRXRD on the SL quality [49]. In Fig. 15, the frequencies of the zone-folded LA phonon lines are plotted versus the average SL period P_{SL}, determined by HRXRD (see Section 5.3.1), for a series of N ML InAs/10 ML GaSb SL's with InSb-like IF's [50]. For comparison, calculated frequencies of zone-folded LA phonons are also shown as a function of the SL period. The calculations are based on the elastic continuum model with the frequency of the $\pm m$ th mode given by [49]

$$\Omega_{\pm m} = (m\pi/P_{SL} \pm \Delta k) v_{aver}. \qquad (1)$$

v_{aver} is the averaged sound velocity of the SL given by $1/v_{aver} = \alpha/v_{InAs} + (1-\alpha)/v_{GaSb}$; α is defined as d_{InAs}/P_{SL} with d_{InAs} as the individual InAs layer width; v_{InAs} and v_{GaSb} are the appropriate sound velocities of bulk InAs and GaSb, respectively [11]. Δk is the momentum transferred in the backscattering Raman experiment, which is approximately twice the momentum of the incident photon. There is excellent agreement between the experimental data and the calculated frequencies,

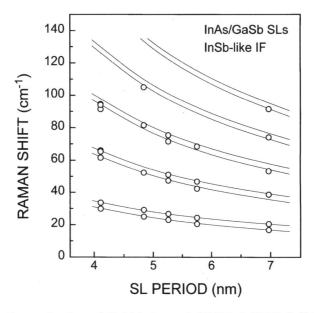

Fig. 15 Frequencies of zone-folded LA phonons in N ML InAs/10 ML GaSb SL's with InSb-like IF's plotted vs. the average SL period determined by HRXRD. Full lines show calculated zone-folded LA phonon frequencies (Ref. [50]).

which demonstrates the consistency of the Raman spectroscopic data and the HRXRD results regarding the SL period.

For InAs/GaSb SL's with GaAs-like IF's, zone-folded LA phonons are observed only up to lower orders than for corresponding SL's with InSb-like IF's [50]. This finding reflects the lower structural quality of the SL's with GaAs-like IF's, which is evident also from the HRXRD analysis (see Section 5.3.1 and Figs. 8 and 9). The frequencies of the zone-folded LA doublets were found to be consistently higher for SL's with GaAs-like than for SL's with InSb-like IF's. This difference has been explained by an increase of the SL sound velocity, compared to the averaged velocity v_{aver} defined above, for the SL's with GaAs-like IF bonds [48,50]. This increase is readily explained by the stronger Ga–As IF bonds [48,50].

For InAs/AlSb heterostructures the situation regarding the possible formation of IF's with both InSb-like and AlAs-like IF bonds is less clear than for InAs/GaSb heterostructures, where controlled growth of both InSb-like and GaAs-like IF's has been demonstrated. Fig. 16 shows a series of room-temperature Raman spectra of three different InAs/AlSb double QW's [36,52]. The spectra were excited at 2.71 eV to provide resonance enhancement for scattering by the InSb-like IF mode [53]. The samples were grown by solid-source MBE on GaAs substrates at a temperature of 410°C using a 1.3 µm thick strain-relaxed AlSb buffer layer. The QW structures were identical except for the shutter sequence used for the growth of the InAs/AlSb IF's. At the heterointerfaces either 1 ML InSb was deposited for the formation of InSb-like IF bonds [9] or 1 or 2 ML AlAs were deposited which in principle should induce the formation of AlAs-like IF bonds [9].

Irrespective of the intended type of IF bonds all samples show scattering by an InSb-like IF mode at 185 cm^{-1} with comparable strength [36,52]. For the sample with 1 ML AlAs deposited at each IF no AlAs-like IF mode is observed at the expected frequencies above the AlSb LO phonon (LO(AlSb)). Instead, only the AlSb LO phonon mode is slightly shifted, to higher frequencies, compared to the AlSb LO phonon in the sample grown with 1 ML InSb at each IF. For the sample grown with 2 ML AlAs at the IF's two modes are resolved in the spectral region of the AlSb LO phonon. The mode at lower frequencies occurs at the same frequency as the AlSb LO phonon in the sample with 1 ML InSb at the IF's. These findings have been interpreted as follows [36,52]. In all three samples InSb-like IF bonds are formed irrespective of the intended type of IF bonds. The addition of AlAs to the InAs/AlSb IF's results in the formation of

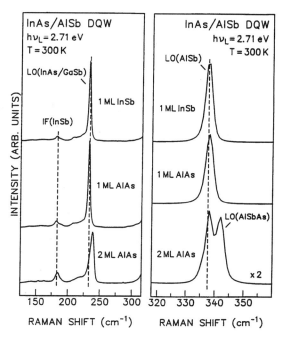

Fig. 16 Room-temperature Raman spectra of three 7.5-nm-wide InAs/AlSb double QW's grown with either 1 ML InSb (top), 1 ML AlAs (middle), or 2 ML AlAs (bottom) at each interface. The spectra were excited at 2.71 eV and recorded in the $[x(y,z)\bar{x}]$ scattering configuration. Note the different energy scales for the left and right portion of the figure (Ref. [52]).

pseudoternary $AlSb_{1-x}As_x$ barriers between and above the two InAs QW's. $AlSb_{1-x}As_x$ is known to show a single-mode behavior with a continuous high-frequency shift of the LO phonon frequency with increasing As content [29]. This mode behavior explains the small high-frequency shift of the LO AlSb phonon observed for the sample with 1 ML AlAs at each IF as well as the appearance of an additional mode at higher frequencies for the sample with 2 ML AlAs per IF. The mode remaining at the frequency found for the AlSb LO phonon in the sample with 1 ML InSb per IF, arises from Raman scattering in the truly binary AlSb buffer layer underneath the InAs QW's, which is not affected by the growth of the IF regions. The high-frequency-shift of the overlapping InAs and GaSb LO phonon lines, the latter arising from the GaSb capping layer, observed for the sample with 2 ML AlAs per IF is most likely caused by strain effects induced by the $AlSb_{1-x}As_x$ barriers [36].

The observation of an InSb-like IF mode in InAs/AlSb heterostructures, irrespective of the intended type of IF bonds, has been explained by a strong exchange interaction between Sb and As at the IF [31]. It has been argued that the Al–As IF bond should be less stable than In–Sb bonds because for AlAs-like IF bonds intermixing at the IF is both thermodynamically favored and driven by a lowering of the residual strain in the IF region [31].

In Raman spectra recorded from InAs/AlSb SL's grown with nominally AlAs-like IF's on either GaSb or InAs buffer layers two closely spaced modes were observed in the spectral range of the AlSb LO phonon [35,54]. The mode at higher frequencies has been assigned to an AlAs-like IF mode [35,54]. However, these SL's showed, like the reference samples grown with InSb-like IF's [35,54], Raman peaks in the frequency range 185–190 cm^{-1} attributed to scattering by an InSb-like IF mode [34]. These findings, together with the before described results, illustrate that control of interfacial bonding is much more difficult to achieve in InAs/AlSb heterostructures than in InAs/GaSb heterostructures.

5.3.3 Far-Infrared Spectroscopy

The heteroepitaxial growth of InAs/AlSb/GaSb heterostructures on GaAs substrates opens a new, quite uncommon possibility for the structural characterization, applying far-infrared (FIR) spectroscopy [55,56]. The optical phonons of these materials are well separated from the reststrahlen band of the GaAs substrate. The typical layer thickness in these quantum wells correspond to an optical thickness on the order of unity in the spectral region of the optical phonons. In the following section it will be shown that optical FIR spectroscopy allows us to gain information about the layer thickness and strain situation.

In Fig. 17 the reflectance spectra of two InAs/AlSb double quantum wells (DQW's) are shown in the spectral region of the optical phonons of the InAs well, the GaSb cap layer, the GaAs substrate, and the AlSb buffer layer. The samples are grown either with InSb-like IF's (upper spectra) or with AlAs-like IF's (lower spectra). Absorbance spectra of the sample with InSb-like IF's measured under two different incident beam angles (0° and 70°) are also included in Fig. 17 covering the spectral region around the AlSb optical phonons (310–345 cm^{-1}). With the angle of incidence near the Brewster angle, the LO-Phonon gets resolved due to the finite p-polarized component of the electric field normal to the sample surface [57].

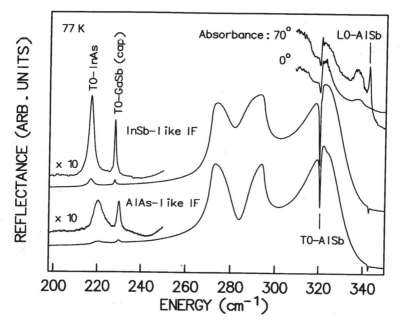

Fig. 17 FIR reflectance (normal incidence) of InAs/AlSb double quantum wells, intentionally grown with AlAs-like and InSb-like interfaces, respectively. The TO phonons of the InAs wells and the GaSb cap are observed. In addition, for the sample with InSb-like interfaces the absorbance spectra recorded at normal incidence and in a Brewster angle geometry are shown in the spectral range of the AlSb LO and TO phonons (Ref. [56]).

In the reflectance spectra of the sample grown with AlAs-like IF's, a broadening and shift to higher energies is observed for both the InAs TO phonon and the TO phonon of the GaSb cap layer as compared to the phonon energies found for the sample with InSb-like IF's. In the case of InAs/AlSb QW's grown on a relaxed AlSb buffer, the InAs wells are under biaxial tensile strain, since the bulk lattice constant of InAs is smaller than that of AlSb. This strain results in a redshift of the InAs TO phonon [55,56] relative to its spectral position in bulk InAs, which is indeed observed for the QW's with InSb-like IF's.

For the intended growth of AlAs-like interfaces it has been found that this growth procedure results in the incorporation of arsenic in the AlSb layers leading to the formation of ternary $AlSb_{1-x}As_x$ barriers [31,36] (see Section 5.3.2). The smaller lattice constant of $AlSb_{1-x}As_x$ compared to that of AlSb results in a less tensile or even compressive strain for the InAs layers, which explains the observed blueshift of the optical phonons.

As discussed above, the spectral position of the optical phonons provides information on the actual strain of the layers. From the full width at half maximum (FWHM) an estimate of the structural quality is gained. Similar information is routinely obtained by performing Raman spectroscopy. In contrast to inelastic light scattering experiments a quantitative analysis of the FIR reflectance or absorbance measurements is much more straight forward to be carried out which allowed us to extract actual layer widths.

From the experimental point of view access to the spectral dependence of the FIR dielectric function of such heterostructures is more easily gained via the measurement of the reflectance as compared to the absorbance [56]. However, the interpretation of reflectance data requires a more sophisticated theoretical modeling of the dielectric function than the evaluation of transmission data. In order to fully understand all the observations, theoretical spectra have to be calculated in the framework of a transfer matrix approach for the propagation of the electromagnetic radiation in a stratified medium, including all the layers deposited onto the substrate. As a result of such a theoretical procedure the actual thickness of the layers can be extracted [48,58] as well as n-type doping levels of contact layers incorporated in the heterostructure [59].

As an example for such a study Fig. 18 shows the reflectance spectra of two series of InAs/GaSb superlattices covering the spectral region around the TO phonons of the SL layers. The SL's consisted of 100 periods with a GaSb layer thickness of 10 ML, which was kept constant throughout all samples. The thickness of the InAs layers was varied in the range between 4 ML and 14 ML. The series of samples were grown with either GaAs-like IF's or InSb-like IF's. The intensity of the TO-phonon signal of the GaSb layers is constant while the intensity of the InAs TO phonon located at $218\,cm^{-1}$ decreases with decreasing layer width.

In Fig. 19 experimental data (full lines) and best ft simulations (dashed lines) of the 14 ML InAs/10 ML GaSb SL's with GaAs-like IF's and InSb-like IF's are shown. The SL grown with GaAs-like IF's shows an additional broad band at about $240\,cm^{-1}$, attributed to the GaAs-like IF mode observed also by Raman spectroscopy (Section 5.3.2). From X-ray and Raman data a significant contamination of the GaSb layers with Arsenic can be excluded. Therefore the strength of this band allows the estimation of the localization length of the IF vibrational mode which was found to be between two and three ML [48].

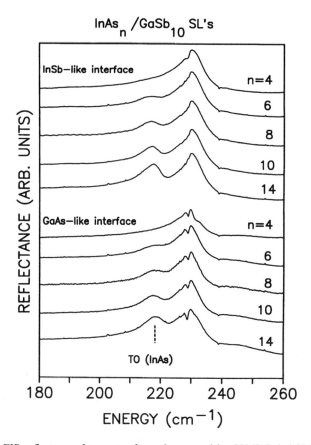

Fig. 18 FIR reflectance of two sets of samples, comprising N ML InAs/10 ML GaSb SLs grown with InSb-like or GaAs-like interfaces. The InAs layer thickness N has been varied from 4 to 14 ML (Ref. [58]).

Similar to what has been observed for the InAs/AlSb QW's the spectral position of the InAs and GaSb TO phonons are shifted relative to their bulk values. These shifts can be related to the internal strain which depends on the type of IF [48]. For example, the spectrally sharp feature in the GaSb TO-phonon band observed for the samples with GaAs-like IF bonds can be explained by a frequency-shift of the GaSb TO phonon of the SL layers relative to the frequency of the TO mode in the GaSb buffer layer due to the strain relaxation of the SL's grown with GaAs-like IF's (see Section 5.3.2). The top part of these SL's are relaxed to a smaller in-plane lattice parameter, which

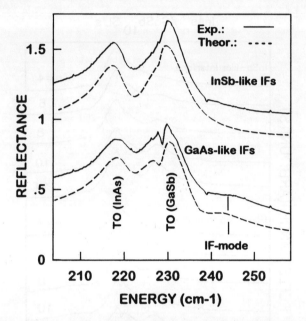

Fig. 19 Comparison between experimental (full lines) and calculated (dashed lines) FIR reflectance spectra of two 14 ML InAs/10 ML GaSb superlattices. Spectra have been offset for clarity. The two upper traces correspond to SL's grown with InSb-like IF's, the two lower curves refer to SL's with GaAs-like IF's. For SL's grown with GaAs-like IF's the contribution of the GaAs-like IF mode has been included in the calculation (Ref. [58]).

induces the frequency-shift of the GaSb TO phonon. This explanation is supported by the observation that the coherently strained SL's grown with InSb-like IF's do not show this sharp feature, except for the sample with the thinnest InAs layers which is known to be partially relaxed.

Besides the information on the strain situation within the SL's obtained from the frequencies of the TO phonons, the actual InAs layer thickness per SL period, determined from a fit to the FIR reflectance spectra can be determined. This thickness is plotted in Fig. 20 versus the nominal InAs layer thickness given by the number of In-containing atomic planes. For a certain nominal InAs layer width in SL's with InSb-like IF bonds the actual width of the InAs layers is smaller by one ML compared to SL's with GaAs-like IF's. Surprisingly, the SL's with GaAs-like IF's show a less intense InAs TO-phonon signal, and thus a smaller actual InAs layer thickness,

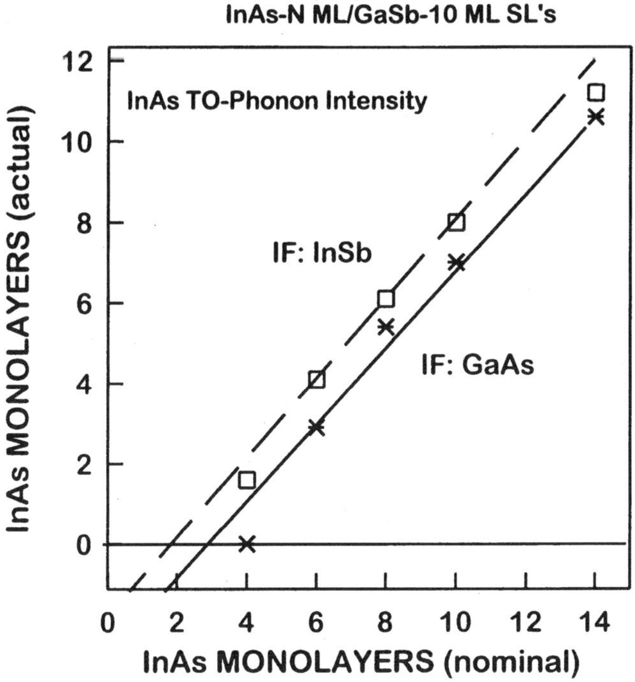

Fig. 20 Actual thickness of the InAs layers in N ML InAs/10 ML GaSb SL's grown with either InSb-like IF's (squares) or GaAs-like IF's (crosses) determined by theoretical fits to the experimental reflectance spectra vs. the nominal thickness (Ref. [48]).

than SL's with InSb-like IF's (see Fig. 18). For both types of IF's the actual InAs layer thickness varies linearly with the nominal InAs layer thickness with an extrapolated actual layer width of zero for a nominal thickness of 2–3 ML. For a given IF type and InAs layer thickness of $\geqslant 6$ ML the actual thickness of the InAs layers, which contribute to the TO-phonon signal, is smaller than the nominal InAs layer thickness by the same amount. For the 4 ML InAs/10 ML GaSb SL's there is an even further reduction of the actual thickness compared to the value expected from the linear extrapolation of the data for $N \geqslant 6$. From the analysis of the HRXRD data (see Section 5.3.1) it is known that the difference between the nominal and the actual SL period is below one ML. Therefore, we have to conclude that certain regions within the individual InAs layer do not contribute to the InAs TO-phonon signal.

The same conclusion has been reached, based on IR absorption data, for InAs/AlSb quantum wells [60,61]. Recently, a possible explanation of this phenomenon has been given [58] on the basis of the assumption of a discontinuous strain distribution within the InAs layers in the vicinity of the IF's. A change of the lattice parameter in the planes facing the IF's results in a shift of the TO-phonon frequency of the InAs layers adjacent to the IF's. Provided that this shift is comparable to the damping of the vibrational mode, the contribution to the reflectance spectrum from the TO-phonon mode in these InAs layers adjacent to the IF's may cancel that from the TO mode in the remainder of the InAs layer. A model of two oscillators spaced by about $5\,\text{cm}^{-1}$ [58] interfering destructively in the reflectivity spectrum explains in particular the absence of any detectable InAs TO-phonon signal in the reflectivity spectra of the 4 ML InAs/10 ML GaSb SL's (see Fig. 18).

A variation of the lattice parameter, and thus of the elastic strain, in the vicinity of the IF's relative to the lattice spacing in the remainder of the layer, has been suggested by Hemstreet *et al.* [17] based on theoretical calculations. These calculations predict in the case of GaAs-like IF bonds an increase of the inter-atomic spacing by 1.8% for the In–As backbonds of the IF. For InSb-like IF bonds a decrease of the length of these backbonds by 1.5% is predicted. These theoretical prediction give support to the above model of two oscillators with slightly different eigenfrequencies [58].

In Fig. 21 the effect of n-type doping on the FIR spectrum of an InAs/GaSb SL is shown. The studied sample (spectrum a) consists of a 150 period InAs/GaSb SL grown on a Be-doped GaSb p-contact layer with the topmost 50 periods of the SL acts as an n-contact due to Si-doping of the InAs layers. Curve (a) shows the spectrum of the doped sample, curve (b) a calculated FIR spectrum fitted to the experimental data, and curve (c) a reference spectrum from a similar but undoped SL. Comparing the spectra from the doped and the undoped SL sample, drastic changes caused mainly by the top n-contact layer are observed.

The contribution of the electron plasma in the n-contact layer causes a high FIR reflectivity near unity. The electromagnetic field interacting with the lattice vibrations undergoes a significant phase shift, which results in derivative-like spectral structures at the TO phonons. From the modeling of such spectra the full structural information on the layer structure can be extracted, as is the case for undoped SL's. In addition, the electron concentration can be evaluated. For the present sample a plasma frequency of $1000\,\text{cm}^{-1}$ has been determined corresponding to an electron concentration of

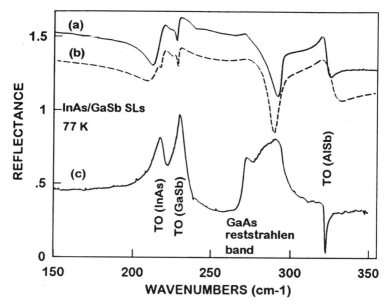

Fig. 21 Curve (a) shows the FIR reflectance of a p-i-n heterostructure, consisting of a p-doped GaSb contact layer, a undoped binary InAs/GaSb SL (100 periods) overgrown with 50 periods of a n-doped SL of same composition. Curve (b) is calculated, curve (c) is from a similar but undoped SL sample. The spectra are shifted by an offset for clarity.

$n = 3 \times 10^{18}$ cm^{-3}. The typical relative error of such a contactless determination of the electron concentration is around 5–10%.

5.4 ELECTRONIC BAND- AND SUBBAND-STRUCTURE

In InAs/AlSb/GaSb heterostructures not only quantization effects modify the band structures of the bulk constituents but also strain effects on the valence and conduction bands, caused by the lattice mismatch between these materials, have to be taken into account. The expected potential profile for the conduction and the valence band of an InAs QW grown on AlSb is drawn in Fig. 22. The InAs layer is subjected to biaxial tensile strain, moving the light hole valence band upwards. The heavy hole (hh)–light hole (lh) splitting is expected to be 190 meV [62]. The hydrostatic component of the strain moves the conduction band downwards, resulting in a band gap (light-hole ⇔ conduction band) of the strained InAs of 287 meV at low temperatures.

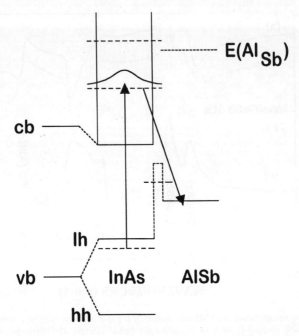

Fig. 22 Band line-up of an 5-nm-thick InAs QW grown on AlSb including the effects of electron confinement, tensile strain and IF states. The single particle picture is shown. The typical electron concentration in such QW's of about $1 \times 10^{12}\,\text{cm}^{-2}$ shift the Fermi level about 100 meV above the lowest electron subband E_1.

The conduction-band discontinuity is 2.0 eV at the Γ-point. In order to calculate the electron confinement correctly, at least a two band model is required because of strong nonparabolicity effects. The confinement in a 7.5-nm-wide QW leads to a position of the first electron subband of 140 meV above the InAs conduction band. The second subband is located by 265 meV further up in energy. There is some scatter of the data in the literature regarding the valence-band offset between InAs and AlSb. Here a value of 100 meV has been taken.

Regarding energies of optical interband transitions, interface states in the forbidden gap may also play a role acting as initial or final states of such transitions. If the InAs QW contains electrons, the single particle picture discussed so far has to be modified. Band filling (Burstein-Moss shift) and band renormalization effects have to be taken into account.

Intentionally undoped InAs/AlSb QW's have been found to contain electrons at concentrations ranging from a few $10^{11}\,\text{cm}^{-2}$ to about

$1 \times 10^{12}\,\text{cm}^{-2}$, depending on the details of the sample structure such as the thickness of the top AlSb barrier layer [63]. Deep donors in the nominally undoped AlSb barriers, InAs/AlSb interface donors, and donor states at the surface of the GaSb capping layer have been identified as sources for these electrons [63]. For thicknesses of the AlSb barrier below 50 nm, the contribution from the surface donors is the dominant one [63]. The mobilities of the electrons in the two-dimensional electron gas (2DEG) formed in the InAs QW were found to depend on the shutter sequence used for the growth of the InAs/AlSb IF region, being significantly higher when 1 ML InSb was deposited at the lower IF [9]. Also the buffer layer sequence influences the 2DEG mobility with low-temperature values exceeding $6 \times 10^5\,\text{cm}^2/\text{Vs}$ achieved at 2DEG concentrations of $1.1 \times 10^{12}\,\text{cm}^{-2}$ when using a GaSb buffer layer followed by a GaSb/AlSb SL [64].

Because of the large conduction-band discontinuity at the Γ-Point of the Brillouin zone of 2 eV across the InAs–AlSb interface the existence of Tamm states has been discussed [65–70], which are intrinsic interface states propagating parallel to the inner surface with an evanescent decay perpendicular to the interface. The energy position of these states has been calculated for InAs/AlSb heterostructures [66]. With an interface thickness of 1 ML InSb these states should be positioned 180 meV above the valence-band maximum of InAs. An increase of the interface thickness to 2 ML should result in a position of 380 meV above the InAs valence-band edge.

Interband transitions across higher lying band gaps, such as the E_1, $E_1 + \Delta_1$, and E_2 gap, can be probed by spectroscopic ellipsometry [71]. As far as InAs/AlSb/GaSb heterostructures are concerned this experimental technique has been applied to InAs/AlSb [42,72] and InAs/GaSb SL's [33] as well as to InAs/AlSb/InAs intraband [73] and InAs/AlSb/GaSb interband tunneling structures [37].

In these pseudomorphically strained heterostructures the higher lying band gaps are affected by the biaxial in-plane strain, similar to the strain effects observed for the fundamental band gap [74]. For pseudomorphic growth along a $\langle 100 \rangle$ direction the k-space degeneracy of the equivalent conduction-band states around the L point (corresponding to the boundaries of the Brillouin zone along the $\langle 111 \rangle$ directions) involved in the E_1 and $E_1 + \Delta_1$ transitions, is not removed [74]. Therefore, these transitions show only strain-induced energy shifts but no splitting.

Similar to the InAs/AlSb heterostructures growth of InAs/GaSb superlattices on a relaxed GaSb buffer leads to biaxially strained InAs layers. As shown in Section 5.3.1 this strain can partially be accommodated by the proper choice of the type of interface. In order to enhance the overlap of the electron and hole wavefunction in the type II InAs/GaSb system the incorporation of In into the GaSb layers has been suggested [7]. Ternary (GaIn)Sb electron barriers are subjected to compressive strain inducing a splitting of the heavy-hole and the light-hole valence bands. The strain-induced energy difference between the two uppermost valence bands easily exceeds 0.1 eV, leading to a strong reduction of the Auger recombination for the p-type material if the SL gap energy is smaller than the valence-subband splitting [75]. For IR detector material with the energy of the fundamental band gap in the 8–12 μm spectral region, the dramatic increase of the minority carrier lifetime should make possible to reach background limited operation at higher temperatures [76,77]. In addition, near room temperature operation of corresponding laser structures can be expected [78].

In the following sub-section 5.4.1 investigations of the electronic properties related to the fundamental band gap will be presented using photoluminescence, calorimetric absorbance, and photocurrent spectroscopy. Measurements of intersubband transitions using inelastic light scattering will be presented in sub-section 5.4.2. Finally, in sub-section 5.4.3, experimental results on interband transitions involving higher lying bands obtained with ellipsometry will be discussed.

5.4.1 Interband Spectroscopy of the Fundamental Energy Gap

5.4.1.1 Infrared Photoluminescence of InAs/AlSb Heterostructures
Photoluminescence provides information on the energy of the effective band gap in type II heterostructures provided that indeed band-to-band transitions are observed. In order to exclude impurity related transitions, it is useful to combine the PL experiment with absorbance measurements. A powerful technique to measure the weak absorbance due to type II optical transitions is calorimetric absorbance spectroscopy (CAS) [79]. A second technique is Fourier-transform photoluminescence excitation (FTPLE) spectroscopy [80,81]. As will be shown below, the first technique provides excellent results even in single QW's, whereas the second technique is restricted to multiple QW's with very high sample quality optimized for emission experiments. However, a drawback of the CAS technique is that the sample has to be cooled down to temperatures of 500 mK or less.

The PL measurements presented in the following were performed by Fourier-transform spectroscopy, using double modulation techniques as described elsewhere [82]. For detection, an InSb photodiode operating at 77 K was used, which defines the low energy cut off. Fig. 23 shows the mid-infrared PL spectra from a series of InAs/AlSb QW samples with varying InAs QW widths. The solid curves indicate spectra recorded for 2.41 eV excitation, while the dotted lines correspond to excitation with a photon energy of 0.94 eV. Dashed vertical lines indicate the low energy onset of the PL as expected from self-consistent two band calculations, assuming a valence-band offset

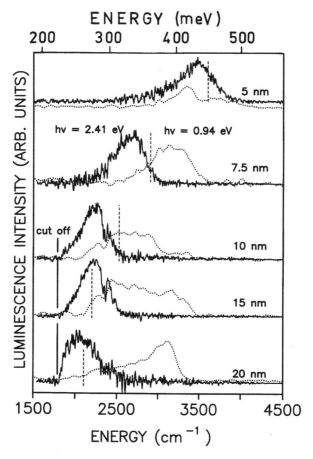

Fig. 23 Spatially indirect PL of InAs/AlSb single quantum wells. The QW width is indicated. The dashed vertical lines correspond to the expected low energy onset of the luminescence. (Solid lines: 2.41 eV excitation, dotted lines: 0.94 eV excitation) (Ref. [83]).

of 100 meV [83]. The spectra show the expected blueshift due to the increasing electron confinement with decreasing QW width. Under visible excitation (2.41 eV) the spectral peaks are all shifted by approximately 80 meV to energies below the expected transition energy. The spectral shape was found to remain unchanged when the excitation density of the 2.41 eV radiation was varied between 10 and 50 W/cm^{-2}. From this observation we conclude that the electron concentration is not changed significantly upon excitation in the visible spectral range.

For excitation at 0.94 eV a broadening and a blueshift of the spectra (dotted lines) is observed, attributed to an increase of the electron concentration in the InAs QW's, pushing the quasi-Fermi level further up into the conduction band. This optically induced change of the electron concentration is known from transport measurements, where an enhancement of the electron concentration of the order of 1×10^{12} cm^{-2} is observed upon IR illumination [84–86]. For the two wider quantum wells (15 and 20 nm) the second subband is calculated to be separated from the first subband by 118 meV and 80 meV, respectively. These values agree very well with the second, higher-energy component of the PL in the above two samples. The broadening of the spectra towards higher energy is limited to a maximum transition energy of about 420 meV. This holds even for the case of 0.94 eV excitation of the thinnest (5 nm) QW sample, where visible excitation leads to PL emission at energies above that limit.

Further measurements, shown in Fig. 24, demonstrate the influence of increasing electron concentration on the PL spectra while the excitation conditions remain constant. Three samples with a constant well width of 15 nm and varying electron concentrations were investigated (0.5×10^{12}, 1×10^{12} and 2.6×10^{12} cm^{-2}). The lowest carrier concentration, was achieved by increasing the AlSb top barrier width [87] from 10 nm to 40 nm, which reduces the electron transfer from the GaSb top layer into the QW. The highest concentration of 2.6×10^{12} cm^{-2} was realized by introducing an additional As doping spike into the 10-nm-thick top AlSb barrier. Excitation was performed at a photon energy of 2.41 eV. The vertical dashed lines in Fig. 24 correspond to the calculated transition energy, where the low energy onset of the PL lines is expected. The sample with the highest electron concentration indeed shows that onset of the PL. The high-energy side of the emission band is limited to 420 meV as already observed in Fig. 23. A decrease of the carrier concentration is accompanied by a redshift of the spectra. Comparing the PL spectra with the calculated

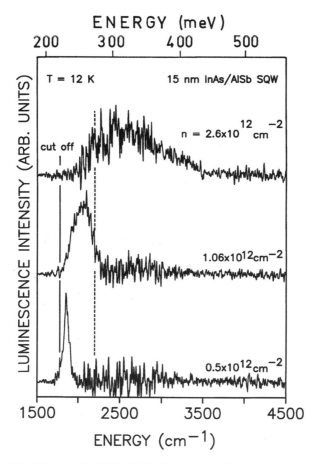

Fig. 24 PL of 15-nm-wide SQW's (2.41 eV excitation). The electron concentration is indicated. Dashed vertical lines correspond to the expected low energy onset (Ref. [83]).

transition energies, a systematic redshift of about 80 meV is observed. This redshift is consistent with the model of interface states positioned 80 meV above the valence-band maximum of AlSb [66]. A further discussion of the influence of IF states on the optical transition will follow below in sub-section 5.4.1.2.

Transport measurements on the present SQW samples show, that in contrast to the QW's with a well width larger than 7.5 nm, the sample with the narrowest QW (5 nm) was highly resistive. This observation is in close agreement with that of Ideshita *et al.* [87] where a decrease of the electron concentration to only $1 \times 10^8 \, \text{cm}^{-2}$ was measured for a 10-nm-wide $Al_{0.5}Ga_{0.5}Sb/InAs/Al_{0.5}Ga_{0.5}Sb$ SQW. These authors attributed the electron accumulation found in wide wells to a deep defect level in the ternary barriers, which is responsible for remote doping of the QW's if the confinement energy is below the energy of that defect level. In narrow QW's the ground state energy of the electrons is raised above the deep level, resulting in high resistivity. This model has been refined by Shen et al. [88], who suggested the cation-on-anion antisite to be the defect responsible for the remote doping. In the case of InAs/AlSb heterostructures this defect (Al_{Sb}) is predicted to produce a deep level which is located about 400 meV above the AlSb valence-band maximum. The cross-over between highly conductive and highly resistive InAs/AlSb QW's was predicted to occur at a well width of 6.3 nm [88], which is in agreement with our observation of a highly resistive 5-nm-thick SQW. This model explains why the high-energy extrema of the PL peaks (dotted lines in Fig. 23) do not exceed a value of 420 meV. At this transition energy the Fermi-level in the QW becomes resonant with the deep level in the AlSb barrier (see insert in Fig. 23). A further high-energy shift of the quasi-Fermi level into the conduction band above the defect energy is not possible, because of an electron transfer into the barrier material populating the deep level.

5.4.1.2 Calorimetric absorbance
In Fig. 25 PL and calorimetric absorbance spectra (CAS) of three samples consisting of two 7.5-nm-wide InAs QW's separated by 5-nm-thick AlSb barriers is shown. The nominal thickness of the InSb-like IF's of the three samples was varied between 1 ML InSb and 3 ML InSb applying monolayer epitaxy. The PL spectra exhibit a significant redshift with increasing IF thickness, despite the nominally identical QW width. We note that two sets of samples have been grown leading to identical PL spectra.

As expected for a type II transition, the CAS spectra show a smooth onset of the absorbance approximately at the spectral position of the PL with an increasing slope at higher energies. A strong enhancement of the absorbance is observed at 600 meV for the two double QW's with 2 and 3 ML InSb IF's. A similar enhancement is observed at 720 meV for the double QW grown with 1 ML InSb IF's.

In an absorbance experiment the spatially direct transitions are expected to dominate the spectra because of the larger oscillator strength compared to the type II transitions across the IF's. The transition energy of the latter are affected by the valence-band offset and interface states, while the spatially direct transitions are only

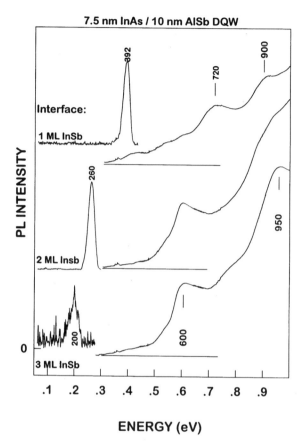

Fig. 25 PL (2.41 eV excitation) and calorimetric absorbance of 7.5-nm-wide InAs DQW's separated by a 5-nm-thick AlSb barrier. The nominal thickness of the InSb-like IF's of the three samples has been varied between 1 ML InSb and 3 ML InSb (Ref. [91]).

sensitive to confinement and strain effects. We have to keep in mind that in the InAs layers the ground state energy of the light holes is increased by confinement (see Fig. 22). This picture becomes even more complicated because of many-body effects related to the high electron concentrations in the InAs layers. In the present structures the Burstein-Moss shift related to a typical 2D electron concentrations of 1×10^{12} cm^{-2} is of the order of 100 meV leading to a blueshift of optical transitions for carriers with $k = k_f$ (k_f: Fermi momentum).

The emission of the three double QW's show up at 392 meV, 260 meV, and 200 meV for, IF widths of 1 ML, 2 ML, and 3 ML, respectively (Fig. 25). A theoretical study predicts a Tamm-state energy of 180 meV above the valence-band maximum of InAs for a 1 ML InSb interface [66], which is expected to be raised to 380 meV upon widening of the InSb-like IF to 2 ML. For a PL transition involving Tamm states as the final state these calculation predict a redshift of 200 meV. Assuming a nonperfect IF with a thickness between 1 and 2 ML for the QW grown with the thinnest IF, the observed peak positions and the redshift of 130 meV are in reasonable agreement with the above model, involving Tamm states as the final state of the emission transition.

Because of the complicated situation outlined above (see Fig. 22) the interpretation of the CAS data can only be preliminary. The onset of the CAS absorbance is almost identically observed for all three samples at 300 meV close to the expected value of 327 meV [89]. At 500 meV the slope of the absorbance spectra increases, most significantly in the spectra of the samples grown with 2 and 3 ML InSb IF's. At this threshold spatially direct transitions set in involving confined light-hole states. The onset of the spatially direct absorbance due to the heavy-hole valence band with the values given above is expected to appear at 600 meV, which may be blueshifted due to band filling effects (Burstein-Moss shift). In the two lower spectra the most significant spectral structures are observed around 600 meV, while in the topmost spectrum a similar structure is located at 720 meV. The energetic difference to the PL line is 330 meV, 340 meV and 400 meV, comparing the sample with 1, 2 and 3 ML InSb IF's, respectively. Taking into account the influence of the interface thickness in order to explain the differences among the PL lines, it is reasonable to assume that in the CAS spectra indeed the heavy hole transitions at 600 meV are observed, providing evidence for an hh–lh splitting of 190 meV.

5.4.1.3 *Optical anisotropy*

Heterostructures grown along the [100] direction are expected to be optically isotropic in the layer plane. An anisotropy can arise in the z-direction because of lifting of the degeneracy of the heavy hole and the light hole valence bands. The hh–lh splitting can be induced either by strain or confinement effects. In InAs/AlSb heterostructures the situation is different because both, the group III and group V elements change across the IF. In contrast to the well known GaAs/(AlGa)As system the interface bonds of the normal (AlSb on InAs) IF are contained in [01−1] planes, in contrast to the bonds of the inverted (InAs on AlSb) IF which are all contained in [011] planes (see Fig. 26). In order to promote the

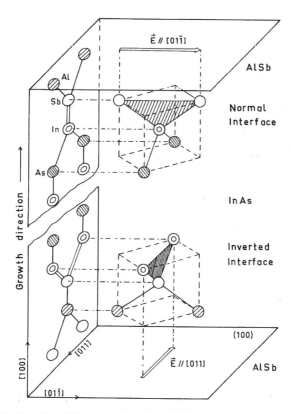

Fig. 26 Configuration of the tetrahedra which build up the normal (AlSb on InAs) and the inverted interface (InAs on AlSb) of a sample grown in [100] direction. The planes containing the InSb-like bonds of the heterointerfaces are drawn shaded. These planes are orthogonal to the (100) sample surface. The projection of these planes on the sample surface is parallel to the [01$\bar{1}$] direction for the normal interface and parallel to the [011] direction for the inverted interface. In addition the projection on a (01$\bar{1}$) plane is shown in the left part of the figure. In this part the projection of the InSb-like bonds is indicated with double lines (Ref. [91]).

InSb-like interface the shutter sequence during growth of the normal interface is inverted compared to the sequence during growth of the inverted interface (see Section 5.2). The inversion of the layer sequence at the interfaces results in a reduction of the SL symmetry from D_{2D} to C_{2v}. Therefore, optical transitions involving the IF's can be expected to translate this microscopic situation into an in-plane anisotropy.

An optical anisotropy in InAs/AlSb heterostructures was first observed by Santos et al. [90] who performed ellipsometry and reflection difference spectroscopy. The strong effect of inequivalent interfaces on the polarization of spatially indirect emission has first been demonstrated on wide InAs layers grown on AlSb [91]. In these heterostructures the contributions of the individual interfaces to the PL spectra could be discriminated using different excitation energies in the PL measurements [91]. Later on, a similar anisotropy has been observed in the PL spectra of type II InP/(InAl)As heterostructures, where the same physical mechanism has been used for explaining the effect [92].

The effect of a strong optical in-plane anisotropy in the PL spectrum of InAs/AlSb type II heterostructures is shown in Fig. 27. For the k-vector of the light propagation perpendicular to the growth surface ($k_\parallel$ growth direction z) a significant dependence of the PL intensity on the in-plane polarization is found (Fig. 27, left). The emission intensity is much stronger for the polarization of the emitted light parallel to the y direction than for polarization parallel to the x direction, where x and y denote two orthogonal (011) type crystallographic orientations. This experimental finding has been confirmed with the emitted light propagation parallel to the x and y direction (see Fig. 27 middle and right). For $k \parallel x$ the emission polarized parallel to the growth plane (y-direction) is much more intense than that polarized perpendicular to the growth

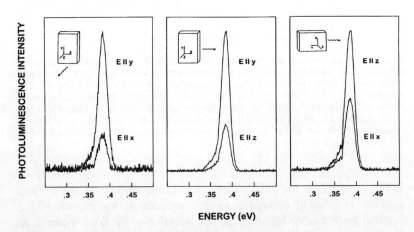

Fig. 27 Polarization resolved PL of an InAs/AlSb DQW consisting of 7.5-nm-wide InAs wells separated by a 5-nm-thick AlSb barrier. Three different geometries are shown. Left part shows emission with k parallel to the growth direction z, middle part k parallel to the layer plane x, right part also shows emission with k parallel layer plane, but the sample has been tilted by 90 degree. (2.41 eV excitation, temperature was 10 K) (Ref. [108]).

plane (z-direction). For $k \parallel y$ the situation is reversed, i.e., the PL polarized parallel to the growth plane (x-direction) is much weaker than that polarized parallel to the z-direction. Therefore, taking the intensity of the emission for polarization parallel to the z-direction as a reference, we conclude that again emission polarized parallel to y is much stronger than emission polarized parallel to x. Thus, the intensity of the different polarized PL components is directly correlated with the crystallographic orientations. We note that this optical anisotropy is found for any InAs/AlSb quantum-well structure grown with thick enough AlSb barriers between the InAs wells to prevent a coupling of the hole wavefunctions throughout the QW structure. Decoupled hole wavefunctions localized at the individual IF's are a necessary condition to distinguish between the radiative recombination across the normal and the inverted interface.

The question arises, which mechanism translates the microscopic symmetry of the interfaces into an orientation of the recombining electron–hole pairs in the layer plane and thus a polarization of the emitted photons. One possibility is the existence of local electric fields established by the IF dipoles oriented along the IF bonds. The tetrahedra building up the interface bonds are rotated by 90 degrees relative to each other for the normal and inverted IF (see Fig. 26). Due to the different orientation of the tetrahedra the projections of the dipole moments of the two interfaces onto the (100) growth surface are also orthogonal to each other. Therefore a local electric field oriented in the layer plane is built up at the individual interfaces according the projection of the interface bonds onto the growth plane.

The spatially indirect PL of a type II system selectively probes electrons and holes with a finite overlap of the wavefunctions across the interfaces. Thus, only electron–hole pairs localized at the interfaces contribute to this emission. These electron–hole pairs are most sensitive to a dipole moment built up by the interface bonds. Even if the total optical anisotropy in the heterostructure is low, the optical anisotropy observed via PL can be high. We only have to assume a different radiative efficiency for the two types of interfaces. The degree of polarization reflects the ratio of the radiative efficiency of the two types of interfaces, which is easily explained by the different structural quality of the two types of interfaces. From Raman experiments there is some evidence, that growth of AlSb on InAs produces the better defined IF [36].

An second explanation might be offered in view of the existence of Tamm states or extended IF states. Wavevectors of these states contain the information of the orientation of the individual IF's. Provided that IF states contribute to the PL process the local symmetry is introduced into the transition matrix element. As outlined above, the spectral position and the redshift of the PL lines shown Fig. 25 provide some evidence, that Tamm states are involved in the emission process.

5.4.1.4 Photocurrent spectroscopy on InAs/GaSb superlattices

The spectral response of the in-plane photocurrent of InAs/GaSb SL's is shown in Fig. 28. The thickness of the GaSb layers was kept constant (10 ML). The thickness of the InAs layers has been varied between 6 and 14 ML. Spectra are shown for two series of samples

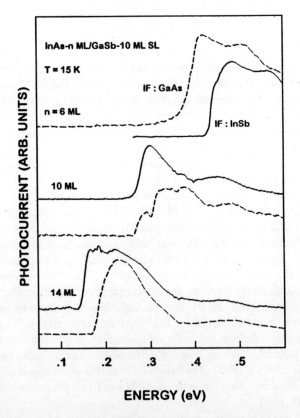

Fig. 28 Photocurrent spectra of N ML InAs/10 ML GaSb SL's grown either with GaAs-like IF's (dashed lines) or with InSb-like IF's (full lines).

grown with either InSb-like IF's or GaAs-like IF's. The band gap shifts from 0.45 eV to a value around 0.1 eV according the decrease of the electron confinement by widening the InAs wells. The onset of the spectra differs for the two types of IF's. The effective band gap of the samples with 6 ML InAs is at a higher energy for the SL grown with InSb-like IF's. The opposite is observed for wider InAs layers. The dependence of the band gap on the InAs layer width and interface type is plotted in Fig. 29. The spectral position of the PL lines of these SL's fit excellently to the band gaps observed in the photocurrent spectra [93,94]. However, the SL's grown with GaAs-like IF's exhibit strongly reduced PL intensities, due to the reduced structural quality evidenced by X-ray diffraction measurements (see Section 5.3.1).

Theoretical considerations [17] suggest a significant change of the valence-band offset between InAs and GaSb induced by different dipoles across the interface generated by the GaAs or the InSb bonds, respectively. Growing the SL's with an InAs layer thickness below 10 ML leads to an enhancement of the relative influence of the interface type. The extrapolation of the present data obtained from coherently strained SL's to a InAs layer thickness of 3 ML leads to a

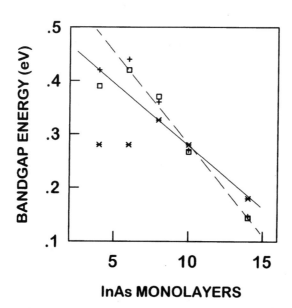

Fig. 29 Energetic position of the effective band gap of InAs/GaSb SL's grown either with GaAs-like IF's (PC: ✶)or with InSb-like IF's (PC: +, PL: ◻). (Ref. [93]).

difference of the band gaps of 100 meV, in excellent agreement with theoretical predictions [17]. The experimental data deviate from this extrapolated dependence for InAs layer width below 6 ML. The strong internal strain leads to relaxation of the SL stack, as shown by X-ray diffraction. In Fig. 29 these relaxed SL films show a lowering of the effective band gap by about 100 meV compared to that expected for coherently strained SL's [93].

5.4.2 Intersubband Spectroscopy by Inelastic Light Scattering

Optical spectroscopy of intersubband transitions in QW's is a powerful method to study the electron subband structure including the effect of nonparabolicity on the subband energies. Because of the small fundamental band gap of InAs of 0.35 eV at 300 K and the large conduction-band offset between InAs and AlSb of 1.35 eV between the Γ-conduction-band minimum in InAs and the X-conduction-band minima in AlSb and of 2.0 eV with respect to the Γ-conduction-band minimum of AlSb[1], quantization energies for electrons in sufficiently narrow InAs/AlSb QW's may become comparable to the InAs band gap energy as shown in Fig. 30. Therefore nonparabolicity of the InAs conduction band has a strong effect on the electron effective mass and thus on the quantization energies [89,95].

The most direct way of intersubband spectroscopy is to perform an IR transmission experiment, with a component of the electric field of the incident light parallel to the quantisation axis to fulfill the polarization rules for intersubband transitions. Such experiments give the energy of the collective intersubband plasmon mode, which is depolarization shifted to higher energies with respect to the bare intersubband transition energy. Intersubband absorption experiments have been performed on InAs/AlSb QW's with an applied magnetic field [96,97]. It has been shown that with the magnetic field oriented parallel to the growth plane, i.e., perpendicular to the quantization axis, IR absorption by both spin-conserving charge-density intersubband excitations and spin-flip intersubband excitations can be observed simultaneously [97]. The observation of the latter type of excitation has been shown to be induced by the bulk inversion asymmetry of InAs [97]. The energy of these spin-density excitations involving spin-flip essentially equal the bare subband spacing.

Inelastic light scattering by electronic intersubband excitations has found widespread use for the study of GaAs/(AlGa)As quantum systems [98]. There the fundamental band gap E_0 and the $E_0 + \Delta_0$

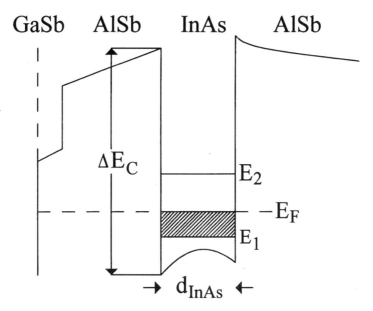

Fig. 30 Schematic conduction-band profile of an intentionally undoped GaSb-capped InAs/AlSb single QW.

band gap of the GaAs QW are accessible by resonant Raman scattering. As these gaps involve the Γ-conduction-band minimum, where the electrons are located, inelastic light scattering by collective charge-density and spin-density excitations [98,99], as well as by true single-particle intersubband transitions [100] can be observed. For InAs QW's, however, only the E_1 and $E_1 + \Delta_1$ band gaps lie in the spectral range accessible in resonant Raman experiments [101]. Thus only coupled longitudinal-optical (LO) phonon-intersubband plasmon excitations, which may couple to the light via the electro-optic, deformation-potential, and Fröhlich mechanisms, can be observed in Raman scattering from InAs QW's [99]. Experimentally both low- and high-frequency coupled intersubband plasmon-phonon modes have been resolved for semiconducting InAs/AlSb QW's [53,102] and for semimetallic InAs/GaSb QW's [103] using optical excitation in resonance with the E_1 gap of InAs.

Figs. 31 and 32 show, respectively, the low-frequency and high-frequency portions of room-temperature Raman spectra recorded from InAs/AlSb single QW's with various well widths. Optical excitation was at 2.6 eV, close to the E_1 gap energy of InAs [101], to resonantly

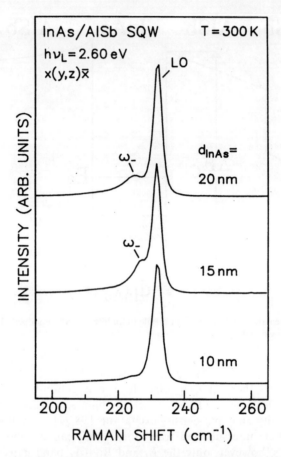

Fig. 31 Low-frequency portions of room-temperature Raman spectra from InAs/AlSb single QW's of different well widths d_{InAs}. The spectra were recorded with the polarization of the scattered light perpendicular to that of the incident light $[x(y,z)\bar{x}]$, where x, y, and z denote [100], [010], and [001] crystallographic directions (Ref. [102]).

enhance scattering by coupled LO-phonon-intersubband plasmon modes [53,102,103]. For well widths $\geqslant 15$ nm the low-frequency coupled LO-phonon-intersubband plasmon mode ω_- and the high-frequency coupled intersubband mode ω_+ are resolved at frequencies around $225\,\text{cm}^{-1}$ and in the range from 1000 to $1400\,\text{cm}^{-1}$, respectively. The ω_+ mode shifts to higher frequencies with decreasing well width reflecting the increase in subband spacing. The scattering strength of both modes decreases with decreasing well width, and for

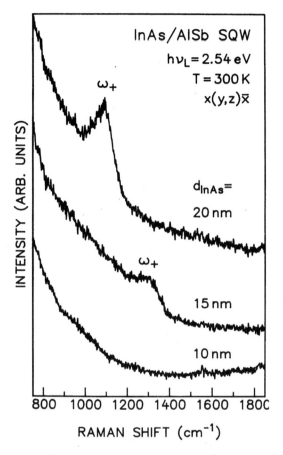

Fig. 32 Same as Fig. 31 for the high-frequency portions of room-temperature Raman spectra from InAs/AlSb single QW's of different well widths d_{InAs} (Ref. [102]).

a width of 10 nm no intersubband mode is observed. The LO phonon signal at frequencies slightly above 230 cm^{-1} arises mostly from the InAs QW with some possible contribution from the GaSb capping layer because of the near coincidence of the LO phonon frequencies in these two materials.

When only the lowest electron subband is significantly populated, the energy spacing between the first and second conduction subband E_{12} and the depolarization shift E_P can be deduced from the energies E_+ and E_- of the intersubband plasmon modes ω_+ and ω_-, when the energies of the LO and transverse optical (TO) phonon in the InAs well are known [98,102]. Intersubband plasmon energies were calculated, taking into account nonparabolicity and finite temperature [104], and fitted to the experimental values for E_+ and E_- to determine the subband spacing E_{12} at the Brillouin zone center ($k_\parallel = 0$) and E_p [102]. The resulting subband spacings are plotted in Fig. 33 versus the nominal width of the InAs QW for samples grown on an AlSb buffer (filled circles) as well as for QW samples grown on a GaSb buffer layer (open triangles) [102]. For comparison, subband spacings at the zone center ($k_\parallel = 0$), calculated for a square InAs QW with finite

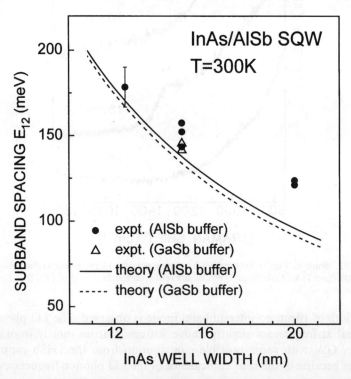

Fig. 33 Subband spacing E_{12} at $k_\parallel = 0$ vs. nominal InAs well width for InAs/AlSb single QW's derived including nonparabolicity effects. Data are shown for samples grown on AlSb buffer layers (filled circles) or GaSb buffer layers (open triangles). Error bars are indicated for the narrowest well width. Theoretical predictions taking into account the effects of strain and nonparabolicity are shown for InAs QW's elastically strained to the in-plane lattice parameter of AlSb (drawn curve) or GaSb (dashed curve) (Ref. [102]).

barrier height and including the effects of strain and nonparabolicity, are also shown for the QW pseudomorphically strained to the in-plane lattice parameter of AlSb (drawn curve) and GaSb (dashed curve) [102]. The theoretical calculations predict, for a given well width, a somewhat smaller subband spacing for QW's grown on a GaSb buffer than for QW's grown on an AlSb buffer because of the different strain situation in the InAs layer [89,102]. However, the scatter of the experimental data is too large to resolve this difference unambiguously [102]. The discrepancy between experiment and theory observed for the samples with a well width of 20 nm might be due to a significant population of the second electron subband which was neglected in the analysis of the Raman data.

Provided that only the lowest electron subband is occupied, the two-dimensional electron concentration N_{2D} can be deduced from the depolarization shift E_p using the relation [98] $E_P^2 = (2e^2/\varepsilon \varepsilon_0) L_{12} E_{12} N_{2D}$. Here, ε is the high-frequency dielectric constant of the QW material, ε_0 the vacuum permittivity, and L_{12} the Coulomb matrix element [98]. The electron concentrations determined this way from the Raman measurements were found to depend critically on the optical power density (2.5 eV photons) used for exciting the Raman spectra. For a 20-nm-wide QW sample grown on an AlSb buffer the N_{2D} was found to increase from 1.1×10^{12} to 2.1×10^{12} cm^{-2} when the excitation power density increased from 200 W/cm^2 to 2 kW/cm^2 [102]. These values have to be compared with the results of room-temperature Hall effect measurements which gave an electron concentration of 1.7×10^{12} cm^{-2} [102]. The lower electron concentration observed for low optical excitation densities, compared to the Hall effect data obtained in the dark, can be explained by a negative photo-effect [84]. Optical excitation with photon energies > 1.55 eV has been found to lead to a drastic decrease in the electron concentration [84]. This negative photo-effect, which results from the spatial separation of photogenerated electron–hole pairs in a built-in space charge field [84], saturates already at low excitation densities when the built-in field is screened by photogenerated carriers. The observed increase in electron concentration with increasing optical excitation density has been attributed to a filling of the InAs conduction band with photogenerated electrons [102]. The subband spacing E_{12}, in contrast, was found to change only very little from 121 to 118 meV upon an increase of the electron concentration from 1.1×10^{12} to 2.1×10^{12} cm^{-2} [102]. This constancy of the subband spacing demonstrates that space charge effects have only a minor influence on E_{12} for the InAs/AlSb QW's because of the large conduction-band offset, which determines the potential profile almost exclusively.

The present example shows that Raman scattering by intersubband plasmon-phonon modes allows an accurate determination of the electron subband spacing in InAs/AlSb QW's. When extracting information on the electron concentration from the Raman spectra, however, care has to be exercised in relating these data to results of transport measurements performed in the dark, because illumination with visible light may cause strong negative and positive photo-effects in InAs/AlSb QW's.

5.4.3 Ellipsometric Spectroscopy of Higher Lying Band Gaps

Fig. 34 shows the imaginary part of the dielectric function of unstrained bulk InAs and for 35 and 11-nm-thick InAs layers

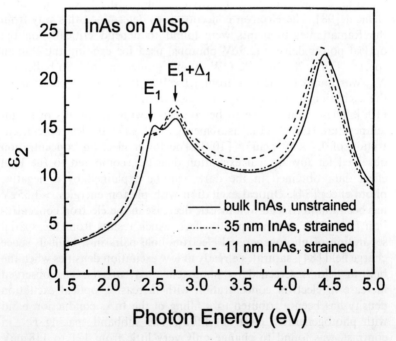

Fig. 34 Imaginary part of the dielectric function (ε_2) of unstrained bulk InAs as well as 35-nm-thick and 11-nm-thick InAs layers, which are under biaxial in-plane tension due to pseudomorphic growth on (100) AlSb. Data were derived from room-temperature ellipsometric measurements.

pseudomorphically strained to the larger in-plane lattice parameter of AlSb. The data were derived from room temperature ellipsometric measurements employing a multilayer model which uses a parametric ansatz for the dielectric function of the InAs layer [73]. The E_1 interband transition is shifted to lower energies by 85 meV for the strained layers whereas the $E_1 + \Delta_1$ transition energy remains constant within the experimental accuracy. The observed behavior is consistent with theory which predicts that for biaxial in-plane tensile strain both inter- and intraband effects lead to a lowering of the E_1 gap energy [74]. For the $E_1 + \Delta_1$ band gap, in contrast, these two contributions have opposite sign and thus effectively cancel each other [74]. The peak around 4.5 eV, which arises from contributions of the E_2 interband transition and transitions across the E'_0 band gap, shows a shift to lower energies by about 40 meV for the strained InAs layers. An analysis similar to that for the E_1 and $E_1 + \Delta_1$ band gap is difficult because of the superposition of different interband transitions.

Fig. 35 shows the imaginary part of the pseudodielectric function for two pairs of N ML InAs/10 ML GaSb SL's with either InSb-like or GaAs-like IF's and InAs layer widths of 4 and 14 ML [33,105]. The data were again derived from room-temperature ellipsometric measurements, Two interband transitions are resolved. They occur for the 14 ML InAs/10 ML GaSb SL's at 2.0 and 2.4 eV, irrespective of the type of IF bonds [33]. These SL's are coherently strained to the in-plane lattice parameter of the strain-relaxed GaSb buffer leaving the InAs layers within the SL under biaxial tension (see Section 5.3.1 and Fig. 11) [48]. The intensities and widths of these interband transitions, however, depend on the type of IF bonds with the transitions being less well resolved for the SL with GaAs-like IF's. The latter observation parallels the larger HRXRD peak width found for the SL's with GaAs-like IF's (see Section 5.3.1 and Fig. 9) indicating a lower structural quality of SL's with GaAs-like IF bonds [48].

The 2 eV interband transition can be assigned to the E_1 band gap of GaSb. The 2.4 eV transition arises from a superposition of contributions from the E_1 band gap of InAs and the $E_1 + \Delta_1$ band gap of GaSb. For the 4 ML InAs/10 ML GaSb SL's these transition energies were found to depend on the type of IF bonds (see Fig. 35) [33]. The transitions are 30–50 meV lower in energy for the SL with InSb-like IF's than for the SL with GaAs-like IF's [33]. This shift of the interband transitions can be related to the fact that both SL's are strain-relaxed with lattice parameters significantly smaller (GaAs-like IF's)

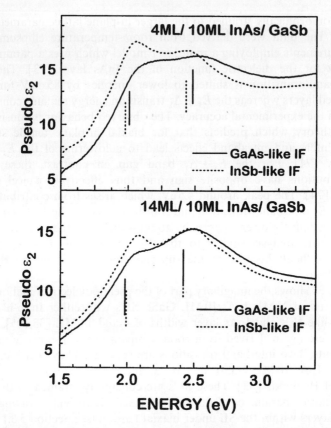

Fig. 35 Imaginary part of the pseudodielectric function of 4 ML InAs/10 ML GaSb SL's (top) and 14 ML InAs/10 ML GaSb SL's (bottom) with either InSb-like or GaAs-like IF's. Critical point energies are indicated by vertical lines. Data were derived from room-temperature ellipsometric measurements (Ref. [105]).

or slightly larger (InSb-like IF's) than that of the GaSb buffer (see Section 5.3.1 and Fig. 11) [48]. As expected, the SL with InSb-like IF's, which has the larger volume of the reduced unit cell, shows the lower transition energies. The energetic position of the 2 eV interband transition averaged over the two types of IF bonds is essentially independent of the InAs layer width [33]. The averaged position of the 2.4 eV transition, in contrast, increases by about 40 meV when the InAs layer width is reduced from 14 to 4 ML [33]. This increase in transition energy has tentatively been explained by quantization effects of the L point electrons in the InAs layers [105].

5.5 RESULTS ON DEVICES

The final step in the characterization of InAs/AlSb/GaSb heterostructures is the testing of devices fabricated from these heterostructures. In the following two examples of devices prepared from optimized epitaxial layer structures will be discussed. The first example deals with p-i-n IR photodiodes, the active region of which consists of an InAs/GaSb SL [93,105]. As the second example results on InAs/AlSb/GaSb resonant interband-tunneling (RIT) diodes employing either AlSb or Al(SbAs) barrier layers will be presented [37,106].

The layer structures used for the preparation of the p-i-n SL photodiodes were grown by solid-source MBE on undoped semi-insulating (100) GaAs substrates using a 1.1 μm thick strain-relaxed GaSb buffer layer. On top of this buffer layer a 0.5 μm thick Be-doped GaSb p-contact layer was grown, followed by 100 periods of an undoped InAs/GaSb SL and a 50 period InAs/GaSb SL with Si-doped InAs layers acting as the n-contact. Both SL's were grown with InSb-like IF's and the temperature for the growth of the SL regions was 410°C. Unpassivated mesa structures were fabricated using standard photolithography and a combination of wet chemical etching and chemically assisted reactive ion etching (CAIBE). Fig. 36 shows the photovoltaic (PV) response and the electroluminescence (EL) spectra of a 8 ML InAs/10 ML GaSb p-i-n diode recorded for various temperatures. The PV response exhibits a well defined onset which is at 0.36 eV (3.4 μm) at low temperatures. The EL spectrum shows a narrow peak at the same energy as the PV onset arising from radiative recombination in the undoped SL region. With increasing temperature the onset of the PV response shifts to lower energies with a cutoff at 0.3 eV (4.1 μm) at room-temperature. The electrically pumped emission persists up to room-temperature. It shows an increasing width and overlap with the PV response spectrum with increasing temperature due to filling of the InAs conduction band with thermally activated carriers.

The InAs/AlSb/GaSb interband-tunneling structures were also grown by solid-source MBE on undoped semi-insulating (100) GaAs substrates. A 1.1 μm thick strain-relaxed AlSb buffer layer followed by a 10-period 2.5 nm GaSb/2.5 nm AlSb smoothing SL were placed between the substrate and the tunneling structure. The layer sequence for the RIT structure started with a 1 μm thick n-type InAs layer doped with Si to a concentration of $1 \times 10^{18} \mathrm{cm}^{-3}$ followed by 50 nm of n-type InAs doped to a level of $2 \times 10^{16} \mathrm{cm}^{-3}$ and a 10-nm-thick undoped InAs spacer. Then the 1.5-nm-wide lower AlSb tunneling

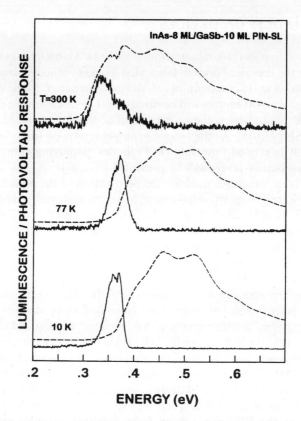

Fig. 36 Photovoltaic response (dashed line) and electroluminescence (full line) spectra of a 8 ML InAs/10 ML GaSb p-i-n SL photodiode recorded at (bottom) 10 K, (middle) 77 K, and (top) 300 K.

barrier was grown followed by a 6.5-nm-thick GaSb QW, the 2.5-nm-thick upper AlSb barrier, another 10-nm-thick undoped InAs spacer layer, 50 nm of InAs with a doping level of 2×10^{16} cm^{-3}, and a 1-μm-thick InAs top contact layer doped to a concentration of 1×10^{18} cm^{-3} (see also the inset of Fig. 37). The growth temperature was 460°C. For structure A the shutter sequence for the growth of the InAs/AlSb IF's corresponded to the deposition of 1 ML of InSb at the IF. For structure B 2 ML of AlAs were deposited at the bottom InAs/AlSb IF and one ML of AlAs was inserted at the top GaSb/AlSb and AlSb/InAs IF's [106], while keeping the total width of the tunneling barriers the same as for structure A. The inset of Fig. 37 shows a schematic band diagram of the RIT diode structure.

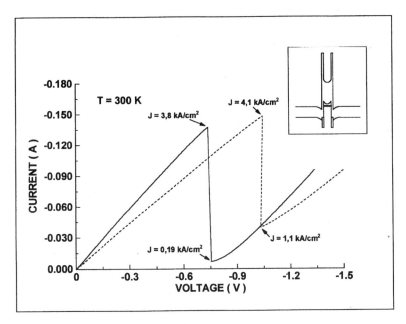

Fig. 37 Room-temperature current–voltage characteristics of InAs/AlSb/GaSb resonant interband-tunneling diodes grown with (full line) and without (dashed line) the deposition of AlAs ML's at the IF's. The inset shows a schematic band diagram of the InAs/AlSb(As)/GaSb resonant interband-tunneling diode (Ref. [37]).

Typical room-temperature current-voltage characteristics of RIT mesa diodes prepared from structure A without and structure B with AlAs hole barriers are shown in Fig. 37 [37]. The polarity of the applied voltage was such that the electron flow was from the bottom to the top electrode. For diode A, a valley current density of $1.1\,\text{kA}/\text{cm}^2$ and a peak-to-valley current ratio (PVR) close to 4 were found at 300 K. For diode B a significantly reduced valley current density of $0.19\,\text{kA}/\text{cm}^2$ and a PVR of 20 were obtained with an only slightly reduced peak current density of $3.8\,\text{kA}/\text{cm}^2$ compared to $4.1\,\text{kA}/\text{cm}^2$ for diode A. These results demonstrate the efficient reduction of the valley current by adding AlAs ML to the AlSb tunneling barriers [106,107]. Detailed magneto-tunneling studies of both types of RIT diodes revealed an almost complete suppression of elastic hole tunneling by the AlSb(As) layers acting as hole barriers, with the remaining valley current mainly caused by inelastic tunneling [106].

As discussed in Section 5.3.2 there is a controversy on whether ML's of AlAs can be formed at the InAs/AlSb heterointerface or whether the insertion of ML's of AlAs results in the formation of pseudo-ternary Al(SbAs). To analyze the present AlSb(As) tunneling barriers by Raman spectroscopy, samples identical to structure A and B were grown except for the lower InAs layers being undoped and the omission of the two topmost n-type InAs layers. Fig. 38 shows low temperature Raman spectra from these samples without and with AlAs ML's added to the AlSb barriers, covering the energy range of the LO phonon of AlSb. With the AlAs ML's added the AlSb LO phonon line shifts from 348.8 to 349.9 cm^{-1} and its width increases from 3.7 to

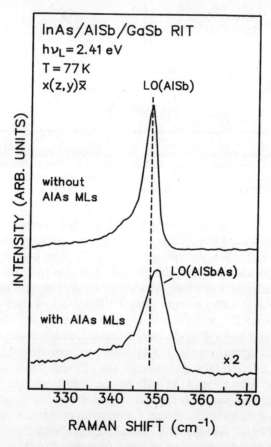

Fig. 38 Low-temperature (77 K) Raman spectra of InAs/AlSb/GaSb resonant interband-tunneling structures grown without (top) and with (bottom) the deposition of AlAs ML's at the interfaces (Ref. [37]).

$6.9\,\text{cm}^{-1}$. This shift and broadening has been taken as a clear proof of the formation of Al(SbAs) barriers [37]. Based on the composition dependence of the LO phonon energy in Al(SbAs) [29], and taking into account strain effects because the AlSb barriers are pseudomorphically strained to the smaller in-plane lattice parameter of InAs, an As content of about 10% has been estimated for the structures grown with AlAs ML's added to the barriers [37]. The formation of $AlSb_{0.9}As_{0.1}$ barriers results in an increase of the valence-band offset with respect to the GaSb QW of about 0.1 eV, compared to binary AlSb barriers. A further consequence is an approximate band line-up of the valence-band edges of the $AlSb_{0.9}As_{0.1}$ barriers and the InAs spacer layers [62] in contrast to the valence-band edge of binary AlSb being 0.1 eV above that of InAs.

5.6 Summary

In this Chapter a review has been given of the growth and characterization of group III arsenide–antimonide heterostructures. Various aspects of molecular-beam epitaxial growth of such heterostructures on GaAs substrates have been covered such as the issues of unintentional As incorporation into the Antimonides, monitoring of growth surface temperature, and controlled formation of different interface bonds. The characterization techniques discussed for the assessment of the structural quality of InAs/AlSb/GaSb heterostructures are HRXRD and phonon spectroscopy by Raman scattering and as well as reflection and absorption measurements. Fourier-transform PL and calorimetric absorption spectroscopy complemented by intersubband Raman spectroscopy and spectroscopic ellipsometry have been introduced as experimental techniques providing information on the band and subband structure of such heterostructures including strain effects. Finally, results on InAs/GaSb SL based IR detectors and emitters and on InAs/AlSb/GaSb resonant interband tunneling diodes have been discussed as examples for device applications.

References

1. A. Nakagawa, H. Kroemer, and J. H. English, *Appl. Phys. Lett.* **54**, 1893 (1989).
2. G. Tuttle and H. Kroemer, *IEEE Trans. Electron Devices* **ED-34**, 2358 (1987); J. D. Werking, C. R. Bolognesi, L. D. Chang, C. Nguyen, E. L. Hu, and H. Kroemer, *IEEE Electron. Device Lett.* **13**, 164 (1992).
3. C. Nguyen, H. Kroemer, and E. L. Hu, *Phys. Rev. Lett.* **69**, 2847 (1992).

4. J. R. Söderström, D. H. Chow, and T. C. McGill, *Appl. Phys. Lett.* **55**, 1094 (1989).
5. L. F. Luo, R. Beresford, and W. I. Wang, *Appl. Phys. Lett.* **55**, 2023 (1989).
6. L. L. Chang, N. Kawai, G. A. Sai-Halasz, R. Ludeke, and L. Esaki, *Appl. Phys. Lett.* **35**, 939 (1979).
7. D. L. Smith and C. Mailhiot, *J. Appl. Phys.* **62**, 2545 (1987); J. L. Johnson, L. A. Samoska, A. C. Gossard, J. L. Merz, M. D. Jack, G. R. Chapman, B. A. Baumgratz, K. Kosai, and S. M. Johnson, *J. Appl. Phys.* **80**, 1116 (1996).
8. R. H. Miles, D. H. Chow, Y.-H. Zhang, P. D. Brewer, and R. G. Wilson, *Appl. Phys. Lett.* **66**, 1921 (1995).
9. G. Tuttle, H. Kroemer, and J. H. English, *J. Appl. Phys.* **67**, 3032 (1990).
10. B. R. Bennett, B. V. Shanabrook, R. J. Wagner, J. L. Davis, and J. R. Waterman, *Appl. Phys. Lett.* **63**, 949 (1993).
11. M. Yano, H. Furuse, Y. Iwai, K. Yoh, and M. Inoue, *J. Cryst. Growth* **127**, 807 (1993).
12. A. Fasolino, E. Molinari, and J. C. Maan, *Phys. Rev. B* **33**, 8889 (1986). *Superlatt. Microstruct.* **3**, 117 (1987).
13. A. Fasolino, E. Molinari and J. C. Maan, *Phys. Rev. B* **39**, 3923 (1989).
14. Y. Liu and B. J. Inkson, *Semicond. Sci. Technol.* **4**, 1167 (1989).
15. D. Kechrakos and J. C. Inkson, *Semicond. Sci. Technol.* **6**, 155 (1991).
16. D. Berdekas and G. Kanellis, *Phys. Rev. B* **43**, 9976 (1991).
17. L. A. Hemstreet, C. Y. Fong, and J. S. Nelson, *J. Vac. Sci. Technol. B* **11**, 1693 (1993).
18. D. M. Symons, M. Lakrimi, R. J. Warburton, R. J. Nicholas, N. J. Mason, P. J. Walker, and M. I. Eremets, *Semicond. Sci. Technol.* **9**, 118 (1994); D. M. Symons, M. Lakrimi, M. van der Burgt, T. A. Vaughan, R. J. Nicholas, N. J. Mason, and P. J. Walker, *Phys. Rev. B* **51**, 1729 (1995).
19. J. R. Meyer, C. A. Hoffman, B. V. Shanabrook, B. R. Bennett, R. J. Wagner, J. R. Waterman, and E. R. Youngdale, *Proceedings of the 22nd Int. Conf. on the Physics of Semiconductors*, Vancouver, Canada (1994), Ed. D. J. Lockwood; (World Scientific, Singapore, 1995), p. 783
20. J. R. Waterman, B. V. Shanabrook, R. J. Wagner, M. J. Yang, J. L. Davis, and J. P. Omaggio, *Semicond. Sci. Technol.* **8**, S106 (1993).
21. S. J. Eglash, H. K. Choi, and G. W. Turner, *J. Cryst Growth* **111**, 669 (1991); J. F. Chen and A. Y. Cho, *J. Electron. Mater.* **22**, 259 (1993); J. F. Klem, J. A. Lott, J. E. Schirber, S. R. Kurtz, and S. Y. Lin, *J. Electron. Mater.* **22**, 315 (1993).
22. T. H. Chiu and W. T. Tsang, *J. Appl. Phys.* **57**, 4572 (1985).
23. T. H. Chiu, W. T. Tsang, J. A. Ditzenberg, S. N. G. Chu, and J. P. van der Ziel, *J. Appl. Phys.* **60**, 205 (1986); S. J. Eglash and H. K. Choi, *Inst. Phys. Conf. Ser.* **120**, 487 (1992).
24. S. Subanna, J. Gaines, G. Tuttle, H. Kroemer, S. Chalmers, and J. H. English. *J. Vac. Sci. Technol. B* **7**, 289 (1989); P. N. Fawcett, B. A. Joyce, X. Zhang, and D. W. Pashley, *J. Cryst. Growth* **116**, 81 (1992); J.M. Kang, M. Nouaoura, L. Lassabatère, and A. Rocher, *J. Cryst. Growth* **143**, 115 (1994).
25. G. D. Kramer, M. S. Adam, R. K. Tsui, and N. D. Theodore, *Inst. Phys. Conf. Ser.* **136**, 727 (1993).
26. J. Schmitz, J. Wagner, H. Obloh, P. Koidl, J. D. Ralston, *J. Electron. Mater.* **23**, 1203 (1994).
27. J. Wagner, J. Schmitz, M. Maier, J. D. Ralston, and P. Koidl, *Solid State Electron.* **37**, 1037 (1994).
28. T. C. McGlinn, T. N. Krabach, M. V. Klein, G. Bajor, J. E. Greene, B. Kramer, S. A. Barnett, A. Lastras, and S. Gorbatkin, *Phys. Rev. B* **33**, 8396 (1986).
29. I. Sela, C. R. Bolognesi, and H. Kroemer, *Phys. Rev. B* **46**, 16142 (1992).
30. B. V. Shanabrook, J. R. Waterman, J. L. Davis, and R. J. Wagner, *Appl. Phys. Lett.* **61**, 2338 (1992).
31. J. Schmitz, J. Wagner, F. Fuchs, N. Herres, P. Koidl, and J. D. Ralston, *J. Cryst. Growth* **150**, 858 (1995); J. Schmitz (unpublished).

32. C. E. C. Wood, K. Singer, T. Ohashi, L. R. Dawson, and A. J. Noreika, *J. Appl. Phys.* **54**, 2732 (1983); H. Yamaguchi and Y. Horikoshi, *Appl. Phys. Lett.* **64**, 2572 (1994).
33. D. Behr, J. Wagner, J. Schmitz, N. Herres, J. D. Ralston, P. Koidl, M. Ramsteiner, L. Schrottke, and G. Jungk, *Appl. Phys. Lett.* **65**, 2972 (1994).
34. I. Sela, C. R. Bolognesi, L. A. Samoska, and H. Kroemer, *Appl. Phys. Lett.* **60**, 3283 (1992).
35. B. R. Bennett, B. V. Shanabrook, and E. R. Glaser, *Appl. Phys. Lett.* **65**, 598 (1994).
36. J. Wagner, J. Schmitz, D. Behr, J. D. Ralston, and P. Koidl, *Appl. Phys. Lett.* **65**, 1293 (1994).
37. J. Wagner, J. Schmitz, H. Obloh, and P. Koidl, *Appl. Phys. Lett.* **67**, 2963 (1995).
38. M. E. Twigg, B. R. Bennett, B. V. Shanabrook, J. R. Waterman, J. L. Davis, and R. J. Wagner, *Appl. Phys. Lett.* **64**, 3476 (1994).
39. M. E. Twigg, B. R. Bennett, and B. V. Shanabrook, *Appl. Phys. Lett.* **67**, 1609 (1995).
40. R. M. Feenstra, D. A. Collins, D. Z.-Y. Ting, M. W. Wang, and T. C. McGill, *Phys. Rev. Lett.* **72**, 2749 (1994).
41. R. M. Feenstra, D. A. Collins, and T. C. McGill, *Superlatt. Microstruct.* **15**, 215 (1994).
42. J. Spitzer, A. Höpner, M. Kuball, M. Cardona, B. Jenichen, H. Neuroth, B. Brar, and H. Kroemer, *J. Appl. Phys.* **77**, 811 (1995).
43. P. M. Thibado, B. R. Bennett, M. E. Twigg, B. V. Shanabrook, and L. J. Whitman, *Appl. Phys. Lett.* **67**, 3578 (1995).
44. A. Y. Lew, E. T. Yu, D. H. Chow, and R. H. Miles, *Appl. Phys. Lett.* **65**, 201 (1994).
45. P. Fewster, *Appl. Surf. Sci.* **50**, 9 (1991).
46. P. van der Sluis, *J. Phys. D: Appl. Phys.* **26**, A188 (1993).
47. E. Koppensteiner, T. W. Ryan, M. Heuken, and J. Söllner, *J. Phys. D: Appl. Phys.* **26**, A35 (1993).
48. N. Herres, F. Fuchs, J. Schmitz, K. M. Pavlov, J. Wagner, J. D. Ralston, P. Koidl, C. Gadaleta, and G. Scamarcio, *Phys. Rev. B* **53**, 15688 (1996).
49. See, e.g., B. Jusserand and M. Cardona, in *Light Scattering in Solids V*, Eds. M. Cardona and G. Güntherodt (Springer, Berlin, 1989), p. 49.
50. J. Wagner, J. Schmitz, N. Herres, J. D. Ralston, and P. Koidl, *Appl. Phys. Lett.* **66**, 3498 (1995).
51. B. V. Shanabrook, B. R. Bennett, and R. J. Wagner, *Phys. Rev. B* **48**, 17172 (1993).
52. J. Wagner, J. Schmitz, R. C. Newman, and C. Roberts, *J. Raman Spectroscopy*, **27**, 231 (1996).
53. J. Wagner, J. Schmitz, J. D. Ralston, and P. Koidl, *Appl. Phys. Lett.* **64**, 82 (1994).
54. B. R. Bennett, B. V. Shanabrook, E. R. Glaser, and R. J. Wagner, *Mat. Res. Soc. Symp. Proc.* **340**, 253 (1994).
55. M. J. Yang, R. J. Wagner, B. V. Shanabrook, W. J. Moore, and J. R. Waterman, C. H. Yang, and M. Fatemi, *Appl. Phys. Lett.* **63**, 3434 (1993).
56. F. Fuchs, J. Schmitz, K. Schwarz, J. Wagner, J. D. Ralston, P. Koidl, C. Gadaleta, and G. Scamarcio, *Appl. Phys. Lett.* **65**, 2060 (1994).
57. D. W. Berreman, *Phys. Rev.* **130**, 2193 (1963).
58. C. Gadaleta, G. Scamarcio, F. Fuchs, and J. Schmitz, *J. Appl. Phys.* **78**, 5642 (1995).
59. C. Gadaleta and F. Fuchs, unpublished.
60. F. Fuchs, J. Schmitz, K. Schwarz, J. Wagner, J. D. Ralston, and P. Koidl, *Appl. Phys. Lett.* **65**, 2060 (1994).
61. M. J. Yang, R. J. Wagner, B. V. Shanabrook, W. J. Moore, J. R. Waterman, M. E. Twigg, and M. Fatemi, *Appl. Phys. Lett.* **61**, 583 (1992).
62. M. P. C. Krijn, *Semicond. Sci. Technol.* **6**, 27 (1991).
63. C. Nguyen, B. Brar, H. Kroemer, and J. H. English, *Appl. Phys. Lett.* **60**, 1854 (1992).
64. C. Nguyen, B. Brar, C. R. Bolognesi, J. J. Pekarik, H. Kroemer, and J. H. English, *J. Electron. Mater.* **22**, 255 (1993).
65. H. Kroemer, C. Nguyen, and B. Brar, *J. Vac. Sci. Technol. B* **10**, 1769 (1992).

66. J. Shen, H. Goronkin J. D. Dow, and S. Y. Ren, *J. Appl. Phys.* **77**, 1576 (1995).
67. W. Shockley, *Phys. Rev.* **56**, 317 (1939).
68. H. M. James, *Phys. Rev.* **76**, 1611 (1949).
69. I. Tamm, *Phys. Z. Sowjetunion* **1**, 733 (1932).
70. H. Ohno, E. E. Mendez, J. A. Brum, J. M. Hong, F. Agullo-Rueda, L. L. Chang, and L. Esaki *Phys. Rev. Lett.* **64**, 2555 (1990).
71. See, e.g., D. E. Aspnes, in *Handbook of Optical Constants*, Ed. E. Palik (Academic Press, New York, 1985), p. 89.
72. J. Spitzer, H. D. Fuchs, P. Etchegoin, A. Höpner, M. IIg, M. Cardona, B. Brar, and H. Kroemer, *Appl. Phys. Lett.* **62**, 2274 (1993).
73. C. M. Herzinger, P. G. Synder, F. G. Cellii, Y.-C. Kao, D. Chow, B. Johs, and J. A. Woollam, *Inst. Phys. Conf. Ser. No.* **141** (IOP, Bristol, 1995), p. 363.
74. F. H. Pollak, in *Semiconductors and Semimetals*, Vol. 32, Ed. T. P. Pearsall (Academic Press, New York, 1990), p. 17.
75. C. H. Grein, P. M. Young, and H. Ehrenreich, *Appl. Phys. Lett.* **61**, 2905 (1992).
76. C. H. Grein, M. E. Flatte, H. Ehrenreich, and R. H. Miles, *J. Appl. Phys.* **77**, 4156 (1995).
77. E. R. Youngdale, J. R. Meyer, C. A. Hoffmann, F. J. Bartoli, C. H. Grein, P. M. Young, H. Ehrenreich, R. H. Miles, and D. H. Dow, *J. Appl. Phys.* **64**, 3160 (1994).
78. P. M. Young, C. H. Grein, H. Ehrenreich, and R. H. Miles, *J. Appl. Phys.* **74**, 4774 (1993).
79. L. Podlowski, A. Hoffmann, and I. Broser, *J. Crystal Growth* **117**, 698 (1992).
80. B. Hamilton and C. Clarke, *Mater. Sci. Forum* **38–41**, 1337 (1989).
81. F. Fuchs, K. Kheng, K. Schwarz, and P. Koidl, *Semicond. Sci. Technol.* **8**, S75 (1993).
82. F. Fuchs, A. Lusson, J. Wagner, and P. Koidl 1989, *Proc. SPIE* Vol. 1145, p. 323 (1989).
83. F. Fuchs, J. Schmitz, J. D. Ralston, P. Koidl, *Appl. Phys. Lett.* **64**, 1665 (1995).
84. Ch. Gauer, J. Scriba, A. Wixforth, J. P. Kotthaus, C. Nguyen, G. Tuttle, J. H. English, and H. Kroemer, *Semicond. Sci. Technol.* **8**, S137 (1993).
85. H. J. Bardeleben, M. O. Manasreh, and C. E. Stutz, *Proc. Int. Conf. on Defects in Semiconductors 1993, Gmunden, Mater. Sci. Forum, Trans Tech Publications*, Aedermannsdorf, Switzerland, (1994), Eds: H. Heiuvich, W. Jautsch, Vol. 143–147, p. 6.
86. D. J. Chadi, *Phys. Rev. B* **47**, 13478 (1993).
87. S. Ideshita, A. Furukawa, Y. Mochizuki, and M. Mizuta, *Appl. Phys. Lett.* **60**, 2549 (1992).
88. J. Shen, J. D. Dow, S. Y. Ren, S. Tehrani, and H. Goronkin, *J. Appl. Phys.* **73**, 8313 (1993).
89. M. J. Yang, P. J. Lin-Chung, R. J. Wagner, J. R. Waterman, W. J. Moore, and B. V. Shanabrook, *Semicond. Sci. Technol.* **8**, S129 (1993), P. J. Lin-Chung and M. J. Yang, *Phys. Rev. B* **48**, 5338 (1993).
90. P. V. Santos, P. Etchegoin, M. Cardona, B. Brar, and H. Kroemer, *Phys. Rev, B* **50**, 8746 (1994).
91. F. Fuchs, J. Schmitz, J. D. Ralston, P. Koidl, R. Heitz, and A. Hoffmann *Proc. ICSMM-7*, Banff (Canada) 1994, *Superlattices and Microstructures* **16**, 35 (1994).
92. J. Böhrer, A. Krost, R. Heitz, F. Heinrichsdorf, L. Eckey, D. Bimberg, and H. Cerva, *Appl. Phys. Lett* **68**, 1072 (1996).
93. F. Fuchs, J. Schmitz, N. Herres, J. Wagner, J. D. Ralston, and P. Koidl, in "Proc. of the 7th Int. Conf. on Narrow Gap Semiconductors", Santa Fe, 1995, *Inst. Phys. Conf. Ser. No.* **144**, IOP Publishing, p. 219 (1995).
94. F. Fuchs, N. Herres, J. Schmitz, K. M. Pavlov, J. Wagner, P. Koidl, and J. H. Roslund, SPIE Vol. 2554, p. 70 (1995).
95. C. Gauer, J. Scriba, A. Wixforth, C. R. Bolognesi, C. Nguyen, B. Brar, and H. Kroemer, *Semicond. Sci. Technol.* **9**, 1580 (1994).

96. A. Simon, J. Scriba, C. Gauer, A. Wixforth, J. P. Kotthaus, C. R. Bolognesi, C. Nguyen, G. Tuttle, and H. Kroemer, *Mater. Sci. Eng. B* **21**, 201 (1993).
97. C. Gauer, A. Wixforth, J. P. Kotthaus, M. Kubisa, W. Zawadzki, B. Brar, and H. Kroemer, *Phys. Rev. Lett.* **74**, 2772 (1995).
98. See, e.g., A. Pinczuk and G. Abstreiter, in *Light Scattering in Solids V*, Eds. M. Cardona and G. Güntherodt (Springer, Berlin, 1989), p. 153.
99. E. Burstein, A. Pinczuk, and D. L. Mills, *Surf. Sci.* **98**, 451 (1980).
100. A. Pinczuk, S. Schmitt-Rink, G. Danan, J. P. Valladares, L. N. Pfeiffer, and K. W. West, *Phys. Rev. Lett.* **63**, 1633 (1989).
101. R. Carle, N. Saint-Criq, J. B. Renucci, M. B. Renucci, and A. Zwick, *Phys. Rev.* **22**, 4804 (1980).
102. J. Wagner, J. Schmitz, F. Fuchs, J. D. Ralston, P. Koidl, and D. Richards, *Phys. Rev. B* **51**, 9786 (1995); J. Wagner, J. Schmitz, D. Richards, J. D. Ralston, and P. Koidl, *Proceedings of the 7th Int. Conf. on Modulated Semiconductor Structures*, Madrid, (1995) *Solid State Electron.* **40**, 281 (1996).
103. Y. B. Li, V. Tsoukala, R. A. Stradling, R. L. Williams, S. J. Chung, I. Kamiya, and A. G. Norman, *Semicond. Sci. Technol.* **8**, 2205 (1993).
104. G. Brozak, B. V. Shanabrook, D. Gammon, D. A. Broido, R. Beresford, and W. I. Wang, *Phys. Rev. B* **45**, 11399 (1992).
105. J. Wagner, F. Fuchs, N. Herres, J. Schmitz, and P. Koidl, *Proceedings of the 3rd Int. Symp. on Long Wavelength Infrared Detectors and Arrays: Physics and Applications, 188th Meeting of the Electrochemical Society*, Chicago, *Electrochem. Soc. Proc.*, Vol. 95-28, 201 (1995).
106. H. Obloh, J. Schmitz, J. D. Ralston, *Int. Symp. on Compound Semiconductors*, San Diego 1994, *Inst. Phys. Conf. Ser. No. 141* (IOP), Bristol (1995), p. 861.
107. S. Tehranl, J. Shen, H. Goronkin, G. Kramer, M. Hoogstra, and T. X. Zhu, *Inst. Phys. Conf. Ser. No. 136*, 209 (1994).
108. F. Fuchs, J. Schmitz, N. Herres, *Proc. 23rd Int. Conf. on the Physics of Semiconductors*, p. 1803, World Scientific Singapore (1996) Eds: M. Scheffler, R. Zimmerwann.

CHAPTER 6

Antimonide-Based Quantum Heterostructure Devices

J. R. MEYER[1], J. I. MALIN[1], I. VURGAFTMAN[1], C. A. HOFFMAN[1] and L. R. RAM-MOHAN[2]

[1] Code 5610, Naval Research Laboratory, Washington, DC 20375, USA; [2] Physics Dept., Worcester Polytechnic Institute, Worcester, MA 01609, USA

6.1.	Introduction	235
6.2.	Multi-Band Finite Element Approach to "Wavelength Engineering"	238
6.3.	IR Electro-Optical Modulators	239
6.4.	Second-Harmonic Generators	250
6.5.	Type-II MWIR Lasers	256
6.6.	Conclusions	267
	References	269

6.1 INTRODUCTION

In recent years, the exploitation of band structure engineering as a means of designing optimized quantum structures with novel properties has led to a revolution in semiconductor electro-optics [1,2]. Tailoring the interband and intersubband optical transitions results in significantly enhanced performance for a broad range of devices, including quantum well diode lasers, electro-optical (EO) modulators, filters, switches, nonlinear optical (NLO) frequency-converters, and IR detectors. Up to this point, nearly all of the devices reported have been based on the 5.7 Å (lattice-matched to GaAs substrates) and 5.9 Å (lattice-matched to InP) III–V heterostructure families, whose growth technologies have been relatively mature for some time. However, there has quite recently been a resurgence of interest in the less studied 6.1-Å family (lattice-matched to GaSb), also known as the antimonides. Innovations in the techniques for MBE growth [3], *in situ* characterization [3,4], and device processing [5–7] have led to rapid advances in the material quality and control over layering definition down to the atomic scale (e.g., the ability to fix the chemical bond type at each heterointerface between constituents with no common cation or anion [4,8,9]).

It will be seen in the following sections that the antimonides possess some unique advantages over the other III–V systems in terms of band structure engineering. These become particularly valuable when one moves from the near-IR ($\lambda < 2\,\mu$m) to the mid-wave and long-wave infrared (MWIR, 3–5 μm, and LWIR, 8–14 μm) spectral regions, where the need for high-performance electro-optic devices has become especially acute for such commercial and military applications as remote chemical sensing (for leak detection, pollution and drug monitoring, chemical process control, etc.), IR spectroscopy, laser surgery, multispectral detection, IR illumination, IR countermeasures, and future ultra-low-loss fiber communications. While some required elements for an integrated EO technology for the MWIR/LWIR are already in place or will soon become available (e.g., one-color detectors [10] and IR fibers [11]), the development of practical sources and modulators lags far behind that of analogous GaAs-based and InP-based quantum well devices for the near-IR and visible spectral regions.

Fig. 1 plots energy gap as a function of lattice constant for the principal antimonide family members, which include the binary compounds InAs, GaSb, and AlSb, a number of alloy combinations such as $Ga_{1-x}Al_xSb$ and $Ga_{1-x}In_xSb$, and even the semimetallic element Sb [12–14]. Collectively, these materials offer a more flexible EO material design than any other quantum heterostructure system, since a uniquely diverse array of key band structure, electronic, and optical properties can be combined in countless layering permutations to suit the needs of a given application. For example, one may specify either narrow-gap or wide-gap, type-I or type-II band alignment in real space, Γ or L conduction minimum in momentum space, large or small interband optical matrix elements, large or small intersubband matrix elements at normal incidence, etc. Even the sensitivity of the valence band offset to interface bond type may be exploited to fabricate materials with band profiles analogous to those of p–n junctions or n–i–p–i structures in the absence of any extrinsic or intrinsic doping [9].

In the following sections, we review several specific device classes for which this remarkable degree of control over the interband and intersubband optical and electronic properties may be used to a particular

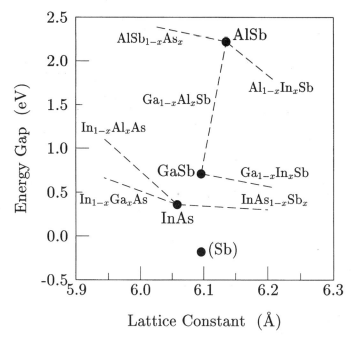

Fig. 1 Ambient-temperature energy gap vs. lattice constant for some of the most prominent binary-compound and ternary-alloy members of the antimonide family, which have lattice constants near 6.1 Å (for [100]-oriented growth). Bowing of the alloy energy gaps has been neglected here. Also shown is elemental Sb, whose lattice constant has the indicated relation to the other family members if growth is along the [111] axis rather than [100].

advantage. These include intersubband-based electro-optical modulators for both normal-incidence and waveguide-mode MWIR and LWIR applications (Section 6.3), IR frequency-conversion devices in which the phase-matching condition can be imposed electrically by varying a bias voltage (Section 6.4), and improved MWIR lasers based on type-II multiple quantum wells as well as a novel type-II interband cascade process (Section 6.5). In several cases, initial experiments have already confirmed the predicted operation and provided quite promising preliminary results. All of these "wavefunction engineered" devices are based on multi-layer configurations incorporating more than the usual two constituents per period. These have been modeled using electronic band structures and electro-optical properties derived from the 8-band finite element method calculation described in Section 6.2.

6.2 MULTI-BAND FINITE ELEMENT APPROACH TO "WAVEFUNCTION ENGINEERING"

Improved growth techniques yielding atomic-level control coupled with rapid advances in device-fabrication technologies have in recent years encouraged a trend toward increasingly complex structural configurations. These employ not only a larger number of material constituents in the same device, but also more intricate substructures (often involving asymmetry) and an expanded menu of parameters which may be modulated. In this section, we briefly outline the analytical tools that are required to design and accurately model devices employing complicated multi-constituent layering geometries.

Whereas standard multiband $k \cdot P$ model calculations are adequate for treating simple, two-constituent quantum wells and superlattices, the boundary conditions for that approach become intractable whenever there is either a larger multiplicity of constituents or a substructure within each period. We therefore employ band structures, wavefunctions, and optical matrix elements calculated using a multiband finite element method (FEM) formalism [2]. Since this 8-band FEM implementation of the $k \cdot P$ model straightforwardly accounts for whatever geometry is imposed on the wavefunctions of the carriers, it allows one to incorporate any III–V or II–VI direct gap semiconducting materials in any combination of layers, and to include the effects of built-in strain and external perturbations such as electric or magnetic fields and hydrostatic pressure. This provides a unique capability to accurately calculate the electronic dispersion relations and optical properties for complicated geometries which would otherwise be intractable without severe approximations.

One begins by writing out the appropriate symmetrized (hermitian) Lagrangian that would generate the coupled differential equations for the envelope functions through a variational procedure. The integral over the physical region of the Lagrangian density, the action integral, is then split up into a number of "cells" or elements, in each of which the physical considerations of the problem hold. The wave functions are assumed to be given locally in each element by fifth-order Hermite interpolation polynomials, which have the property that the expansion coefficients correspond to the values of the wave function and its derivatives at select points, called nodes, in the element. The global wave functions are constructed by joining the locally-defined interpolation functions and matching the function and its derivative across the element boundary for each of the bands included in the analysis.

The spatial dependence of the wavefunctions, manifested through the interpolation polynomials, is next integrated out, leaving the action integral dependent on the unknown nodal values of the wavefunction. The usual variational principle is then implemented as a variation of the nodal values under which the action integral is a minimum. This *nodal variational principle* leads to a "Schrödinger equation" for the nodal values. The integration of the action integral is performed element by element, giving rise to element matrices which are then overlaid into a global matrix in a manner consistent with the boundary conditions. This results finally in a generalized eigenvalue problem [15,16], which may be solved for the eigenenergies and wave functions with a standard diagonalizer.

The FEM capability allows us to go beyond conventional bandgap or band-structure engineering, to a more general approach that may be designated *wavefunction engineering* [2]. This signifies a fundamental redesign of the electronic states at the quantum mechanical level, so as to optimize the band mixing and spatial distribution of the electron and hole wavefunctions to achieve a desired set of properties. The tailoring may be accomplished, e.g., by changing the geometrical placement of different materials in the heterostructure so as to maximize the desired localization or shape of the wavefunctions. The expanded flexibility naturally has broad implications for optoelectronic device design. By tailoring the wavefunctions we can control optical selection rules, optical matrix elements, carrier lifetimes, overlap integrals, tunneling currents, electro-optical and nonlinear optical coefficients, and so on. Such manipulations will be exploited repeatedly in the device configurations discussed below.

6.3 IR ELECTRO-OPTICAL MODULATORS

As the emerging IR technologies noted in the introduction become more sophisticated, EO switches, limiters, intensity and phase modulators, beam steering components, etc. will increasingly be required for incorporation as key elements. Unfortunately, the stage of development for MWIR and LWIR modulators is currently quite primitive when compared to the rapid advances occurring in the performance of near-IR EO devices based on the excitonic quantum-confined Stark effect (QCSE) [17–20]. QCSE intensity modulators employ the shift with applied electric field of the excitonic absorption edge in a GaAs-based or InP-based heterostructure with energy gap E_g near the photon energy $\hbar\omega$ [17]. Phase modulators may similarly be constructed, since from the Kramers–Kronig relations the variation of the absorption coefficient must be accompanied by a modulation of the refractive index [21]. While the QCSE could in principle be extended to longer wavelengths by employing quantum wells with a smaller energy gap, that approach is unworkable in practice because narrower-gap semiconductors tend to have much broader fundamental absorption edges and display far weaker exciton lines due to the smaller electron mass.

As a possible alternative mechanism for the modulation of MWIR and LWIR wavelengths, intersubband interactions in asymmetric quantum structures have been suggested [22,23], since the magnitudes and widths of the intersubband resonances tend to be comparable to those for exciton transitions in the near-IR. However, practical devices have yet to be developed despite the experimental confirmation of quite large EO coefficients [24–31]. This is due in large part to the polarization selection rule which yields vanishing interactions between the intersubband system and normal-incidence radiation whenever the electrons populate the isotropic Γ-valley in a GaAs-based or InP-based quantum well, since coupling to the beam then requires a component of the optical electric field along the quantization axis [32]. A second fundamental limitation which has received far less attention is the near impossibility of designing a low-insertion-loss IR phase modulator if the EO response relies on intersubband processes alone [33]. As will be seen below, this follows directly from the relation between the resonant absorption coefficient and the resonant contribution to the refractive index.

Here we discuss an antimonide-family EO modulator design [33] which removes both of these limitations. It has been pointed out that when the electrons populate highly-anisotropic L or X valleys and the symmetry axes of the elliptical constant-energy surfaces are tilted with respect to the confinement axis, the in-plane and growth-direction motions become coupled and strong normal-incidence intersubband interactions can occur [31,35]. Large absorption coefficients have been confirmed experimentally for the case of L-valleys in antimonide quantum wells [36–40], and normal-incidence optical devices have been proposed [33,41–45].

For EO applications, this property can be used to greatest advantage if we specifically design the device so as to set up a competition between Γ-valley states and L-valley states having nearly the same energy. In this form of wavefunction engineering, both the valley-identity and the spatial profiles of the wavefunctions are crucial. The structure consists of an asymmetric double quantum well (ADQW) whose most distinctive feature is that an applied electric field (F) can transfer virtually the entire electron population from one well whose conduction-band minimum is at the Γ-point (where the intersubband transition matrix elements vanish at normal incidence) to a second well whose minimum is at the L-point (where the intersubband transitions are strongly coupled to the optical field at normal incidence) [46]. Thus by varying F, we can radically alter the polarization selection rules, optical matrix elements, intersubband resonance energies, and in-plane electron effective mass. It will be seen below that the latter is important because it enables strong tuning of the plasma contribution to the refractive index and thereby provides an ideal non-resonant mechanism for low-loss phase modulation.

Fig. 2 illustrates the Γ and L band edge profiles, quantized energy levels, and L-valley wavefunctions for a GaSb-$Ga_{1-x}Al_xSb$-$Ga_{1-x}In_xSb$-AlSb ADQW (layer thicknesses $d_1 - d_2 - d_3 - d_B$), in which the $Ga_{1-x}Al_xSb$ barrier layer separating the two wells (d_2) is thin enough to permit sub-nanosecond interwell tunneling times following reversal of the applied voltage. Energy levels and matrix elements for the Γ-valley subbands were derived from the full 8-band FEM calculation while the L-valley properties were obtained from a 1-band version of the algorithm, which provides a good approximation since interband interactions are relatively weak at the L point. The 1-band calculation employed L-point quantization masses for each constituent material: $m_{zz} = 3m_t m_l/(m_t + 2m_l)$ for [100] growth [47], where m_t and m_l are the transverse and longitudinal masses.

Note that in the structure of Fig. 2, the GaSb well on the left (ℓ) has its conduction-band minimum at the L-point while the $Ga_{1-x}In_xSb$ well on the right (r) has its minimum at the Γ-point (the Γ-valley well may alternatively be composed of InAs). The structure is designed such that the lowest quantized levels in the two wells lie close enough in energy that the net ground state for the ADQW can be modulated by reversing the direction of the applied electric field from positive [Fig. 2(a)] to negative [Fig. 2(b)]. It should also be noted that most of the electrons remain in the L well whenever the two bands are evenly aligned in energy, since the L-valley density of states is more than 20 times greater due to its 4-fold degeneracy and larger density-of-states mass.

Following use of the FEM formalism to derive dispersion relations for all of the quantized energy levels, the relative populations of the two wells as a function of applied field may be evaluated using Fermi–Dirac statistics. The net sheet concentration per period is

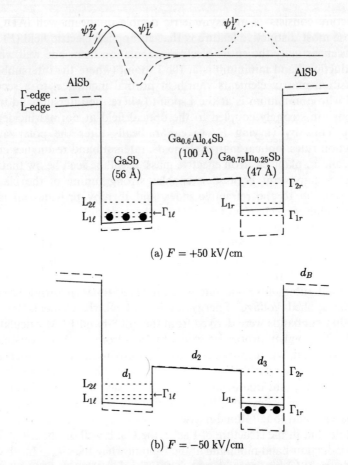

Fig. 2 Γ-point (dashed) and L-point (solid) conduction band profiles, quantized energy levels, and wavefunctions (L-point only) for an AlSb-GaSb-$Ga_{0.6}Al_{0.4}$Sb-$Ga_{0.75}In_{0.25}$Sb-AlSb ADQW at applied fields of (a) $+50$ kV/cm and (b) -50 kV/cm.

given by

$$N_s = \sum_{B=L,\Gamma} N_{sB}, \quad (3.1)$$

where

$$N_{sB} = \sum_{w=\ell,r} \sum_j \frac{2\nu_B}{(2\pi)^2} \int_0^\infty d^2k_\parallel f_0[E_{jB}^w(k_\parallel, F)], \quad (3.2)$$

w is the left or right well, j is the subband index, the valley degeneracies are $v_L = 4$ and $v_\Gamma = 1$, $k_\parallel$ is the in-plane wavevector, $E^w_{jB}(k_\parallel, F)$ is the field-dependent energy of a given state, and

$$f_0 = \frac{1}{\exp[(E^w_{jB} - E_F)/k_B T] + 1}, \qquad (3.3)$$

is the Fermi distribution function. The Fermi energy $E_F(F)$ is self-consistently adjusted to assure such that $N_{s\Gamma}(F) + N_{sL}(F)$ is equal to the net doping concentration, which remains fixed as the field is varied.

Before Eqs. (3.1)–(3.3) may be evaluated, we must first determine the relation between the actual electric field in the active region (F_{act}) and the applied field (F), which is opposed by an internal field (F_{int}) induced by the charge separation. For definiteness, we assume that the ionized donors responsible for the n-type doping are uniformly distributed throughout the active double well and spacer region, which has a total thickness of $d_1 + d_2 + d_3$. From elementary capacitance theory, the internal field is then

$$F_{int} = \frac{2\pi(N_{s\ell} - N_{sr})e}{\kappa}, \qquad (3.4)$$

where $N_{s\ell}$ and N_{sr} are the total electron concentrations present in the left and right wells (summed over L and Γ states) under given external bias conditions and κ is the net static dielectric constant. The voltage drop across each period of the structure is then $V \approx \mathcal{N}[F_{act}(d_1/2 + d_2 + d_3/2) + (F_{act} - F_{int})(d_3/2 + d_B + d_1/2)]$ and the average applied field F is:

$$F = F_{act} - F_{int}\left[1 - \frac{d_1/2 + d_2 + d_3/2}{d}\right], \qquad (3.5)$$

where $d \equiv d_1 + d_2 + d_3 + d_B$ is the total period of the ADQW, $\mathcal{N}$ is the number of periods, F_{int} is taken from Eq. (3.4), and it should be remembered that F_{act} and F_{int} have opposite signs. For a doping level of $N_s \approx 1 \times 10^{12}\,\text{cm}^{-2}$, complete population transfer requires that approximately half of the applied field ($F \approx 50\,\text{kV/cm}$) be devoted to overcoming the internal field, i.e., to the second term of Eq. (3.5), while for somewhat heavier doping levels (e.g., $N_s > 2 \times 10^{12}\,\text{cm}^{-2}$) internal fields tend to dominate the voltage requirements.

The shift of the relative energy difference between the L-valley (left well) and Γ-valley (right well) minima, which is responsible for modulating their relative populations, is determined primarily by the product of the local field (F_{act}) and the distance separating the centers of the two wells ($\approx d_1/2 + d_2 + d_3/2$). While this consideration appears to favor a very thick spacer layer, two factors limit the optimum value of d_2 to the order of ≈ 100 Å. First, if the spacer layer is too thick, the interwell transfer following a field reversal becomes too slow since the tunneling time varies exponentially with d_2. And second, a large d_2 ultimately limits the fill factor and hence the net absorption coefficient, since there can be fewer double-well repeats per unit thickness if the spacer comes to represent the dominant fraction of each total period.

The Γ–L EO modulator may be operated either cooled or uncooled (for which somewhat higher voltages must be applied to achieve the same degree of modulation [48]). Although ambient-temperature operation is essential for some applications, $T_{op} = 77$ K is often convenient because many IR systems are cooled anyway, e.g., if the modulator is to be used as an absorption filter for enabling multi-spectral LWIR and/or MWIR detection. For the example of $d_2 = 100$ Å, $N_s = 1 \times 10^{12}$ cm^{-2} electrons per period, and $T_{op} = 77$ K, a relatively-modest applied field of ± 50 kV/cm is sufficient to switch from $< 4\%$ to $> 94\%$ of the electrons populating L-states in the left well, and slightly higher fields produce almost complete emptying and filling [33]. This controllable transfer of electrons from L states to Γ states provides an effective and flexible mechanism for modulating both the IR absorption coefficient and the refractive index.

The absorption coefficient associated with intersubband transitions can be derived using the standard perturbation solution for dipole transitions in an optical field. Ignoring depolarization corrections, that yields [49]

$$\alpha_{\text{isb}}^{ij}(\hbar\omega) = \frac{\gamma^2 \alpha_0^{ij}}{(\hbar\omega - E_{ij})^2 + \gamma^2}, \quad (3.6)$$

where i and j are the initial and final subband indices, $E_{ij} \equiv E_j - E_i$ is the field-dependent intersubband splitting, and we have phenomenologically introduced a Lorentzian broadening parameter γ (half of the full-width at half maximum) which can in practice be dominated by either collisions, nonuniformity, or nonparabolicity [50–52]. In cgs units, the peak absorption is given by

$$\alpha_0^{ij} = \frac{4\pi n e^2 (N_s^i - N_s^j) \omega z_{ij}^2}{(d_1 + d_3)\kappa c \gamma}, \quad (3.7)$$

where n is the refractive index, c is the speed of light, N_s^i is the electron density populating a given subband, and we have followed the common practice of normalizing to the well thickness (here the sum of the two well thicknesses). The $\hat{z}$-component of the dipole matrix element is related to the inverse mass tensor components $(1/m_{mn})$ through the relation [53]:

$$z_{ij} \equiv \langle i|z|j \rangle \propto \left(\frac{\hat{e}_x}{m_{zx}} + \frac{\hat{e}_y}{m_{zy}} + \frac{\hat{e}_z}{m_{zz}} \right), \qquad (3.8)$$

where $\hat{e}$ is the unit polarization vector of the optical electric field. Thus for an isotropic electron valley (i.e., Γ) with $1/m_{zx} = 1/m_{zy} = 0$ we have $z_{ij} \propto \cos\theta_z$, where θ_z is the angle between $\hat{e}$ and the $\hat{z}$-axis, and z_{ij} must vanish at normal incidence. However, for a highly anisotropic electron valley (i.e., L) for which $1/m_{zx} = (2^{1/2}/3)(m_l - m_t)/3m_t m_l$ and $m_{zy} = 0$ for [100] growth, z_{ij} can be substantial even for $\hat{e}$ in the $\hat{x}-\hat{y}$ plane. In what follows, we may therefore ignore any contribution to the normal-incidence absorption coefficient by Γ-valley electrons and take $\alpha_{\text{isb}} \propto N_{\text{sL}}$.

An important consequence is that rather than simply inducing a shift of the resonance wavelength as for a conventional QCSE or intersubband modulator, here the field actually turns the absorption line on and off [28,48]. This is illustrated in Fig. 3, which indicates a high contrast ratio (over 20:1) between the positive-field (solid curve) and negative-field (dashed curve) normal-incidence peak absorption coefficients. The switching fields are somewhat higher here than in the example discussed above due to the larger doping density of $N_s = 2 \times 10^{12}$ cm^{-2} that has been assumed.

Preliminary experimental data confirming the effectiveness of this EO mechanism are given in Fig. 4, which shows results for a Γ-L ADQW modulator fabricated and tested at Columbia University [54]. The structure was the same as in Fig. 2, except that the Ga$_{1-x}$In$_x$Sb Γ well was replaced by a 12-Å-thick InAs quantum well [48] and the GaSb well thickness (d_1) was reduced to 30 Å for operation at MWIR rather than LWIR wavelengths. The device contained 25 periods, and the 77 K absorption spectra (relative to the zero-field result) shown in the figure correspond to various bias voltages between 0 (bottom curve, for which most electrons occupy Γ-valley states in the InAs wells), and 14 V (top curve, for which most of the electrons occupy L-valley states in the GaSb wells). The maximum absorption modulation of 3200 cm^{-1} may be expected to increase substantially once narrower linewidths are achieved.

We next show that the same Γ-L ADQW configuration is also suitable for phase modulation devices. It is well known, e.g., from the Kramers–Kronig relations, that the resonant intersubband absorption

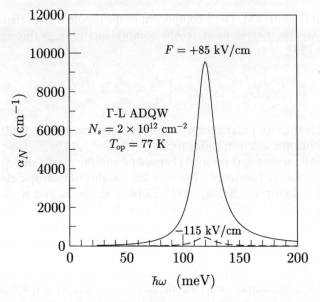

Fig. 3 Normal-incidence absorption coefficient vs. photon energy at $T = 77\,\text{K}$ and applied fields of $+85\,\text{kV/cm}$ (solid curve) and $-115\,\text{kV/cm}$ (dashed curve) for the ADQW structure shown in Fig. 2. $\gamma = 10\,\text{meV}$ is assumed here and below.

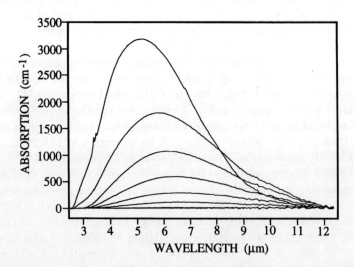

Fig. 4 Experimental absorption spectra at normal incidence ($T = 77\,\text{K}$) for a Γ–L EO modulator at various applied voltages between 0 (bottom curve) and 14 V (top curve). All α_N are relative to the value at zero field. (From Ref. [54]).

must be accompanied by a resonant contribution to the refractive index. This has the approximate form [23]

$$n_{\text{isb}}^{ij}(\hbar\omega) \approx -\frac{c\alpha_0^{ij}}{2\omega}\frac{(\hbar\omega - E_{ij})\Gamma}{(\hbar\omega - E_{ij})^2 + \Gamma^2}, \quad (3.9)$$

which has the characteristics that it changes sign at the resonance energy, has its largest magnitude one broadening parameter away from resonance ($\hbar\omega = E_{ij} \pm \Gamma_{ij}$), and decreases much more gradually than the absorption coefficient far away from resonance [as $(\hbar\omega - E_{ij})^{-1}$ in contrast to $\alpha_{\text{isb}}^{ij} \propto (\hbar\omega - E_{ij})^{-2}$]. As in the case of the absorption coefficient, n_{isb} in the Γ–L ADQW can be turned on and off by the electric field rather than simply shifted. For $N_s = 1 \times 10^{12}$ cm^{-2}, the dashed curve in Fig. 5 represents the normal-incidence modulation of n_{isb} corresponding to a field variation ΔF of -100 kV/cm ($+40 \rightarrow -60$ kV/cm).

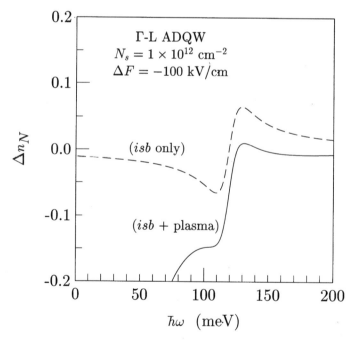

Fig. 5 Field-reversal ($+40 \rightarrow -60$ kV/cm) shift of the normal-incidence refractive index vs. photon energy at $T = 77$ K for the ADQW from Fig. 2. The dashed curve is the resonant intersubband contribution alone, while the solid curve also includes the free-carrier plasma contribution.

While the index shift is seen to be strong, a serious limitation of modulators based solely on the intersubband interactions is that the resonant Δn is largest in regions where there is also significant absorption loss [see Eqs. (3.6) and (3.9)]. If we define L_π to be the propagation length required to induce a phase shift of π, the figure of merit

$$L_\pi \alpha = \frac{\pi c \alpha}{\omega |\Delta n|} \tag{3.10}$$

must be kept as small as possible if insertion losses are to be minimized. For a conventional stepped well or ADQW in which the main effect of the applied field is to shift the transition energy by an amount ΔE_{12} (without inducing any intervalley transfer), it is apparent from Eqs. (3.6) and (3.9) that the ratio $\alpha_{isb}/\Delta n_{isb}$ will be minimized well away from resonance, where

$$L_\pi^{isb} \alpha_{isb}^{12} \approx \frac{2\pi\gamma}{\Delta E_{12}}. \tag{3.11}$$

This implies that for realistic values of γ (at least $\approx 5\,\text{meV}$) and ΔE_{12} (less than $\approx 40\,\text{meV}$), it becomes nearly impossible to design a modulator with $L_\pi^{isb} \alpha_{isb}^{12} \ll 1$ [33].

On the other hand, a crucial advantage of the Γ–L ADQW structure is that the interwell transfer modulates not only n_{isb}, but also the plasma contribution to the refractive index [33]. This has the form [55]

$$n_p(F) = -\sum_B^{\Gamma L} \frac{2\pi n N_{sB}(F) e^2}{\kappa \omega^2 d m_\parallel^B}, \tag{3.12}$$

where $m_\parallel^B$ is the in-plane effective mass for Γ-valley or L-valley electrons. In conventional Γ–Γ asymmetric designs, $\Delta n_p \approx 0$ because the total density is fixed by the doping level and the field modulation of the energy levels and wavefunctions is not accompanied by any appreciable modification of the Γ-valley in-plane mass. However, in the Γ–L ADQW the plasma contribution to the index shift can be quite large, since the effect of the field reversal is to transfer the carriers from L states with a heavy in-plane mass $[1/\langle (m_\parallel^L)^{-1}\rangle_{avg} \to m_{zz}^L \approx 0.2 m_0]$ in one well to Γ states with a much lighter mass $(m_\parallel^\Gamma \approx 0.05 m_0)$ in the other well. The solid curve in Fig. 5 represents the net index change due to the combined contributions by the resonant intersubband and plasma processes.

Besides increasing the net $|\Delta n|$ at all photon energies below the resonance energy, the plasma mechanism is particularly attractive for phase modulation devices because it remains large in regions where the resonant intersubband absorption is weak (compare to Fig. 3). Thus the Γ–L ADQW can yield $L_\pi \alpha \ll 1$ as long as one moves far away from the intersubband resonances. To extend the low-insertion-loss region to larger $\hbar\omega$, it is necessary only to increase the resonance energy E_{12}^L by employing thinner quantum wells. For example, $E_{12}^L = 180$ meV yields $L_\pi \alpha < 0.3$ for operation at $\hbar\omega = 120$ meV [33].

We finally consider whether it is possible to employ the Γ–L ADQW structure as a low-insertion-loss phase modulator when the beam propagates in the plane. For TE waveguide modes ($\theta_z = 90°$) the operation is quite analogous to that discussed above, since the optical polarization again lies in the plane as for normal incidence. Furthermore, a similar device is also suitable for TM modes ($\theta_z = 0°$), although the nature of the modulation becomes qualitatively different because L-valley states then provide the only plasma contribution to the dielectric constant. In the TM case $n_p \to 0$ when the electrons occupy Γ-valley states, because the growth-direction mass in the quantum well is effectively infinite and the optical electric field along that axis does not interact with plasma oscillations in the plane. However, voltage-induced transfer to the anisotropic L-valleys results in a significant n_p, since the off-diagonal mass components couple the in-plane motion to growth-direction fields. The insertion loss will again be quite small as long as the intersubband resonance energies are tuned far away from $\hbar\omega$.

Note that because the conduction-band offset between GaSb and AlSb is on the order of 1 eV for both Γ and L states [56,57], the same basic scheme may be employed for EO modulators spanning IR wavelengths between 1.5 and 20 μm. The large band offset also favors a high fill factor, since even a relatively thin AlSb barrier (e.g., ≈ 70 Å) will be quite effective in preventing leakage due to tunneling from one period to the next. We finally note that the large L-point density of states is advantageous in terms of enabling heavy doping levels accompanied by modest Fermi energies. Sheet densities per well of 1.6×10^{12} cm^{-2} have already been demonstrated [37], and considerably larger N_s are probably feasible. Were such high electron concentrations employed in a Γ-valley system, the Fermi level would exceed E_{12} and states in the second subband would be populated even in the absence of excitation.

An example of an application requiring normal-incidence operation is an EO-tunable absorption filter, which in combination with a broad-band IR focal plane array would permit multi-spectral detection, since one or more modulators could be used to transmit or block selected sub-regions of the LWIR or MWIR atmospheric windows. Angular incidence is not feasible for this application due to severe constraints related to compactness, acceptance angle, and non-deviation of the beam. A second example is any configuration incorporating a Fabry–Perot etalon. In high-performance QCSE devices for the near-IR, one frequently places the EO medium within a resonant cavity in order to significantly enhance the modulation per unit active-layer thickness [18]. Ideally, this is accomplished with monolithically-grown Bragg mirrors, i.e., quarter-wave stacks layered above and below the active region. Reflectivities exceeding 0.999 have been obtained using high-index/low-index combinations such as AlGaAs/AlAs and AlGaInP/AlInP [58], and longer-wavelength antimonide-family mirrors consisting of GaSb/AlSb and $Ga_{1-x}Al_xSb$/AlSb have also been reported [59,60]. Etalon effects will significantly enhance the modulation magnitude in Γ–L EO devices, as well as the normal-incidence nonlinear optical response of the L-valley second-harmonic generators described in the next section.

We emphasize again that the utility of any EO modulator based on Γ-valley intersubband processes alone is severely restricted. Since the interaction strength is reduced by $\cos^2\theta_z$ even in cases where angular incidence may be tolerated, only in waveguide-mode applications with TM polarization does this selection rule not represent a significant limitation. On the other hand, the Γ–L intervalley-transfer mechanism provides a strong response for both surface incidence at arbitrary angle and waveguide propagation with arbitrary polarization. Quantitative calculations indicate that the EO coefficients are comparable to those associated with the excitonic QCSE, and typically the spectral bandwidth is much greater. It follows that most of the near-IR device geometries now being implemented may be mimicked at longer wavelengths if we substitute Γ–L intersubband interactions for the QCSE. This holds for both intensity and phase modulators operating either at normal-incidence or in waveguide mode (including Mach–Zehnder configurations [20].

6.4 SECOND-HARMONIC GENERATORS

The primary obstacle now blocking the progress of such important MWIR technologies as chemical sensing and IR countermeasures is the lack of inexpensive, convenient, and reliable sources capable of emitting high powers at non-cryogenic operating temperatures. While a number of researchers have addressed this need by investigating intersubband second-harmonic generation (SHG) or sum and difference frequency mixing in GaAs-based and InP-based heterostructure systems, in this section we point out several advantages of employing antimonide intersubband nonlinear processes instead. As an alternative type of source, the next section will discuss two new MWIR lasers based on type-II interband transitions in antimonide quantum wells.

If depletion and saturation are neglected, the SHG conversion efficiency is given (in cgs units) by [61]

$$\eta = \frac{128\pi^3}{n\kappa c^3}[\omega_1\chi^{(2)}_{2\omega}L']^2\frac{\sin^2(\pi L'/L_c)}{(\pi L'/L_c)^2}I_1, \qquad (4.1)$$

where ω_1 is the frequency of the pump beam, I_1 is its intensity, L_c is the coherence length (discussed below), $L' = L/\sin\theta_z$ is the propagation length, and L is the total thickness of the active material. As in the case of the absorption coefficient discussed in the preceding section, the SHG coefficient ($\chi^{(2)}_{2\omega}$) vanishes for Γ-valley states whenever the optical polarization is in the plane ($\propto \cos^3\theta_z$). However, it is perhaps not surprising that intersubband interactions in the anisotropic L-valleys of antimonide-based quantum wells can, at least under some conditions, yield a strong nonlinear response at normal incidence [62]. For a three-level system [61,63],

$$\chi^{(2)}_{2\omega} \approx \sum_{v=1}^{4}\frac{n^2e^3N^v_s z^v_{12}z^v_{13}z^v_{23}}{d\kappa(\hbar\omega - E_{12} - i\gamma)(2\hbar\omega - E_{13} - i\gamma)}, \qquad (4.2)$$

where N^v_s is the electron density in the vth L-valley and z^v_{ij} is the dipole matrix element corresponding to transitions between subbands i and j in that valley. Recall from the discussion following Eq. (3.8) that while the matrix elements for Γ-valley electrons are proportional to $\cos\theta_z$, they do not vanish at normal incidence when the electrons are in an L-valley.

It should first be emphasized that the normal-incidence $\chi^{(2)}_{2\omega}$ is *not* large for growth along the [100] axis [45,63]. In that case the four L-valleys are degenerate and equally populated and the alternating signs of the various z^v_{ij} produce canceling contributions, hence the net $\chi^{(2)}$ vanishes. One must therefore grow along a lower-symmetry axis in order to make $|z^v_{ij}|$ vary with v. The equality of the quantization masses is also then removed, which in turn breaks the energy level degeneracy and produces unequal populations in the four valleys [64].

For example, growth along [511] leads to $\chi^{(2)}_{2\omega} \approx 4 \times 10^{-8}$ m/V (normalized to the entire period rather than just the quantum well thicknesses) for a GaSb-Ga$_{1-x}$Al$_x$Sb-AlSb stepped well structure optimized for the doubling of CO$_2$ laser radiation. While this is somewhat less than the largest values reported for GaAs-based and InP-based intersubband systems [49,65-67], it is nonetheless more than two orders of magnitude larger than in bulk GaAs, and the convenience of normal incidence and the potential for significantly enhancing the conversion efficiency by placing the nonlinear medium in a Fabry–Perot cavity are attractive.

It has also been pointed out that the incorporation of an L-valley "momentum-space reservoir" should significantly improve the saturation properties of Γ-valley intersubband processes in antimonide-based SHG devices [2,68,69]. The function of the reservoir, which is optically inactive because the L-valley transition energies are tuned far away from resonance, is to remove electrons from excited states in the upper subbands and return them to the lowest subband via intervalley scattering on a time scale much faster than conventional Γ-valley intersubband relaxation times. This inhibits saturation at high pump intensities and increases the maximum η. At a pump-beam wavelength of 10.6 μm, conversion efficiencies up to 33% are predicted [69] for waveguide geometry, although the excitation levels required to achieve that value are quite high ($I_1 > 100$ MW/cm^2).

We next consider an antimonide-based second-harmonic generation device which allows one to "tune in" the crucial phase-matching condition electrically, simply by varying an applied bias voltage. While intersubband processes in GaAs-Al$_x$Ga$_{1-x}$As and In$_{1-x}$Ga$_x$As-In$_{1-x}$Al$_x$As stepped and double quantum wells have yielded large $\chi^{(2)}_{2\omega}$ [49,66,67], the net conversion efficiencies reported thus far have never reached 1% [67]. This performance has been limited in part by short phase coherence lengths

$$L_c = \frac{\lambda}{4\delta n}, \qquad (4.3)$$

where $\delta n \equiv n(2\omega_1) - n(\omega_1)$ is the difference between the refractive indices at the pump and second-harmonic frequencies. As the beams propagate, any index mismatch will lead to a gradual loss of phase coherence that produces an oscillation of the direction of power flow between the two beams rather than a one-way conversion of pump intensity to second-harmonic intensity. Some form of phase matching will therefore be required if intersubband-based frequency-conversion devices are to be realized with L_c long enough to be of practical interest. To date, most techniques for phase matching have relied on either birefringence [70,71], which applies only to restricted classes of nonlinear materials, or quasi-phase matching [72], in which the optical medium is modulated to reverse the sign of $\chi^{(2)}$ every coherence length. However, a third alternative is to exploit the large intersubband contribution to the refractive index [73] and its high degree of tunability with applied field as a means of compensating for the bulk-like mismatch related to phonon processes (δn_b) [74]. That is, we can electrically impose the phase-matching condition: $\delta n = \delta n_b + \delta n_{isb}(F) \to 0$ by adjusting F [69,75].

Fig. 6 illustrates an antimonide-family ADQW designed to exploit this possibility. We consider waveguide geometry, since it is impractical

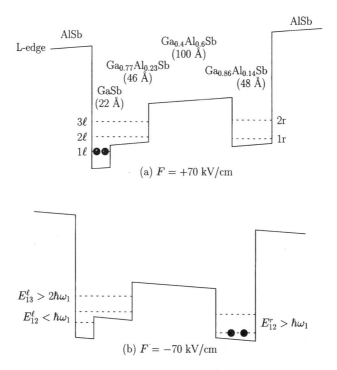

Fig. 6 Conduction band profiles and quantized energy levels for an L–L ADQW, at applied electric fields of (a) + 70 kV/cm and (b) − 70 kV/cm (adequate for 300 K operation). The left well is stepped in order to induce the asymmetry required for second-harmonic generation.

to achieve path lengths long enough for optimum conversion efficiency using surface incidence [69] when the active quantum wells must be grown by MBE or MOCVD (although a Fabry–Perot alternative will be discussed briefly below). Since intervalley transfer is not required for this application, the compositions have been chosen such that both quantum wells have their minima at the L-point. As in the Γ–L ADQW, however, we retain the capability of transferring electrons from one well to the other by varying the applied electric field. Note also that the left well is stepped in order to insure a large value of z^{ℓ}_{13} (which vanishes due to symmetry in a square well).

The L–L ADQW layer thicknesses and compositions are designed such that for a pump-beam photon energy of $\hbar\omega_1 = 120\,\text{meV}$: $E^{\ell}_{12} < \hbar\omega_1 < E^{\text{r}}_{12}$ while $E^{\ell}_{13} < 2\hbar\omega_1$. Eq. (3.9) may then be used to obtain the dependences of n_{isb} on photon energy illustrated in Fig. 7 for positive (solid curve) and negative (dashed curve) applied fields. We emphasize first that for photon energies near $\hbar\omega_1$, reversal of the bias polarity leads to an especially large modulation of the index. Note also that as long as the electrons are in the left well ($F > 0$) we have $n_{\text{isb}}(\omega_1) < 0$ and $n_{\text{isb}}(2\omega_1) > 0$, implying that δn_{isb} is positive. On the other hand, when

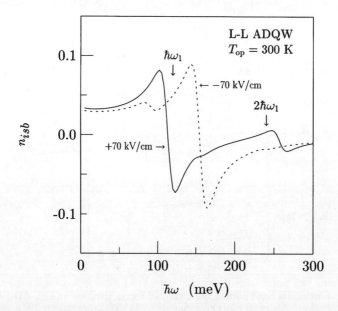

Fig. 7 Intersubband contribution to the refractive index vs. photon energy at two applied biases for the L–L ADQW from Fig. 6 ($N_s = 1 \times 10^{12}\,\text{cm}^{-2}$).

the electrons are in the right well ($F < 0$) the signs of n_{isb} at the two frequencies are reversed and δn_{isb} is negative. The result is that we can use the applied bias to tune n_{isb} to whatever positive or negative value is required to compensate δn_b and achieve phase matching.

Quantitative results for these trends are illustrated in Fig. 8, which plots the dependences of $n_{\text{isb}}(\omega_1)$, $n_{\text{isb}}(2\omega_1)$, and δn_{isb} F. As expected from the discussion above, the indices at the pump and second-harmonic frequencies both vary monotonically with increasing F, but in opposite directions. Since the relative shift spanning the range -0.07 to $+0.07$ significantly exceeds typical values for the bulk mismatch ($0.02 < \delta n_B < 0.05$), we can "tune in" whatever compensation value is required to achieve phase matching under given operating conditions. Furthermore, this configuration is quite forgiving in terms of fabrication tolerances (layer thicknesses, doping levels, waveguide properties, etc.), prior knowledge of the bulk-like index dispersion, and theoretical uncertainties, as well as variations of δn_{isb} with intensity near and above

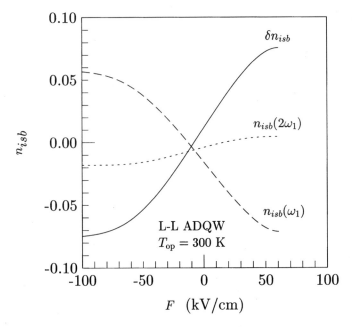

Fig. 8 Intersubband contribution to the refractive index vs. applied field at the fundamental ($\hbar\omega_1 = 117$ meV, dashed curve) and doubled (234 meV, dotted curve) photon energies, and the mismatch between the two (solid curve), for the L-L ADQW from Fig. 6 ($N_s = 1 \times 10^{12}$ cm^{-2}).

saturation [75]. The operational bandwidth is also expanded since the bias voltage can be readjusted for each pump wavelength of interest, and the same device can achieve phase matching of either second harmonic generation or non-degenerate frequency mixing. A final advantage is that the second-harmonic output may be rapidly modulated with a high on/off contrast ratio, since the phase mismatch can be electrically tuned for maximum or minimum conversion efficiency [69].

A more detailed analysis of bias-controlled intersubband frequency-mixing devices may be found in Ref. [69]. That work demonstrates that the highest conversion efficiencies should be attainable in voltage-tunable structures having two regions with distinct ADQW layer thicknesses: one active set of wells devoted to second-harmonic generation along with a second set whose sole function is to impose phase matching.

As mentioned above, an alternative approach to assuring a long effective path length for high conversion efficiency is to operate at normal incidence (using off-axis growth) and place the active region in a Fabry–Perot etalon, e.g., using Bragg mirrors. This option is not available if one considers only Γ-valley intersubband processes in GaAs-based and InP-based quantum wells, since then the angular incidence required to produce a non-vanishing SHG is incompatible with the etalon geometry. Fabry–Perot frequency-conversion devices may also naturally take advantage of bias tuning to impose phase-matching and assure that the coherence length is longer than the effective path length.

6.5 TYPE-II MWIR LASERS

Nearly all of the interband semiconductor lasers developed previously have employed active-region constituents having a type-I band alignment, since large optical matrix elements are required if appreciable gain is to be generated. However, it is known that type-II InAs-Ga$_{1-x}$In$_x$Sb superlattices can display strong interband absorption as long as the layers are thin enough to allow significant interpenetration of the electron and hole wavefunctions [76,77]. It follows that gain should also be attainable from this and other wavefunction-engineered type-II configurations as long as the wavefunction overlap is sufficiently large. In fact, it has been pointed out that MWIR lasers based on type-II antimonides should have significant advantages [2,78–80] over any of the analogous type-I devices studied to date. This section will discuss recent theoretical and experimental progress toward the development of high-performance MWIR type-II quantum well lasers (T2QWLs). We also describe a novel type-II interband cascade laser (T2ICL), which combines the advantages of the bipolar T2QWL with the electron recycling feature of the intersubband-based unipolar quantum cascade laser (QCL) demonstrated by Faist et al. [81].

For visible and near-IR wavelengths, there is an established and relatively mature diode laser technology employing GaAs-based and InP-based active gain regions. More recently, InP-based and antimonide lasers emitting out to $\lambda \approx 2\,\mu m$ have imroved to the point where their ambient-temperature performance approaches that of the shorter-wavelength devices [82,83]. However, the development of analogous III–V devices with non-cryogenic operating temperatures and high output powers at MWIR wavelengths has proven to be much more challenging [84–92] (current IV–VI [93] and II–VI [94] MWIR lasers are similarly inadequate). This rapid decline in performance beyond $2\,\mu m$ may be attributed primarily to losses associated with Auger recombination, although other factors such as poor electrical confinement can also contribute. The increasing importance of the Auger process at longer wavelengths becomes obvious if we examine the functional dependence of the Auger coefficient (γ_3) on emission wavelength as estimated by setting $\hbar\omega = E_g$. Experimental values for a variety of III–V systems [95–101] (points) and HgCdTe [102] (curve) at 300 K are plotted in Fig. 9. The dashed curve represents an approximate boundary between the radiative and Auger regimes, i.e., for γ_3 values below the curve emission into the lasing mode dominates the recombination at high injection levels, while above the curve non-radiative Auger losses dominate. Note that while all of the data points corresponding to $\lambda \leqslant 2.2\,\mu m$ fall below the boundary, all γ_3 reported to date for $\lambda \geqslant 3.0\,\mu m$ lie above. Thus the dramatic degradation of laser performance with wavelength between 2 and $3\,\mu m$ (Fig. 1) is precisely what is expected as a consequence of non-radiative Auger losses.

For the different type-I narrow-gap III–V systems (open points), the variation of γ_3 with λ is seen to roughly follow the $Hg_{1-x}Cd_xTe$ alloy curve. In contrast, however, the data for type-II quantum wells [101,100] (filled circles) fall as much as an order of magnitude below that trend (over two orders of magnitude Auger suppression has been obtained [103] at 77 K). This is because intervalence resonances can be removed [78] and the T2QW band structure can be engineered so as to frustrate the simultaneous conservation of energy and momentum which must occur in any CHHH Auger event (in which the

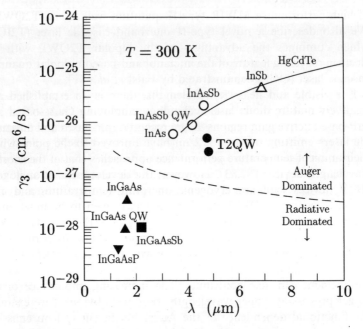

Fig. 9 Experimental Auger coefficients vs. wavelength at 300 K for a number of III–V binaries, alloys, and quantum wells [95–101], the dependence on alloy composition for $Hg_{1-x}Cd_xTe$ [102] (solid curve), and the approximate boundary between the Auger-dominated and radiative-dominated regimes at high injection levels (dashed curve).

recombination of an electron and heavy hole, CH, is accompanied by the excitation of a second hole to a lower valence-band state, HH). In fact, Grein *et al.* have predicted on the basis of detailed Auger calculations that optimized type-II MWIR structures will actually be radiative-limited up to 300 K [78]. This projection naturally has important implications for the ultimate feasibility of high-performance MWIR lasers operating at ambient temperatures.

Researchers at Hughes have recently demonstrated stimulated emission in the 3–5 μm range from type-II diode lasers containing active regions consisting of 4-1/2-period InAs-$Ga_{0.75}In_{0.25}$Sb superlattices [79,104,105]. While these devices have displayed quite promising characteristics, a superlattice active region is ultimately non-optimal due to its energy dispersion along all three coordinate axes [106]. It is well known that quantum well lasers with quasi-2D electron and hole populations tend to significantly outperform [107–109] double heterostructure lasers with 3D bulk carriers once a given fabrication technology has matured. This is primarily due to the much higher gain per injected carrier at threshold for the more concentrated 2D density of states [110]. While the holes in a type-II InAs-Ga$_{1-x}$In$_x$Sb superlattice generally have minimal dispersion along the growth axis (i.e., they are nearly quasi-2D), the strong penetration of the electron wavefunctions into the thin Ga$_{1-x}$In$_x$Sb barriers leads to a nearly isotropic electron mass ($m_{nz}/m_{n\parallel} \approx 1.2$ from the FEM calculation). Thus electrons distributed over a broad range of k_z participate in the spontaneous emission, while only those near the zone center actually contribute to the gain at the lasing frequency.

Fig. 10 illustrates the Γ-valley conduction, valence, and split-off band profiles, along with the energy levels and wavefunctions calculated using the FEM formalism for a wavefunction-engineered four-constituent type-II multiple quantum well (InAs/Ga$_{1-x}$In$_x$Sb/InAs/AlSb) which preserves the large optical matrix elements of the InAs/Ga$_{1-x}$In$_x$Sb superlattice but has 2D dispersion relations for both electrons and holes [80]. Note that while the electron ground state is split into symmetric (E1S) and antisymmetric (E1A) levels due to the double InAs layers in each period, the energy separation of 120 meV is large enough to insure that virtually all of the injected electrons will occupy E1S levels under typical operating conditions. We find that even though the electron wavefunctions (solid curves) have their maxima in the InAs layers and the hole wavefunctions (dashed curves) are centered on the Ga$_{1-x}$In$_x$Sb, their overlap is sufficient to yield interband optical matrix elements more than 70 % as large as those in typical type-I heterostructures.

Furthermore, CHHH Auger recombination is suppressed because the energy gap does not resonate with any valence intersubband transitions involving H1 near its maximum (it falls approximately halfway between H1–H2 and H1–H3). In particular, the split-off band is now far too low to provide a final hole state for CHHS Auger processes, which often dominate the Auger lifetime (τ_A) in type-I structures containing InAs or InAs-rich alloys [84,99]. Hence *all* multi-hole Auger processes are energetically unfavorable, and even CHCC Auger events (in which the CH recombination is accompanied by an electron transition to a higher-energy conduction-band state) are reduced by the small in-plane effective mass for holes near the band extremum ($\approx 0.075 m_0$). The reduced density of states which accompanies the small in-plane hole mass also serves to decrease the threshold carrier density required to achieve population inversion. The structure illustrated in Fig. 10 has AlSb cladding layers, which provide excellent

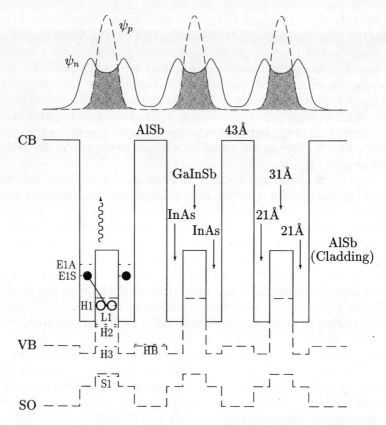

Fig. 10 Γ-valley conduction, valence, and split-off band profiles for the four-constituent T2QWL with the indicated layer thicknesses (lattice-matched to the AlSb optical cladding layers). Also shown are 8-band FEM results for the electron (solid) and hole (dashed) wavefunctions, along with energy extrema for the various conduction and valence subbands.

optical confinement as well as a nearly exact lattice match to the active quantum-well region. We finally point out that electrical confinement ceases to be an issue in antimonide-based type-II heterostructures, because the AlSb cladding layers provide large offsets for both the conduction and valence bands.

The gain above threshold is naturally closely related to the absorption coefficient below threshold, apart from dependences on the electron and hole occupation factors. For a system with 2D dispersion relations, the gain is given by [111]

$$g(\hbar\omega) = \frac{4\pi^2 n e^2}{\hbar\kappa c\hbar\omega d} \frac{2}{(2\pi)^2} \int_0^{2\pi/a} d^2k_\parallel \frac{(\gamma/\pi)|P_{CV}(k_\parallel)|^2(f_e + f_h - 1)}{[E_e(k_\parallel) - E_h(k_\parallel) - \hbar\omega]^2 + \gamma^2}, \quad (5.1)$$

where we take $\gamma \approx 4$ meV based on experimental transport results for n-type and p-type InAs-Ga$_{1-x}$In$_x$Sb superlattices [112] and the electron and hole quasi-Fermi energies (E_{Fe} and E_{Fh}) which govern the Fermi distribution functions [$f_i(k_\parallel, E_{Fi})$] are related to the injected sheet carrier concentration per period through Eqs. (3.2) and (3.3). The concentration is in turn related to the injection current by the expression

$$j(N_s) = \mathcal{N} N_s e \left(\frac{1}{\tau_R(N_s)} + \frac{1}{\tau_{NR}(N_s)} \right) + j_L \equiv j_R + j_{NR} + j_L, \quad (5.2)$$

where $\mathcal{N}$ is again the total number of periods in the active region, τ_R is the radiative lifetime, $\tau_{NR} \approx \tau_A$ is the nonradiative lifetime, and j_L is the leakage current resulting from inadequate electrical confinement. The radiative lifetime due to spontaneous emission processes has the form [111]

$$\frac{1}{\tau_R} = \frac{4n^3 e^2}{\hbar^4 \kappa c^3 N_s} \int_0^\infty \hbar\omega \, d(\hbar\omega) \frac{2}{(2\pi)^2} \int_0^{2\pi/a} d^2k_\parallel \frac{(\gamma/\pi)|P_{CV}(k_\parallel)|^2 f_e f_h}{[E_e(k_\parallel) - E_h(k_\parallel) - \hbar\omega]^2 + \gamma^2}. \quad (5.3)$$

Energy-level and matrix-element dispersion relations from the 8-band FEM algorithm were used in Eqs. (5.1)–(5.3) to calculate radiative lifetimes and gain spectra for the type-II quantum well laser from Fig. 10 [80]. Defining g_{max} to be the maximum in $g(\hbar\omega)$ corresponding to a given injection current density, the solid curve in Fig. 11 plots the predicted dependence of g_{max} on j_R at an operating temperature of 300 K (the dashed curve will be discussed below). The calculation assumes $\mathcal{N} = 12$, for which the optical confinement factor representing the overlap of the optical mode with the active region is $\approx 15\%$. Due to the 2D density of states, the current required to achieve a given gain is slightly lower and the gain curve is much steeper for the T2QWL than for an analogous type-II superlattice structure with 3D electron dispersion and the same energy gap. The larger $\partial g/\partial N_s$ in the quantum well leads to a significantly smaller linewidth enhancement factor (≈ 1.7, vs. ≈ 4.2 in the superlattice) and hence reduced tendency toward filamentation. The arrow on the curve indicates that the net parasitic losses due to reflection at the facets and free carrier absorption in the cladding layers and active region are estimated to be on the order of

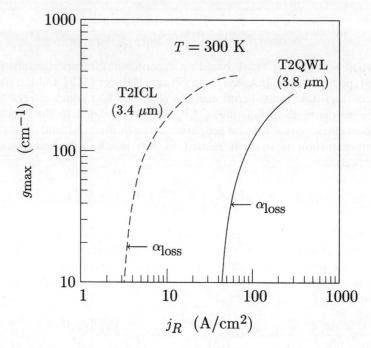

Fig. 11 Maximum gain vs. radiative contribution to the current density at an operating temperature of 300 K, for the type-II quantum well laser (solid curve, which assumes a 12-period structure with active-region optical confinement factor ≈ 0.15) and the type-II interband cascade laser (dashed curve, which assumes a 25-period structure with optical confinement factor ≈ 0.79).

40 cm^{-1} for this structure. The threshold current required to achieve transparency is then ≈ 60 A/cm^2, which would be attractive even when compared to typical results for near-IR GaAs-based and InP-based quantum well lasers operating at ambient temperature. The dependence on T yields a favorable characteristic temperature $T_0 \equiv j_{th}(T)/j'_{th}(T)$ of 350 K. While the analysis has thus far been based on the theoretical prediction [78] that optimized InAs-Ga$_{1-x}$In$_x$Sb structures are radiative-limited up to 300 K, it is also useful to examine the opposite "worst-case" limit, in which we take the Auger coefficient to be no better than the best value already observed for an MWIR type-II structure at 300 K (see Fig. 9) [101]. This upper bound for γ_3 leads to $j_{NR} \approx 1200$ A/cm^2, which is less attractive than the j_R from Fig. 11 but is nonetheless a factor of ≈ 50 lower than an extrapolation from the best current MWIR ($\lambda \geqslant 3$ μm) diode thresholds [88]. The

temperature dependence then yields $T_0 = 70$ K, which considerably exceeds the values in the 17–30 K range reported to date for III–V interband diodes operating at $T \geqslant 120$ K and emitting at $\lambda \geqslant 3.2$ μm.

The first T2QWLs have quite recently been grown at University of Houston and characterized at NRL [113]. Broad-area laser bars were cleaved from MBE-grown multiple quantum wells having the four-constituent design and layer thicknesses illustrated in Fig. 10. Although electrically-pumped diode sources will ultimately have much broader utility, these preliminary devices were optically pumped by pulses from -switched Nd:YAG ($\lambda_{pump} = 1.06$ μm) and Ho:YAG ($\lambda_{pump} = 2.06$ μm) lasers. At 87 K, the photoluminescence spectrum exhibited a broad peak at 4.17 μm, while above the lasing threshold the spectrum narrowed by a factor of 15. The peak lasing wavelength of 3.86 μm at 81 K red-shifted to 4.04 μm at 270 K, due to the temperature shift of the energy gap. Peak output powers were 650 mW at 81 K and 10 mW at 270 K.

Fig. 12 illustrates the observed threshold intensity for lasing (I_{th}) as a function of temperature (points). While the value* of 3 kW/cm^2 at 81 K is somewhat higher than that observed by Le *et al.* for pulsed optical pumping ($\lambda_{pump} = 2.1$ μm) of an InAs$_x$Sb$_{1-x}$GaSb double heterostructure laser emitting at ≈ 4.06 μm (77 K) [89] here the increase with increasing temperature is much more gradual. The slope of the dependence yields $T_0 = 96$ K for $T_{op} < 170$ K and $T_0 = 35$ K for 170 K $\leqslant T_{op} \leqslant 270$ K, with much lower I_{th} than earlier results for the temperature range $T_{op} > 170$ K which is accessible with a TE-cooler. Lasing continued up to $= 285$ K, which is more than 70 K higher than any previously-reported T_{max} for an optically-pumped or electrically-pumped III–V interband laser with $\lambda \geqslant 3.2$ μm

A quantitative analysis [113] of the threshold data presented in Fig. 12 implies an Auger lifetime considerably shorter than that obtained at the same carrier concentrations in the earlier T2QW Auger experiments [46,101] (filled circles in Fig. 9). Significant further improvements may therefore be expected from MWIR laser structures with layer thicknesses better optimized for Auger suppression.

A new type of semiconductor laser that has justly been heralded as a revolutionary breakthrough is the quantum cascade laser demonstrated by researchers at AT&T Bell Labs [84,114,115]. In a staircase of In$_{1-x}$Ga$_x$As-In$_{1-x}$Al$_x$As coupled quantum wells lattice-matched to an InP substrate, unipolar electron injection produces a subband

---

*Note added in proof: Subsequent experiments have determined that the high thresholds for this early device were due to poor facet quality.

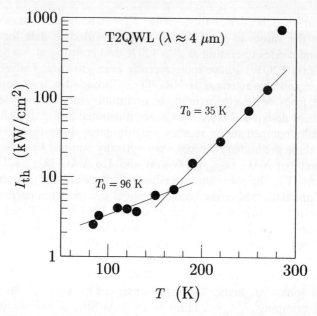

Fig. 12 Threshold pump intensity vs. temperature. The straight-line fits corresponds to characteristic temperatures of 96 K for $T_{op} < 170$ K and 35 K for 170 K $< T_{op} < 270$ K. (From Ref. [113]).

population inversion and lasing results from stimulated intersubband transitions. Moderate cw powers have been demonstrated (e.g., 17 mW at $T = 50$ K) [114], and $T_{max} = 320$ K was recently reported for $\lambda \approx 5\,\mu$m in pulsed mode [115]. A key element is the "electron recycling" which accompanies the cascade nature of the injection. That is, in contrast to conventional diode lasers in which a maximum of one photon can be emitted for every electrically-injected electron and hole, in the QCL each electron entering the device can in principle produce an additional photon for each period. However, the threshold current is unavoidably high since the non-radiative relaxation time associated with intersubband optical phonon processes is orders of magnitude shorter than the radiative lifetime [116], a relationship which is inherent to the QCL design.

Here we discuss an alternative configuration that is unique to the antimonides, the type-II interband cascade laser [117,118]. The T2ICL retains the electron recycling advantage of the cascade geometry while effectively eliminating the phonon relaxation path. As a consequence, the threshold current can potentially be reduced by orders of magnitude since only Auger recombination remains as a significant non-radiative decay mechanism, and we have seen above that Auger losses can be strongly suppressed in type-II devices.

A typical T2ICL schematic is illustrated in Fig. 13, although it should be emphasized that the concept is quite flexible and numerous alternative configurations are possible [118]. The staircase device consists of many active regions separated from one another by injection regions, each of which serves as the collector for one active region and the emitter for the next. The injection region may be composed of either a graded quaternary InAlAsSb alloy or a digitally-graded InAs–AlSb superlattice [79]. With an appropriate bias such as that illustrated in Fig. 13, electrons are injected from the left and tunnel through the first AlSb barrier into the E1 band of the InAs QW in the active region. Since the subsequent GaInSb and AlSb barriers prevent tunneling into the next injection region, the electrons may escape only by making radiative or non-radiative transitions to the first GaInSb hole QW. The radiative efficiency will be high, since the phonon relaxation process which dominates in the intersubband QCL

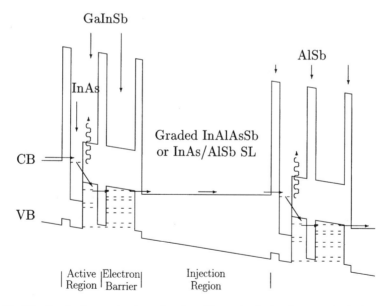

Fig 13. Conduction and valence band profiles and calculated energy levels for the type-II interband cascade laser (T2ICL). The injection is unipolar, with electrons entering at the left and leaving at the right, creating an additional photon at each step of the staircase (From Ref. [118]).

has been eliminated, and Auger losses can be strongly suppressed as in the T2QWL. Once the electrons have made optical transitions to the valence band, they tunnel in less than 100 ps into the second GaInSb QW, whose thickness is designed such that free carrier absorption due to intervalence transitions is minimized at the lasing wavelength. The electrons finally transfer to the conduction band of the next injection region through elastic interband tunneling, [119, 120], a crucial step which can only be accomplished in type-II antimonide-based structures. A cascade of photons results as the electrons descend the staircase in this manner, since there is a separate radiative transition at each active region of the device.

Whereas it is apparent from Eq. (5.2) that the injection current density needed to achieve a given population inversion in a conventional QW laser is proportional to the total number of carriers in *all* periods of the structure ($\mathcal{N} N_s$), for cascade configurations it is proportional only to the sheet concentration in a single period (N_s), since through interband tunneling each period provides the injected carriers required for the next. Detailed FEM calculations [118] analogous to those discussed above for the T2QWL yield the maximum gain as a function of j_R at 300 K represented by the dashed curve in Fig. 11. The predicted threshold current for the T2ICL is more than 3 orders of magnitude lower than the QCL result of $\approx 10{,}000\,\text{A/cm}^2$ at the same temperature, and is ≈ 300 times lower than the best QCL result to date for 77 K. Furthermore, it is considerably lower than the analogous theoretical threshold for the T2QWL (solid curve), since j_{th} is now independent of the quantum well multiplicity $\mathcal{N}$. Note also that while the magnitude of the gain is reduced by the relatively low fill factor in the cascade structure (due to the thick injection region in each period), this is more than offset by the higher optical confinement factors which will be practical since additional periods do not carry a penalty of additional injection current. The confinement factor is 79 % in the present example of a 25-period (1 µm thick) T2ICL lattice-matched to AlSb cladding layers. The larger confinement factor in the active region also reduces the net losses due to free carrier absorption, which are expected to be greater for that portion of the beam occupying the cladding layer.

We emphasize again that numerous variations on the interband cascade laser are possible. For example, while the type-II band alignment at the interface between one period and the next is a defining characteristic that must be present in order to enable the interband tunneling step which returns electrons to the conduction band, it is not essential that the active region have a type-II alignment. One may also consider a type-I interband cascade laser (T1ICL) [118], in which the wavefunctions for both electrons and holes taking part in the lasing transitions have their maxima in the same quantum well (e.g., $InAs_xSb_{1-x}$). One can even design near-IR T1ICL devices emitting at wavelengths of 1 μm or shorter [121].

6.6 CONCLUSIONS

We have illustrated just a few of the new opportunities for IR electro-optical devices based on antimonide quantum heterostructures. While preliminary antimonide optical devices have been under investigation for several years, e.g., type-II-superlattice LWIR detectors [76, 122,123], normal-incidence QWIPs employing L-valley intersubband transitions [35–39,124], and more recently type-II superlattice lasers [79,104,105], our focus in this chapter has been on exploiting "wavefunction engineering" concepts to significantly broaden the range and flexibility of the device structures. By employing quantum heterostructures with more complex multi-layer configurations than the simple two-constituent active regions used in most of the earlier work [125], the spatial distribution, band admixture, and overlap of the electron or hole wavefunctions may be manipulated so as to offer a remarkable degree of control over the energy levels, optical matrix elements, tunability with applied electric and optical fields, and other important properties relevant to device performance. This approach becomes especially powerful when combined with the diversity of the antimonide family of materials.

We have described Γ–L ADQW structures in which an applied bias transfers virtually the entire electron population from Γ states in one quantum well to L states in another. This leads to a high degree of control over and tunability of the intersubband optical interactions in either normal-incidence or waveguide-mode geometry, and in fact provides all of the elements required for a wide range of EO devices for the MWIR/LWIR that are entirely analogous to near-IR modulators based on the excitonic QCSE. These include normal-incidence and waveguide-mode intensity and phase modulators, Fabry–Perot devices, total internal reflection switches, and beam-steering devices, all of which can be adapted to wavelengths throughout the MWIR and LWIR. Modulation of either the absorption coefficient or the refractive index is particularly effective in the Γ–L process, because rather than simply shifting the optical resonances to a different energy, the applied field can actually *eliminate* them if desired. The large non-resonant plasma contribution to the index shift, which has no analog in the excitonic regime, is especially attractive since it enables the design of EO phase modulators with minimal insertion loss. The first Γ–L ADQW intensity modulator was recently demonstrated experimentally at Columbia [54].

Further possibilities emerge when additional substructures (steps, etc.) are incorporated into one or both wells of the ADQW design. For example, we considered second-harmonic generation in a waveguide-mode L–L ADQW for which the phase-matching condition can be tuned in electrically by transferring electrons between the two wells in such a way that the intersubband contribution to the refractive index fully compensates the bulk-like contribution. As long as the growth is along an axis other than [100], antimonide quantum wells with L-valley minima also yield a substantial intersubband $\chi^{(2)}$ at normal incidence. This contrasts the vanishing normal-incidence SHG for previously-studied systems with Γ-valley conduction-band minima, and opens the possibility for a significant enhancement of the conversion efficiency through placement of the active region in a Fabry–Perot cavity.

Finally, we have outlined the considerable potential for high-performance MWIR lasers based on type-II multiple quantum wells. The four-constituent T2QWL is unique in combining the advantages of (1) large conduction and valence band offsets for excellent carrier confinement, (2) potential for the strong suppression of non-radiative Auger losses (significant reduction of $\gamma_3(300\,\text{K})$ in type-IIs has already been confirmed experimentally [100,101]), (3) strong overlap of the electron and hole wavefunctions to assure large optical matrix elements, and (4) a 2D density of states for both electrons and holes to assure optimum gain per injected carrier. Detailed gain calculations yield $j_{\text{th}} < 100$ A/cm^2 for ambient-temperature operation if the Auger rate can be suppressed to the extent predicted theoretically [78]. Preliminary experimental characterization of the first optically-pumped T2QWLs ($\lambda \approx 4\,\mu\text{m}$) yielded a maximum operating temperature (285 K), which exceeded all previous results for interband III–V lasers emitting at wavelengths beyond 3 μm [113]. We have also discussed the operation of an interband cascade laser, which retains the "photon recycling" advantage of a cascade geometry while eliminating the phonon relaxation path which reduces by orders of magnitude the quantum efficiency near threshold of the intersubband QCL. Ultra-low j_{th} are predicted for the T2ICL, since the required current is reduced by a factor of the quantum well multiplicity from that for a conventional diode laser. As has been noted previously with respect to the intersubband QCL [81], the emission photon energies for both of the type-II laser configurations (T2QWL and T2ICL) are controlled almost entirely by quantum confinement rather than by the energy gaps of the constituents, and can in principle be tuned from zero to more than 1 eV.

References

1. F. Capasso, *Surf. Sci.* **142**, 513 (1984).
2. L. R. Ram-Mohan and J. R. Meyer, *J. Nonlinear Opt. Phys. Mat.* **4**, 191 (1995).
3. R. H. Miles, D. H. Chow, Y.-H. Zhang, P. D. Brewer, and R. G. Wilson, *Appl. Phys. Lett.* **65**, 1921 (1995).
4. P. M. Thibado, B. R. Bennett, M. E. Twigg, B. V. Shanabrook, and L. J. Whitman, *Appl. Phys. Lett.* **67**, 3578 (1995).
5. M. Inoue, K. Yoh, and A. Nishida, *Semicond. Sci. Technol.* **9**, 966 (1994).
6. T. Utzmeier, T. Schlosser, K. Ensslin, J. P. Kotthaus, C. R. Bolognesi, C. Nguyen, and H. Kroemer, *Solid-State Electron.* **37**, 575 (1994).
7. Y. Chen, D. M. Symons, M. Lakrimi, A. Salesse, G. B. Houston, R. J. Nicholas, N. J. Mason, and P. J. Walker, *Superlatt. Microstruct.* **15**, 41 (1994).
8. G. Tuttle, H. Kroemer, and J. H. English, *J. Appl. Phys.* **67**, 3032 (1990).
9. J. R. Meyer, C. A. Hoffman, B. V. Shanabrook, B. R. Bennett, R. J. Wagner, J. R. Waterman, and E. R. Youngdale, *Proc. 22nd Int. Conf. Phys. Semicond.* (Vancouver, 1994), p. 783.
10. *Quantum Wells and Superlattices for Long Wavelength Infrared Detectors*, Ed. M. O. Manasreh, (Artech House, Boston, 1993).
11. J. S. Sanghera, V. Q. Nguyen, P. C. Pureza, F. H. Kung, R. Miklos, and I. D. Aggarwal, *J. Lightwave Tech.* **12**, 737 (1994).
12. T. D. Golding, J. A. Dura, W. C. Wang, A. Vigliante, S. C. Moss, H. C. Chen, J. H. Miller, Jr., C. A. Hoffman, and J. R. Meyer, *Appl. Phys. Lett.* **63**, 1098 (1993).
13. J. R. Meyer, C. A. Hoffman, T. D. Golding, J. T. Zborowski, and A. Vigliante, *Proc. 22nd Int. Conf. Phys. Semicond.* (Vancouver, 1994), p. 1612.
14. Particularly intriguing is the potential for a "Super-Type-II" band alignment, i.e., an overlap of hundreds of meV between the conduction band minimum in InAs and the valence band maximum of Sb.
15. K.-J. Bathe, *Finite Element Procedures in Engineering Analysis* (Prentice-Hall, New Jersey, 1982).
16. W. H. Press, S. A. Teukolsky, W. T. Vetterling, and B. R. Flannery, *Numerical Recipes* (Cambridge University Press, 1992).
17. D. A. B. Miller, D. S. Chemla, T. C. Damen, A. C. Gossard, W. Wiegmann, T. H. Wood, and C. A. Burrus, *Phys. Rev. Lett.* **53**, 2173 (1984).
18. R.-H. Yan, R. J. Simes, and L. A. Coldren, *IEEE J. Quant. Electron.* **27**, 1922 (1991).
19. T. K. Woodward, J. E. Cunningham, and W. Y. Jan, *J. Appl. Phys.* **78**, 1411 (1995).
20. M. Fetterman, C.-P. Chao, and S. R. Forrest, *IEEE Phot. Tech. Lett.* **8**, 69 (1996).
21. D. S. Chemla, D. A. B. Miller, P. W. Smith, A. C. Gossard, and W. Wiegmann, *IEEE J. Quant. Electron.* **20**, 265 (1984).
22. P. F. Yuh and K. L. Wang, *IEEE J. Quant. Electron.* **25**, 1671 (1989).
23. D. A. Holm and H. F. Taylor, *IEEE J. Quant. Electron.* **25**, 2266 (1989).
24. S. R. Parihar, S. A. Lyon, M. Santos, and M. Shayegan, *Appl. Phys. Lett.* **55**, 2417 (1989).

25. Y. J. Mii, R. P. G. Karunasiri, K. L. Wang, M. Chen, and P. F. Yuh, *Appl. Phys. Lett.* **56**, 1986 (1990).
26. N. Vodjdani, B. Vinter, V. Berger, E. Bockenhoff, and E. Costard, *Appl. Phys. Lett.* **59**, 555 (1991).
27. E. Martinet, F. Luc, E. Rosencher, Ph. Bois, and S. Delaitre, *Appl. Phys. Lett.* **60**, 895 (1992).
28. V. Berger, E. Dupont, D. Delacourt, B. Vinter, N. Vodjdani, and M. Papuchon, *Appl. Phys. Lett.* **61**, 2072 (1992).
29. E. B. Dupont, D. Delacourt, and M. Papuchon, *IEEE J. Quant. Electron.* **29**, 2313 (1993).
30. E. B. Dupont, D. Delacourt, and M. Papuchon, *Appl. Phys. Lett.* **63**, 2514 (1993).
31. F. Capasso, C. Sirtori, and A. Y. Cho, *IEEE J. Quant. Electron.* **30**, 1313 (1994).
32. B. F. Levine, *J. Appl. Phys.* **74**, R1 (1993).
33. J. R. Meyer, C. A. Hoffman, F. J. Bartoli, and L. R. Ram-Mohan, *Appl. Phys. Lett.* **67**, 2756 (1995).
34. C.-L. Yang and D.-S. Pan, *J. Appl. Phys.* **64**, 1573 (1988).
35. H. Xie, J. Piao, J. Katz, and W. I. Wang, *J. Appl. Phys.* **70**, 3152 (1991).
36. E. R. Brown, S. J. Eglash, and K. A. McIntosh, *Phys. Rev. B* **46**, 7244 (1992).
37. L. A. Samoska, B. Brar, and H. Kroemer, *Appl. Phys. Lett.* **62**, 2539 (1993).
38. Y. Zhang, N. Baruch, and W. I. Wang, *Appl. Phys. Lett.* **63**, 1068 (1993).
39. B. Brar, L. Samoska, H. Kroemer, and J. H. English, *J. Vac. Sci. Technol. B* **12**, 1242 (1994).
40. G. Ru, Y. Zheng, and A. Li, *J. Appl. Phys.* **77**, 6496 (1995).
41. H. Xie, W. I. Wang, J. R. Meyer, C. A. Hoffman, and F. J. Bartoli, *J. Appl. Phys.* **74**, 2810 (1993).
42. H. Xie, W. I. Wang, J. R. Meyer, C. A. Hoffman, and F. J. Bartoli, *J. Appl. Phys.* **74**, 1195 (1993).
43. H. Xie and W. I. Wang, *Appl. Phys. Lett.* **63**, 776 (1993).
44. W. Xu, Y. Fu, M. Willander, and S. C. Shen, *Phys. Rev. B* **49**, 13760 (1994).
45. M. J. Shaw and M. Jaros, *Phys. Rev. B* **50**, 7768 (1994).
46. ADQW structures in which an electric field was used to transfer the electron population from one well to another were also discussed by N. Vodjdani, B. Vinter, V. Berger, E. Bockenhoff, and E. Costard, *Appl. Phys. Lett.* **59**, 555 (1991). However, in that work both wells had Γ-valley minima.
47. H. Xie, W. I. Wang, J. R. Meyer, and C. A. Hoffman, and L. R. Ram-Mohan, *J. Nonlinear Opt. Phys. Mat.* **4**, 337 (1995).
48. J. R. Meyer, C. A. Hoffman, F. J. Bartoli, and L. R. Ram-Mohan, *Proc. Narrow Gap Semicond. 1995*, *Inst. of Phys. Conf. Series* **144**, p. 330.
49. E. Rosencher and Ph. Bois, *Phys. Rev. B* **44**, 11315 (1991).
50. P. von Allmen, M. Berz, G. Petrocelli, F.-K. Reinhart, and G. Harbeke, *Semicond. Sci. Technol.* **3**, 1211 (1988).
51. H. Asai and Y. Kawamura, *J. Appl. Phys.* **68**, 5890 (1990).
52. M. Zaluzny, *Phys. Rev. B* **43**, 4511 (1991).
53. C.-L. Yang, D.-S. Pan, and R. Somoano, *J. Appl. Phys.* **65**, 3253 (1989).
54. Q. Du, J. Alperin, and W. I. Wang, *Appl. Phys. Lett.* **67**, 2218 (1995).
55. M. Ziman, *Princples of the Theory of Solids*, 2nd Edition (Cambridge University Press, 1972), Chapter 7.
56. C. Alibert, A. Joullie, A. M. Joullie, and C. Ance, *Phys. Rev. B* **27**, 4946 (1983).
57. J. Menendez, A. Pinczuk, D. J. Werder, J. P. Valladares, T. H. Chiu, and W. T. Tsang, *Solid State Commun.* **61**, 703 (1987).
58. P. L. Gourley and M. E. Warren, *J. Nonlinear Opt. Phys. Mat.* **4**, 27 (1995).
59. B. Lambert, Y. Toudic, Y. Rouillard, M. Baudet, B. Guenais, B. Deveaud, I. Valiente, and J. C. Simon, *Appl. Phys. Lett.* **64**, 690 (1994).
60. G. Tuttle, J. Kavanaugh, and S. McCalmont, *IEEE Phot. Tech. Lett.* **5**, 1376 (1993).

61. Y. R. Shen, *The Principles of Nonlinear Optics* (Wiley, New York, 1984).
62. H. Xie, W. I. Wang, J. R. Meyer, and L. R. Ram-Mohan, *Appl. Phys. Lett.* **65**, 2048 (1994).
63. M. Zaluzny and V. B. Bondarenko, *Appl. Phys. Lett.* **68**, 1872 (1996).
64. L. R. Ram-Mohan, J. R. Meyer, H. Xie, and W. I. Wang, *Appl. Phys. Lett.* **68**, 1873 (1996).
65. P. Boucaud, F. H. Julien, D. D. Yang, J. -M. Lourtioz, E. Rosencher, P. Bois, and J. Nagle, *Appl. Phys. Lett.* **57**, 215 (1990).
66. C. Sirtori, F. Capasso, D. L. Sivco, A. L. Hutchinson, and A. Y. Cho, *Appl. Phys. Lett.* **60**, 151 (1992).
67. Z. Chen, M. Li, D. Cui, H. Lu, and G. Yang, *Appl. Phys. Lett.* **62**, 1502 (1993).
68. J. R. Meyer, C. A. Hoffman, F. J. Bartoli, E. R. Youngdale, and L. R. Ram-Mohan, *IEEE J. Quant. Electron.* **31**, 706 (1995).
69. I. Vurgaftman and J. R. Meyer, *IEEE J. Quant. Electron.* **32**, 1334 (1996).
70. J. A. Giordmaine, *Phys. Rev. Lett.* **8**, 19 (1962).
71. P. D. Maker, R. W. Terhune, M. Nisenoff, and C. M. Savage, *Phys. Rev. Lett.* **8**, 21 (1962).
72. M. M. Fejer, G. A. Magel, D. H. Jundt, and R. L. Byer, *IEEE J. Quant. Electron.* **28**, 2631 (1992).
73. G. Almogy, M. Segev, and A. Yariv, *Opt. Lett.* **19**, 1192 (1994).
74. Placement of the SHG medium in a waveguide adds a further contribution to δn, which is treated in detail in Ref [69].
75. J. R. Meyer, C. A. Hoffman, F. J. Bartoli, and L. R. Ram-Mohan, *Appl. Phys. Lett.* **67**, 608 (1995).
76. D. L. Smith and C. Mailhiot, *J. Appl. Phys.* **62**, 2545 (1987).
77. R. H. Miles, D. H. Chow, J. N. Schulman, and T. C. McGill, *Appl. Phys. Lett.* **57**, 801 (1990).
78. C. H. Grein, P. M. Young, and H. Ehrenreich, *J. Appl. Phys.* **76**, 1940 (1994).
79. R. H. Miles, D. H. Chow, T. C. Hasenberg, A. R. Kost, and Y.-H. Zhang, *Inst. Phys. Conf. Ser.* **144**, 31 (1995).
80. J. R. Meyer, C. A. Hoffman, F. J. Bartoli, and L. R. Ram-Mohan, *Appl. Phys. Lett.* **67**, 757 (1995).
81. J. Faist, F. Capasso, D. L. Sivco, C. Sirtori, A. L. Hutchinson, and A. Y. Cho, *Science* **264**, 553 (1994).
82. J. S. Major, Jr., J. S. Osinski, and D. F. Welch, *Electron. Lett.* **29**, 2112 (1993).
83. H. K. Choi and G. W. Turner, *Appl. Phys. Lett.* **67**, 332 (1995).
84. M. Aidaraliev, N. V. Zotova, S. A. Karandashev, B. A. Matveev, N. M. Stus', and G. N. Talalakin, *Fiz. Tekh. Poluprov.* **27**, 21 (1993) [*Semicond.* **27**, 10 (1993)].
85. S. J. Eglash and H. K. Choi, *Appl. Phys. Lett.* **64**, 833 (1994).
86. S. R. Kurtz, R. M. Biefeld, L. R. Dawson, K. C. Baucom, and A. J. Howard, *Appl. Phys. Lett.* **64**, 812 (1994).
87. A. N. Baranov, A. N. Imenkov, V. V. Shertstnev, and Yu. P. Yakovlev, *Appl. Phys. Lett.* **64**, 2480 (1994).
88. H. K. Choi, S. J. Eglash, and G. W. Turner, *Appl. Phys. Lett.* **64**, 2474 (1994).
89. H. Q. Le, G. W. Turner, J. R. Ochoa, and A. Sanchez, *Electron. Lett.* **30**, 1944 (1994).
90. Y.-H. Zhang, R. H. Miles, and D. H. Chow, *IEEE Selected Topics in Quant. Electron.* **1**, 749 (1995).
91. H. K. Choi, G. W. Turner, and H. Q. Le, *Appl. Phys. Lett.* **66**, 3543 (1995).
92. H. K. Choi and G. W. Turner, *Appl. Phys. Lett.* **67**, 332 (1995).
93. Z. Feit, D. Kostyk, R. J. Woods, and P. Mak, *IEEE Phot. Tech. Lett.* **7**, 1403 (1995).
94. H. Q. Le, J. M. Arias, M. Zandian, R. Zucca, and Y.-Z. Liu, *Appl. Phys. Lett.* **65**, 810 (1994).

95. H. Bruhns and H. Kruse, *Phys. Stat. Sol.* (b) **97**, 125 (1980).
96. A. P. Mozer, S. Hausser, and M. H. Pilkuhn, *IEEE J. Quant. Electron.* **21**, 719 (1985).
97. P. Brosson, J. Benoit, A. Joullie, and B. Sermage, *Electron. Lett.* **23**, 417 (1987).
98. S. Hausser, G. Fuchs, A. Hangleiter, K. Streubel, and W. T. Tsang, *Appl. Phys. Lett.* **56**, 913 (1990).
99. J. R. Lindle, J. R. Meyer, C. A. Hoffman, F. J. Bartoli, G. W. Turner, and H. K. Choi, *Appl. Phys. Lett.* **67**, 3153 (1995).
100. E. R. Youngdale, J. R. Meyer, C. A. Hoffman, B. R. Bennett, J. R. Waterman, B. V. Shanabrook, and R. J. Wagner, unpublished data.
101. J. R. Lindle, J. R. Meyer, C. A. Hoffman, and R. H. Miles, unpublished data.
102. J. Bajaj, S. H. Shin, J. G. Pasko, and M. Khoshnevisan, *J. Vac. Sci. Technol. A* **1**, 1749 (1983).
103. E. R. Youngdale, J. R. Meyer, C. A. Hoffman, F. J. Bartoli, C. H. Grein, P. M. Young, H. Ehrenreich, R. H. Miles, and D. H. Chow, *Appl. Phys. Lett.* **64**, 3160 (1994).
104. T. C. Hasenberg, D. H. Chow, A. R. Kost, R. H. Miles, and L. West, *Electron. Lett.* **31**, 275 (1995).
105. D. H. Chow, R. H. Miles, T. C. Hasenberg, A. R. Kost, Y.-H. Zhang, H. L. Dunlap, and L. West, *Appl. Phys. Lett.* **67**, 3700 (1995).
106. The Hughes active regions actually had quantized rather than continuous electron dispersion along the growth axis because they usually contained only 4–1/2 periods (4 InAs and 5 $Ga_{1-x}In_xSb$). Their gain properties would therefore be expected to fall somewhere between the superlattice and quantum well results.
107. N. Holonyak, Jr., R. M. Kolbas, R. D. Dupuis, and P. D. Dapkus, *IEEE J. Quant. Electron.* **16**, 170 (1980).
108. W. T. Tsang, *Appl. Phys. Lett.* **39**, 786 (1981).
109. H. Sakaki, *Solid State Commun.* **92**, 119 (1994).
110. W. T. Tsang, in *Semiconductors and Semimetals*, Vol. 24, Ed. R. Dingle (Academic, New York, 1987), Chapter 7.
111. A. Yariv, *Quantum Electronics*, 3rd edition (Wiley, New York, 1989).
112. C. A. Hoffman, J. R. Meyer, E. R. Youngdale, F. J. Bartoli, and R. H. Miles, *Appl. Phys. Lett.* **63**, 2210 (1993). Also unpublished data.
113. J. I. Malin, J. R. Meyer, C. L. Felix, C. A. Hoffman, L. Goldberg, F. J. Bartoli, C.-H. Lin, P. C. Chang, S. J. Murry, R. Q. Yang, and S.-S. Pei, *Appl. Phys. Lett.* **68**, 2976 (1996).
114. J. Faist, F. Capasso, C. Sirtori, D. L. Sivco, A. L. Hutchinson, and A. Y. Cho, *Electron. Lett.* **30**, 865 (1994).
115. J. Faist, F. Capasso, C. Sirtori, D. L. Sivco, J. N. Baillargeon, A. L. Hutchinson, S.-N. G. Chu, and A. Y. Cho, *Appl. Phys. Lett.* **68**, 3680 (1996).
116. J. Faist, F. Capasso, C. Sirtori, D. L. Sivco, A. L. Hutchinson, S.-N. G. Chu, and A. Y. Cho, *Appl. Phys. Lett.* **64**, 1144 (1994).
117. R. Q. Yang, *Superlatt. Microstruct.* **17**, 77 (1995).
118. J. R. Meyer, I. Vurgaftman, R. Q. Yang, and L. R. Ram-Mohan, *Electron. Lett.* **32**, 45 (1996).
119. J. R. Soderstrom, D. H. Chow, and T. C. McGill, *Appl. Phys. Lett.* **55**, 1094 (1989).
120. L. F. Luo, R. Beresford, and W. I. Wang, *Appl. Phys. Lett.* **55**, 2023 (1989).
121. J. R. Meyer and L. R. Ram-Mohan, unpublished analysis.
122. D. H. Chow, R. H. Miles, J. R. Soderstrom, and T. C. McGill, *Appl. Phys. Lett.* **56**, 1418 (1990).
123. R. H. Miles and D. H. Chow, in *Long Wavelength Infrared Detectors*, Ed. M. Razeghi, (Gordon and Breach, Newark, in press).
124. E. R. Brown and S. J. Eglash, *Phys. Rev. B* **41**, 7559 (1990).
125. An exception has been the more complex layering configurations employed in antimonide-family resonant tunneling devices, e.g., in Refs. [119] and [120].

CHAPTER 7

Mid-Infrared Strained Diode Lasers

G. G. ZEGRYA

A.F. Ioffe Physical-Technical Institute,
Academy of Sciences of Russia, St. Petersburg, Russia

7.1.	Introduction	274
7.2.	Wave Functions and Carrier Spectra in Strained Quantum Well Heterostructures	275
7.3.	Gain and Spontaneous emission rate in Strained Quantum Well Heterostructure Lasers	283
	7.3.1. Gain	283
	7.3.2. Spontaneous emission rate	285
7.4.	Nonradiative Transitions and Carrier Leakage in Strained Quantum Well Heterostructure Lasers	286
	7.4.1. Intraband absorption at the heteroboundaries	286
	7.4.1.1. Intraband absorption by electrons	287
	7.4.1.2. Intraband absorption by holes	291
	7.4.2. New mechanism of Auger recombination in quantum wells	292
	7.4.2.1. Matrix element of Auger transition	295
	7.4.2.2. Mechanism of Auger recombination in type-I heterostructures with strained quantum wells	299
	7.4.2.3. Mechanism of Auger recombination in type-II heterostructures with quantum wells	308
	7.4.3. Non-threshold mechanism of Auger recombination	325
	7.4.4. Carrier leakage	329
7.5.	Threshold Characteristics of Mid-Infrared Strained Quantum Well Lasers	331
	7.5.1. Threshold condition	331
	7.5.2. Threshold carrier density	332
	7.5.3. Auger current	335

7.5.4.	Mechanism of suppression of Auger recombination current	340
7.5.5.	Leakage current. Two mechanisms	344
7.5.6.	Analysis of the Auger current dependence on the quantum well parameters and on strain	346
7.6.	Comparison with Experiments	351
7.7.	New Approach to Creation of Mid-Infrared Lasers Operating at Room Temperature	356
7.7.1.	Auger engineering in quantum well laser structures	356
7.7.2.	Mechanism of suppression of intraband absorption processes	362
7.8.	Conclusions	364
	Acknowledgments	366
	References	366

7.1 INTRODUCTION

Semiconductor lasers emitting in the mid-infrared (IR) range (2–5 µm) have been intensely investigated recently [1–6]. The interest in this type of lasers is aroused by outlooks for their extensive application to different branches of science and engineering.

First of all, mid-infrared lasers as components of semiconductor gas analyzers are of great importance in environmental protection and ecology monitoring because of their wavelength range (2–5 µm) where a number of industrial toxic gases have strong absorption lines [7–9]. Moreover, in this wavelength range there are three atmospheric transmission windows, which enables widespread use of lasers emitting in a spectral window as elements of laser radars and for illumination [9]. Application of mid-infrared lasers to medicine is also very promising [10].

A number of materials, such as the III–V semiconductor compounds [1,4,8,9,11] II–VI [6,12] and lead-salts, [5,13,14] have been used in creating mid-infrared lasers. For example, lead-salt lasers with emission wavelengths of $\sim 4\,\mu m$ can operate in pulsed mode at the temperatures up to 290 K [13] and in CW mode up to 200 K [14]. In Ref. [9] the advantages of the compounds III–V over other materials that might be used in creating mid-infrared lasers were analyzed.

There are two different approaches to creating mid-infrared lasers based on III–V alloys. They are aimed at developing lasers based on intraband transitions (quantum cascade lasers) [11,16] and lasers using interband transitions [1–4,8,9,15]. Lasers with interband transitions usually have lower limits of operating temperature than lead-salt lasers. For example, at, $\lambda \sim 4\,\mu m$, the InAsSb/InAlAsSb quantum-well lasers operate in pulsed mode up to 165 K and in CW mode up to 123 K [9]. A maximum operating temperature of about 180 K was reported for mid-infrared lasers based on the InAsSb(P)/InAsSbP system in Ref. [15].

Mid-infrared III–V lasers based on interband transitions are studied below. Up to now, there has been no lasers of this kind emitting in the wavelength range of $\lambda \geqslant 4\,\mu m$ at room temperature. The purpose of this chapter is to study fundamental physical processes governing the operation of mid-infrared lasers based on strained quantum wells. We will also study the limits of their operation at room temperature. We will describe a variety of new effects due to the carrier–interface interaction. The most important of these are: a new non-threshold channel of Auger recombination (Section 7.4.2), intraband absorption (Section 7.4.1), and leakage (Section 7.4.4). These and other processes discussed in the present chapter are to be taken into account in creating mid-infrared diode lasers. And finally, we will propose a new fundamental approach to development of mid-infrared lasers operating at room temperature.

7.2 WAVE FUNCTIONS AND CARRIER SPECTRA IN STRAINED QUANTUM WELL HETEROSTRUCTURES

To investigate elementary processes of recombination in strained quantum wells it is first necessary to find wave functions of carriers. Analysis shows that the wave functions of carriers in the strained semiconductor heterostructures are to be calculated in the framework of the multiband approximation [17–19]. This is particularly important, as we will see below, for the evaluation of Auger recombination rate, since when calculating the matrix element of this process, it is necessary to take account of the interconversion between light and heavy holes in their interaction with a heterobarrier [20–23].

Different approaches are used to describe band states in semiconductor heterostructures with vertical heterobarriers [24,25]. The resulting equations are too complicated to be used in analyzing recombination processes.

We will use Kane multiband model [17]. The basis chosen within the model gives qualitatively and quantitatively precise results both for carrier spectrum (allowing for the interconversion between light and heavy holes) and for macroscopic quantities such as radiative recombination rate, Auger recombination rate etc.

Semiconductors based on III–V compounds have the symmetry group T_d. With the spin–orbit interaction neglected, in most of these semiconductors the wave functions are transformed according to the non-degenerate representation Γ_1 in the conduction band and to the triply degenerate representation Γ_{15} in the valence band. Accordingly, within the model, the basis wave functions of the bottom of the conduction band and the top of the valence band are chosen in the form of $|s\rangle$- and $|\mathbf{p}\rangle$-functions (the X axis is perpendicular to the heterobarrier). The wave functions of electrons and holes are the result of superposition of the basis states,

$$\psi = u(\mathbf{r})|s\rangle + \mathbf{v}(\mathbf{r})|\mathbf{p}\rangle, \qquad (1)$$

where $u(\mathbf{r})$ and $\mathbf{v}(\mathbf{r})$ are smooth envelopes to the Bloch functions [26]. When there is a biaxial stress, the set of equations for the envelopes can be represented as follows:

$$\left[E - \frac{E_g}{2} - V_c(x) - d_c(x)\right]u - \gamma\hat{\mathbf{k}}\mathbf{v} = 0, \qquad (2)$$

and

$$\left[E + \frac{E_g}{2} + V_v(x) - d_h(x) - \delta(x) + \frac{\hbar^2 \hat{\mathbf{k}}^2}{2m_{hh}}\right]v_x - \gamma\hat{k}_x u = 0, \qquad (3)$$

$$\left[E + \frac{E_g}{2} + V_v(x) - d_h(x) + \delta(x) + \frac{\hbar^2 \hat{\mathbf{k}}^2}{2m_{hh}}\right]\mathbf{v}_\| - \gamma\hat{\mathbf{k}}_\| u = 0. \qquad (4)$$

Here E is the energy of the particles measured from the centre of the bandgap of the narrow-gap semiconductor, E_g is the bandgap of the narrow-gap semiconductor without strain, m_{hh} is the heavy hole effective mass of the bulk material, $\hat{\mathbf{k}} = -i\nabla$, $V_c(x)$ and $V_v(x)$ are the heterobarrier heights for electrons and holes respectively in the absence of strain; $d_h = (d_x + d_\|)/2$, $\delta = (d_x - d_\|)/2$, and the Kane matrix element γ calculated with account of spin–orbit interaction takes the form

$$\gamma^2 = \frac{\hbar^2}{2m_c} \frac{E_g(E_g + \Delta_{so})}{E_g + 2\Delta_{so}/3}, \qquad (5)$$

where m_c is the electron effective mass, Δ_{so} is the spin–orbit splitting constant. We will consider the case of strained quantum well and non-strained barriers. Then

$$[d_c(x), d_x(x), d_\|(x)] = \begin{cases} d_c, d_x, d_\|, & \text{for } -a/2 < x < a/2, \\ 0, & \text{for } x > a/2, \ x < -a/2, \end{cases} \quad (6)$$

where a is the width of the quantum well (Fig. 1).

The values of d_c, d_x and $d_\|$ are expressed in terms of the deformation potential constants of the conduction and valence band [27]

$$d_c = a_c \operatorname{tr}(\varepsilon) \equiv a_c(\varepsilon_{xx} + 2\varepsilon_\|),$$
$$d_x = a_v \operatorname{tr}(\varepsilon) - 2b(\varepsilon_\| - V\varepsilon_{xx}), \quad (7)$$
$$d_\| = a_v \operatorname{tr}(\varepsilon) + b(\varepsilon_\| - \varepsilon_{xx}).$$

Here a_c, a_v and b are the hydrostatic and uniaxial deformation potentials,

$$\varepsilon_\| = \varepsilon_{yy} = \varepsilon_{zz} = (a_B - a_W)/a_W,$$
$$\varepsilon_{xx} = -\frac{2\sigma}{1-\sigma}\varepsilon_\|, \quad (8)$$

where a_B and a_W are the lattice constants in the barrier and well materials respectively, σ is Poisson's ratio, $\sigma = C_{12}/(C_{11} + C_{12})$, and C_{ij} are the elastic stiffness constants. In what follows we will study the quantum wells with a vertical barrier. Therefore, $V_c(x)$ and $V_v(x)$ are equal to

$$[V_c(x), V_v(x)] = \begin{cases} 0, & \text{for } -a/2 < x < a/2, \\ V_c, V_v, & \text{for } x > a/2, \ x < -a/2. \end{cases} \quad (9)$$

According to Eqs. (2)–(4), the energy spectrum E splits into an electron branch E_c, light-hole branch E_{hl} and heavy-hole branch E_{hh}.

The electron and hole wave functions that are determined from the Kane equations (2)–(4) must filled the following boundary conditions at the heterointerfaces. At $x = -a/2$ and $a/2$ the electron wave functions have continuous components u and v_x:

$$u^> = u^<, \quad v_x^> = v_x^<, \quad (10)$$

where the superscripts "<" and ">" refer to the values of these components to the left and right of the heterointerface respectively.

The electron wave function components v_y and v_z parallel to the heterojunction are discontinuous. Note that within the chosen multi-gap Kane model, we neglected the term $\hbar^2 \hat{k}^2/(2m_{hh})$ in finding the wave functions of electrons and deducing the boundary conditions (11) in

Eqs. (2)–(4) [19]. Including this term in the electron wave functions would result in an excess of accuracy within the method of envelopes of the wave functions. However, taking into account this term is of primary importance for holes, since it describes the interconversion between light and heavy holes in their interaction with the heterobarrier. The components v_x, v_y, v_z, and the derivatives $\partial v_y/\partial x$, $\partial v_z/\partial x$, are continuous at the heterointerfaces $x = -a/2, a/2$, while the derivative $\partial v_x/\partial x$, and the component u has a discontinuity:

$$\mathbf{v}^> = \mathbf{v}^<, \quad \frac{\partial v_y^>}{\partial x} = \frac{\partial v_y^<}{\partial x}, \quad \frac{\partial v_z^>}{\partial x} = \frac{\partial v_z^<}{\partial x},$$

$$\frac{\partial v_x^>}{\partial x} - \frac{\partial v_x^<}{\partial x} = i\gamma \frac{2m_{\text{hh}}}{\hbar^2}(u^> - u^<). \tag{11}$$

The boundary conditions for the wave functions of electrons (10) and holes (11) are derived from the Kane equations (2)–(4) by integrating them with respect to the transverse coordinate x and with regard to the continuity of the x-component of the flux density j_x.

It is well-known [28] that wave functions of particles (electrons and holes) in a symmetrical quantum well are either odd or even functions of the coordinates. Consequently, the components of the envelopes u and $\mathbf{v}$ of the wave functions of electrons and holes localized in quantum wells can also be classified by their parity. We will exhibit the wave functions of carriers for states of the same parity.

The wave function of electrons localized in the quantum well $(-a/2 < x < a/2)$ has the form

$$\psi_c(\mathbf{r}) = A \exp(i\mathbf{q}\rho) \begin{bmatrix} \cos(kx) \\ i\lambda_{x-} k \sin(kx) \\ \lambda_{\|-} q_y \cos(kx) \\ \lambda_{\|-} q_z \cos(kx) \end{bmatrix}. \tag{12}$$

Here $\mathbf{k} = (k_x, \mathbf{q}) \equiv (k, \mathbf{q})$ is the quasi-momentum of electrons, $\lambda_{\|-} = \gamma/(E_c + E_g/2 - d_\|)$, $\lambda_{x-} = \gamma/(E_c + E_g/2 - d_x)$, since $d_x, d_\| \ll E_g$, $\lambda_{x-} \cong \lambda_{\|-} = \lambda_-$, λ_- is the characteristic wavelength of electron, corresponding to an energy of the order of E_g; at $\mathbf{k} = 0$ we have $\lambda_- \approx \gamma/E_g \equiv \lambda_g = \hbar/\sqrt{2m_c E_g}$;

$$E_c = \frac{d_c}{2} + \left[\frac{E_g^2}{4} + d_c \frac{E_g}{2} + \gamma^2(k^2 + q^2)\right]^{1/2} \tag{13}$$

is the electron energy, ρ is the coordinate in the plane of the quantum well, A is the normalization factor. For $x > a/2$ (the subbarrier region) we obtain

$$\psi_c(\mathbf{r}) = B \exp(i\mathbf{q}\rho - \kappa_c x) \begin{bmatrix} 1 \\ i\lambda_+ \kappa_c \\ \lambda_+ q_y \\ \lambda_+ q_z \end{bmatrix}, \qquad (14)$$

where $B = A \cos(ka/2)$, $\lambda_+ = \gamma/(E_c + E_g/2 + V_v)$, κ_c^{-1} is the characteristic damping length for the components of the electron wave function beneath the barrier,

$$\kappa_c^2 = q^2 - \frac{E_c - V_c - E_g/2}{\gamma \lambda_+},$$

$$A^2 = \frac{2}{a + 1/\kappa_c}. \qquad (15)$$

An expression similar to Eq. (14) occurs for the subbarrier part of the wave function in the region $x < -a/2$. Applying the boundary conditions (10) we deduce the dispersion equation for even electron levels in the quantum well

$$\text{tg}\frac{ka}{2} = \frac{\kappa_c}{k}\frac{\lambda_+}{\lambda_-}. \qquad (16)$$

Now we consider the states of light and heavy holes. According to Eqs. (2)–(4), the wave functions of holes in the region within the quantum well ($-a/2 < x < a/2$) take the form

$$\psi_h(\mathbf{r}) = H \exp(i\mathbf{q}\rho) \begin{bmatrix} q/P_h B_x^W \cos(P_h x) \\ iq \sin(P_h x) \\ -(P_h + B_x^W \cdot B_c^W/P_h)(q_y/q)\cos(P_h x) \\ -(P_h + B_x^W \cdot B_c^W/P_h)(q_z/q)\cos(P_h x) \end{bmatrix}$$

$$+ L \exp(i\mathbf{q}\rho) \begin{bmatrix} A_x^W \cos(P_l x) \\ iP_l \sin(P_l x) \\ q_y(A_x^W/A_\parallel^W)\cos(P_l x) \\ q_z(A_x^W/A_\parallel^W)\cos(P_l x) \end{bmatrix}, \qquad (17)$$

where

$$A_x^W = \left[E_h + \frac{E_g}{2} - d_h - \delta + \frac{\hbar^2}{2m_{hh}}(P_1^2 + q^2) \right] \Big/ \gamma,$$

$$A_\parallel^W = \left[E_h + \frac{E_g}{2} - d_h + \delta + \frac{\hbar^2}{2m_{hh}}(P_1^2 + q^2) \right] \Big/ \gamma,$$

$$B_x^W = \left[E_h + \frac{E_g}{2} - d_h - \delta + \frac{\hbar^2(P_h^2 + q^2)}{2m_{hh}} \right] \Big/ \gamma, \quad (18)$$

$$B_c^W = \left[E_h - \frac{E_g}{2} - d_c \right] \Big/ \gamma,$$

$q = (q_y^2 + q_z^2)^{1/2}$, P_l and P_h are the x-components of the wave vector of the light and heavy holes respectively, L and H are the amplitudes of the wave functions corresponding to the light and heavy holes. It is important that the amplitudes L and H are to be calculated with reference to the interconversion between light and heavy holes using the boundary conditions (11). The subbarrier part of the hole wave function in the region $x > a/2$ takes the form

$$\psi_h(\mathbf{r}) = \tilde{H} \exp(i\mathbf{q}\rho - \kappa_h x) \begin{bmatrix} 0 \\ -iq \\ \kappa_h(q_y/q) \\ \kappa_h(q_z/q) \end{bmatrix} + \tilde{L} \exp(i\mathbf{q}\rho - \kappa_l x) \begin{bmatrix} -A_B \\ i\kappa_l \\ -q_y \\ -q_z \end{bmatrix}, \quad (19)$$

where $A_B = [E_h + E_g/2 + V_v + (\hbar^2/2m_{hh})(q^2 - \kappa_l^2)]/\gamma$, $\tilde{L}$ and $\tilde{H}$ are the amplitudes of the wave functions of light and heavy holes with account of their interconversion in the subbarrier region. The hole wave function in the region $x < -a/2$ has the form analogous to Eq. (19). The dispersion equation for holes is derived by substituting the wave functions (17) and (19) into the boundary conditions (11):

$$\left\{ P_h \operatorname{ctg} P_h \frac{a}{2} \left(1 + \frac{B_x^W \cdot B_c^W}{P_h^2} + \frac{q^2}{P_h^2} \frac{B_x^W}{A_B} \right) + \kappa_h \frac{q^2 + P_h^2 + B_x^W B_c^W}{q^2 - \kappa_h^2} \right\}$$

$$\times \left\{ -P_l \operatorname{tg} P_l \frac{a}{2} \left[1 - \frac{q^2}{q^2 - \kappa_h^2} \left(1 - \frac{A_x^W}{A_\parallel^W} \right) \right] + \kappa_l \frac{A_x^W}{A_B} \right\}$$

$$= q^2 \left\{ 1 - \frac{\kappa_l}{P_h} \frac{B_x^W}{A_B} \operatorname{ctg} P_h \frac{a}{2} - \frac{q^2 + P_h^2 + B_x^W B_c^W}{q^2 - \kappa_h^2} \right\}$$

$$\times \left\{ \frac{A_x^W}{A_\parallel^W} - \frac{A_x^W}{A_B} - \frac{\kappa_h P_l}{q^2 - \kappa_h^2} \left(1 - \frac{A_x^W}{A_\parallel^W} \right) \operatorname{tg} P_l \frac{a}{2} \right\}. \quad (20)$$

At $q=0$ Eq. (20) splits into two equations corresponding to the noninteracting light and heavy holes.

In what follows a theoretical study of the Auger recombination processes in quantum wells will be of particular interest. Therefore, it is appropriate to make the following remark. According to Eq. (20), interconversion between the light and heavy holes takes place at $a \in q \neq 0$ (see Fig. 2). In the temperature range $T > E_{1h}$, where E_{1h} is the energy of the first size-quantization level of heavy holes (here the temperature T is taken in energy units), the main contribution to Auger recombination rate is made by the values $q > \pi/a$. Within this range of q the influence of the spin–orbit interaction on the spectrum and the hole wave functions can be neglected [21]. Analysis shows that allowing for the influence of spin–orbit interaction on the process of Auger recombination leads only to a quantitative change in the recombination rate by γ (see Eq. (5)), and to a qualitative and quantitative change of the overlap integral between the initial and final electron states (see Eq. (77)).

Note that in the framework of the Kane model, apart from the states of heavy holes considered above, there exist heavy hole states polarized in the plane of the heterobarrier (y, z) [29]. However, these states take no part in the process of Auger recombination, since the overlap integral between the electrons and the specified holes is equal to zero.

Further, when studying the mechanism of Auger recombination we will restrict ourselves to the case of shallow quantum wells with the excitation energy of the Auger electron (approximately E_g) exceeding the depth of the quantum well (in Section 7.7 the opposite case will be considered). In this case an excited Auger electron passes into a state belonging to the continuous spectrum. The wave function of high-speed Auger electrons is the result of superposition of the incident and reflected waves in the region $x < -a/2$ and of the transmitted wave in the region $x > a/2$ [30]:

$$\psi_f(\mathbf{r}) = A_f \exp(i\mathbf{q}\rho) \begin{bmatrix} \exp(i\bar{k}_f(x-a/2)) + r\exp(-i\bar{k}_f(x-a/2)) \\ \lambda_+ \bar{k}_f [\exp(i\bar{k}_f(x-a/2)) - r\exp(-i\bar{k}_f(x-a/2))] \\ \lambda_+ q_y [\exp(i\bar{k}_f(x-a/2)) + r\exp(-i\bar{k}_f(x-a/2))] \\ \lambda_+ q_z [\exp(i\bar{k}_f(x-a/2)) + r\exp(-i\bar{k}_f(x-a/2))] \end{bmatrix}, \quad (21)$$

$$\psi_f = A_f t \exp\left(i\mathbf{q}\rho + \bar{k}_f\left(x - \frac{a}{2}\right)\right) \begin{bmatrix} 1 \\ \lambda_+ \bar{k}_f \\ \lambda_+ q_y \\ \lambda_+ q_z \end{bmatrix}. \quad (22)$$

Here r and t are the amplitudes of reflection and transmission of an electron over the barrier, and $\bar{k}_f$ is the x-component of the quasi-momentum of a high-speed Auger electron over the barrier:

$$\bar{k}_f^2 + q^2 = \frac{E_f - V_c - E_g/2}{\gamma \lambda_+}, \tag{23}$$

A_f is the normalization factor,

$$A_f = \{1 + \lambda_+^2 (\bar{k}_f^2 + q^2)\}^{-1/2}. \tag{24}$$

The wave function of highly excited electrons within the quantum well ($-a/2 < x < a/2$) can be represented as follows:

$$\psi_f(\mathbf{r}) = A \exp(i\mathbf{q}\rho)$$

$$\times \begin{bmatrix} d_+ \exp(ik_f(x-a/2)) + d_- \exp(-ik_f(x-a/2)) \\ \lambda_- k_f [d_+ \exp(ik_f(x-a/2)) - d_- \exp(-ik_f(x-a/2))] \\ \lambda_- q_y [d_+ \exp(ik_f(x-a/2)) + d_- \exp(-ik_f(x-a/2))] \\ \lambda_- q_z [d_+ \exp(ik_f(x-a/2)) + d_- \exp(-ik_f(x-a/2))] \end{bmatrix}, \tag{25}$$

where d_- and d_+ are the amplitudes of the reflected and transmitted waves in the region of the quantum well, k_f is the x-component of the quasi-momentum of a highly excited Auger electron in the region of the quantum well,

$$k_f^2 + q^2 \cong (E_f - E_g/2 - d_c)(E + E_g/2)/\gamma^2. \tag{26}$$

In Eq. (26) we neglected d_x and $d_\parallel$ in comparison with E_g, since $(d_x, d_\parallel) \ll E_g$. The amplitudes r, d_+, d_- and t are calculated with the use of the boundary conditions (11) and the expression for A_4:

$$r = 2i\delta_v e^{i\varphi} \sin(k_f a), \qquad d_+ = e^{i\varphi}(1 - \delta_v),$$

$$d_- = \delta_v e^{i\varphi}, \qquad t = e^{i\varphi}, \tag{27}$$

$$\varphi = a(k_f - \bar{k}_f), \qquad \delta_v = \frac{2(V_c - d_c) + V_v + d_x}{8E_g}.$$

In deriving Eq. (27) we used fact that $\delta_v \ll 1$.

The expressions for the wave functions allow one to study elementary processes of carrier recombination in strained semiconductor heterostructure lasers from fundamental principles.

7.3 GAIN AND SPONTANEOUS EMISSION RATE IN STRAINED QUANTUM WELL HETEROSTRUCTURE LASERS

7.3.1 Gain

The gain $g(\omega)$ will be estimated with the aid of the density operator formalism. The polarization $\mathbf{P}(t)$ is given as follows by the one-particle density operator ρ [31]:

$$\mathbf{P}(t) = en \sum_{mm'} \int\int \mathbf{r}_{m'm}(\mathbf{p},\mathbf{p}')\rho_{mm'}(\mathbf{p},\mathbf{p}',t)\,d\mathbf{p}\,d\mathbf{p}'. \tag{28}$$

Here e is the elementary charge, n is the two-dimensional concentration of electrons, $r_{m'm}$ is the dipole matrix element of the interband transition, $\rho_{mm'}$ are the matrix elements of the density operator ρ.

Further we will consider only four components of the density operator: ρ_{vv} representing the state of an electron in the valence band, ρ_{cc} representing the state of the electron in the conduction band, ρ_{vc} describing transition from the conduction band ("c"-state) into the valence band ("v"-state), ρ_{cv} describing transition from "v" into "c". We consider direct interband transitions for which $\mathbf{r}_{vv} = \mathbf{r}_{cc} = 0$. The matrix element of the interband transition between the electron and hole levels may be presented in the form

$$r_{vc}(\mathbf{q}_c, \mathbf{q}_v) = r_{cv}(\mathbf{q}_c, \mathbf{q}_v) = r_{vc}(\mathbf{q})\delta_{\mathbf{q}_c - \mathbf{q}_v, 0}, \tag{29}$$

where $\mathbf{q}$ is the longitudinal momentum of charge carriers. Taking into account these notes, we obtain the following expression for the polarization vector:

$$\mathbf{P}(t) = n \sum_{n_e, n_h} \int \mathbf{r}_{vc}(\mathbf{q})[\rho_{vc}(\mathbf{q},t) + \rho_{cv}(\mathbf{q},t)]\,d\mathbf{q}. \tag{30}$$

The non-diagonal components of the density matrix can be estimated from the following set of equations:

$$i\hbar\frac{\partial \rho_{vc}}{\partial t} = \hbar\omega_{vc}\rho_{vc} - \frac{i\hbar}{T_{vc}}\rho_{vc} - e\mathbf{E}[\mathbf{r},\rho]_{vc}, \tag{31}$$

$$i\hbar\frac{\partial \rho_{cv}}{\partial t} = \hbar\omega_{cv}\rho_{cv} - \frac{i\hbar}{T_{cv}}\rho_{cv} - e\mathbf{E}[\mathbf{r},\rho]_{cv}, \tag{32}$$

where $\omega_{cv} = (E_c - E_v)/\hbar$, E_c and E_v are the electrons and hole energies respectively, $\mathbf{E}$ is the electric field of the wave. Since ρ is a Hermitian operator, $T_{vc} = T_{cv} = \tau$. The constant τ is called longitudinal dipole–dipole relaxation time. It is related to the linewidth of the optical transition. To solve Eqs. (31) and (32) for ρ_{cv} and ρ_{vc}, we use the approximation of the first harmonic of electromagnetic field. Thus, we obtain

$$\rho_{vc}(\mathbf{q}, t) = -\frac{e}{\hbar} D(q) \frac{\mathbf{r}_{cv}(\mathbf{q}) \mathbf{E} e^{i\omega t}}{\omega_{cv} - \omega + i/\tau}, \qquad (33)$$

$$\rho_{cv}(\mathbf{q}, t) = -\frac{e}{\hbar} D(q) \frac{\mathbf{r}_{cv}(\mathbf{q}) \mathbf{E} e^{-i\omega t}}{\omega_{cv} - \omega + i/\tau}, \qquad (34)$$

where $D(\mathbf{q}) = \rho_{cc} - \rho_{vv}$ is the difference in population between the conduction and the valence bands. Expressing D in terms of the occupation probability of the levels f_c and f_v we deduce

$$D = \frac{2}{(2\pi)^2 n} [f_c(q) - f_v(q)]. \qquad (35)$$

Finally, according to Eqs. (30), (33)–(35), we derive the following expression for the polarization vector:

$$\mathbf{P}(t) = -\frac{4e^2}{\hbar} \mathbf{E} \sum_{n_c, n_h} \int \frac{d^2q}{(2\pi)^2} (\mathbf{r}_{cv} \cdot \mathbf{e})^2 [f_c(q) - f_v(q)]$$

$$\times \frac{(\omega_{cv} - \omega) \cos \omega t + 1/\tau \sin \omega t}{(\omega_{cv} - \omega)^2 + 1/\tau^2}. \qquad (36)$$

Then we use the definition of the dielectric permittivity χ:

$$\mathbf{P}(t) = \chi(\omega) \mathbf{E}. \qquad (37)$$

The imaginary part of the dielectric permittivity is $\varepsilon''(\omega) = 4\pi \operatorname{Im} \chi(\omega)$. The gain $g(\omega)$ is related to $\varepsilon''(\omega)$ by

$$g(\omega) = -\frac{\omega}{c} \frac{\varepsilon''}{\sqrt{\varepsilon_\infty}}, \qquad (38)$$

where ε_∞ is the dielectric permittivity at the optical frequencies. Substituting the expression for $\varepsilon''(\omega)$ into Eq. (38) and using the relations (36) and (37), we derive the following expression for the gain $g(\omega)$:

$$g(\omega) = \frac{16\pi}{\sqrt{\varepsilon_\infty}} \frac{e^2}{\hbar c} \hbar \omega \sum_{n_c, n_h} \int \frac{d^2q}{(2\pi)^2} (\mathbf{r}_{cv} \mathbf{e})^2 [f_c(q) + f_h(q) - 1]$$

$$\times \frac{\delta_\tau}{(E_c - E_h - \hbar\omega)^2 + \delta_\tau^2}, \qquad (39)$$

where $\delta_\tau = \hbar/\tau$. In Eq. (39) we took into account the fact the $f_v(q) = 1 - f_h(q)$ where $f_h(q)$ is the distribution function of holes. The dipole matrix element $\mathbf{r}_{cv}$ is convenient to be represented as

$$\mathbf{r}_{cv} = \vec{\mathcal{P}}_{cv}/\hbar\omega, \quad (40)$$

$$\vec{\mathcal{P}}_{cv} = \int dx\, \mathbf{j}_{cv}, \quad (41)$$

where $\mathbf{j}_{cv}$ is the probability flux density:

$$\mathbf{j}_{cv} = i\gamma(u_c^* \mathbf{v}_v + u_v \mathbf{v}_c^*). \quad (42)$$

The final expression for the gain $g(\omega)$ is derived from Eqs. (39) and (40):

$$g(\omega) = \frac{8\pi}{\sqrt{\varepsilon_\infty}} \frac{e^2}{hc} \frac{1}{\hbar\omega} \sum_{n_c, n_h} \int q\, dq\, |\vec{\mathcal{P}}_{cv}|^2 [f_c(q) + f_h(q) - 1] L(\omega, q), \quad (43)$$

where

$$L(\omega, q) = \frac{1}{\pi} \frac{\delta_\tau}{(E_c - E_h - \hbar\omega)^2 + \delta_\tau^2},$$

$$L(\omega, q) \to \delta(E_c - E_h - \hbar\omega), \quad \text{for } \delta_\tau \to 0, \quad (44)$$

$|\vec{\mathcal{P}}_{cv}|^2 = |\vec{\mathcal{P}}_{cv}^\parallel|^2/2$ is the expression for TE-mode, $|\vec{\mathcal{P}}_{cv}|^2 = |\vec{\mathcal{P}}_{cv}^x|^2$ is the one for TM-mode. In deriving this equation we used the fact that the radiation matrix element is, according to Eqs. (41) and (42), equal to $\vec{\mathcal{P}}_{cv} = (\mathcal{P}_{cv}^x, \mathcal{P}_{cv}^\parallel)$.

7.3.2 Spontaneous Emission Rate

The specific spectral density $\Phi(\omega)$ of emission due to recombination of non-equilibrium two-dimensional electrons and holes can be related to the absorption coefficient $\alpha(\omega)$ by the following equation:

$$\frac{\Phi(\omega)}{\alpha(\omega)} = \frac{\hbar\omega^3}{\pi^2 v^2} \frac{1}{e^{(\hbar\omega - \Delta F)/T} - 1}, \quad (45)$$

where $v = c/\sqrt{\varepsilon_\infty}$, $\Delta F = F_e - F_h$ is the difference between the Fermi quasi-levels of electrons and holes.

In thermodynamic equilibrium the spectral emission intensity $\Phi(\omega)$ is equal to that of black emission, then from Eq. (45) follows the Kirchhoff law [32].

On the basis of Eq. (45), the radiative recombination rate summed over all frequencies can be represented by the integral

$$R_{ph} = \int_0^\infty \frac{1}{\hbar\omega} \Phi(\omega)\,d\omega = \frac{\varepsilon_\infty}{\pi^2 c^2} \int_0^\infty \frac{\alpha(\omega)\omega^2\,d\omega}{\exp[(\hbar\omega - \Delta F)/T] - 1}. \quad (46)$$

In a special case of $\Delta F = 0$ expression (46) goes over into the Van Roosbroeck and Shockley formula [33]. The absorption factor can be found from Eq. (43). Substituting the expression for $\alpha(\omega)$ into Eq. (46), integrating over ω with account of $L(\omega, \mathbf{q}) = \delta[E_c(q) - E_h(q) - \hbar\omega]$ yield the following:

$$R_{ph} = \frac{8\varepsilon_\infty}{\pi\sqrt{\kappa_0}} \frac{e^2}{\hbar c} \frac{1}{\hbar^3 c^2} \sum_{n_c, n_h} \int q\,dq \left[|\mathscr{P}_{cv}^x|^2 + \frac{1}{2}|\mathscr{P}_{cv}^{\|}|^2 \right] f_c(E_c) f_h(E_h)(E_c - E_h). \quad (47)$$

7.4 NONRADIATIVE TRANSITIONS AND CARRIER LEAKAGE IN STRAINED QUANTUM WELL HETEROSTRUCTURE LASERS

7.4.1 Intraband Absorption at the Heteroboundaries

It is accepted that in semiconductor heterostructures, as well as in bulk semiconductors, the intraband absorption by free carriers is weak because the absorption is possible when not only an electron and a photon but also some other particle (a phonon or impurity atom) participates in the process. Thus, in semiconductor heterolasers radiation is mainly lost on the resonator mirrors.

Nevertheless, in semiconductor heterostructures based on narrow-gap semiconductors there is a considerable intraband absorption at high temperatures. This absorption leads to suppression of lasing at high temperatures [4]. As it will be shown below, with decreasing active layer width the intraband absorption becomes stronger to reach few tens of cm^{-1}.

This section considers the mechanism of absorption by free carriers without the third particle (a phonon or an impurity atom). We will not consider the absorption by holes in the valence band of semiconductors with $E_g \approx \Delta_{so}$. The intraband absorption by an electron without the third particle generally prohibited by the momentum and energy conservation law, becomes possible if the electron interacts with a heterobarrier. This can be proved by the following. The wave function of an electron near the heterointerface ($x < 0$) and within the region of the heterobarrier ($x > 0$) is a wave packet comprising waves with different values of the quasi-momentum k_x, amidst which, with a nonzero probability $a(k_x)$ (the X axis is perpendicular to the heterobarrier), there are quasi-momenta corresponding to the final quasi-momentum k_{fx} of the excited electron. Consequently, the absorption of a photon $\hbar\omega$ by the electron does not demand momentum transfer, i.e., this transition is resonant. It is important that the quasi-momenta of electrons that have absorbed a photon $\hbar\omega$ are mostly perpendicular to the heterointerface, i.e., aligned with the X-axis, due to the different behavior of the parallel $\mathbf{q}$ and perpendicular k_x components of the quasi-momentum in absorbing a photon: $\mathbf{q}$ stays constant, while k_x changes from $k_{ix} \approx (2m_c E_{0c}/\hbar)^2$ in the initial state to $k_{fx} \approx (2m_c\omega/\hbar)^{1/2}$ in the final state, where E_{0c} is the energy of a first size-quantization electron level. As a result, $|\mathbf{q}_f| \ll k_{fx}$.

We will see that despite the fact that the law of conservation of the longitudinal component of the momentum ($\mathbf{q}_i = \mathbf{q}_f$) is observed, the absorption of radiation polarized in the plane of the heterobarrier is nonzero [34]. This originates from the fact that the electron wave function contains the contribution from P-states of the valence band (see Section 7.2, Eq. (21)), equal to $\gamma q/E_g$. Then the matrix element of the intraband transition M_{if} related to absorption of light polarized along the heteroboundary plane, is proportional to the Kane matrix element γ raised to the third power:

$$M_{if} \propto \gamma^2 \frac{\gamma q}{E_g}. \qquad (48)$$

This means that electrons in narrow-gap semiconductors are described by Kane's model. This model, as well as Dirac's, has a relativistic correction factor of the order of $\gamma^2 q^2/E_g^2$. The third power of γ in the matrix element is the result of the allowance for the term in Kane's Hamiltonian describing the non-parabolicity and proportional to $\hat{P}^4$, where $\hat{P}$ is the momentum operator.

7.4.1.1 Intraband absorption by electrons

In this section we will focus on the influence of the interface upon the process of intraband absorption by electrons [35]. We consider the case of the heterostructure with the vertical barrier V_c where an electron occurs in the state "f" of the continuous spectrum as the result of absorbing the photon $\hbar\omega > V_c$. Optical intraband transitions between the localized quantum well states are studied in Ref. [36]. The total number of photons $\hbar\omega$ per unit time absorbed in the transition from the initial state "i" into the final state "f" is equal to

$$W_{if} = \frac{2\pi}{\hbar} \int\int |M_{if}(\mathbf{k}_i, \mathbf{k}_f)|^2 f(\mathbf{k}_i)[1 - f(\mathbf{k}_f)]\delta(E_f - E_i - \hbar\omega) \frac{d^3 k_i}{(2\pi)^3} \frac{d^3 k_f}{(2\pi)^3}, \tag{49}$$

where $f(\mathbf{k})$ is the Fermi–Dirac distribution function of the electron, E_i is the initial and E_f is the final energies of electron states, M_{if} is the matrix element of the intraband transition

$$M_{if} = \int \psi_i^*(\mathbf{r}) \hat{H}' \psi_f(\mathbf{r}) d\mathbf{r}. \tag{50}$$

Assuming the light is polarized in the plane of the heterointerface, the matrix of electron–photon scattering Hamiltonian takes the form

$$\hat{H}' = \frac{e}{c} A_0 \begin{bmatrix} 0 & 0 & (\gamma^*/\hbar)e_y & (\gamma^*/\hbar)e_z \\ 0 & 0 & 0 & 0 \\ (\gamma/\hbar)e_y & 0 & 0 & 0 \\ (\gamma/\hbar)e_y & 0 & 0 & 0 \end{bmatrix}, \tag{51}$$

where A_0 is the amplitude of the vector potential of the wave, e_y, e_z are the components of the unit polarization vector in the direction of $\mathbf{A}$. Let us substitute the explicit expressions for the wave functions (12), (14), (21) and (22) along with $\hat{H}'$ from Eq. (51) into Eq. (50). After integrating over the volume we obtain the following expression for the matrix element of intraband absorption of the light polarized along the plane of the heterointerface:

$$M_{if}^{\parallel} = -\frac{e}{c} A_{0z} A_{ki} A_{kf}^* (2\pi)^2 \delta(\mathbf{q}_i - \mathbf{q}_f) 2\cos\beta_i e^{i\beta_i} d_f \frac{\gamma^2}{\hbar}$$

$$\times \frac{\gamma q_{fx}}{8E_g^2} \left\{ \frac{\gamma \kappa_c}{E_g} \frac{3(V_c - d_c) + 2(V_v + d_x)}{E_g} \right.$$

$$\left. - \frac{i\gamma k_{fx}^+}{E_g} \frac{3(V_c - d_c) + 5(V_v + d_x)}{E_g} \right\}. \tag{52}$$

Here $2\beta_i$ is the phase shift originating from the reflection of the electron from the interface, d_f is the amplitude of electron passing above the barrier, $\mathbf{k}_f^+ \equiv (k_{fx}^+, \mathbf{q}_+)$ and $\mathbf{k}_f^- \equiv (k_{fx}^-, \mathbf{q}_f)$ are the wave vectors of the electrons excited by the light in wide- and narrow-gap materials respectively,

$$\tg \beta_i = -\frac{\kappa_c}{k_{ix}} \frac{\lambda_+}{\lambda_-},$$

$$d_f = \frac{2k_{fx}^-}{[k_{fx}^- + k_{fx}^+(\lambda_+/\lambda_-)]}. \tag{53}$$

Similar considerations show that the matrix element of absorption of the light polarized perpendicularly to the heterointerface is equal to

$$M_{ij}^\perp = -i(e/c)A_{0x}A_{k_i}A_{k_f}^*(2\pi)^2 \delta(\mathbf{q}_i - \mathbf{q}_f) 2\cos\beta_i e^{i\beta i} d_f \gamma^2/\hbar$$

$$\times \frac{3(V_c - d_c) + V_v + d_x}{4E_g^2} \frac{k_{fx}^+ - 2i\kappa_c}{k_{jx}^+ - i\kappa_c}. \tag{54}$$

It is useful to compare the expressions for $M_{if}^\parallel$ and $M_{if}^\perp$. From Eqs. (52) and (53) we have

$$\frac{|M_{if}^\parallel|}{|M_{if}^\perp|} \cong \frac{3(V_c - d_c) + 5(V_v + d_x)}{3(V_c - d_c) + V_v + d_x} \frac{\gamma^2 q_{fz} k_{fx}^+}{2E_g^2} \sim \left(\frac{T}{E_g}\right)^{1/2} < 1. \tag{55}$$

In deriving Eq. (55) we used the fact that $\kappa_c \ll k_{fx}^+$. Note that above mechanism of absorption is immediately connected with the interaction between charge carriers and the abrupt heterobarrier. This can be proved by the direct dependence of the matrix element of intraband transitions M_{if} on the barrier height V_c. The matrix element $M_{ij}^\perp$ was studied in Ref. [37].

Substituting the expression for the matrix element $M_{if}^\perp$ from Eq. (54) into Eq. (49) and integrating over the initial and final states, we find the total number of absorbed photons W_{if}. The absorption factor is therefore equal to

$$\alpha_i^\perp \approx \frac{16}{3} \frac{(2\pi)}{\sqrt{\varepsilon_\infty}} \frac{e^2}{\hbar c} \lambda_\omega^2 \lambda_g \frac{\Gamma}{a^2} n \left[\frac{3(V_c - d_c) + V_v + d_x}{4E_g}\right]^2 \frac{\varepsilon_\perp}{V_c}$$

$$\times \left(\frac{E_g + \Delta_{so}}{E_g + (2/3)\Delta_{so}}\right)^{3/2} \frac{E_g/2 + \hbar\omega}{[(E_g + V_c + \hbar\omega)(\hbar\omega - V_c)]^{1/2}}, \tag{56}$$

where $\lambda_\omega = (\hbar/2m_c\omega)^{1/2}$, $\varepsilon_\perp$ is the size-quantization energy of electrons, Γ is the optical confinement factor. Eq. (56) is valid while $\hbar\omega > V_c$ holds. If the energy of a photon $\hbar\omega$ is close to the heterobarrier height V_c, the absorption factor has a root singularity. The obtained result is of major practical importance. The very fact of non-monotonicity of the absorption factor as a function of frequency ω can be a basis for an optical method of determining the offsets V_c and V_v in heterostructures. Since the absorpton factor has an acute maximum at the frequency $\omega = V_c/\hbar$, we can determine V_c by analyzing the relation $\alpha_i(\omega)$.

Below, we will be particularly interested in the case $\hbar\omega \approx E_g$. Assuming in Eq. (56) $\hbar\omega = E_g \gg V_c$, we get

$$\alpha_i^\perp \approx \frac{8\sqrt{2}\,2\pi}{\sqrt{\varepsilon_\infty}} \frac{e^2}{\hbar c} \lambda_g^3 n_c \frac{\Gamma}{a^2} \left[\frac{3(V_c - d_c) + V_v + d_x}{4E_g}\right]^2$$

$$\times \frac{\varepsilon_\perp}{V_c} \left(\frac{E_g + \Delta_{so}}{E_g + (2/3)\Delta_{so}}\right)^{3/2}. \tag{57}$$

Now we give another explicit expression for the absorption factor of radiation polarized along the plane of the quantum well. Substituting Eq. (52) into Eq. (49) and summing over the initial and final states, at $\hbar\omega = E_g$ we obtain for $\alpha_i^\parallel$

$$\alpha_i^\parallel \approx \frac{6\sqrt{2}\,2\pi}{\sqrt{\varepsilon_\infty}} \frac{e^2}{\hbar c} \lambda_g^3 n_c \frac{\Gamma}{a^2} \left[\frac{3(V_c - d_c) + 5(V_v + d_x)}{4E_g}\right]^2$$

$$\times \frac{\varepsilon_\perp}{V_c} \frac{T}{E_g} \left(\frac{E_g + \Delta_{so}}{E_g + (2/3)\Delta_{so}}\right)^{7/2}. \tag{58}$$

It is evident that with our approximation we have $\alpha_i^\parallel \ll \alpha_i^\perp$. Their ratio is of the order of

$$\frac{\alpha_i^\parallel}{\alpha_i^\perp} \sim \frac{T}{E_g} \ll 1. \tag{59}$$

Analysis shows that account for the spin–orbit interaction considerably affects the magnitude of the absorption factor $\alpha_i^\parallel$. In this case the ratio (59) assume the form [37]

$$\frac{\alpha_i^\parallel}{\alpha_i^\perp} \sim \left(\frac{\Delta_{so}}{E_g + (2/3)\Delta_{so}}\right)^2. \tag{60}$$

The above developed theory of intraband absorption in quantum well heterostructures is based on the interaction between charge carriers and an abrupt heterobarrier. The phonon-free mechanism of intraband absorption can significantly prevail over the phonon-involving mechanism of absorption in heterostructures based on narrow-gap semiconductors with $E_g \leqslant 0.5$ eV. As follows from Eq. (57), for $\hbar\omega = E_g$ the absorption factor decreases with increasing E_g as $\alpha_i^\perp \propto E_g^{-7/2}$. When $E_g < 0.5\,\text{eV}$, $\alpha_r^\perp > 10\,\text{cm}^{-1}$.

It is interesting to apply the obtained results to estimate the intraband absorption in mid-infrared diode lasers. Substituting the parameters typical for InAs into Eq. (57), we find that at practical values of strain we have $\alpha_i^\perp \approx 50\,\text{cm}^{-1}$. Thus, output losses can predominate over the losses on the resonator mirrors. This will lead to a sharper temperature dependence of the threshold concentration and, consequently, to a sharper temperature dependence of the threshold current density for mid-infrared lasers.

The described mechanism of the intraband absorption refers to both type-I and type-II heterostructures. The difference between these heterostructures is shown below (see Section 7.4.2). We note that the expressions obtained for the intraband absorption factor (Eqs. (56)–(58)) are valid for type-II heterostructures, too, if we change V_v to $-|V_v|$ (see (91)). The dependence of $\alpha_i^\perp$ on V_c and V_v is monotonous for type-I heterostructures, $\alpha_i^\perp$ rises with increasing V_c and V_v. However, the dependence of $\alpha_i^\perp$ on these parameters is appreciably different for type-II heterostructures: it has a non-monotonous run and comes to a minimum at $|V_v|/V_c \approx 3$. Thus, for type-II heterostructures the intraband absorption is suppressed when a certain relationship between the parameters exists. As we will see below, in type-II heterostructures it is also possible to suppress the Auger recombination processes (Section 7.4.2).

It is remarkable that the conditions under which the suppression of intraband absorption and Auger recombination takes place, turn out to be the same: $|V_v|/V_c \cong 3$.

7.4.1.2 Intraband absorption by holes

To conclude this section we briefly discuss the process of intraband absorption by holes. We will consider the process when, due to absorbing a photon $\hbar\omega$, a hole undegoes a transition from its initial state "i" into the so-band. We will take into account the fact that $\hbar\omega > \Delta_{so}$ and $\hbar\omega - \Delta_{so} \gg T$. The matrix element of absorption of the radiation polarized perpendicularly to the quantum well plane equals

$$M_{if}^\perp \approx -\frac{ie}{\hbar c} A_{0x}(2\pi)^2 \delta(\mathbf{q}_i - \mathbf{q}_f) \frac{V_v}{k_{fx}^2} v_{ix}^*\left(\frac{a}{2}\right) v_{fx}\left(\frac{a}{2}\right), \tag{61}$$

where $v_{ix}(a/2)$ and $v_{fx}(a/2)$ are the values of the x-component of the holes wave function in the initial and final states at the heterointerface

$x = a/2$. For example, for $v_{ix}(a/2)$, according to Section 7.2, we have

$$v_{ix}^*\left(\frac{a}{2}\right) \approx -\frac{iH_i}{\sqrt{q_i^2 + k_{xi}^2}} \frac{\varepsilon_{hi}}{V_v}\left[q_{hi} + \left(\frac{2m_c V_v}{\hbar^2}\right)^{1/2}\right], \qquad (62)$$

where $\varepsilon_{hi} = (\hbar^2/2m_{hh})(q_{hi}^2 + k_{xi}^2)$, H_i is the amplitude of the heavy hole wave function (see Eq. (17)).

Substituting Eq. (61) into (49) and taking account of the explicit expressions for v_{ix} and v_{fx}, we obtain after summing over the initial and final states

$$\alpha_h^\perp \approx \frac{32}{3\sqrt{\varepsilon_\infty}} 2\pi \frac{e^2}{\hbar c} \frac{V_v}{\hbar\omega} \frac{\langle\varepsilon_{hi}\rangle}{\hbar\omega - \Delta_{so}} \lambda_{\omega-\Delta}^3 p \frac{\Gamma}{a^2}\left[\left(\frac{\langle\varepsilon_{hi}\rangle}{V_v}\right)^{1/2} + \left(\frac{m_c}{m_h}\right)^{1/2}\right]^2, \qquad (63)$$

where p is the two-dimensional concentration of holes, $\langle\varepsilon_{hi}\rangle = T$ holds for a broad quantum well and $\langle\varepsilon_{hi}\rangle = \varepsilon_{h1} \approx (\hbar^2/2m_{hh})(\pi/a)^2$ for a narrow one, $\lambda_{\omega-\Delta} = \hbar/[2m_{so}(\hbar\omega - \Delta_{so})]^{1/2}$, m_{so} is the hole effective mass in the so-band.

Analysis shows that the intraband absorption by holes can also be efficient. There exist conditions under which it dominates over the absorption by electrons.

7.4.2 New Mechanism of Auger Recombination in Quantum Wells

The space restrictions imposed on carriers in the quantum wells by abrupt heterobarriers affect not only wave the functions and spectrum of carriers but also, as stated above (Section 7.4.1), the recombination processes. Up to now in the works studying the Auger recombination processes in heterostructures and quantum wells, their mechanisms have been assumed the same as those in bulk semiconductors [38–42], i.e., the processes are of threshold nature and the Auger recombination rate is an exponential function of temperature, $G \propto \exp(-E_{th}/T)$, where E_{th} is the threshold energy.

Nevertheless, the very existence of the heterobarrier appreciably influences the electron–electron interaction in semiconductor quantum structures. This interaction is fundamental: the heterobarrier lifts the restrictions imposed on the electron–electron collision processes by the laws of momentum and energy conservation. This gives rise to non-threshold channels of Auger recombination, weakly dependent on temperature.

In this section we will study a new mechanism of Auger recombination in heterostructures, the so-called non-threshold Auger process first predicted in Ref. [19]. The rate of this non-threshold process displays a weak power dependence on temperature, while the process itself becomes a prevailed mechanism of nonradiative recombination of non-equilibrium charge carriers both in heterostructures and in quantum wells. Let us give a qualitative description of the new mechanism of Auger recombination.

(1) The restrictions imposed on the Auger recombination processes by the law of momentum conservation are lifted for an electron that is located close to heterobarrier or within the subbarrier region, since the electron wave function in this region is a wave packet comprising waves with different values of quasi-momenta, amidst which there are quasi-momenta corresponding to the final momentum of the excited Auger electron.

(2) Energy transmission from a recombining electron–hole pair to a high-speed Auger electron is resonant. The electron acquires a momentum necessary for a transition into the final highly excited state to occur from the interaction with the heterobarrier, rather than with another electron. The matrix element of electron–electron interaction does not contain the law of momentum conservationfor the component perpendicular to the heterointerface. The main contribution to the matrix element of Coulomb interaction is given by small transmitted momenta, i.e., the electron–electron interaction is long-range for the Auger recombination and, as a result ofthis, the Auger transition rate does not exhibit threshold (exponential) dependence on temperature, which is distinctive for bulk semiconductors.

(3) The ejection of high-speed Auger electrons takes place mainly along the axis perpendicular to the heterointerface, i.e., the axis of maximum spacial heterogeneity. The electrons are ejected above the barrier in a narrow angle range $\Delta\theta \sim (T/E_g)^{1/2} \ll 1$ (Fig. 1).

The Auger recombination rate is calculated in the framework of first-order perturbation theory in the electron–electron interaction:

$$G = \frac{2\pi}{\hbar} \frac{1}{S} \sum_{1,2,3,4} |M|^2 \delta(E_1 + E_2 - E_3 - E_4)$$
$$\times f_c(E_1) f_c(E_2) f_h(E_3) [1 - f_c(E_4)], \tag{64}$$

where $f_c(E_i)$ is the Fermi–Dirac distribution function of the i-th particle ($i = 1, 2, 3, 4$), E_1 and E_2 are the energies of the initial states, E_3

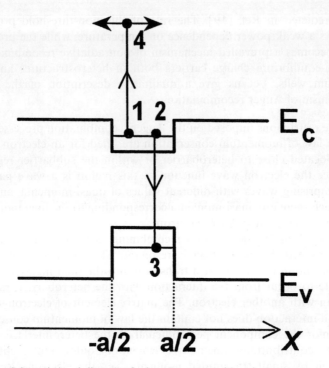

Fig. 1 Schematic band diagram of a heterostructure with a single quantum well. The numbers "1" and "2" denote the initial states of the particles, the numbers "3" and "4" the final states. The transitions from the initial to final states in the Auger recombinition process are shown by arrows.

and E_4 are the energies of the final states, S is the heterojunction area, and M is the matrix element of the electron–electron interaction. Summation in Eq. (64) is performed over the initial ("1" and "2") and final ("3" and "4") states of the particles. After being averaged over the particle spin states the square of the matrix element assumes the following form:

$$\langle |M|^2 \rangle = |M_\mathrm{I}|^2 + |M_\mathrm{II}|^2 - M_\mathrm{I} M_\mathrm{II}^*, \tag{65}$$

$$M_\mathrm{I} = \int \psi_2^*(\mathbf{r}) \psi_3(\mathbf{r}) \frac{e^2}{\kappa_0 |\mathbf{r} - \mathbf{r}'|} \psi_1^*(\mathbf{r}') \psi_4(\mathbf{r}') \, d^3r \, d^3r'. \tag{66}$$

Here κ_0 is the static permittivity of the medium, the expression for M_II is obtained from Eq. (66) by the $1 \rightleftharpoons 2$ substitution.

7.4.2.1 Matrix element of Auger transition

The key point in calculating the Auger recombination rate is to determine the Auger transition matrix element. Expanding the Coulomb potential in the Fourier integral, we obtain

$$M_I = \frac{4\pi e^2}{\kappa_0} \int \frac{d^3q}{(2\pi)^3} \frac{1}{q^2} I_{14}(\mathbf{q}) I_{23}(-\mathbf{q}), \tag{67}$$

$$I_{ij}(\mathbf{q}) = \int \psi_i^*(\mathbf{r}) \psi_j(\mathbf{r}) e^{i\mathbf{q}\mathbf{r}} d^3r. \tag{68}$$

Here I_{ij} is the overlap integral between the particle states "i" and "j".

As shown earlier [19,21,30,34], the overlap integrals entering in the Auger recombination matrix element should be calculated in the framework of the multiband Kane model taking into account the non-parabolicity of the carrier spectrum dispersion. For the overlap integral I_{14} between the states of the localized "1" and excited "4" electron, this is important for the following reasons. First, the wave function of an electron excited highly in the conduction band contains an essential contribution from $|\mathbf{P}\rangle$-states of the valence band (i.e. $u_4 \sim v_{4x}$). Second, the specified overlap integral consists of the contributions from three intervals of integration over x: two intervals of subbarrier motion of the electron "1" ($x < -a/2$, $x > a/2$, see Fig. 1) and a quantum well interval ($-a/2 < x < a/2$). The contributions from the subbarrier motion interval and that from the quantum well interval compensate each other when being summed, which results in appearance in the Auger recombination matrix element of an additional small factor of the order of $\tilde{V}/E_g$, where $\tilde{V} = 3(V_c - d_c)/4 + (V_c + d_x)/4$ is the effective barrier height for electrons (with the strain taken into account). Note that in the II-type heterostructures at definite structure parameters an analogous compensation results in suppressing the Auger recombination processes [30,35,43].

When calculating the electron–hole overlap integral I_{24}, it is necessary to take into account the light-heavy hole interconversion and an essentially non-parabolic dependence of the hole spectrum on the longitudinal momentum components. This is can be done in the framework of the Kane model. It is noteworthy that the dependence of the overlap integral I_{23} on the transferred momentum $|\mathbf{q}_1 - \mathbf{q}_3|$ is essentially nonlinear. Fig. 3 shows I_{23} versus the momentum transferred. For the case of a single non-strained quantum well (see Fig. 3(a)), the linear dependence of I_{23} on the transferred momentum takes place only when the momentum is small, $q < \pi/a$; in the region $q > \pi/a$, I_{23}

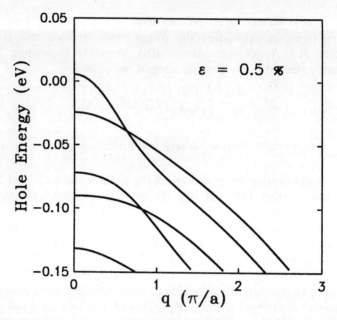

Fig. 2 Spectrum of holes in a type-I heterostructure versus longitudinal momentum. The spectrum was calculated from Eq. (20). The following parameters were taken for the calculation: $a = 80\,\text{Å}$, $E_g = 0.4\,\text{eV}$, $\Delta_{so} = 0.38\,\text{eV}$, $V_c = 0.2\,\text{eV}$, $V_v = 0.15\,\text{eV}$, $m_h = 0.41\,m_0$, $m_c = 0.03\,m_0$.

is a slowly decreasing function of this momentum. Fig. 3(b) shows the same dependence for a single strained quantum well structure [44]. In this case the dependence of the overlap integral on transferred momentum has essentially a non-monotonous run. Thus, strain has an appreciable effect (both qualitative and quantitative) on the transferred-momentum dependence of the overlap integral. In Ref. [42] the Auger recombination rate was calculated using an approximate expression for the electron–hole overlap integral $|I_{23}|^2 = \alpha|\mathbf{q}_2 - \mathbf{q}_3|^2/E_g$, where α is some fitting parameter. As shown in the present paper, this approach is not valid in the general case and it does not allow analysis of the dependence of the Auger recombination rate dependence on quantum well parameters and strain. Thus, to calculate the overlap integrals appearing in the Auger recombination matrix element it is necessary to use the carrier wave function derived in the framework of the multiband Kane model.

From the general form of the Kane equations (2)–(4) we can derive the scalar product of two electron wave functions $\psi_i^* \psi_j$, comprising

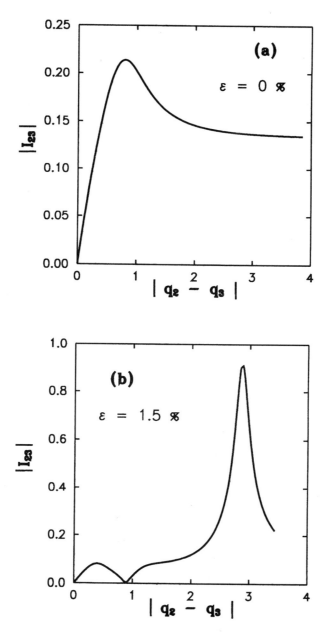

Fig. 3 The electron–hole overlap integral $I_{23}(q_x = 0)$ versus the transferred momentum $|\mathbf{q}_2 - \mathbf{q}_3|$ for an unstrained (a) and strained (b) InAlAsSb/GaSb quantum well; $E_g^{\text{eff}} = 0.4\,\text{eV}$, $a = 80\,\text{Å}$, the band parameters of the structure are taken from Ref. [53].

columns of four components of the envelopes **u** and **v**:

$$\psi_i^*\psi_j = u_i^*u_j + v_i^*v_j = \frac{1}{E_i - E_j}\,\text{div}\,\mathbf{j}_{ij}, \tag{69}$$

where, according to Eq. (42), the probability flux density **j**, is equal to

$$\mathbf{j}_{ij} = i\gamma(u_i^*v_j + u_jv_i^*). \tag{70}$$

Substituting Eqs. (69) and (70) into Eq. (68) and integrating in the plane of the quantum well, we obtain the following expression for I_{14}:

$$I_{14} = \frac{\gamma}{E_4 - E_1}\{(\mathbf{q}_4 - \mathbf{q}_1)\mathbf{I}_\| - q_x I_x\}\delta_{\mathbf{q}_\| + \mathbf{q}_4 - \mathbf{q}_1, 0} \equiv \tilde{I}_{14}(q)\delta_{\mathbf{q}_\| + \mathbf{q}_4 - \mathbf{q}_1, 0}, \tag{71}$$

where

$$\mathbf{I}_\| = \int_{-\infty}^{+\infty} e^{iq_x x}(u_4\mathbf{v}_{1\|}^* + u_1^*\mathbf{v}_{4\|})\,dx, \tag{72}$$

$$I_x = \int_{-\infty}^{+\infty} e^{iq_x x}(u_4 v_{1x}^* + u_1^* v_{4x})\,dx. \tag{73}$$

Here we took into account the fact that $\mathbf{v} = (v_x, \mathbf{v}_\|)$, $\delta_{q,0}$ is the Kronecker delta. We substitute Eq. (71) for I_{14} into Eq. (67) and with account of the explicit form of I_{23} we obtain for the matrix element

$$M_I = \frac{4\pi e^2}{\kappa_0}\int dx\,\psi_3(x)\psi_2^*(x)\int\frac{dq_x}{2\pi}\frac{e^{-iqx}\tilde{I}_{14}(q_x)}{(\mathbf{q}_1 - \mathbf{q}_4)^2 + q_x^2}\,\delta_{\mathbf{q}_1 + \mathbf{q}_2, \mathbf{q}_3 + \mathbf{q}_4} \tag{74}$$

$$\equiv \tilde{M}_I\delta_{\mathbf{q}_1 + \mathbf{q}_2, \mathbf{q}_3 + \mathbf{q}_4}.$$

The matrix element M_I can be calculated exactly by analytical methods. The following scheme can be conveniently used in calculations. First, we calculate I_x and $\mathbf{I}_\|$. Then we integrate Eq. (74) with respect to q. Then by integrating with respect to q_x and using the residue theorem, we conclude that there are two types of poles in the complex plane of q_x: (i) the poles that correspond to a small momentum transferred at the Coulomb interaction, $q_x = |\mathbf{q}_1 - \mathbf{q}_4| \cong 1/\lambda_T$, $\lambda_T = (\hbar^2/2m_h T)^{1/2}$; (ii) the poles that correspond to a large transferred momentum, $q_x \approx Q$, where $Q = 1/(\sqrt{2}\lambda_g)$, $\lambda_T \gg \lambda_g$. Thus, the matrix element of the Auger recombination transition comprises two parts:

$$M = M^{(1)} + M^{(2)} \equiv [\tilde{M}^{(1)} + \tilde{M}^{(2)}]\delta_{\mathbf{q}_1 + \mathbf{q}_2, \mathbf{q}_3 + \mathbf{q}_4}, \tag{75}$$

where $M^{(1)}$ and $M^{(2)}$ are the contributions that correspond respectively to the small and large transferred momenta. The effective Coulomb long-range interaction (small transferred momentum) in the Auger recombination process is characteristic only of heterostructures because of the interaction of carriers with the heteroboundary [19]. In bulk semiconductors, the only contribution to the matrix element corresponds to a large transferred momentum [18,19].

In type-I heterostructures, the main contribution to the matrix element is given by the term $M^{(1)}$ [19,35], which corresponds to a small transferred momentum; in this case $M^{(2)}/M^{(1)} \sim (T/V_c)\sqrt{m_h/m_c} \ll 1$. In type-II heterostructures the ratio between $M^{(1)}$ and $M^{(2)}$ substantially depends on the heterostructure parameters (the heterobarrier heights V_c and V_v, strain and quantum well widths for electrons a and holes b). Depending on the ratio V_v/V_c, the following cases are possible in type-II heterostructures: $M^{(1)} \gg M^{(2)}$, $M^{(1)} \sim M^{(2)}$ and $M^{(1)} \ll M^{(2)}$.

A further analysis is convenient to perform separately for heterostructures of I and II type, since the mechanisms of Auger recombination are appreciably different for type-I and type-II heterostructures [35]. This difference will be described in detail below.

7.4.2.2 Mechanism of Auger recombination in type-I heterostructures with strained quantum wells

In type-I heterostructures the offsets of the conduction and valence bands at the interface between two materials have opposite directions (Fig. 1). Below we will use in calculating the Auger recombination rate the following relationship between the parameters: $T \ll (V_c, V_v) \ll E_g$. This relationship takes place for the majority of semiconductor heterostructures. However, we will use a precise numerical calculation of the Auger recombination rate in comparing theoretical results with experimental data.

Now we turn to the calculation of the matrix element M.

A. *Small momentum transferred at Coulomb interaction of charge carriers.* We consider the overlap integral of electrons I_{14} (see Eq. (71)) at small q. First, we calculate I_x and $I_\parallel$ using Eqs. (72) and (73) along with explicit expressions for the wave functions and Kane equations from Section 7.2. Then we express v_{1x} and v_{4x} from the Kane equation and substitute them into Eq. (73). Integrating by parts, we obtain

$$I_x = \frac{i\gamma}{4E_g^2}[3(V_c - d_c) + V_v + d_x]\left\{u_{c1}^*\left(-\frac{a}{2}\right)u_{c4}^Q\left(-\frac{a}{2}\right)e^{iqa/2}\right.$$
$$\left. - u_{c1}^*\left(\frac{a}{2}\right)u_{c4}^Q\left(\frac{a}{2}\right)e^{-iqa/2}\right\}. \tag{76}$$

Here $q = \pm i|\mathbf{q}_1 - \mathbf{q}_4|$, $u_{c1}(-a/2)$ and $u_{c4}^Q(-a/2)$ are the values of the components of the envelopes of the wave functions at the heterointerface $x = -a/2$; $u_{c1}(a/2)$ and $u_{c4}^Q(a/2)$ are their values at $x = a/2$. We remind that the components u and v_x for electrons are continuous.

We can calculate $I_\parallel$ in a similar way. Here the neglect of the spin–orbit interaction leads to $|I_\parallel|/|I_x| \approx (TV_c/E_g^2)^{1/2} \ll 1$ [21,30,35]. However, allowing for the spin–orbit interaction qualitatively and quantitatively changes the relation between $I_\parallel$ and I_x. Kane equations with allowance for the spin–orbit interaction lead to $v_\parallel/v_x \sim \Delta_{so}/(E_g + (2/3)\Delta_{so})$ [20]. Consequently,

$$\frac{|I_\parallel|}{|I_x|} \simeq \frac{1}{3} \frac{\Delta_{so}}{E_g + (2/3)\Delta_{so}}. \tag{77}$$

In this case the contribution from $I_\parallel$ into the overlap integral I_{14} turns out to be important.

Calculating I_x, according to Eq. (73), we subdivided the integration over x into three parts: the quantum well region $(-a/2 < x < a/2)$ and the two barrier regions ($x > a/2$ and $x < -a/2$). As stated above, the contributions to the matrix element from the quantum well region and from two barrier regions are of the same order, but of different sign. As a result, these contributions cancel each other when being summed, so that a small term of the order of $(\widetilde{V}/E_g)^{1/2}$ appears in the expression for I_x. Thus, the main (with respect to $1/Q$) term in I_x is proportional to $1/Q^3$. As a result, we find that the contribution of I_x to the overlap integral I_{14} is determined by the following expression:

$$\widetilde{I}_{14}^x(q) = \frac{i\gamma^2 q}{4E_g^2(E_1 - E_4)}[3(V_c - d_c) + V_v + d_x]\left[u_{c1}^*\left(-\frac{a}{2}\right)u_{c4}^Q\left(-\frac{a}{2}\right)e^{iqa/2}\right.$$

$$\left. - u_{c1}^*\left(\frac{a}{2}\right)u_{c4}^Q\left(\frac{a}{2}\right)e^{-iqa/2}\right]. \tag{78}$$

As in the case of a single heterobarrier, [19,35] we have obtained an important result: the overlap integral I_{14} and therefore the matrix element of the Auger transition are proportional to the heterobarrier height for electrons and holes, which results from the interaction of charge carriers with heterobarriers and from the absence of the law of quasi-momentum conservation for the transverse component.

Then we substitute the expression obtained for $\widetilde{I}_{14}^x$ into Eq. (74). Performing integration with respect to q we get

$$\tilde{M}_{\mathrm{I}} = \frac{4\pi e^2}{\kappa_0} \int_{-\infty}^{+\infty} \psi_2^*(x)\psi_3(x) J(x)\, dx, \qquad (79)$$

$$J(x) = \frac{\gamma^2 \tilde{V}}{2E_g^3} \left\{ u_{c1}^*\left(\frac{a}{2}\right) u_{c4}^Q\left(\frac{a}{2}\right) \exp\left[-P_{14}\left|x-\frac{a}{2}\right|\right] \mathrm{sign}\left(x-\frac{a}{2}\right) \right.$$
$$\left. - u_{c1}^*\left(-\frac{a}{2}\right) u_{c4}^Q\left(-\frac{a}{2}\right) \exp\left[-P_{14}\left|x+\frac{a}{2}\right|\right] \mathrm{sign}\left(x+\frac{a}{2}\right) \right\}, \quad (80)$$

where

$$P_{14} = |\mathbf{q}_1 - \mathbf{q}_4| \sim 1/\lambda_T; \qquad \tilde{V} = \frac{3}{4}(V_c - d_c) + \frac{1}{4}(V_v + d_x),$$

$$\mathrm{sign}\, x = \begin{cases} 1, & \text{for } x > 0, \\ -1, & \text{for } x < 0. \end{cases}$$

In Eq. (80) we used the fact that $E_4 - E_1 = E_g$. Then, we integrate Eq. (79) with respect to x. Expressing the product $\psi_2^*(x)\psi_3(x)$ in terms of the flux density (see Eqs. (69) and (70)) and performing integration by parts, we obtain the final expression for the matrix element of Auger recombination at small transferred momenta

$$\tilde{M}_{\mathrm{I}}^{(1)} = \frac{4\pi e^2}{\kappa_0} \frac{\gamma^3 \tilde{V}}{E_g^4} u_{c2}^*\left(\frac{a}{2}\right) u_{c4}^Q\left(\frac{a}{2}\right) \left[u_{c1}^*\left(\frac{a}{2}\right)\right]' \left(1 + \frac{\Delta_{\mathrm{so}}}{3E_g + 2\Delta_{\mathrm{so}}}\right)$$
$$\times \frac{1}{[2a(P^2 + P_h^2)]^{1/2}} \frac{P_h}{[P_h^2 + P^2 (A_x^W/A_\parallel^W)^2]^{1/2}} \left(\frac{5}{2} \frac{A_x^W}{A_\parallel^W} - \frac{3}{2}\right), \quad (81)$$

where $[u_{c1}^*(a/2)]' = (\partial u_{c1}^*/\partial x)_{x=a/2}$. In calculating $\tilde{M}_{\mathrm{I}}^{(1)}$ we used the explicit form of the hole wave function component v_{h3}. In Eq. (81) we took account of the interconversion between the light and heavy holes interacting with a heterobarrier.

Now we turn to the calculation of $M_{\mathrm{I}}^{(2)}$.

B. *The case of large transferred momenta.* Using the expression (68), we can express I_{23} as follows:

$$I_{23} = -\frac{1}{q^2} \int_{-\infty}^{+\infty} dx\, e^{-iqx} \frac{d^2}{dx^2} (u_{c2}^* u_{h3} + v_{c2}^* v_{h3}). \qquad (82)$$

While deriving Eq. (82) we integrated the expression for I_{23} with respect to x by parts twice. Then, using the Kane equations (2)–(4), we express the second derivatives of u and v. However, we should only retain the terms proportional to $\partial V_c/\partial x$, $\partial V_v/\partial x$, $\partial d_c/\partial x$ and $\partial d_x/\partial x$. It is these terms that contain singularity. As a result we obtain for I_{23}

$$I_{23} = \frac{i(V_c + V_v - d_c + d_x)}{\gamma Q^2} \left[u_{c2}^*\left(-\frac{a}{2}\right) v_{h3x}\left(-\frac{a}{2}\right) e^{iqa/2} \right.$$
$$\left. - u_{c2}^*\left(\frac{q}{2}\right) v_{h3x}\left(\frac{a}{2}\right) e^{-iqa/2} \right]. \tag{83}$$

Substituting I_{23} from Eq. (83) into Eq. (74) for $\tilde{M}_I$ and changing $q \to Q$ we come to

$$\tilde{M}_I^{(2)} = \frac{i(V_c + V_v - d_c + d_x)}{\gamma Q} \int\int \frac{dq}{2\pi} dx \, (u_{c2}^* u_{c4}^Q + v_{c2}^* v_{c4}^Q) e^{iqx}$$
$$\times \left[u_{c2}^*\left(-\frac{a}{2}\right) v_{h3x}\left(-\frac{a}{2}\right) e^{iqa/2} - u_{c2}^*\left(\frac{q}{2}\right) v_{h3x}\left(\frac{a}{2}\right) e^{-iqa/2} \right]. \tag{84}$$

Taking account of $\int_{-\infty}^{+\infty}(dq/2\pi) e^{iqx} = \delta(x)$, we obtain the final expression for $\tilde{M}_I^{(2)}$

$$\tilde{M}_I^{(2)} = \frac{4\pi e^2}{\kappa_0} \frac{2i(V_c + V_v - d_c + d_x)}{\gamma Q^4} \left[u_{c1}^*\left(\frac{a}{2}\right) \right]^2 u_{c4}^Q\left(\frac{a}{2}\right) v_{h3x}\left(\frac{a}{2}\right). \tag{85}$$

Comparing the expression (85) for $\tilde{M}_I^{(2)}$ with (81) for $\tilde{M}_I^{(1)}$, we get

$$\frac{\tilde{M}^{(2)}}{\tilde{M}^{(1)}} \sim \frac{T}{V_c} \sqrt{\frac{m_h}{m_c}}. \tag{86}$$

Thus, in type-I semiconductor heterostructures, the contribution to the matrix element of the Auger transition from large transferred momenta for $V_c \gg T$ is smaller than that from small transferred momentum. However, the exact numerical calculation of the Coulomb matrix element M (see Eq. (75)) given above shows that for real heterostructures it is important to take into account both $\tilde{M}^{(1)}$ and $\tilde{M}^{(2)}$.

C. *Auger recombination rate.* Now we substitute the square of the absolute value of the matrix element $|M|^2$ (see Eq. (65)) into (64). Then using (74), (75) and (81) we perform summation over the initial and final states. Below, for the sake of a qualitative analysis, we will consider that only the first size-quantization level of electrons is occupied. As was noted above, for heavy holes, $T > E_{0h}$, so the summation over the size-quantization levels is changed to integration with respect to P_h. Integration is also to be performed over the states of the highly excited electron. As a result, we derive the following expression for the Auger recombination rate:

$$G = \frac{3}{\sqrt{2}\hbar\gamma} \int \frac{d^2q_1}{(2\pi)^2} \int \frac{d^2q_2}{(2\pi)^2} \int \frac{d^2q_3}{(2\pi)^2} \int \frac{dP_h}{2\pi}$$
$$\times \{|\tilde{M}_I|^2 + |\tilde{M}_{II}|^2 - \tilde{M}_I\tilde{M}_{II}^*\}|_{E_4=3E_g/2} f_c(E_1)f_c(E_2)f_h(E_3). \quad (87)$$

Then, we should integrate over q_1, q_2 and q_3, using a polar coordinate system. Thus, we get for the Auger recombination rate

$$G = \frac{8\pi^2}{3\sqrt{2}} \frac{E_B}{\hbar} n^2 p \lambda_g^4 \frac{\tilde{V}^2}{E_g^2} \frac{E_{0c}}{T} \frac{m_c}{m_h} \frac{\sin^2 ka}{(1+1/\kappa_h a)(1+1/\kappa_c a)^2} \frac{\lambda_g^3}{a^3} J_d, \quad (88)$$

where

$$J_d = \frac{3}{2\pi\sqrt{\pi}} \int\int dx\, d^2\rho\, e^{-(x^2+\rho^2)} \frac{x^2((5/2)(A_x^W/A_\parallel^W) - 3/2)^2}{(x^2+\rho^2)[x^2+\rho^2(A_x^W/A_\parallel^W)^2]}, \quad (89)$$

$E_B = m_c e^4/(2\hbar^2\kappa_0^2)$ is the Bohr electron energy,

$$x = P_h/P_T, \quad \rho^2 = q^2/P_T^2, \quad P_T = (2m_h T)^{1/2}/\hbar.$$

If there is no strain, $J_d = 1$. Then Eq. (88) goes over into an expression for the Auger recombination rate for a non-strained quantum well [21,35]. In the limit $(V_c = V_v) \to \infty$ in the absence of strain Eq. (88) goes over into the expression for the Auger recombination rate obtained in Ref. [45].

As follows from (88), the Auger recombination rate displays a power dependence on temperature. Besides, for the type-I heterostructures with strained quantum wells the Auger rate considerably depends on the quantum well width a, the heterobarrier heights V_c and V_v for electrons and holes respectively and on the value of elastic strain. Its dependence on V_c and V_v is determined by the dependence on V_c and V_v of the overlap integral $I_{14}(q)$ of the electrons in the initial state "1" and the final state "4" for small transferred momentum. As shown above, the specified overlap integral contains three parts: two parts of subbarrier motion of electron ($x < -a/2$ and $x > a/2$) and a part of the quantum well ($-a/2 < x < a/2$). The contribution from the subbarrier parts and the quantum well part cancel each other when being summed, the resulting overlap integral I_{14} decreasing as $\tilde{V}/E_g$ [19,35]. This means that the new mechanism of Auger recombination being considered in this paper is due to the interaction of charge carriers with heterointerfaces [19].

The dependence of the Auger recombination rate on the strain value ε is determined by the dependence of the overlap integrals I_{14} and I_{23} on ε (see Eqs. (78), (81)). With increasing ε the effective value of the heterobarrier height decreases; hence the overlap integral of electron and highly excited electron decreases, too. The overlap integral I_{23} appreciably changes depending on strain: for small transferred momenta $q < \pi/a$, I_{23} falls as ε rises and for large transferred momenta $q > \pi/a$, it depends on ε non-monotonously (see Fig. 3(b)). As a result, the Auger recombination rate decreases with increasing strain ε. Fig. 4 shows the Auger recombination rate versus the strain ε, the possible range of ε is determined by the parameters of quantum well. For ε corresponding to real structures, the Auger recombination rate drops with increasing strain ε. The decrease of the quantum well width makes this dependence steeper (Fig. 4(a)).

Fig. 5 shows the Auger recombination rate against the quantum well width. This dependence is not monotonous, having a maximum at a certain value of a. With increasing temperature, the value of the maximum shifts to the region of great a (see Fig. 5(b)). The non-monotonicity of G as a function of a is connected with a non-monotonous run of the overlap integral I_{14} as a function of a.

Fig. 6 plots the Auger recombination rate against V_c. At $V_v = \text{const}$ the Auger rate is a rising function of V_c coming to saturation at $V_c \sim E_g$. As follows from Eq. (88) for the Auger recombination rate, G considerably depends on the bandgap width E_g: $G \propto E_g^{-11/2}$. The obtained results are important for analysis of the threshold characteristics of quantum well lasers.

As previously noted, as a result of the Auger process having a threshold nature, the coefficient of the Auger recombination in bulk semiconductors

$$C_A = G/(n^2 p) \tag{90}$$

appears to be an exponential function of temperature: $C_A \propto \exp(-E_{th}/T)$. In quantum wells, however, the Auger process is non-threshold with the Auger coefficient being weakly dependent on temperature. Exact analysis shows that over a wide range of the quantum well widths a the Auger coefficient only varies by a factor of two (see Section 7.4.3). Thus, the dependence of C_A on temperature is determined by the temperature dependence of the bandgap E_g and the electron–hole overlap integral I_{23}. Fig. 7 plots the Auger coefficient as a function of temperature for the case of E_g being temperature dependent (solid curve) and E_g temperature independent, at different values of the strain [44]. In both

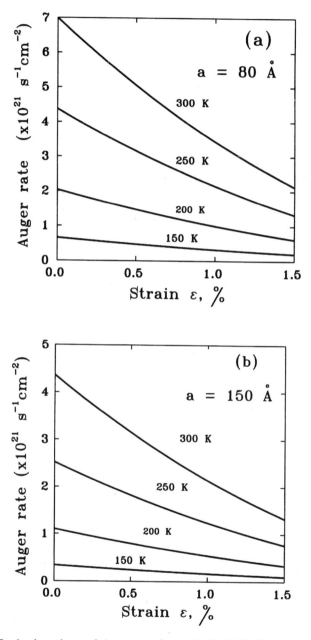

Fig. 4 Strain dependence of Auger rate for an InAlAsSb/GaSb quantum well at various temperatures for different quantum well widths: (a) $a = 80$ Å and (b) $a = 150$ Å.

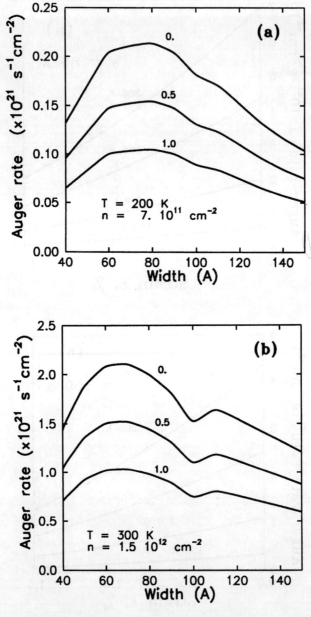

Fig. 5 Auger rate for an InAlAsSb/GaSb quantum well with various compressive strains ($\varepsilon = 0.5$ and 1%) versus well width: (a) $T = 200\,\text{K}$, $n = 7 \times 10^{11}\,\text{cm}^{-2}$; (b) $T = 300\,\text{K}$, $n = 1.5 \times 10^{12}\,\text{cm}^{-2}$.

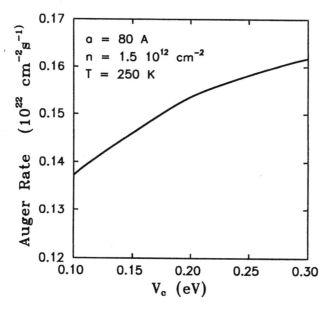

Fig. 6 Auger rate versus conduction band offset for an InAlAsSb/GaSb quantum well.

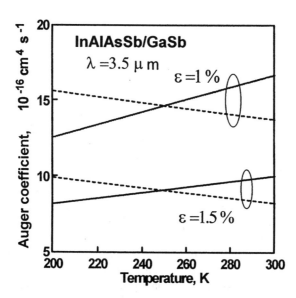

Fig. 7 Temperature dependence of Auger coefficient for an InAlAsSb/GaSb quantum well with various strains, $\lambda = 3.5\,\mu m$.

cases the Auger coefficient is a power function of temperature. A more detailed analysis of C_A as a function of T is performed in Section 7.4.3.

The results obtained in this section are important for the analysis of threshold characteristics of mid-infrared lasers based on quantum wells (see Sections 7.4–7.6).

Thus, we have shown that in type-I heterostructures with quantum wells the Auger recombination process becomes non-threshold owing to the interaction of charge carriers with heterointerfaces, while the Auger recombination rate is a power function of temperature.

We will only consider one of the possible Auger recombination processes in quantum wells, namely, CHCC process. It is worth stressing that, along with the CHCC Auger process, we have to take into account the CHHS Auger process, i.e., the Auger process involving hole–hole interaction, as a result of which one of the holes passes into the conduction band and the other passes into the so-band absorbing the energy and momentum. It is important that this Auger process also appears to be non-threshold. Moreover, in a number of semiconductor heterostructures, for which $E_g - \Delta_{so} \gg T$ holds, this process may be the main channel of nonradiative Auger recombination at certain quantum well parameters. For a qualitative analysis an approximate expression for the CHHS Auger process rate may be used:

$$G^{CHHS} \approx \frac{8}{\pi} \frac{E_B}{\hbar} \left(\frac{E_{0c}}{E_g - \Delta_{so}} \right)^{5/2} \left(\frac{m_c}{m_{so}} \right)^{5/2} \lambda_g^2 \lambda_{E_g - \Delta}^2 n p^2, \quad (91)$$

where $\lambda_{E_g - \Delta} = \hbar/[2m_c(E_g - \Delta_{so})]^{1/2}$. In analyzing the Auger recombination current (Section 7.5.3) we will also take account of the CHHS process.

7.4.2.3 Mechanism of Auger recombination in type-II heterostructures with quantum wells

In type-I heterostructures the offsets of the conduction band and valence band at the interface between two materials have opposite directions (Fig. 1). Type-II heterostructures have the following distinguishing features [46]: (i) the offsets of the conduction (V_c) and the valence band (V_v) have the same direction (Fig. 8) and different signs: $V_c > 0$, $V_v = -|V_v| < 0$; (ii) in contrast to the case of type-I heterostructures, here the electrons and holes are spatially separated, so that the recombination is possible only via tunneling through the heterobarrier. It will be shown below that in type-II heterostructures the Auger recombination rate is also a power function of temperature, but the

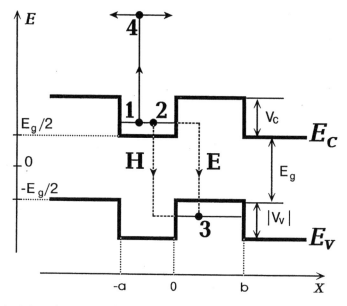

Fig. 8 Schematic representation of the band diagram of a type-II heterostructure with quantum wells: "1" and "2" are the initial states of the particles, "3" and "4" are the final states, H and E are two possible channels for the recombination of electron "2" and hole "3".

Auger recombination mechanisms in heterostructures of types I and II are fundamentally different.

The purpose of the present section is to investigate theoretically the Auger recombination processes of nonequilibrium carriers in type-II heterostructures.

Type-II heterostructures have two important channels for Auger recombination: (1) an Auger process involving two electrons and one hole (the CHCC process); (2) an Auger process involving one electron and one hole with passage of a second hole into the spin-split-off band (the CHHS process). In the present work we restrict ourselves to consideration of the CHCC process (Fig. 8), since it dominates over the CHHS process if $(E_g - \Delta_{so})/E_g > m_c/m_{so}$ holds, where E_g is the effective bandgap width (see Fig. 8).

We note that in type-II heterostructures, unlike those of type I, the CHCC Auger process involves two channels (Fig. 8): (1) an electron tunneling through the heterobarrier and recombining with a hole in the quantum well (the E channel); (2) a hole tunneling through the heterobarrier and recombining with an electron in the quantum well (the H channel). The contributions from these two channels to the Auger recombination matrix element are of the same order, owing to the interconversion of light and heavy holes in their interaction with the heterointerface. It turns out that only the E channel is important in the radiative recombination process.

To calculate the Auger recombination rate G it is first necessary to find the wave function of carriers in a type-II heterostructure. Fig. 8 shows the band diagram of such a structure with quantum wells. To the left of the point $x = 0$ (in the region $x < 0$) there is a quantum well of width a for electrons, and in the region $x > 0$ there is a quantum well of width b for holes. We shall henceforth assume that nonequilibrium electrons and holes are trapped in these wells. We note that the effective bandgap width E_g (i.e., the minimal energy separation between the electron and hole states) is smaller than the bandgap widths for both semiconductors (see Fig. 8). As previously noted, the wave functions of carriers must be calculated in the multiband approximation. It will be shown below that this is especially important for type-II heterostructures, since there are two channels for the recombination of electrons and holes (the E and H channels). These two channels interfere with each other due to the interconversion of light and heavy holes. Therefore, the Auger recombination matrix element must be calculated with account of the complicated structure of the valence band, particularly the interconversion of light and heavy holes when they interact with the heterointerface. The wave functions of electrons and holes are the result of superposition of the basis states (Eq. (1)). The set of equations for the envelopes of Bloch functions is derived in Section 7.2 (Eqs. (2)–(4)). The band offsets $V_c(x)$ and $V_v(x)$

$$V_c(x) = \begin{cases} 0, & -a < x < 0, \\ V_c, & x < -a,\ x > 0, \end{cases}$$

$$V_v(x) = \begin{cases} 0, & 0 < x < b, \\ V_v, & x > b,\ x < 0. \end{cases} \tag{92}$$

We note that V_c and V_v have different signs in type-II heterojunctions: $V_c > 0$, $V_v = -|V_v| < 0$. The electron and hole wave functions in type-II heterostructures are to be calculated with allowance for (91).

In type-II heterostructures the Auger transition matrix element also comprises two parts, according to Eq. (75).

In type-II heterostructures, along with the long-range part of the Coulomb interaction matrix element that corresponds to small momentum transfers, the short-range part that corresponds to large momentum transfers must be taken into account. Since the electrons and holes in type-II heterostructures are spatially separated, there are two channels for Auger recombination, viz., E and H (Fig. 8). As will be shown below, these two channels interfere with each other to decrease the matrix element $M^{(1)}$. In addition, $M^{(1)}$ is significantly smaller than $M^{(2)}$ when a certain relationship exists between V_c and V_v (see Eq. (98)). Thus, in type-II heterostructures the contributions to the Auger recombination rate are made by both the long-range part of the Coulomb interaction (small momentum transfers) and short-range part of the Coulomb interaction (large momentum transfers).

In calculating $M^{(1)}$ and $M^{(2)}$ below, we shall use the following relationships between the parameters: $T \ll (V_c, |V_v|) \ll E_g$. Such a relationship between the parameters is observed for most of semiconductor heterostructures. In addition, for the sake of simplicity we will disregard the influence of strain on the matrix element and the Auger recombination rate. This is possible, since the value of elastic strain does not affect the Auger recombination rate in type-II heterostructure so much as the parameters V_c and $|V_v|$. It is then convenient to calculate $M^{(1)}$ and $M^{(2)}$ separately [30,43].

A. *Small momentum transfers.* Let us proceed to the calculataion of $M^{(1)}$. We consider the electron overlap integral $I_{14}(q)$ at small q. We expand I_{14} in powers of $1/Q$. Then, according to (71)–(73), we obtain

$$\tilde{I}_{14} \cong -\frac{\gamma q}{E_g} I_x, \qquad (93)$$

$$I_x \cong -\frac{i\gamma}{4E_g^2}[3(V_c - d_c) - |V_v| + d_c][u_1(0)u_4(0) - e^{-iqa}u_1(-a)u_4(-a)]$$

$$+ \frac{\kappa_c(3V_c - |V_v|)}{2Q^2 E_g}[u_1(0)v_{x4}(0) + e^{-iqa}u_1(-a)v_{x4}(-a)]$$

$$+ \frac{iq[5(V_c - d_c) - |V_v| + d_x]}{2Q^2 E_g}[u_1(0)v_{x4}(0) - e^{-iqa}u_1(-a)v_{x4}(-a)]. \qquad (94)$$

Here $q = \pm i|\mathbf{q}_1 - \mathbf{q}_4|$, $u_1(0)$, $v_{x4}(0)$ and $u_4(0)$ are the values of the components of the envelope wave function at the heterointerface $x = 0$; $u_1(-a)$, $v_{x4}(-a)$, and $u_4(-a)$ are the values of the components at $x = -a$. We recall that the components u and v_x for electrons are continuous. In (93) we took into account that $E_4 - E_1 \cong E_g$.

To calculate I_x according to (73), we divide the integration with respect to x into three regions: the region of the well ($-a < x < 0$) and two barrier regions ($x > 0$ and $x < -a$). It turns out that the contributions to the matrix element from the region of the quantum well and from two barrier regions are of the same order, but of different sign. As a result, they cancel one another, so that a small parameter of the order $\sqrt{V/E_g}$ (where $V = 3V_c - |V_v|$, see (93) appears in the exprssion for I_x. Thus, the leading (with respect to $1/Q$) term in I_x is proportional to $1/Q^3$. In (93) we retained the next order in $1/Q$, i.e., $1/Q^4$, since the main term, which is proportional to $1/Q^3$, has the multiplier $(3V_c - |V_v|)$, which may be close to zero. It should be stressed that such a situation does not occur in type-I heterostructures, and it is sufficient to restrict the calculation of I_x to the main term is proportional to $1/Q^3$. In type-I heterostructures the main term in $1/Q$ is proportional to $(3V_c + V_v)$, where $V_v > 0$.

We substitute the expression for $\tilde{I}_{14}$ (Eqs. (93) and (94)) into (74). Performing the integration with respect to q, we obtain

$$\tilde{M}_I = \frac{4\pi e^2}{\kappa_0} \int_{-\infty}^{+\infty} \psi_2^*(x) \psi_3(x) J(x)\, dx, \tag{95}$$

$$J(x) = \frac{\gamma^2}{8E_g^2} u_1^*(0) u_4(0) \exp(-p_{14}|x|) \left\{ 3[(V_c - d_c) - |V_v| + d_x] \right.$$

$$\left. \times \left(1 + i\frac{\kappa_c}{Q}\right) \operatorname{sign} x + i[5(V_c - d_c) - |V_v| + d_x] \frac{p_{14}}{Q} \right\}, \tag{96}$$

where $p_{14} = |\mathbf{q}_1 - \mathbf{q}_4| \sim 1/\lambda_T$. We note that in (95) we neglect the terms which are proportional to $\exp(-p_{14}a) \ll 1$. This is correct when $p_{14}a \approx \sqrt{2m_h T a^2/\hbar^2} \cong \sqrt{T\pi^2/E_{1h}} \gg 1$. Henceforth we assume that $E_{1h} < T \ll |V_v|$.

Performing integration with respect to x in (94), we obtain the final expression for the matrix element $M^{(1)}$ at small momentum transfers. Expressing the product $\psi_2^*(x)\psi_3(x)$ in (94) terms of the flux density (see Eqs. (69) and (70)) and integrating by parts, we obtain

$$\tilde{M}_I^{(1)} = \frac{4\pi e^2}{\kappa_0} \frac{i\gamma^2}{8E_g^2} \frac{u_1(0) u_4(0)}{p_{14}} \left\{ (3V_c - |V_v|)\left(1 + i\frac{\kappa_1}{Q}\right) \right.$$

$$\times [J_h + J_e - 2p_{14} u_2^*(0) v_{3x}(0)] + i(5V_c - |V_v|)$$

$$\left. \times \frac{p_{14}}{Q} [J_h - J_e + 2i(\mathbf{q}_3 - \mathbf{q}_2) u_2^*(0) \mathbf{v}_{3\|}(0)] \right\}, \tag{97}$$

where

$$J_e = \int_{-\infty}^{0} \left\{ p_{14}^2 [u_2^* v_{3x} + u_3 v_{2x}^*] + i(\mathbf{q}_3 - \mathbf{q}_2) \frac{\partial}{\partial x} (u_2^* v_{2\parallel} + u_3 v_{2\parallel}^*) \right\} dx;$$

J_h is the same as J_e, except that it is taken in the integration limits from 0 to ∞.

When M is calculated, it is important to take into account the interconversion of light and heavy holes, accompanying their interaction with the heterointerface. The matrix element M consists of two parts: the contribution from the region $x < 0$ and the contribution from the region $x > 0$. These two contributions correspond to two channels of recombination of spatially separated electrons and holes (the E and H channels, see Fig. 8). The E channel corresponds to the recombination of an electron tunneling through the heterobarrier with holes which transform into each other when being reflected from the heterointerface. In the case of the H channel, an electron trapped in the well recombines with holes that transform into each other when tunneling through the heterobarrier. For quantum wells in which the size-quantization energy of electrons and holes is much smaller than the height of the heterobarriers ($E_{0e} \ll V_c$, $E_{0h} \ll |V_v|$), we find that the contributions from the E and H channels to $M^{(1)}$ are of the same order, but of different sign. Thus, there is a destructive interference between these two channels for Auger recombination. Such interference produces an additional small parameter of the order of $[Tm_h/V_c m_c]^{3/2} < 1$ in the matrix element $M^{(1)}$ for type-II heterostructures in comparison with the Auger recombination matrix element for type-I heterostructures [19,42]. As a result, the Auger recombination matrix element for small momentum transfers equals

$$\tilde{M}_1^{(1)} = \frac{4\pi e^2}{\kappa_0} \frac{\gamma^3}{8E_g^4} u_1^*(0) u_2(0) u_4(0) H \varepsilon_h \sin \frac{P_h b}{2}$$

$$\times \left\{ (3V_c - |V_v|) \left(1 + \frac{i\kappa_1}{Q}\right) \left[\frac{m_h}{m_c} \frac{\mathbf{q}_3 \cdot (\mathbf{q}_3 - \mathbf{q}_2)}{q_3 V_e} + 2 \frac{q_3}{|P_1|} \sqrt{\frac{2m_c}{|V_v|\hbar^2}} \right] \right.$$

$$\left. + i \frac{p_{14}}{Q} (5V_c - |V_v|) \frac{m_h}{m_c} \frac{\mathbf{q}_3 \cdot (\mathbf{q}_3 - \mathbf{q}_2)}{q_3 V_c} \right\}. \tag{99}$$

Here $\varepsilon_h = \hbar^2(P_h^2 + q_3^2)/2m_h$ is the energy of a heavy hole measured from the valence-band edge of the semiconductor in the quantum well region $0 < x < b$;

$$u_1(0) = u_2(0) = A \cos \frac{ka}{2}, \qquad u_4(0) = A_4 t.$$

Let us proceed to the calculation of $M^{(2)}$.

B. *Large momentum transfers.* In type-I semiconductor heterostructures the contribution to the Auger transition matrix element $M^{(2)}$ from large momentum transfers is small in comparison with that from small momentum transfers (see Eq. (86)). As will be shown below, in type-II heterostructures this is not valid: $M^{(1)}$ and $M^{(2)}$ may be of the same order, and under certain conditions the contribution $M^{(2)}$ from large momentum transfers may significantly exceed $M^{(1)}$. This difference between the mechanisms of Auger recombination in heterostructures of types I and II is attributed to the magnitude of the overlap integral I_{14} of the electrons in the initial "1" and final "4" states at small momentum transfers.

Using expression (74) for M_I, we represent $\tilde{M}_I^{(2)}$ in the form

$$\tilde{M}_I^{(2)} = \frac{4\pi e^2}{\kappa_0} \left[\int_{-\infty}^{0} \psi_2^*(x) \psi_3(x) J_{14}^{(e)} \, dx + \int_{0}^{\infty} \psi_2^*(x) \psi_3(x) J_{14}^{(h)} \, dx \right], \quad (100)$$

where

$$J_{14}^{(h)} = \int_{0}^{\infty} \frac{dq}{2\pi} I_{14}(q) \frac{e^{-iqx}}{q^2 + p_{14}^2}, \quad (101)$$

$J_{14}^{(e)}$ is the same as $J_{14}^{(h)}$, except for it is taken in the range from $-\infty$ to 0. When $J_{14}^{(e)}$ and $J_{14}^{(h)}$ are calculated using the residue theorem, only the poles corresponding to large momentum transfers $|q| \approx Q$ should be taken into account, and the pole $q = \pm i p_{14}$ should be disregarded, since it is already taken into account in $M^{(1)}$. Performing integration in Eq. (100) with respect to q, we obtain

$$J_{14}^{(h)} = \frac{BA_4 t}{(\bar{k}_4 + i\kappa_c)^2 + p_{14}^2} [1 + \gamma^2 \lambda_{-}^{(1)} \lambda_{-}^{(4)} (\mathbf{q}_1 \mathbf{q}_4 - i\kappa_c \bar{k}_4)] \exp(ix\bar{k}_4 - \kappa_c x). \quad (102)$$

Here $\lambda_{-}^{(i)} = \gamma/(E_i + |V_v| + E_g/2)$, where $i = 1, 4$.

According to Eq. (100), the expression for $J_{14}^{(e)}$ has the form

$$J_{14}^{(e)} = \int_{\infty}^{0} \frac{dq}{2\pi} I_{14}(q) \frac{e^{-iqx}}{q^2 + p_{14}^2}. \quad (103)$$

Integrating with respect to q in (102) by the use of the residue theorem and discarding the poles corresponding to small momentum transfers at $q = \pm ip_{14}$, which are already taken in $M^{(1)}$, we obtain the following expression for $J_{14}^{(e)}$

$$J_{14}^{(e)} = \frac{A_4 A}{2} \sum_{v=-1,1} \frac{d_+ \exp[il(v)] + d_- \exp[-il(v)]}{p_{14}^2 + (Q + vk)^2}$$
$$\times [1 + \lambda_-^{(1)}\lambda_-^{(4)}(\mathbf{q}_1\mathbf{q}_4 - vkk_4)]. \quad (104)$$

where $l(v) = vkx_a + \kappa_4 x$.

We substitute (101) and (103) into (99). Integration with respect to x must be performed next. The integral has the following form:

$$\int_0^\infty f(x) e^{i\xi x} dx = -\frac{1}{i\xi} f(+0) + \frac{1}{(i\xi)^2} f'(+0) - \frac{1}{(i\xi)^3} f''(+0) + \cdots. \quad (105)$$

A relation similar to (104) holds in the case of integration from $-\infty$ to 0. The expression (104) is obtained using integration by parts. Here we use $f(\pm\infty) = 0$ and note the $f(+0)$ is the value of f at the point $x = +0$. In our case $\xi \approx Q$ holds; therefore, Eq. (104) is nothing more than an expansion of $M^{(2)}$ in powers of $1/Q$. As in the case of small momentum transfers, we restrict ourselves to the first two non-vanishing terms in the expansion in powers of $1/Q$ in $M^{(2)}$. As a result, the integral with respect to x (from $-\infty$ to $+\infty$) equals

$$\int_{-\infty}^\infty f(x) e^{i\xi x} dx = -\frac{1}{i\xi}[f(+0) - f(-0)] + \frac{1}{(i\xi)^2}[f'(+0) - f'(-0)]$$
$$-\frac{1}{(i\xi)^3}[f''(+0) - f''(-0)] + \cdots, \quad (106)$$

where $f(-0)$ is the value of the function f at the point $x = -0$. It was assumed in (104) and (105) that $f(x)$ is a smooth function separately in $[-\infty, 0]$ and $[0, +\infty]$; both the function $f(x) \equiv u_2^*(x)u_3(x) + \mathbf{v}_2^*(x)\mathbf{v}_3(x)$ and its derivatives $f'(x)$, f'', ... have finite discontinuities at the heterointerface ($x = 0$). The jumps in the function and its derivatives are calculated directly, using the Kane equations (2)–(4) and the boundary conditions (11). As a result, we obtain for $\tilde{M}_1^{(2)}$

$$\tilde{M}_1^{(2)} = \frac{4\pi e^2}{\kappa_0} u_1(0) u_2(0) u_4(0) \frac{1}{i\gamma Q^4} \left\{ (V_c - |V_v|) + v_{3x}(0) \right.$$
$$+ \varepsilon_h H \sin\left(\frac{P_h b}{2}\right) \frac{1}{Q}\left[\frac{m_h}{m_c} \frac{\mathbf{q}_2 \cdot \mathbf{q}_3}{q_3} \sqrt{\frac{2m_h|V_v|}{\hbar^2}} + \frac{2m_c}{\hbar^2}(V_c^{3/2}|V_v|^{-1/2})\right.$$
$$\left.\left. + |V_v|^{1/2} V_c^{1/2} + 2|V_v| - 8V_c\right] \right\}. \quad (107)$$

Here $v_{3x}(0)$ is the value of the component of the hole wave function at the heterointerface $x=0$. In the approximation $\varepsilon_h \ll |V_v|$ for states of a parity for which v_{3x} is odd, we have

$$v_{3x}(0) = iH \sin\left(\frac{P_h b}{2}\right)\left[\frac{q_3}{|p_1|}\sqrt{\frac{2m_c \varepsilon_h^2}{\hbar^2 |V_v|}} + \frac{\varepsilon_h}{|V_v|} q_3 \left(1 + \frac{2m_c \varepsilon_h}{\hbar^2 |P_1|^2}\right)\right]. \quad (108)$$

Expressions (75), (99), and (108) completely specify the Auger transition matrix element M_I. As we have already noted above, M_{II} is obtained from M_I by means of the interchange of indices $1 \leftrightarrow 2$.

It follows from the expressions we obtained for $M^{(1)}$ and $M^{(2)}$ that in type-II heterostructures the relationship between $M^{(1)}$ and $M^{(2)}$ greatly depends on the parameters of heterostructure, viz., on the heights of the heterobarriers V_c and V_v and the widths of the quantum wells for electrons a and holes b (Fig. 8). The following cases are possible, depending on the relationship between V_c and V_v: $M^{(1)} \gg M^{(2)}$; $M^{(1)} \sim M^{(2)}$; $M^{(1)} \ll M^{(2)}$.

C. *Auger recombination rate.* To calculate the rate G, the square of the absolute value of the matrix element $|M|^2$ (see (65)) must be substituted into the expression (64), and the summation must be performed over the initial and final states of the particles. We shall assume below that only the ground state of the electron size-quantization is filled. As was noted above, for heavy holes we have $T > E_{0h}$; therefore, we replace the summation over the hole size-quantization levels by integration with respect to P_h. Integration should also be performed over the states of highly excited electron. Then for G we have

$$G = \frac{2\pi}{\hbar} \int \frac{d^2 q_1}{(2\pi)^2} \int \frac{d^2 q_2}{(2\pi)^2} \int \frac{d^3 q_3}{(2\pi)^2} \int \frac{dP_h}{2\pi} \int \frac{dk_4}{2\pi} \{|\tilde{M}_I|^2 + |\tilde{M}_{II}|^2 - \tilde{M}_I \tilde{M}_{II}^*\}$$
$$\times \delta(E_1 + E_2 - E_3 - E_4) f_c(E_1) f_c(E_2) f_h(E_3). \quad (109)$$

In deriving (109) we sum over $\mathbf{q}_4$ using the Kronecker delta corresponding to the conservation law for the longitudinal momentum component $\mathbf{q}$ (see Eq. (74)). Then in (108) we perform integration with respect to the momentum k_4 of the highly excited electron using the energy of δ-function. We use the explicit expressions for the energies of electrons and holes, in which it is important to take into account the nonparabolic character of the spectrum of highly excited electron, to represent the argument of the δ-function in the form

$$E_1 + E_2 - E_3 - E_4 \cong \frac{3}{2} E_g - \sqrt{\frac{E_g^2}{4} + \gamma^2 k_4^2}. \quad (110)$$

In (109) we took into account that $|\mathbf{k}_1| \cong |\mathbf{k}_2| \cong \sqrt{m_c/m_h}|\mathbf{k}_3| \ll |\mathbf{k}_4|$. As a result, the integration with respect to k_4 gives an expression identical to (87). We perform the ensuing integration with respect to q_1, q_2, and q_3 using a polar coordinate system. Calculation of the integrals is a simple, but tedious procedure. As we have already noted, heavy holes polarized in the plane of the heterointerface make no contribution to the Auger recombination rate. Therefore, in the expression for the rate we must require that the concentration of the holes that participate in the recombination process be equal to half the total concentration of holes. As a result, the final expression for the Auger recombination rate has the following form:

$$G = \tilde{G}\left(g_1 + \frac{V_c}{E_g}g_2 + \frac{T}{E_g}g_3\right) = G_1 + G_2 + G_3. \quad (111)$$

Here

$$\tilde{G} = 32\sqrt{2}\pi^2 \frac{E_B}{\hbar} \frac{T}{E_g} \frac{m_c}{m_h} n^2 p \lambda_g^4 \frac{\kappa_c^4 \lambda_g^5}{b} \frac{\cos^4(ka/2)}{(1+\kappa_c a)^2}, \quad (112)$$

$$g_1 = \frac{(2V_c - |V_v|)^2}{|V_v|V_c} + \sqrt{\pi}\left(\frac{m_h}{m_c}\right)^{3/2}\left(\frac{T}{|V_v|}\right)^{1/2} \frac{3V_c - |V_v|}{4V_c} \frac{2V_c - |V_v|}{V_c}$$

$$+ \left(\frac{m_h}{m_c}\right)^3 \frac{T}{V_c}\left(\frac{3V_c - |V_v|}{4V_c}\right)^2, \quad (113)$$

$$g_2 = \frac{(3V_c - |V_v|)^2}{16V_c^2} + \left[\frac{2V_c}{|V_v|} + \left(\frac{m_h}{m_c}\right)^{3/2}\left(\frac{\pi T}{|V_v|}\right)^{1/2} + \left(\frac{m_h}{m_c}\right)^3 \frac{T}{2V_c}\right]$$

$$+ \left(\frac{m_h}{m_c}\right)^2 \frac{T}{8V_c^3}(3V_c - |V_v|)(5V_c - |V_v|)\left[\left(\frac{V_c}{|V_v|}\right)^{1/2} + \frac{3}{8}\left(\frac{m_h}{m_c}\right)^{3/2}\left(\frac{T}{V_c}\right)^{1/2}\right]$$

$$+ \frac{1}{16}\left(\frac{m_h}{m_c}\right)^4\left(\frac{T}{V_c}\right)^2\left(\frac{5V_c - |V_v|}{V_c}\right)^2, \quad (114)$$

$$g_3 = \frac{1}{8}\left(\frac{m_h}{m_c}\right)^2\left[\left(\sqrt{\frac{m_h|V_v|}{m_c V_c}} + \frac{3V_c - |V_v|}{2V_c}\right)^2 + \left(\frac{5V_c - |V_v|}{2V_c}\right)\left(\frac{\pi m_h T}{m_c V_c}\right)^{1/2}\right.$$

$$\left.\times \left(\sqrt{\frac{m_h|V_v|}{m_c V_c}} + \frac{3V_c - |V_v|}{2V_c}\right) + \left(\frac{5V_c - |V_v|}{2V_c}\right)^2 \frac{m_h}{m_c} \frac{T}{V_c}\right], \quad (115)$$

Three terms in (111) are the result of expansion of the matrix element in powers of $1/Q$. The g_1 and g_2 terms originate from the $M^{(1)}$ part of M corresponding to small momentum transfers in the Coulomb interaction of the electrons, and g_3 originates from the $M^{(2)}$ part of M corresponding to large momentum transfers.

D. *Radiative recombination rate.* Substituting the explicit expression for the components of the electron and hole wave functions (u and v) into (41) and performing integration with respect to x, we obtain

$$\mathscr{P}_x^{cv} = \gamma B \left\{ \frac{Hq_c}{\kappa_c^2 + P_h^2} \left[P_h \cos \frac{P_h b}{2} - \kappa_c \sin \frac{P_h b}{2} \right] \right.$$

$$+ \frac{LP_1}{\kappa_c^2 + P_1^2} \left[P_1 \cos \frac{P_1 b}{2} - \kappa_c \sin \frac{P_1 b}{2} \right]$$

$$\left. - \frac{\tilde{H}q_c}{\kappa_h^2 + k^2} (\kappa_h + \kappa_c) + \frac{\tilde{L}\kappa_1}{\kappa_1^2 + k^2} \right\}$$

$$\equiv \mathscr{P}_x^E + \mathscr{P}_x^H, \tag{116}$$

$$|\mathscr{P}_\parallel|^{cv} = i\gamma \left\{ -\frac{HP_h}{\kappa_c^2 + P_h^2} \left[\kappa_c \cos \frac{P_h b}{2} + P_h \sin \frac{P_h b}{2} \right] \right.$$

$$+ \frac{Lq_c}{\kappa_c^2 + P_1^2} \left[\kappa_c \cos \frac{P_1 b}{2} + P_1 \sin \frac{P_1 b}{2} \right]$$

$$\left. + \frac{\tilde{H}\kappa_h}{\kappa_h^2 + k^2} (\kappa_h + \kappa_c) - \frac{\tilde{L}q_c}{\kappa_1^2 + k^2} (\kappa_1 + \kappa_c) \right\}$$

$$\equiv \mathscr{P}_\parallel^E + \mathscr{P}_\parallel^H. \tag{117}$$

Here $\mathscr{P}_{x,\parallel}^E$ and $\mathscr{P}_{x,\parallel}^H$ correspond to two channels E and H for electron–hole recombination (see Fig. 8). The terms proportional to H and L in (116) and (117) correspond to $\mathscr{P}_{x,\parallel}^E$, and the terms proportional to $\tilde{H}$ and $\tilde{L}$ correspond to $\mathscr{P}_{x,\parallel}^H$. We recall that the E channel corresponds to the recombination of a tunneling electron with a hole in the quantum well, and that the H channel corresponds to the recombination of a tunneling hole with an electron in the quantum well.

As we know, the contributions to the radiative recombination rate in the case of non-degenerate electron statistics are made by $q \leq q_T$, where $q_T = 1/\lambda_T$. It follows from (116) and (117) that we have $|\mathscr{P}_x| \ll |\mathscr{P}_\parallel|$ when $q \leq q_T$:

$$\frac{|\mathscr{P}_x|}{|\mathscr{P}_\parallel|} \approx \frac{m_c}{m_h} \sqrt{\frac{V_c}{|V_v|}} \ll 1. \tag{118}$$

Therefore, the main contribution to the radiative recombination rate is made by $\mathbf{P}_\parallel$. As follows from (116) and (117), when $q_c = q_v = 0$, we have $P_x = 0$, and $\mathbf{P}_\parallel$ is nonzero:

$$|\mathscr{P}_\parallel| \cong HB \sin\left(\frac{P_h b}{2}\right) \frac{P_h^2}{\kappa_c^2 + P_h^2}. \tag{119}$$

Here we took into account that the holes are found in a size-quantization level of heavy holes and $\kappa_c > P_h$. It is noteworthy that $\mathbf{P}_\parallel$ is weakly dependent on q_c and q_v. In addition, $\mathscr{P}_\parallel(q_c = q_v = 0) \approx \mathscr{P}_\parallel(q_c = q_v = q_T)$. Thus, the optical transition matrix element M_R is determined by $\mathscr{P}_\parallel$ and scarcely depends on q_c and q_v. Therefore, in a further calculation of the radiative recombination rate we can set $q_c = q_v = 0$ in the expression for $\mathscr{P}_\parallel$. We recall that the Auger transition matrix element is strongly dependent on the longitudinal components of the electron momentum (q_2) and the hole momentum (q_3) and that when $q_2 = q_3 = 0$ holds, the Auger transition matrix element equals zero (see Eqs. (67), (71), and (74)).

It should also be stressed that the contributions of the two recombination channels (E and H) to $p_\parallel$ differ significantly:

$$\frac{\mathscr{P}_\parallel^H}{\mathscr{P}_\parallel^E} \approx \frac{m_c}{m_h} \ll 1. \tag{120}$$

A conclusion of fundamental importance follows from (120): the optical transition matrix element M_R and, therefore, the radiative recombination rate in type-II heterostructures are determined by the E channel for electron–hole recombination. As was shown above, in the case of Auger recombination, the contributions of the E and H channels to the Auger transition matrix element are of the same order, but of different sign, and when they are summed, they compensate each other. Hence, a destructive interference of the E and H channels takes place for Auger recombination with a resultant decrease in its rate.

Some comments on the physical meaning of the interference mechanism of two channels (E and H) in the Auger recombination rate and on the predominance of the E channel over the H recombination channel in the case of radiative recombination are appropriate.

As we have already noted, the Auger recombination matrix element for small momentum transfers is proportional to the difference between the longitudinal components of the electron and hole momenta

$\mathbf{q}_3 - \mathbf{q}_2 = \mathbf{q}_1 - \mathbf{q}_4$ (see Eqs. (67), (72), and (74)). When $\mathbf{q}_3$ is nonzero, interconversion of light and heavy holes occurs upon their interaction with the heterointerface. As a result, the hole incident on the heterointerface tunnels through the heterobarrier as a light hole. This results in a significant increase in the part of the Auger recombination matrix element corresponding to the H channel, which is of the same order as the matrix element corresponding to the E channel, but of opposite sign.

Such destructive interference of the E and H channels does not occur in the case of radiative recombination, since, as we have already noted, $\mathscr{P}_{cv}$ is weakly dependent on the longitudinal momenta of electrons (q_c) and holes (q_v). Also, since the interconversion of light and heavy holes does not take place at $q_v = 0$, this process is not important for optical transitions. In this case the H channel of electron–hole recombination is ineffective, since it corresponds to the tunneling of a hole with a heavy mass. As a result, in the case of optical transitions, the ratio between $\mathscr{P}_\parallel^H$ and $\mathscr{P}_\parallel^E$ is given by (120).

We substitute (116) and (117) into (47). Performing summation over the c and v states of the particles and integrating with respect to ω, for the radiative recombination rate we obtain

$$R \cong 2\pi \frac{\varepsilon_\infty}{\sqrt{\kappa_0}} \frac{e^2}{\hbar c} \frac{E_g}{\hbar} \frac{E_g}{2m_c c^2} \frac{\hbar^2 np}{2m_c V_c} \frac{T}{V_c} \frac{m_h}{m_c}. \qquad (121)$$

We note that the radiative recombination process of electrons and holes is just as effective in type-II heterostructures as in those of type I. For comparison, we present the ratio of the radiative recombination rates $R \equiv R_\mathrm{II}$ and R_I in type-II and type-I heterostructures, respectively, for the same heterostructure parameters:

$$R_\mathrm{II}/R_\mathrm{I} \cong \left(\frac{T m_h}{V_c m_c}\right)^2 \leqslant 1. \qquad (122)$$

As follows from (111), the Auger recombination rate is a power function of temperature. In addition, G is highly dependent on the parameters a, b, V_c, and V_v.

As we have already noted, G_1 and G_2 in (111) originate from the part $M^{(1)}$ of the matrix element M, which corresponds to small momentum transfers in the Coulomb interaction of electrons, while G_3 originates from the part $M^{(2)}$ of M, which corresponds to large momentum transfers. We note that, as follows from (111)–(115), when $V_c > |V_v|$ holds, we have $G_1 > G_2 \gg G_3$. It should be specially emphasized that under the condition $3V_c \approx |V_v|$, i.e., $(V_c - |V_v|)/V_c \ll 1$, we have $G_3 \ll (G_2, G_3)$.

Under another condition $3V_c \approx |V_v|$ we have $G_2 \ll (G_1, G_3)$. This means that G_1 and G_2 have a minimum at the values of $|V_v|/V_c$ just indicated (Fig. 9). Therefore, the Auger recombination rate has a minimum at certain values of $|V_v|/V_c$: $G^{min} \approx G_3$. Thus, the minimum value of the Auger recombination rate G^{min} is determined by the Coulomb interaction matrix element $M^{(2)}$ for large momentum transfers. This means that there is an effective suppression of the Auger recombination processes in type-II heterostructures, since, as we have already noted, when $V_c > |V_v|$, $G_3 \ll (G_1, G_2)$.

Such an effective suppression of the Auger recombination rate in type-II heterostructures is attributed to the behavior of the overlap integral $I_{14}(q)$ of the electron in the initial state "1" and the final state "4" for a small momentum transfer q. This overlap integral has contributions from three regions: two regions of subbarrier motion of the electron ($x < -a$ and $x > 0$) and the region of the quantum well ($-a < x < 0$). When the contributions from these three regions are summed, it is found in the case of both type-I and type-II heterostructures that the contributions from the subbarrier regions and the region of the quantum well cancel each other. The resultant overlap integral is diminished by a factor of $(3V_c + V_v)/E_g$. We note that for type-I heterostructures $V_v > 0$, and for type-II heterostructures $V_v = -|V_v| < 0$. Consequently, we find that in type-II heterostructures there is a strong decrease in the overlap integral $I_{14}(q)$ of the electron in the initial and final states when the momentum transfer q is small under the condition $3V_c \cong |V_v|$. This is a result of the cancellation of the contributions to $I_{14}(q)$ of the regions indicated above.

Fig. 9 presents the dependence of the logarithm of the Auger recombination rate on the ratio $|V_v|/V_c$ calculated from (111) with account of the exact expressions for g_1, g_2, and g_3 (see (112)–(115)). We have obtained a result of fundamental importance: the Auger recombination rate has a minimum when $3V_c \cong |V_v|$. In addition, the ratio of the Auger recombination rate $G^{min} \equiv G(3V_c \sim |V_v|)$ at the minimum and the rate is small, when $V_c \geqslant |V_v|$: $G^{min}/G(V_c \geqslant |V_v|) \ll 1$. The obtained result is of fundamental importance, since it demonstrates the possibility of suppressing the Auger recombination processes in type-II heterostructures. Such an effective suppression of these processes is attributed to the nature of the Coulomb interaction between electrons in an Auger transition. Under some conditions ($|V_v| < V_c$) the Coulomb interaction between electrons is mainly an effective long-range interaction (small momentum transfers), producing a large value for the Auger recombination rate. Under another condition ($3V_c \cong |V_v|$) the

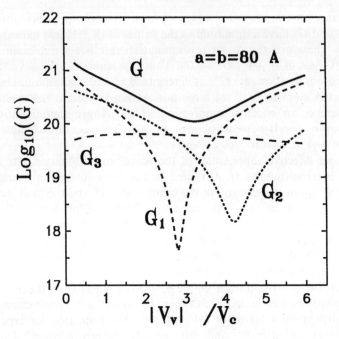

Fig. 9 Logarithm of the Auger recombination rate $\log G$ versus $|V_v|/V_c$ at $T = 290$ K. The dashed curves correspond to the contributions G_1, G_2, and G_3 to the total rate G (solid curve). The following parameters, which are characteristic of an InGaAsSb/GaSb structure (see Ref. [47]), were used in the calculation: $E_g = 0.6$ eV, $V_c = 0.25$ eV, $m_c = 0.04\,m_0$, $m_h = 0.4\,m_0$, $a = b = 80$ Å, $n = p = 1.2 \times 10^{12}$ cm^{-2}.

Coulomb interaction between the electrons has a predominantly short-range character (large momentum transfers), causing a sharp decrease in the Auger recombination rate.

Suppression of the Auger recombination process in type-II heterostructures is of fundamental importance for creating optoelectronic devices with improved characteristics. The Auger recombination processes are known to cause a decrease in the internal quantum efficiency of semiconductor quantum-well lasers and an increase in the threshold current density at high temperatures [21]. The mechanism of suppressing Auger recombination processes in type-II heterostructures studied in this section makes it possible, in particular, to solve the problem of long-wavelength lasers ($\lambda > 4$ μm), viz., to raise their operating temperature to room and higher temperatures.

It is interesting to compare the minimum Auger recombination rate in type-II heterostructures with the rate in type-I heterostructures G_1.

Using the expression for G_1 from Eq. (88), we have for identical values of the heterostructure parameters

$$\frac{G^{\min}}{G_I} \sim \left(\frac{T}{V_c}\frac{m_h}{m_c}\right)^3 \frac{V_c}{E_g} \ll 1. \tag{123}$$

Therefore, in type-II heterostructures, unlike type-I heterostructures, a significant decrease in the rate of the Auger recombination processes is possible. This conclusion, in particular, can be significant in developing optoelectronic devices.

Let us proceed to an analysis of the radiative recombination rate in type-II heterostructures. As has already been noted, the radiative recombination rates in type-II ($R \equiv R_{II}$) and type-I (R_I) heterostructures are of the same order: $R_{II} \leqslant R_I$ (see Eq. (122)).

In the case of quantum wells based on narrow-gap semiconductors [21], the Auger recombination rate in type-I heterostructures (G_1) is known to significantly exceed the radiative rate (R_I) at high temperatures. Therefore, in these semiconductor structures the internal quantum efficiency η is much less than unity: $\eta = R_I/(G_1 + R_I) \ll 1$.

In type-II heterostructures the situation can be exactly opposite. The ratio between the radiative and Auger recombination rates in these heterostructures has a sharp maximum as a function of $|V_v|/V_c$ (Fig. 10). It is noteworthy that for type-II heterostructures even with a small effective bandgap width E_g, the maximum value of R_{II}/G_{II} can be greater than unity. Consequently, by selecting the optimal parameters for the heterostructure (at which R/G is maximum), we can achieve the maximum internal quantum efficiency η for long-wavelength lasers based on type-II heterostructures.

The mechanism for suppressing the Auger recombination processes, theoretically predicted in Refs. [30,43,35], was detected experimentally when a laser of a new type employing a single InAs/GaSb type-II heterojunction was fabricated [47,48]. Suppression of the Auger recombination processes in type-II heterostructures with a superlattice based on In(As, Sb) was also discovered in another experimental study [49]. We summarize the main results of this section.

(1) It has been shown that the Auger-recombination process is always threshold-free in type-II heterostructures and that its rate is a power function of temperature.
(2) The fundamental difference between the mechanisms of the Auger recombination of nonequilibrium carriers in type-I and type-II heterostructures has been demonstrated:

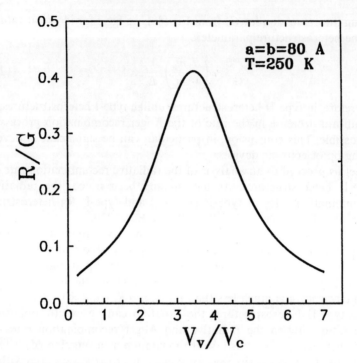

Fig. 10 Ratio of the radiative recombination rate R to the Auger recombination rate G versus ratio $|V_v|/V_c$ at $T = 290$ K. The heterostructure parameters are the same as in Fig. 9.

(i) in type-II heterostructures there are two channels for Auger recombination, viz., E and H, which interfere with each other to a significant degree;

(ii) G_{II} and G_I depend on the heterostructure parameters V_c and V_v in a essentially different manner: G_{II} has a minimum as a function of $|V_v|/V_c$, while G_I is an increasing function of V_v/V_c;

(iii) the Auger recombination rate is diminished in type-II heterostructures in comparison with type-I heterostructures;

(iv) in quantum wells for which $E_{1h} < T < E_{1c}$ holds, the temperature dependence of C_{AII} differs from that of C_{AI}: $C_{AII} \propto T^2$ and $C_{AI} \propto 1/T$. It follows from this analysis that in the temperature range where the recombination of nonequilibrium carriers is determined by the nonradiative channel, the lifetime $\tau = n/G$ is greater in type-II structures than in type-I structures.

(3) Another fundamental conclusion of the present work is the following: the radiative recombination rate in type-II heterostructures is

of the same order as in type-I heterostructures with the same parameters.
(4) It has been shown that in type-II heterostructures the internal quantum efficiency η can be significantly higher than the quantum efficiency in type-I structures with the same parameters.
(5) We predict a considerable decrease in the value of threshold current density of lasers based on type-II heterostructures in comparison with lasers based on type-I heterostructures due to the mechanism for suppressing the Auger recombination rate when the ratio of the barrier heights V_v and V_c ($3|V_v| \cong V_c$) has a certain value. We shall not analyze in detail the influence of the recombination processes studied on the operation of optoelectronic devices here. The purpose of the present study is to investigate the mechanisms of radiative and Auger recombination in type-II heterostructures and to elucidate the main features distinguishing them from the recombination mechanisms in type-I heterostructures.

7.4.3 Non-Threshold Mechanism of Auger Recombination

In Section 7.4.2 we developed the theory of a new non-threshold Auger recombination process in heterostructures with quantum wells. This mechanism was first proposed in Ref. [19].

It is noteworthy that the Auger recombination mechanism in quantum wells fundamentally differs from that in bulk semiconductors.

In homogeneous semiconductors the Coulomb interaction matrix element of Auger recombination contains only the part that corresponds to large momentum transfers due to the requirements of the energy and momentum conservation laws. The consequences of this are the following:

(1) a very indirect course of the transition of an electron from the initial to the highly excited final state in **k** space;
(2) a threshold (exponential) dependence of the Auger recombination rate on temperature.

As previously shown in Ref. [19], in type-I semiconductor heterostructures the main contribution to the Coulomb interaction matrix element is made by the long-range part, which corresponds to small momentum transfers. This feature of the behavior of the Auger transition matrix element in the presence of heterointerface is fundamentally related to the interaction of carriers with the heterointerface.

One result of such an interaction is the absence of a conservation law for the transverse (x) component of the quasi-momentum of the particles. The consequences of this are the following:

(1) a threshold-free character of the Auger recombination process in quantum structures, where the rate is a power function of the temperature, rather than an exponential function, as in a bulk semiconductor;
(2) a resonant character of the Auger transition, i.e., the transition of an electron from the initial state to a highly excited state takes place as a result of the only energy transfer from the recombining electron–hole pair without momentum transfer; the electron acquires the momentum needed for the transition to the highly excited final state from the interaction with the heterointerface.

Note that in quantum well heterostructures the main contribution to the temperature dependence of the Auger recombination rate is determined by that of the carrier concentration, while the Auger recombination coefficient C_A (see Eq. (90)) scarcely depends on temperature. Fig. 11 plots the Auger recombination rate G as a function of temperature for a type-I quantum well at different values of the quantum well width: $a = 80$ Å (Fig. 11(a)) and $a = 150$ Å (Fig. 11(b)). A numerical calculation shows that G depends on temperature mainly in terms of the concentration of non-equilibrium electrons and holes n and p and the Auger coefficient C_A.

The weak temperature dependence of the Auger recombination rate originates from the Auger recombination mechanism being non-threshold in quantum wells. We recall that in bulk semiconductors the Auger recombination coefficient is an exponential function of temperature: $C_A^{3D} \propto \exp(-E_{th}/T)$. An exact analysis has shown that the specified coefficient is a power function of temperature over a wide range of quantum well parameters. Here the temperature dependence of C_A is determined by that of the bandgap width and the electron–hole overlap integral I_{23}. Fig. 12 shows the Auger recombination coefficient versus temperature at different values of the quantum well width: $a = 80$ Å (Fig. 12(a)) and $a = 150$ Å (Fig. 12(b)). Over a wide temperature interval from 50 to 300 K the Auger coefficient has a non-monotonous run, having a maximum at a certain temperature. The position of the maximum of C_A scarcely depends on the quantum well width. However, with increasing quantum well width a, the temperature dependence of the Auger coefficient becomes sharper. At $a = 150$ Å the maximum value of C_A for $T = 200$ K is approximately

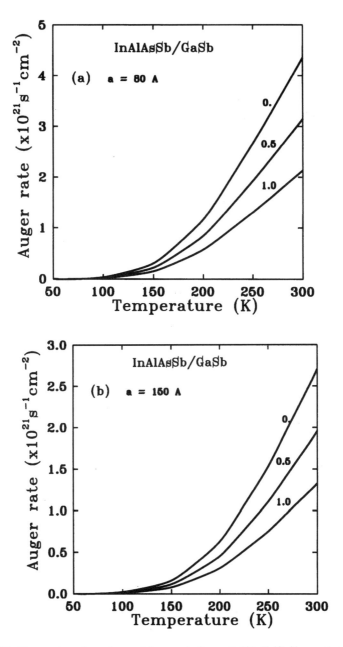

Fig. 11 Temperature dependence of Auger rate for an InAlAsSb/GaSb quantum well with various strains, (a) $a = 80$ Å, (b) $a = 150$ Å.

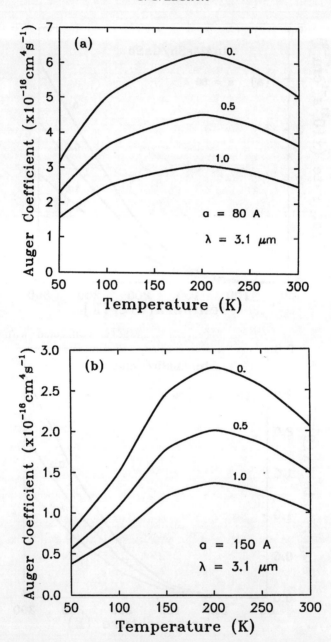

Fig. 12 Temperature dependence of Auger coefficient for an InAlAsSb/GaSb quantum well with various strains, $\lambda = 3.1\,\mu\text{m}$: (a) $a = 80\,\text{Å}$, (b) $a = 150\,\text{Å}$.

three times as much as the maximum value of C_A for $T = 50\,\text{K}$. Such a behavior of C_A with increasing quantum well width is attributed to the change in the overlap integral I_{23}. As the quantum well broadens, the distance between size-quantization levels decreases: $E_{h,n+1} - E_{h,n} \ll T$. The spectrum of holes becomes quasi-discrete (nearly continuous). Therefore, the electron–hole overlap integral I_{23} explicitly depends on temperature: $|I_{23}| \propto T^{-1/2}$. Another interesting result is that as the strain increases, the maximum of C_A is smoothened out, moreover, C_A falls with increasing ε (Fig. 12).

7.4.4 Carrier Leakage

It is known that there is a considerable contribution to the threshold current of mid-infrared lasers from leakage at high temperatures [38,39]. No consistent theory of the mechanism of the leakage in long-wavelength lasers has been developed yet.

The purpose of this section is to propose two mechanisms of carrier leakage out of a quantum well that become important at high temperatures.

The first mechanism of carrier leakage is connected with the process of intraband absorption studied in Section 7.4.1. As we did in Section 7.4.1, we consider here the intraband absorption of the frequency $\omega \approx E_g/\hbar$ in a quantum well with $(V_c, V_v) < E_g$. As a result of absorbing a photon $\hbar\omega \cong E_g$, an electron (or a hole) transits from the initial state characterized by the momentum $\mathbf{k}_i$ into the final state with the momentum $\mathbf{k}_f$. As shown in Section 7.4.1, the longitudinal component of the quasi-momentum $\mathbf{q}_i = \mathbf{q}_f$ is conserved in the process. However, the law of conservation of the transverse components k_{ix} and k_{fx} does not hold, i.e., an electron that has absorbed a photon $\hbar\omega$ changes its momentum by interacting with the heterointerface. In the final state the transverse component of the momentum is equal to $k_{fx} \approx (2m_c\omega/\hbar)^{1/2} \gg k_{ix}$ holds. Thus, the electrons that have absorbed a photon $\hbar\omega$, "escape" from the heterointerface, since $k_{fx} \gg |\mathbf{q}_f|$. These electrons are not trapped by the quantum well, since they move away from the heterointerface to a distance appreciably greater than the diffusion length during the energy relaxation time $\tau_\varepsilon \sim 10^{-12}\,\text{s}$. As shown in Section 7.4.1, the probability of this intraband absorption process essentially depends on the frequency ω when $\hbar\omega \approx E_g$ holds. In this case we have for the absorption coefficient $\alpha_i^\perp \propto \omega^{-7/2}$. Consequently, this mechanism of carrier leakage influences the threshold characterisics of mid-infrared lasers. Moreover, this absorption process influences the temperature dependence of the gain $g(\omega)$ and, as a result, the temperaure dependence of the threshold concentration of the non-equilibrium charge carriers n_{th} and p_{th} (see Section 7.5.2).

Thus, intraband absorption at heterointerfaces leads to two effects adversely affecting the operation of mid-infrared lasers: (i) carriers ionize out of the quantum well, which leads to a downward shift of the Fermi quasi-level and to an increase in the threshold concentration and (ii) threshold carrier concentration is a non-linear function of temperature, which, in turn, leads to a steep temperature dependence of the threshold current density (see Section 7.5.2).

The second mechanism of leakage is connected with Auger recombination processes. As in the case of intraband absorption, each Auger recombination event is accompanied by ionization of electrons (holes) out of the quantum well. Electrons "escape" from the heterointerface, since the quasi-momentum component transverse to the heterointerface k_{fx} is essentially greater than the longitudinal component q_f: $k_{fx} \approx (2m_c E_g)^{1/2}/\hbar \gg q_f \approx (2m_c T)^{1/2}/\hbar$. It is worth emphasizing that the Auger recombination process becomes important in mid-infrared lasers, since its rate displays a sharp dependence on the bandgap width: $G \propto E_g^{-11/2}$. As a result, the process of Auger recombination gives the main contribution to the value of the threshold current density of mid-infrared lasers (see Section 7.5.3). In addition, Auger ionization of the carriers out of the quantum well causes a downward shift of the Fermi quasi-level, which leads to a more sharp temperature dependence of the threshold concentration of charge carriers.

As we see, there occurs an essential delocalization of the charge carriers out of the quantum well, owing to the processes of Auger recombination and intraband absorption. This causes a sharper temperature dependence of the threshold concentration of charge carriers, which, in its turn, leads to a considerable increase of the leakage and Auger recombination currents, suppresses lazing at high temperatures. Just such a behavior is seen in experiment.

It is therefore evident that creation of mid-infrared lasers operating at high temperature (up to room temperature) requires the laser structure to be optimized in order to suppress the processes of Auger recombination and intraband absorption. In Section 7.7 a new fundamental approach to creation of mid-infrared lasers operating at high temperatures will be proposed. We will see that within the approach there exists a mechanism of suppression of the leakage current, Auger recombination, and intraband absorption.

7.5 THRESHOLD CHARACTERISTICS OF MID-INFRARED STRAINED QUANTUM WELL LASERS

7.5.1 Threshold Condition

For calculating the threshold current it is necessary to calculate the threshold concentration of carriers. The threshold concentration is determined by the condition that concentrations of electrons n_{th} and holes p_{th} are equal and by the threshold condition:

$$\tilde{g}_{th} = \Gamma g_{th} = \alpha_i + \frac{1}{2}\ln\left(\frac{1}{R}\right). \quad (124)$$

Here the modal gain $\tilde{g}_{th}$ is expressed in terms of the local gain coefficient g (see Eq. (43)) and the gain confinement factor Γ; α_i being the value of internal losses.

Below, we will take account of internal losses connected with absorption on the heterointerface in analyzing threshold characteristics of mid-infrared lasers (see Section 7.4.1). These internal losses considerably influence the dependence of the gain on temperature and carrier concentration, moreover, they influence the temperature dependence of the threshold concentration. Fig. 13 plots the modal gain $\tilde{g}_{th}$ versus

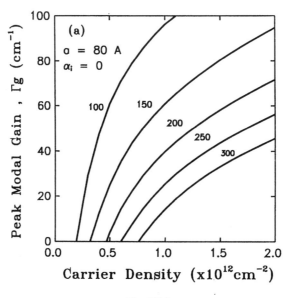

Fig. 13(a)

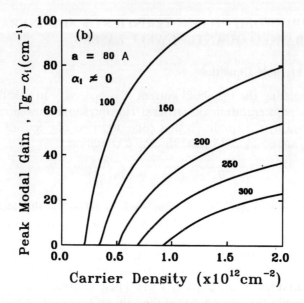

Fig. 13 Peak modal gain versus carrier density for unstrained InAlAsSb/GaSb quantum wells with $a = 80$ Å at various temperatures: (a) $\alpha_i = 0$; (b) $\alpha_i \neq 0$.

the carrier concentration without taking α_i into account (Fig. 13(a)) and with allowance for α_i (Fig. 13(b)). The internal losses connected with absorption at heterointerfaces therefore affect the dependence of $\tilde{g}_{th}$ on the non-equilibrium carrier concentration. Fig. 14 plots the mode gain $\tilde{g}_{th}$ against temperature for two cases: $\alpha_i = 0$ and $\alpha_i \neq 0$. The presence of the intraband absorption sharpens the temperature dependence of $\tilde{g}_{th}$ in comparison with that in the absence of intraband absorption ($\alpha_i = 0$).

7.5.2 Threshold Carrier Density

The intraband losses α_i are considerable for mid-infrared lasers, so that for long enough lasers ($L > 400\,\mu$m) we have $\alpha_i > \ln(1/R)/L \equiv \alpha^*$. In this case the threshold carrier concentration n_{th} determined by relation (124) depends on the quantum well parameters, such as heterobarrier heights for electrons and holes V_c and V_v, the quantum well width a, and the value of strain. The dependence of n_{th} on the quantum well parameters is attributed to the dependence of α_i on V_c, V_v, a, d_c and d_x (see Section 7.4.1, Eq. (57)).

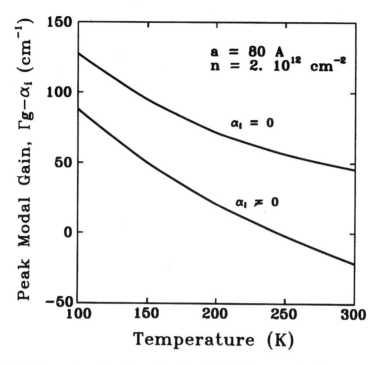

Fig. 14 Temperature dependence of peak modal gain for an unstrained quantum well with different internal losses: $a = 80\,\text{Å}$, $n = 2 \times 10^{12}\,\text{cm}^{-2}$.

Fig. 15 plots the threshold concentration n_{th} as a function of temperature for two cases: (a) $\alpha_i = 0$ and (b) $\alpha_i \neq 0$. The dependence of n_{th} on T is almost linear in the case (a); in (b) n_{th} is a non-linear function of temperature, increasing with temperature. Fig. 16 shows the threshold concentration versus the heterobarrier height for electrons V_c. As V_c increases, n_{th} slowly increases, too. This behavior of the threshold concentration is caused by the dependence of α_i on V_c: at small values of V_c we have $\alpha_i \propto V_c^2$; then as V_c rises further, α_i becomes a linear function of V_c: $\alpha_i \propto V_c$ (see Eq. (57)). Therefore, $\alpha_i^\perp$ increases with increasing V_c. As a result, the threshold concentration becomes higher. Fig. 17 presents the dependence of the threshold concentration on the strain value ε. As ε increases, the threshold concentration falls down, since the internal losses α_i become lower (see Eq. (60)). The decrease of $\alpha_i^\perp$ at higher ε is attributed to the effective decrease of the quantum well depth for electrons V_c: $V_c \rightarrow V_c - d_c$.

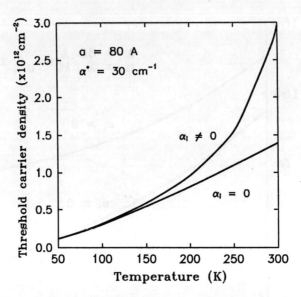

Fig. 15 Temperature dependence of threshold carrier density for an unstrained InAlAsSb/GaSb quantum well with various internal losses. External losses are $\alpha^* = 30\,\text{cm}^{-2}$.

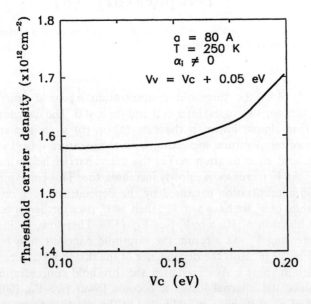

Fig. 16 Threshold carrier density versus conduction band offset for an InAlAsSb/GaSb quantum well: $a = 80\,\text{Å}$, $T = 250$ K, $V_v = V_c + 0.05\,\text{eV}$, $\alpha_i \neq 0$.

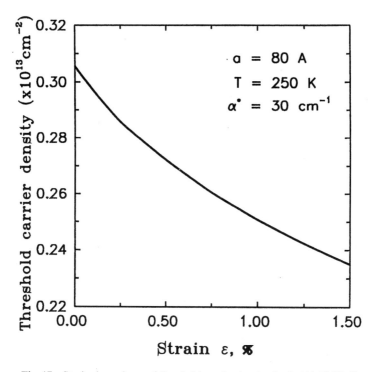

Fig. 17 Strain dependence of threshold carrier density for InAlAsSb/GaSb.

7.5.3 Auger Current

It is known that Auger recombination processes contribute mostly to the threshold current density J_{th} in mid-infrared diode lasers. It is therefore necessary to know the dependence of the Auger recombination current J_A on the quantum well parameters in order to be able to optimize laser structure by decreasing the threshold current and increasing the internal quantum output.

The purpose of this section is to study the dependence of J_A on the quantum well parameters, such as the heterobarrier heights V_c and V_v, the quantum well width a, the strain value ε. In addition, we will study J_A as a function of temperature and non-equilibrium carrier concentration. We will perform a qualitative analysis of the dependence of J_A on the quantum well parameters using model parameters.

The Auger recombination current is determined by the rate of the processes, CHCC and CHHS, so that we have:

$$J_A = e(G^{CHCC} + G^{CHHS}). \tag{125}$$

Let us now consider the contribution from the CHCC Auger recombination process to the value of J_A. Fig. 17 shows the temperature dependence of J_A for two values of the quantum well width ($a = 80$, 150 Å). Qualitative and quantitative analyses show that the Auger recombination rate G^{CHCC} turns out to be a power function of temperature. We remind that up to now J_A has been assumed to be an exponential function of temperature for lasers based on quantum wells (Refs. [38,39]). The performed analysis has shown that this exponential dependence does not take place under any conditions. The narrower the quantum well, the steeper becomes the temperature dependence of the Auger current J_A, since the decrease of a increases the Auger recombination coefficient (see Eqs. (88) and (90)). We emphasize that the main contribution to the temperature dependence of J_A is given by the temperature dependence of the concentration of non-equilibrium electrons n and holes p. As noted above, in the absence of internal losses ($\alpha_i = 0$) the dependence of the threshold carrier concentration is approximately linear: $n_{th} \propto T$. In this particular case the Auger current J_A depends on temperature as $J_A \propto T^3$ (see Fig. 18). For real situations the interband absorption is nonzero: $\alpha_i \neq 0$ close to the generation

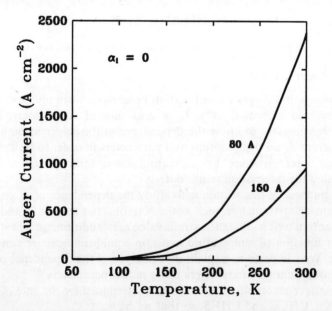

Fig. 18 Temperature dependence of Auger current for an InAlAsSb/GaSb quantum well with various well widths, internal losses are $\alpha_i = 0$.

threshold of mid-infrared lasers, which appreciably affects the temperature dependence of threshold concentration: n_{th} is not a linear function of T (see Fig. 15). Here the temperature dependence of the Auger current J_A is steeper (see Fig. 19).

Thus, the analysis we performed shows a fundamental contribution from the intraband absorption (see Section 7.4.1) to the temperature dependence of the Auger current J_A for mid-infrared lasers.

According to Eq. (88), for $n = p$, the Auger recombination current depends on the carrier concentration approximately as $J \propto n^3$. Fig. 20 shows J_A as a function of n for two different values of the quantum well width. The dependence of J_A on n is approximately cubic for both cases.

Now we proceed to analysis of the dependence of the Auger current J_A on the strain value ε. It is rather difficult to study the dependence of J_A on ε in a general case. The performed analysis has shown that for most cases J_A is a decreasing function of ε, though there may be

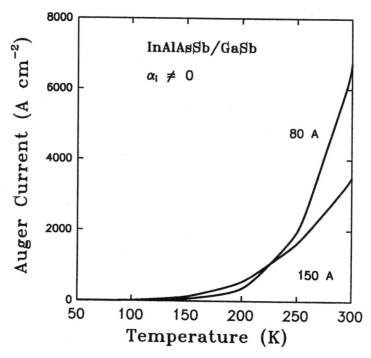

Fig. 19 Temperature dependence of Auger current for an InAlAsSb/GaSb quantum well with various well widths. Internal losses are $\alpha_i \neq 0$.

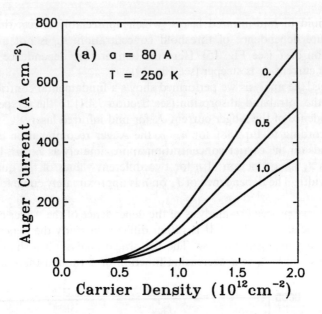

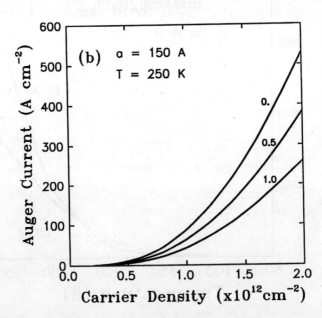

Fig. 20 Auger current versus carrier density for an InAlAsSb/GaSb quantum well with various strains at $T = 250$ K: (a) $a = 80$ Å, (b) $a = 150$ Å.

the opposite situation: over a certain range of ε the current J_A may rise with ε.

Nevertheless, a quantitative analysis of the dependence of J_A on ε can only be carried out numerically for real (not model) parameters of the heterostructure. Fig. 21 shows the dependence of J_A on ε for the case of occupied main size-quantization levels [44]. Then the Auger current J_A falls down with increasing ε. We will consider the dependence of J_A on ε again in Section 7.6 while analyzing the dependence of the threshold current density on the strain value ε and temperature T for certain laser structures.

In conclusion of this section it is appropriate to make the following remark. As is well-known, semiconductor structures with strained layers were first proposed chiefly in order to suppress the process of nonradiative Auger recombination in mid-infrared lasers [50]. However, up-to-date mid-infrared strained diode lasers ($\lambda \geqslant 4\,\mu$m) based on III–V can only operate at low temperatures ($T < 200$ K). Thus, using

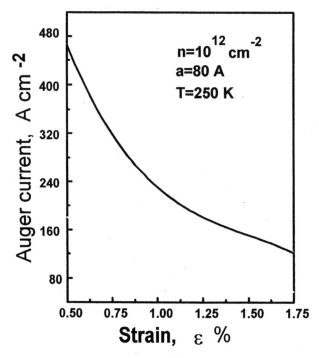

Fig. 21 Strain dependence of Auger current for an InAlAsSb/GaSb quantum well: $n = 10^{12}\,\text{cm}^{-2}$, $a = 80$ Å.

quantum wells with strained layers in creating long-wavelength lasers ($\lambda \geqslant 4\,\mu\text{m}$) does not increase their operating temperature up to room temperature. It is, in fact, natural, since, as we saw, at real values of the strain ε it is impossible to make the Auger recombination current even one order of magnitude smaller. However, creation of mid-infrared lasers based on III–V that could operate at room temperature requires suppression of the Auger recombination current by, at least, two orders of magnitude, so that J_A be smaller than the radiative recombination current J_R. It is of fundamental importance that the suppression of the Auger recombination current fails to solve the problem of increasing the maximum operation temperature T_1 of mid-infrared lasers. As was shown above, the temperature dependence of the threshold current density is essentially affected by the internal losses α_i. Thus, suppression of the two processes is required to increase the maximum operating temperature T_1 of mid-infrared lasers: (i) Auger recombination processes and (ii) intraband absorption processes (α_i). It is impossible to suppress both of the specified processes within the frame of strained laser structures at the same time. Thus, this approach [50] does not make it possible to create mid-infrared diode lasers based on III–V operating at room temperature.

However, in Section 7.7 we will propose a new fundamental approach to creation of mid-infrared lasers operating at high temperature which enables, among all, simultaneous suppression of the Auger recombination and intraband absorption processes.

In the next chapter we will consider one of possible mechanisms of Auger recombination current suppression in type-II quantum well heterostructures by an example of traditional heterostructures.

7.5.4 Mechanism of Suppression of Auger Recombination Current

The theoretical analysis performed in Section 7.4 showed that the mechanism of Auger recombination is considerably different for type-I and type-II heterostructures, with the dependences of the Auger recombination rate on the quantum well parameters (V_x, V_v, a and b) differing both qualitatively and quantitatively. Still more fundamental difference between type-I and type-II heterostructures is the possibility of an essential suppression of Auger recombination for type-II heterostructures [43]. The suppression of Auger recombination processes, as will be shown below, leads to a sharp decrease of the Auger recombination current in comparison with type-I heterostructures. It is important that the conditions under which the Auger recombination current J_A is at a minimum do not affect the radiative current J_R. The mechanism of the suppression of the Auger recombination current for type-II heterostructures is considered in detail in Section 7.4.2.

In this section we will study qualitatively and quantitatively the Auger recombination current J_A and the radiative recombination current J_R for type-II heterostructures with quantum wells. We will analyze the dependence of J_A and J_R on the parameters of the quantum well for electrons and holes respectively.

Fig. 22 shows the Auger current J_A and radiative current J_R against the ratio of the heterobarrier heights for electrons and holes: $|V_v|/V_c$ [51,52]. The Auger current J_A is a non-monotonous function of the ratio $|V_v|/V_c$ having a minimum at $|V_v|/V_c \approx 3$. The minimum value of the Auger current may be of the order of the radiative current J_R. The radiative current is a slowly rising function of $|V_v|/V_c$. It is evident that the non-monotonous run of the dependence of J_A on the ratio $|V_v|/V_c$ affects the dependence of the internal quantum efficiency

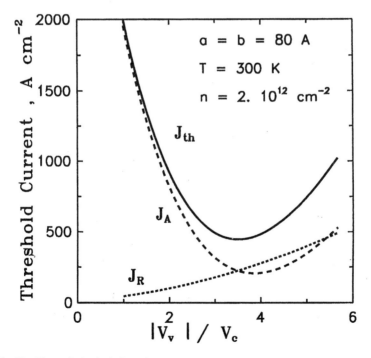

Fig. 22 Theoretical calculation of the threshold current as a function of valence band to conduction band offset ratio for a quantum well laser based on type-II heterostructure.

$\eta = J_R/(J_R + J_A)$ on the quantum well parameters. Fig. 23 plots the internal quantum efficiency versus the ratio $|V_v|/V_c$. At the minimum of J_A ($J_A = J_A^{\min}$) η has a maximum. This is to be taken into account in designing mid-infrared lasers based on type-II heterostructures.

It is interesting to study the dependence of the threshold current $J_{th} = J_{A,th} + J_{R,th}$ on temperature T for different values of the ratio $|V_v|/V_c$. Fig. 24 plots this dependence for $|V_v|/V_c = 1$ and $|V_v|/V_c = 3$. Under the conditions needed for the Auger current to be suppressed, the threshold current J_{th} scarcely depends on temperature. At high temperature and at $|V_v|/V_c \approx 3$ the threshold current J_{th} is approximately one order of magnitude less than it is at $|V_v|/V_c \cong 1$. Fig. 25 plots the internal quantum efficiency η against temperature for the two cases just specified. At $|V_v|/V_c \approx 3$ the internal quantum efficiency η may be of the order of or more than 10% at high temperatures.

Another important result that we have obtained is connected with the intraband absorption factor $\alpha_i^\perp$ (see Section 7.4.1). For type-II

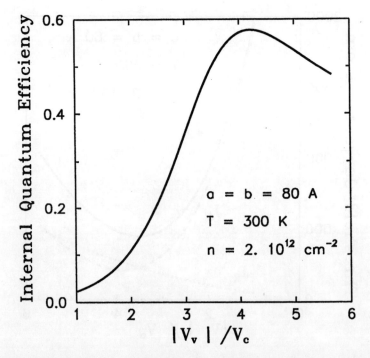

Fig. 23 Theoretical calculation of internal quantum efficiency as a function of valence band to conduction band offset ratio.

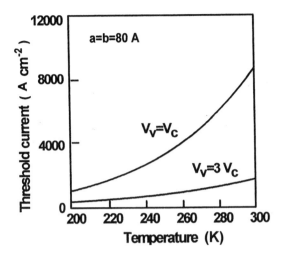

Fig. 24 Temperature dependences of threshold current for a quantum well laser based on type-II heterostructure, calculated for different valence band to conduction band offset ratios.

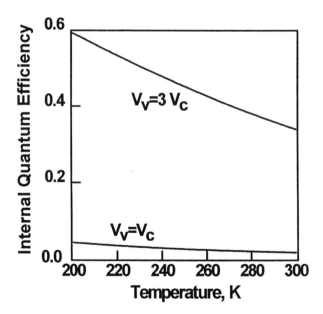

Fig. 25 Temperature dependence of internal quantum efficiency for a quantum well laser based on type-II heterostructure, calculated for different valence band to conduction band offset ratios.

heterostructures the intraband absorption factor is determined by expression (57) provided that V_v is changed for $-|V_v|$. $\alpha_i^\perp$ has a minimum of $3V_c \cong |V_v|$, this minimal value being equal to

$$\alpha_{i\min} \approx \frac{8\sqrt{2}\,2\pi}{\sqrt{\varepsilon_\infty}} \frac{e^2}{\hbar c} \lambda_{E_g}^3 n_c \frac{\Gamma}{a} \frac{\varepsilon_\perp^2}{E_g^2} \frac{V_v^2}{E_g^2} \frac{\varepsilon_\perp}{V_c} \left(\frac{E_g + \Delta_{so}}{E_g + (3/2)\Delta_{so}}\right)^{3/2}. \qquad (126)$$

According to Eq. (58), we have for type-II heterostructures

$$\frac{\alpha_{i\min}^\perp}{\alpha_i^\parallel} \sim \frac{\varepsilon_\perp}{T} \frac{\varepsilon_\perp}{E_g} \ll 1. \qquad (127)$$

Thus, in type-II heterostructures the intraband absorption is considerably suppressed under the condition $3V_c \approx |V_v|$.

Consequently, under the condition necessary for suppression of the Auger recombination processes, the intraband absorption processes are also suppressed. This result shows the advantages of type-II heterostructures over those of type-I to be taken into consideration when creating mid-infrared lasers based on semiconductor compound III–V.

7.5.5 Leakage Current. Two Mechanisms

In Section 7.4.4 we considered two mechanisms of leakage of non-equilibrium carriers out of the quantum well: (i) leakage caused by intraband absorption and (ii) leakage caused by Auger recombination. We assume that each mechanism leads to the leakage of electrons and holes out of the quantum well.

It is evident that the processes of carrier ionization in the quantum well contribute to the value of the leakage current J_L. As the temperature and the non-equilibrium carrier concentration rise, the Auger recombination rate and the intraband absorption factor and, consequently, the leakage current J_L rise, too. We suppose that the carrier ionization in the quantum well caused by Auger recombination is more efficient than that caused by intraband absorption. We also assume that an excited Auger electron moves away into the contact, not being trapped back by the quantum well. Under these conditions the leakage current is determined by the Auger recombination current:

$$J_L \cong \beta J_A, \qquad (128)$$

where the factor β describes the fraction of the Auger electrons reaching the contact. The maximum value of β evidently equals unity: $\beta_{\max} = 1$. It is necessary to take account of the contribution of the leakage current into the threshold current density for mid-infrared lasers:

$$J_{th} = J_{R,th} + J_{A,th} + J_{L,th}. \tag{129}$$

Fig. 26 presents the dependence of the threshold current density J_{th} on temperature for type-I heterostructure with allowance for the leakage current for two different quantum well widths: $a = 80$ Å (Fig. 26(a)) and $a = 150$ Å (Fig. 26(b)) For estimation of J_L we assumed that $\beta = 1/2$. An analysis has shown that the leakage current qualitatively and quantitatively affects the temperature dependence of the threshold current density for mid-infrared lasers. This is to be taken into account in optimizing laser structures.

As shown above (Section 7.5.4), under certain conditions the suppression of Auger recombination becomes possible in type-II heterostructures. It is natural to expect that under the same conditions the leakage current is suppressed in mid-infrared lasers based on type-II heterostructures. Fig. 27 plots J_{th} against temperature for mid-infrared

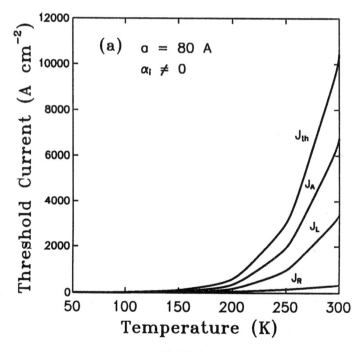

Fig. 26(a)

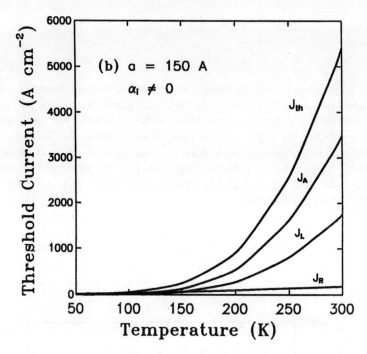

Fig. 26 Temperature dependence of threshold current, calculated for a quantum well laser with $\alpha_i \neq 0$: (a) $a = 80$ Å, (b) $a = 150$ Å.

type-II lasers with account of the contribution from J_L; the laser structure was chosen to provide $3V_c \approx |V_v|$. It is this kind of laser structure that enables considerable suppression of the leakage current.

It is noteworthy that there are conditions under which the carriers that are excited out of the quantum well by the Auger recombination or intraband absorption processes recombine in the subbarrier region without reaching the contact. The carriers recombining in the subbarrier regions also contribute to the leakage current. Complete suppression of the leakage current requires laser structures with deep quantum wells, i.e., $V_c > E_g$. We will study this problem in detail in Section 7.7.

7.5.6 Analysis of the Auger Current Dependence on the Quantum Well Parameters and on Strain

In Section 7.5.3 we performed a qualitative analysis of the dependence of the Auger current J_A on the quantum well parameters (V_c, V_v, a, ε) using model parameters. A quantitative analysis of this dependence is

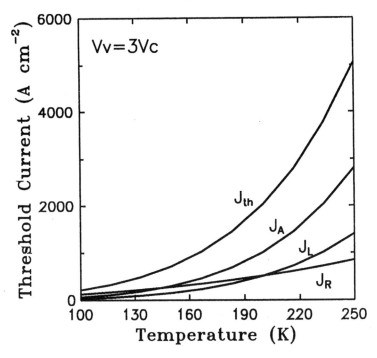

Fig. 27 Temperature dependence of threshold current, calculated for a quantum well laser based on type-II heterostructure, $|V_v|/V_c = 3$.

only possible if the parameters are predefined. This is especially important in optimizing laser structure.

In this section we will generalize the results of Section 7.5.3 and study the Auger recombination current near the generation threshold along with its dependence on the quantum well parameters. First of all, it is interesting to demonstrate the dependence of $J_{A,th}$ on the wavelength λ of the laser emission (Fig. 28). As was previously noted, the Auger recombination rate sharply decreases when the bandgap width E_g rises. As a result, $J_{A,th} \propto \lambda^{11/2}$, where λ is the laser wavelength. Over the range of the wavelengths $3 < \lambda < 4\,\mu\text{m}$ the Auger current $J_{A,th}$ increases by one order of magnitude. This is therefore to be taken into account when creating lasers of this wavelength range. While estimating the dependence of $J_{A,th}$ on λ we used the parameters of the quantum well based on $\text{In}_x\text{Al}_{1-x}\text{As}_{1-y}/\text{GaSb}$ [53].

The dependence of the Auger current on the quantum well width a was rather unexpected. In the above we analyzed the dependence of

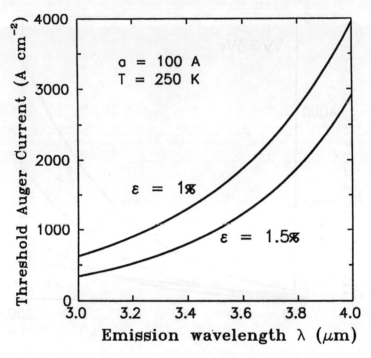

Fig. 28 Threshold Auger current versus the emission wavelength for a InAlAsSb quantum well laser with various strains, $n = p = 2.5 \times 10^{12}\,\text{cm}^{-2}$, $a = 100\,\text{Å}$, $T = 250\,\text{K}$.

the Auger recombination rate on a (see Section 7.4.2). It is appropriate to emphasize here that the Auger current $J_{A,\text{th}}$ depends on a through a corresponding dependence of the Auger coefficient C_A and the threshold concentration n_{th}. The Auger coefficient as a function of a has a non-monotonous run, being mainly determined by the quantum-well-width dependence of the overlap integral I_{14} between the electrons in the initial "1" and final "4" states. At a certain value of $a = a_{\max}$ the coefficient $C_A(a_{\max})$ has a maximum, since $C_A \propto |I_{14}|^2/a^n$, and the overlap integral is $I_{14} \propto \sin k_x a$ the coefficient C_A decreases approximately as $1/a^7$. On the other hand, the threshold concentration also falls down as a increases. As a result, the Auger current is a decreasing function of the quantum well width a for $a > a_{\max}$ at the generation threshold (Fig. 29) [44]. It is worth noting that while a rises from 30 to 100 Å the Auger current J_A decreases approximately by one order of magnitude. We consider the obtained result very important, so as it is necessary to take it into account in creating mid-infrared lasers.

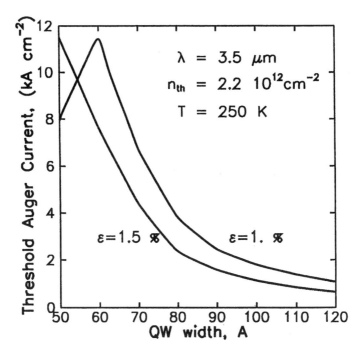

Fig. 29 Threshold Auger current versus quantum well width, calculated for an InAlAsSb quantum well laser with various strains: $\lambda = 3.5\,\mu\text{m}$, $T = 250\,\text{K}$, $n = 2.2 \times 10^{12}\,\text{cm}^{-2}$.

Thus, analysis of the dependence of J_A on a shows that structures with narrow quantum wells ($a < 50\,\text{Å}$) cannot be used for creating long-wavelength lasers. It is evident that for every laser wavelength there exists an optimum width of the quantum well a_{opt} providing a minimal value of $J_{A,\text{th}}$. The problem of optimizing the laser structure must be solved by a self-fitting procedure, with reference to the dependence on the quantum well width and other laser characteristics: the gain g and the intraband absorption factor α_i.

As was previously noted, the non-threshold mechanism of the Auger recombination is caused by interaction of electrons (holes) with the heterointerface. This interaction results in the Auger recombination coefficient C_A being a rising function of the heterobarrier heights for electrons V_c and holes V_v. Moreover, as V_c grows, the threshold concentration also rises, since the increase in V_c leads to higher internal losses α_i (see Fig. 16). Thus, the Auger current rises as V_c rises and the current $J_{A,\text{th}}$ scarcely depends on V_c when $V_c \leqslant E_g$ holds (Fig. 30).

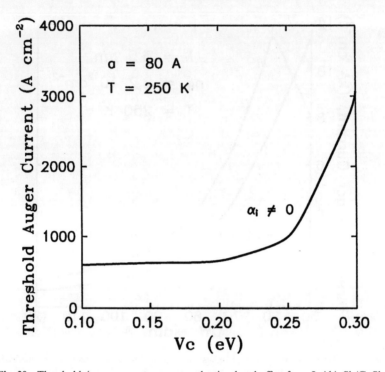

Fig. 30 Threshold Auger current versus conduction band offset for a InAlAsSb/GaSb quantum well laser: $a = 80$ Å, $T = 250$ K, $\alpha_i \neq 0$.

We have already mentioned that the analysis of the dependence of J_A on the strain ε is complicated. Nevertheless, it has shown that the dependence of the current J_A on the strain value ε is sharper for narrow quantum wells than for wide ones. Fig. 31 plots $J_{A,\text{th}}$ against ε for two different values of the quantum well width: $a = 80$ and 100 Å [44]. Here we took account of the threshold concentration also being strain-dependent (see Fig. 17).

Thus, we have shown that the qualitative and quantitative analysis of the dependence of the Auger current on the quantum well parameters cannot be carried out without studying the Auger recombination mechanism microscopically. Such an analysis of the new Auger recombination mechanism from the fundamental principles was performed for the first time in Ref. [19]. In addition, we have demonstrated a fundamental difference between the Auger recombination mechanisms in heterostructures and bulk semiconductors. We believe that the obtained results reveal unambiguously the fundamental reasons why

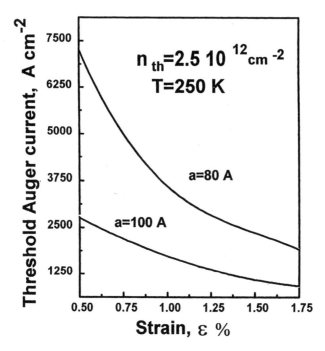

Fig. 31 Strain dependence of threshold Auger current for an InAlAsSb/GaSb quantum well laser with various well widths, $n_{th} = 2.5 \times 10^{12}\,\text{cm}^{-2}$.

mid-infrared lasers ($\lambda \geqslant 4\,\mu\text{m}$) fail to operate at room temperature. Moreover, while analyzing the Auger recombination mechanisms in type-I and II heterostructures, we demonstrated the advantages of type-II heterostructures over those of type I as materials for creating mid-infrared lasers based on semiconductor compounds III–V.

7.6 COMPARISON WITH EXPERIMENTS

It deserves some interest to apply our theoretical results to an analysis of the threshold characteristics of mid-infrared lasers. A most important of these are the threshold current density J_{th}. There have been no works concerned with the microscopic theory of temperature dependence of the threshold current density up to now. As a rule, in works analyzing the temperature dependence of the threshold current density in semiconductor lasers (mainly lasers with quantum wells) an empirical law is used (see, e.g., Refs. [38,39]):

$$J_{th} = J_0 \exp(T/T_0), \tag{130}$$

where T_0 is the characteristic temperature: $T_0^{-1} = d\ln(J_{th})/dT$. The characteristic temperature T_0 is different for different laser structures. Since T_0 strongly depends on temperature, its definition as a logarithmic derivative of J_{th} with respect to temperature is valid for a very narrow range of temperatures, in which $T_0 =$ const.

A theoretical analysis has shown from the fundamental principles that the temperature dependence of J_{th} (cf. Eq. (130)) cannot be exponential under any conditions in semiconductor lasers. Thus, expression (130) lacks strict physical substantiation, the use of T_0 as a laser characteristic being contradictory. On the other hand, depending on the value of T_0 the conclusion about radiative versus non-radiative recombination relationship in laser structures is made. The best characteristics in this sense is the internal quantum efficiency $\eta = 1/(1 + G/R)$, where R is the radiative recombination rate and G is the total nonradiative recombination rate.

Below, we will study the threshold current density in mid-infrared lasers and analyze various experimental results. In Ref. [9] an experimental analysis of the temperature dependence of J_{th} for InAsSb/InAlAsSb multi quantum-well (MQW) diode lasers emitting at 3.9 μm was performed. The experimental value of J_{th} at 80 K corresponds to 78 A/cm^2, at 165 K $J_{th} = 3.5$ kA/cm^2. Fig. 32 presents the temperature dependence of J_{th}, calculated within the frame of our model for the laser structure just specified (solid curve). In the same figure experimental values of J_{th} are plotted. We note that the sharp increase of the threshold current (by a factor of 45), in the temperature range from 80 to 165 K is attributed to the following. As the temperature rises, the intraband absorption coefficient α_i increases, the threshold concentration n_{th} increasing, too. The dependence of n_{th} on temperature is essentially non-linear (see Fig. 15): $n_{th} \propto T^\alpha$. The power α is a rising function of the temperature taking values in the interval $1 \leqslant \alpha < 2$. For $T > 80$ K the main contribution to J_{th} is given by the Auger recombination current $J_{A,th}$. In this case $n_{th} \propto T^{5/3}$; consequently, $J_{A,th} \propto T^5$. Then $(d\ln J_{th}/dT)^{-1} = T_0 \approx 33$ K. Let us now consider 3.9 μm InAsSb/AlAsSb double-heterostructure diode lasers (DH lasers) [54] and compare them with the quantum well laser stated above. When $T < 80$ K holds, the threshold current density of DH lasers is less than that of MQW lasers based on InAsSb/InAlAs. However at high temperature we come to the opposite relationship:

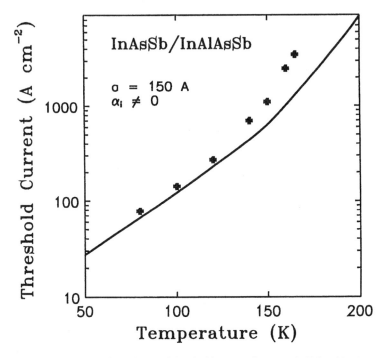

Fig. 32 Temperature dependence of threshold current for an InAsSb/InAlAsSb quantum well laser with $a = 80$ Å and $\alpha_i \neq 0$.

$J_{th}(T = 165\,K) \approx 3.5\,kA/cm^2$, and $J_{th}^{DH}(T = 165\,K) \cong 7\,kA/cm^2$. Thus, for $T > 80\,K$ J_{th} of the DH lasers is approximately twice as much as that of the MQW lasers, for, when $T > 80\,K$ holds, the main contribution to J_{th} is given by the Auger recombination current J_A. As shown in Section 7.5.3, the Auger recombination current is a power function of temperature for MQW lasers and an exponential function for DH lasers. Thus, the exponential rise of J_{th} with temperature (DH lasers) is sharper than the power dependence of J_{th} on T (QW lasers), the difference between the magnitudes of J_{th}^{QW} and J_{th}^{DH} being caused by the different mechanisms of Auger recombination in quantum wells and DH structures.

We now consider long-wavelength lasers based on AlGaAsSb/InGaAsSb [55,56]. In Ref. [55] double-heterostructure diode lasers emitting at 3 μm were studied. For this type of lasers at 40 K we have $J_{th} \approx 9\,A/cm^2$. The limiting operating temperature T_l of such a laser equals 255 K; at this temperature $J_{th} \approx 13.4\,kA/cm^2$, $T_0 = 28\,K$. It is evident that at $T > 100$ K the main contribution to J_{th} for this laser is given by the Auger current. As noted above, the Auger current is an exponential function of the temperature for DH lasers. A theoretical estimation of T_0 at 200 K gives approximately 27 K.

The laser studied in Ref. [56], is based on quantum wells (4QWs). Its emission wavelength is 2.78 μm. The limiting operating temperature is $T_1 = 288$ K. The threshold current density at this temperature equals 10 kA/cm². Our theoretical estimation gives $J_{th} \approx 9.8$ kA/cm². The parameters needed for estimation were obtained from Ref. [56]. It is noteworthy that for long-wavelength lasers the internal losses α_i increase with the number of quantum wells rise. Thus, in analyzing the number-of-well dependence of J_{th}, the dependence of α_i on the number of wells is to be taken into account.

Let us consider InAsSb/InAlAs strained quantum-well lasers emitting at 4.5 μm [57]. The InAsSb/InAlAs lasers have operated pulsed mode up to 85 K, with the threshold current density of 350 A/cm² at 50 K, and $J_{th} = 1.95$ kA/cm² at 85 K. At $T = 85$ K for the MQW laser InAsSb/InAsAsSb with the wavelength $\lambda = 3.9$ μm we have the threshold current close to 100 A/cm². Such a sharp increase of the threshold current density by more than one order of magnitude with the wavelength increasing from 3.9 to 4.5 μm can be explained by the following two reasons. One is the sharp dependence of the Auger recombination coefficient C_A on the wavelength: $C_A \propto \lambda^{11/2}$. The other important reason is the increase of the threshold concentration with increasing rise of the wavelength λ (Fig. 33). Let us give a more detailed explanation of this statement. As was shown in Section 7.4.1, the internal losses corresponding to the intraband absorption depend on the emission wavelength. When the quantum energy is $\hbar\omega = E_g$, the internal losses are $\alpha_i^\perp \propto \lambda^{7/2}$. The fact that α_i increases with λ, results in the following: for Eq. (124) to stay valid it is necessary to increase $\tilde{g}_{th}$, i.e. it is necessary to increase the threshold concentration n_{th}. Therefore, as λ rises, $\alpha_i(\lambda)$ increases in a considerably sharper manner than the gain $\tilde{g}(\lambda)$. Consequently, the increase of the wavelength leads to a rise in the threshold concentration n_{th}. On the other hand, the threshold current for this type of lasers is known to be mainly determined by the Auger current: $J_{th} \approx J_{A,th}$. The Auger current $J_{A,th} \propto n_{th}^3$, thus, a small increase of n_{th} induced by the increase of α_i leads to a steep rise in J_{th} which is seen in experiment [56–58]. Fig. 34 plots an experimental temperature dependence of J_{th} for InAsSb/InAlAs strained quantum-well lasers emitting at 4.5 μm [51] (points). The same figure shows a

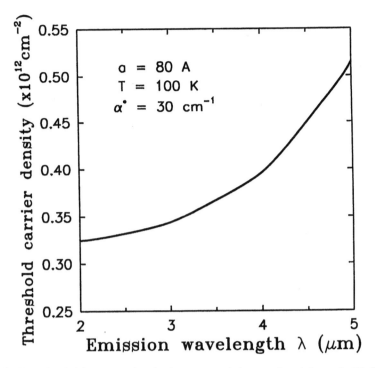

Fig. 33 Threshold concentration density versus emission wavelength for an InAlAsSb/GaSb quantum well laser.

theoretical dependence of J_{th} on T calculated in the framework of our model (solid curve).

To conclude the section we note that the limiting operating temperature T_l of mid-infrared lasers is affected by two processes: (i) Auger recombination (Section 7.4.2) and (ii) intraband absorption (Section 7.4.1). As shown in the present chapter, the mechanisms of Auger recombination and intraband absorption in quantum wells differ fundamentally from the corresponding mechanisms in bulk semiconductors. Both Auger recombination coefficient and internal loss factor depend considerably on the quantum well parameters, temperature, and emission wavelength λ. Creation of mid-infrared lasers operating at high temperature requires simultaneous suppression of the Auger recombination and intraband absorption processes under the same conditions.

One of possible mechanisms of such suppression will be discussed in the following section.

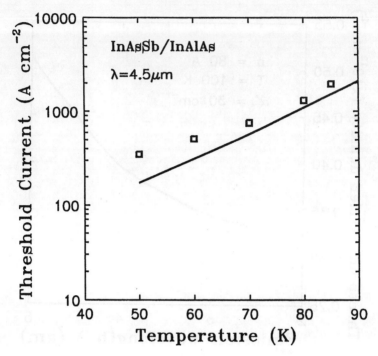

Fig. 34 Comparison of theoretically calculated (solid line) and experimental (dashed line) temperature dependences of threshold current for an InAsSb/InAlAsSb quantum well laser: $a = 150$ Å, $\alpha_i \neq 0$, $\lambda = 4.5\,\mu m$.

7.7 NEW APPROACH TO CREATION OF MID-INFRARED LASERS OPERATING AT ROOM TEMPERATURE

7.7.1 Auger Engineering in Quantum Well Laser Structures

In the preceding sections the main processes governing the operation of semiconductor lasers based on quantum wells were studied. The processes of radiative and Auger recombination in lasers on quantum wells were described in detail. The dependence of the Auger and radiative recombination rates and the internal loss factor α_i on the quantum well parameters (the heterobarrier heights for electrons and holes V_c and V_v, the quantum well width a), temperature, emission wavelength and value of elastic strain ε was analyzed. We have paid much attention to studying the processes of the intraband absorption in quantum well lasers.

This detailed analysis has demonstrated that the main factors that prevent mid-infrared lasers based on semiconductors III–V from operating at room temperature are: (i) considerable prevalence of the Auger recombination rate over that of the radiative recombination: the radiative recombination rate drops with increasing emission wavelength as $R \propto \lambda^{-2}$, while the Auger rate, on the contrary, increases sharply: $G \propto \lambda^{11/2}$; (ii) considerable prevalence of the internal loss factor α_i over the loss on mirrors α^*; α_i rises sharply as the wavelength grows: $\alpha_i \propto \lambda^{7/2}$; (Fig. 35) (iii) great leakage current corresponding to carrier delocalization in the quantum well and connected with two processes: Auger recombination and intraband absorption.

The three reasons just specified lead to the two fundamental consequences: (i) a steep fall of the internal quantum output with increasing wavelength and the temperature; (ii) suppression of laser generation owing to the impossibility of the threshold condition (120) to be satisfied, since at a certain temperature T_1 and a preset

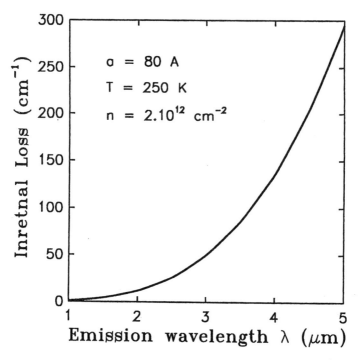

Fig. 35 Internal loss versus emission wavelength for a InAlAsSb/GaSb quantum well laser: $a = 80 \text{ Å}$, $T = 250 \text{ K}$, $n = 2 \times 10^{12} \text{ cm}^{-2}$.

wavelength λ we have $\tilde{g}(T_1, \lambda) - \alpha_i(T_1, \lambda) < \alpha^*$. The greater λ, the less is T_1, i.e. the limiting operating temperature decreases as the wavelength λ rises (see Fig. 36). Violation of the generation condition (124) induces the transition of the system from absolute instability to convection instability, the latter meaning that the active medium can only amplify radiation incident from the outside.

It is evident that creation of mid-infrared lasers operating at room temperature requires: (i) essential suppression of the Auger recombination processes at high temperatures to make it less or of the same order of magnitude as the radiative recombination rate; (ii) essential suppression of the internal losses for α_i to be smaller than α^*: $\alpha_i \ll \alpha^*$; (iii) leakage suppression.

The purpose of the present section is to propose a new fundamental approach to creation of mid-infrared diode lasers operating at room temperature [59]. We will see below that within the new approach it is possible to: (i) considerably suppress the Auger recombination rate

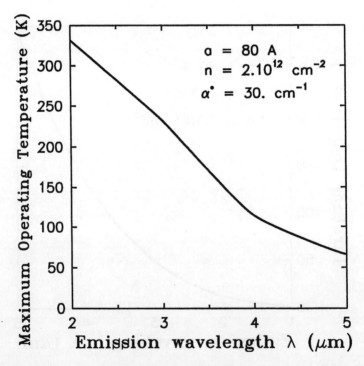

Fig. 36 Maximum operating temperature versus emission wavelength, calculated for an InAlAsSb/GaSb quantum well laser.

(approximately by three orders of magnitude) [52]; (ii) suppress the intraband absorption process (by two orders of magnitude); (iii) completely suppress the leakage current [59].

The essence of the new approach is the control over the Auger recombination rate (Auger engineering) and over the processes of intraband absorption. We consider deep quantum wells for both electrons and holes. The depth of the quantum wells is enough for the distance between the nearest size-quantization levels for electrons (namely, the ground and the first excited levels) to be more than the effective bandgap width $\tilde{E}_g$ (Fig. 37).

$$E_{1c} - E_{0c} > \tilde{E}_g \equiv E_{0c} + E_g + E_{0h}, \qquad (131)$$

where E_{1c} is the energy of the first excited size-quantization level for electrons.

Let us now consider the CHCC Auger recombination process in such a deep quantum well under condition (131): two electrons localized on the basic level (in the states "1" and "2") interact, one of them recombining with a hole (state "3") and the other passing to the highly excited state "4" (see Fig. 37). Henceafter the highly excited electron gets between the size-quantization levels, i.e. still remains on the basic size-quantization level. This requires the longitudinal component of the momentum to change from $q_1 \sim q_T = \sqrt{2m_c T}/\hbar$ to $q_4 \approx Q = \sqrt{2m_c \tilde{E}_g}/\hbar$. The mechanism of this Auger recombination process is analogous to that in bulk semiconductors. We will later prove that this process is threshold, the Auger recombination rate depending on temperature exponentially.

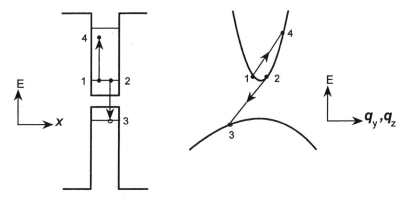

Fig. 37 CHCC threshold Auger process in a deep quantum well, shown schematically: (a) coordinate space, (b) momentum space.

Now we proceed to a more detailed description of the Auger recombination process in a deep quantum well [60,61]. As a result of the Coulomb interaction of electrons in the states "1" and "2", one of them recombines with a hole passing into the state "3" while the other electron passes into the sate "4", remaining on the basic size-quantization level. Such an Auger transition is possible under the condition that the electron–electron interaction is strong, since the transmitted momentum q is of the order of Q. The longitudinal component of the quasi-momentum is conserved in the process: $\mathbf{q}_1 + \mathbf{q}_2 = \mathbf{q}_3 + \mathbf{q}_4$. Since it is only the ground size-quantization levels of electrons and holes that participate in the Auger recombination process, the rate of the process, according to Eq. (64), equals

$$G = \frac{2\pi}{\hbar} \frac{1}{S} \int \int \int \frac{d^2 q_1}{(2\pi)^2} \frac{d^2 q_2}{(2\pi)^2} \frac{d^2 q_3}{(2\pi)^2} |\tilde{M}_\text{I}|^2 f_c(E_1) f_c(E_2) f_\text{h}(E_3)$$

$$\times \delta\{E_1(\mathbf{q}_1) + E_2(\mathbf{q}_2) - E_3(\mathbf{q}_3) - E_4(\mathbf{q}_1 + \mathbf{q}_2 - \mathbf{q}_3)\}. \tag{132}$$

In deriving (132) we took acount of Eq. (74) and integrated over $\mathbf{q}_4$, using δ-function, which expresses the law of the longitudinal component of the momentum conservation.

The integrand in Eq. (132) contains two steep dependences: the energy delta-function and a Bolzmann distribution function:

$$f_c(E_1) f_c(E_2) f_\text{h}(E_3) \propto \exp\{-[\varepsilon_1(\mathbf{q}_1) + \varepsilon_2(\mathbf{q}_2) + \varepsilon_\text{h}(\mathbf{q}_3)]/T\},$$

$$\varepsilon_{1,2}(\mathbf{q}_{1,2}) = E_{1,2}(\mathbf{q}_{1,2}) - \frac{E_\text{g}}{2}; \qquad \varepsilon_\text{h}(\mathbf{q}_3) = -\frac{E_\text{g}}{2} - E_\text{h}(\mathbf{q}_3), \tag{133}$$

where $\varepsilon(\mathbf{q}_{1,2})$ and on $\varepsilon_\text{h}(\mathbf{q}_3)$ are the kinetic energies of electrons and holes, the relation between which is determined by the laws of energy and momentum conservation. The region where the kinetic energy of the initial electrons and holes (134) is at a minimum under the condition that the conservation laws are obeyed, makes the main contribution to the integral:

$$\varepsilon_1(\mathbf{q}_1) + \varepsilon_2(\mathbf{q}_2) + \varepsilon_\text{h}(\mathbf{q}_\text{h}) = \min,$$

$$\varepsilon_1(\mathbf{q}_1) + \varepsilon_2(\mathbf{q}_2) + \varepsilon_\text{h}(\mathbf{q}_\text{h}) + E_\text{g} = \varepsilon_4(\mathbf{q}_1 + \mathbf{q}_2 - \mathbf{q}_3). \tag{134}$$

The conditional extremum (135) with respect to $\mathbf{q}_{1,2}$ and $\mathbf{q}_3$ can be calculated by the method of indeterminate Lagrange multipliers for arbitrary dispersion relations. Next, expanding the integrand near the extremum ($\mathbf{q}_{10}$, $\mathbf{q}_{20}$, $\mathbf{q}_{30}$) in the exponential (134) up to second-order infinitesimals, we reduce the problem to calculation of Gaussian integrals. The minimum value of the energy in Eq. (135) gives the kinetic threshold of the reaction (i.e., of the Auger process). From the condition of extremality it follows that all the particle speeds are equal and directed along the same line:

$$\mathbf{v}_1 = \mathbf{v}_2 = -\mathbf{v}_3. \qquad (135)$$

Close to the extremum the quasi-momenta are related as follows:

$$|\mathbf{q}_{10}| \approx |\mathbf{q}_{20}| \approx \frac{m_c}{m_h^*}|\mathbf{q}_{30}|,$$

$$|\mathbf{q}_{30}| \approx 2\left(\frac{m_c E_g}{\hbar^2}\right)^{1/2} \equiv \sqrt{2Q}, \qquad (136)$$

where m_h^* is the two-dimensional effective hole mass. Therefore, the close to the extreme point the quasi-momentum of holes is significantly greater than that of electrons: $|\mathbf{q}_{30}| \gg (|\mathbf{q}_{10}|, |\mathbf{q}_{20}|)$.

As a result, for the Auger recombination rate we obtain

$$G = \frac{3\sqrt{\pi}}{\hbar^2}|\tilde{M}_1(\mathbf{q}_{10}, \mathbf{q}_{20}, \mathbf{q}_{30})|^2 \left(\frac{m_c}{E_g}\right)^{1/2}\left(\frac{\hbar^2}{2m_h T}\right)^{1/2} n^2 p \exp\left(-\frac{E_{th}}{T}\right), \qquad (137)$$

where $E_{th} = 2m_c \tilde{E}_g / m_h^*$. The two-dimensional hole mass m_h^* may be considerably (2–3 times) smaller than m_h owing to the interconversion of the light and heavy hole states in the quantum well (see Fig. 2) The threshold energy for such structures can therefore appear to be somewhat greater than the threshold energy for bulk semiconductors. The value of $\tilde{M}_1$ is determined by expression (74). According to (74), after integrating over q_x we deduce for $\tilde{M}_1$

$$\tilde{M}_1 = \frac{4\pi e^2}{\kappa_0}\frac{1}{2q_{30}}I_{13}(iq_{30})I_{24}(-iq_{30}). \qquad (138)$$

The overlap integral I_{ij} entering in Eq. (138) is defined by (68). For I_{24} we have

$$I_{24}(-iq_{30}) \approx |A|^2/q_{30}, \qquad (139)$$

where $|A|^2 = 2/(\alpha + 1/\kappa_c)$. In a similar way, for I_{13} we get

$$I_{13}(iq_{30}) = \frac{\gamma}{E_{c1} - E_{h3}} U_{c1}^*\left(\frac{a}{2}\right)\left(\frac{2}{a}\right)^{1/2}\frac{P_h}{q_{30}}\frac{A_\parallel^W}{A_x^W}. \qquad (140)$$

Thus, the final expression for $|\tilde{M}_1|^2$ assumes the form

$$|\tilde{M}_1|^2 = \left(\frac{4\pi e^2}{\kappa}\right)^2 \left(\frac{2}{a + 1/\kappa_c}\right)^3 \frac{2}{a} \frac{\gamma^2}{\tilde{E}_g q_{30}^4} \cos^2\frac{ka}{2} \frac{P_h^2}{q_{30}^2} \left(\frac{A_\parallel^W}{A_x^W}\right)^2. \quad (141)$$

Here

$$\cos^2\frac{ka}{2} \approx \frac{k_{1x}^2}{k_{1x}^2 + \kappa_c^2} \approx \frac{E_{0c}}{V_c}.$$

Substituting Eq. (141) into (137) we derive for the Auger recombination rate

$$G \approx 12(2\pi)^{5/2} \frac{E_B}{\hbar} \frac{E_{0c}}{V_c} \left(\frac{m_h}{m_c} \frac{T}{E_g}\right)^{1/2} \frac{\lambda_g^8}{a^4} n^2 p \exp\left(-2\frac{m_c}{m_h^*} \frac{\tilde{E}_g}{T}\right) \left(1 - \frac{m_h^*}{m_c} \frac{\delta}{\tilde{E}_g}\right). \quad (142)$$

Thus, we obtained an exponential suppression of the Auger recombination process in deep quantum wells ($V_c > E_g$) compared to shallow ones ($V_c \ll E_g$)).

It is interesting to compare what we derived for the Auger rate in a deep quantum well G^B (see Eq. (142)) with the expression for a shallow quantum well, G^S. Discarding the strain, we obtain

$$\frac{G^B}{G^S} \approx \frac{E_g}{V_c^B} \frac{E_g}{E_{0c}} \frac{\lambda_g}{a} \frac{T}{V_c^S} \frac{m_h}{m_c} \left(\frac{m_h}{m_c} \frac{T}{E_g}\right)^{1/2} \exp\left(-\frac{2m_c}{m_h^*} \frac{\tilde{E}_g}{T}\right), \quad (143)$$

where V_c^B and V_c^S are the depths of deep and shallow quantum wells respectively. The ratio (143) is always less than 1. Assume $E_g = 0.4\,\text{eV}$, $V_c^B \cong 0.5\,\text{eV}$, $V_c^S \cong 0.2\,\text{eV}$, $T = 300\,\text{K}$, $a = 50\,\text{Å}$, $m_h^* = m_h/2$, we derive $G^B/G^S \sim 2 \times 10^{-3}$, $m_h^* = m_h/2$. This ratio appears to be even smaller, since we have used an approximate expression for G^S.

We have thus shown that in deep quantum wells there can be considerable suppression of the Auger recombination rate and consequent rise in the internal quantum efficiency η.

7.7.2 Mechanism of Suppression of Intraband Absorption Processes

However, the most valuable result is that under the conditions needed for the Auger recombination rate to be less than the radiative recombination rate such a deep quantum well enables considerable suppression of the intraband absorption process.

The mechanism of the intraband absorption is appreciably different. The absorption of a photon is only possible when phonons or impurities participate, i.e. the process only occurs in second-order perturbation theory.

In the process of scattering of an electron on a LO-phonon and with account of [62] we deduce for the intraband absorption factor α_i^f

$$\alpha_i^f = \frac{8\pi}{\sqrt{\varepsilon_\infty}} \alpha_{\rm ef} \frac{e^2}{\hbar c} \lambda_\omega^2 n \frac{\Gamma}{a} \left(\frac{\omega_0}{\omega}\right)^{3/2}, \qquad (144)$$

where $\alpha_{\rm ef}$ is the electron–phonon interaction constant (Froehlich constant), moreover, $\alpha_{\rm ef} \ll 1$; $\hbar\omega_0$ is the energy of LO-phonon. For III–V compounds we have $0.01 < \alpha < 0.1$. We should compare the expression obtained for α_i^f (144) with that for the intraband absorption factor in a shallow quantum well $\alpha_i^\perp$ (see (57)). Neglecting the strain and assuming $V_c^S = V_v^S$, $\hbar\omega = E_g$, we have

$$\frac{\alpha_i^f}{\alpha_i^\perp} \approx \frac{\alpha_{\rm ef}}{4\sqrt{2\pi}} \frac{\hbar\omega_0}{V_c^S} \frac{(\hbar\omega_0 E_g)^{1/2}}{\varepsilon_\perp} \frac{\alpha}{\lambda_g}. \qquad (145)$$

This ratio is always much less than unity. Using the same parameters as in estimating the relationship (143), and with account of the fact that $\alpha_{\rm ef} \sim 3 \times 10^{-2}$, we obtain $\alpha_i^f/\alpha_i^\perp \approx 1 \times 10^{-3}$. In such a structure the internal losses are therefore much less than those on mirrors, $\tilde{\alpha}_i^f \ll \alpha^*$. Assuming $\alpha^* = 30\,{\rm cm}^{-1}$, we have that at $T = 300\,{\rm K}$ the threshold concentration equals $2 \times 10^{12}\,{\rm cm}^{-2}$. At the same value of the parameters the Auger coefficient, according to Eq. (142), is $C_A^B \equiv G^B/n^2p = 2.4 \times 10^{-17}\,{\rm cm}^4/{\rm s}$. Thus, the Auger recombination rate equals $G^B \approx 1.92 \times 10^{20}\,{\rm cm}^{-2}{\rm s}^{-1}$. Consequently, the Auger recombination current equals $J_{A,{\rm th}} \approx 31\,{\rm A}/{\rm cm}^2$. Under the same conditions we have $J_{R,{\rm th}} \approx 80\,{\rm A}/{\rm cm}^2$, and the leakage current is almost zero. The threshold current density $J_{\rm th}$ of lasers based on quantum well does not therefore exceed $120\,{\rm A}/{\rm cm}^2$. Then, $\eta \geqslant 50\%$. This means that such a laser can operate at room and even higher temperatures.

Fig. 38 shows the temperature dependence of the Auger current J_A and the radiative current for lasers based on deep quantum wells. The radiative current varies linearly with temperature; the Auger current is an exponential function of temperature. Nevertheless, at $T = 300\,{\rm K}$ we have $J_{A,{\rm th}} < J_{R,{\rm th}}$.

We have only considered the case of intraband absorption by holes and of CHCC Auger process. All the obtained conclusions are, however, valid for the case of holes.

Above, we discussed the suppression of the Auger recombination and intraband absorption processes for type-I heterostructures.

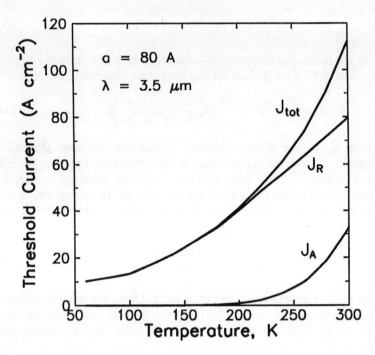

Fig. 38 Temperature dependence of threshold current, calculated for a deep quantum well: $\lambda = 3.5\,\mu\text{m}$, $a = 80\,\text{Å}$.

It is important that the materials most appropriate for creating mid-infrared lasers based on deep quantum wells are the type-II heterostructures with quantum wells [61]. These structures enable independent control of the distance between two size-quantization levels for electrons and holes. It is therefore possible to suppress the processes of Auger recombination and intraband absorption for electrons and holes simultaneously in type-II heterostructures.

7.8 CONCLUSIONS

A theory of elementary processes governing the operation of mid-infrared lasers based on strained quantum wells with strained layers based on III–V semiconductor compounds has been proposed from fundamental principles. The influence of heterointerfaces on the Auger recombination and intraband absorption mechanisms in quantum wells has been demonstrated. Analytical expressions for the Auger recombination rate and the intraband absorption factor in the quantum wells based on heterostructures of the type I and II have been obtained. This enabled us to analyze qualitatively the dependence of threshold characteristics of mid-infrared lasers on the quantum well parameters and the value of strain. This analysis proved to be efficient in studying the limits of operation of mid-infrared lasers at high temperatures. The main purpose of the present chapter was to perform a qualitative microscopic analysis of the processes in long-wave lasers affecting the limiting operating temperature.

It has been shown that the limiting operating temperature is mainly affected by the process of intraband absorption. The existence of this absorption caused by the interaction of electrons (holes) and heterointerfaces, results in a non-linear dependence of non-equilibrium charge carriers on temperature at the generation threshold. The non-linearity of this dependence originates in that both the Auger and radiative recombination processes are steep power functions of temperature.

We have formulated the condition determining the limiting operating temperature, i.e., the temperature at which the laser generation is suppressed.

A big part of the chapter is devoted to studying Auger recombination mechanisms in type-I and type-II heterostructures with quantum wells. It has been shown that the Auger recombination mechanisms in the two types of heterostructures are considerably different. The mechanism of suppression of the Auger recombination and intraband absorption processes in type-II heterostructures has been studied.

We have demonstrated the advantages of the type-II heterostructures over the heterostructures of type I.

To compare theoretical results with experimental data a numerical (computer) calculation of the threshold current was performed. We have used exact expressions for the Auger recombination current.

In the end of the chapter we proposed a new fundamental approach (Auger engineering) to creation of mid-infrared diode lasers operating at room temperature. Within the approach simultaneous suppression of the Auger recombination and intraband absorption processes is possible, so that the Auger recombination current does not exceed the radiative recombination rate.

Thus, in the present paper we discussed in detail the main physical processes governing the operation of the mid-infrared lasers; studied the limits of their operation at room temperature; proposed a new approach to creation of the mid-infrared lasers operating at room temperature.

Acknowledgments

I owe a debt of gratitude to D. Z. Garbuzov, M. I. Dyakonov, R. F. Kazarinov, M. P. Michailova, V. I. Perel, R. A. Suris and Yu. P. Yakovlev for discussous and productive criticism. I also wish to thank my colleagues: A. D. Andreev, L. E. Asryan, M. F. Bryzhina, N. A. Gun'ko and A. S. Polkovnikov for their invaluable help in creation and designing of this paper.

References

1. A. E. Bochkarev, L. M. Dolginov, A. E. Drakin, P. G. Eliseev, and B. N. Sverdlov, *Sov. J. Quantum Electron.* **18**, 1362 (1988).
2. A. N. Baranov, T. N. Danilova, B. E. Dzhurtanov, A. N. Imenkov, S. G. Konnikov, A. M. Litvak, V. E. Usmanskii, and Yu. P. Yakovlev, *Sov. Tech. Phys. Lett.* **14**, 727 (1988).
3. H. K. Choi and S. J. Eglash, *Appl. Phys. Lett.* **59**, 1165 (1991).
4. M. Aidaraliev, G. G. Zegrya, N. V. Zotova, S. A. Karandashev, B. A. Matveev, N. M. Stus', and G. N. Talalakin, *Sov. Phys. Semicond.* **26**, 138 (1992).
5. D. Partin, *Superlatt. Microstruct.* **1**, 131 (1985).
6. P. Becla, *J. Vac. Sci. Technol.* **A6**, 2725 (1988).
7. G. Herzberg, *Molecular Spectra and Molecular Structure*, Vol. 2, Infrared and Raman Spectra of Polyatomic Molecules, Van Nostrand, New York (1945), Chap. 9, p. 306.
8. M. W. Prairie and D. L. McDaniel, *SPIE* **2214**. Space Instrumentation and Dual-Use Technologies, Orlando, 207–218 (April 1994).
9. H. K. Choi and G. W. Turner, *Proc. Soc. Photo-Opt. Instrum. Eng.* **2382**, 236 (1995).
10. S. D. Mancha, A. Keipert, and M. W. Prairie, *SPIE* **2214**. Space Instrumentation and Dual-Use Technologies, Orlando, 197–206 (April 1994).
11. J. Faist, F. Capasso, D. L. Sivco, C. Sirtori, A. L. Hutchinson, and A. Y. Cho, *Electron. Lett.* **30**, 865 (1994).
12. J. M. Arias, M. Zandian, R. Zucca, and J. Singh, *Semicond. Sci. Technol.* **8**, S255 (1993).
13. B. Spanger, U. Schiessel, A. Lambrecht, H. Bottner, and M. Tache, *Appl. Phys. Lett.* **53**, 2582 (1988).
14. Z. Feit, D. Kostyk, R. J. Woods, and P. Mak, *Appl. Phys. Lett.* **58**, 343 (1991).
15. A. H. Baranov, A. N. Imenkov, V. V. Sherstnev, and Yu. P. Yakovlev, *Appl. Phys. Lett.* **64**, 2480 (1994).
16. R. F. Kazarinov and R. A. Suris, *Sov. Phys. Semicond.* **5**, 707 (1971).
17. E. O. Kane, *J. Phys. Chem. Solids* **1**(2), 249 (1957).
18. B. L. Gelmont, *Sov. Phys. JETP* **48**(2), 268 (1978).
19. G. G. Zegrya and V. A. Kharchenko, *Sov. Phys. JETP* **74**(1), 173 (1992).
20. R. A. Suris and G. G. Zegrya, *Semicond. Sci. Technol.* **7**, 347 (1992).
21. G. G. Zegrya, A. D. Andreev, N. A. Gun'ko, and E. V. Frolushkina, *Proc. SPIE* **2399**, 307 (1995).
22. G. G. Zegrya, P. Voisin, D. K. Nelson, A. N. Starukhin, and A. N. Titkov, *Nanostructures: Physics and Technology, International Symposium* (St. Petersburg, Russia), 101 (1994).
23. G. G. Zegrya and A. D. Andreev, *International Semiconductor Conference (Sinaia, Romania), CAS'95 Proceedings*, 253 (1995).
24. M. G. Burt, *J. Phys. Condens. Matter* **4**, 6651 (1992).

25. B. A. Foreman, *Phys. Rev. B* **49**, 1757 (1994).
26. R. A. Suris, *Sov. Phys. Semicond.* **20**, 1258 (1986).
27. G. L. Bir and G. E. Pikus, *Simmetry and Strain-Induced Effects in Semiconductors* (Wiley, New York, 1974).
28. L. D. Landau and E. M. Lifshitz, *Quantum Mechanics* (Pergamon, New York, 1977).
29. A. V. Sokol'skii and R. A. Suris, *Sov. Phys. Semicond.* **21**, 529 (1987).
30. G. G. Zegrya and A. D. Andreev, *Sov. Phys. JETP* **82**, 328 (1996).
31. R. H. Pantell and H. E. Puthoff, *Fundamentals of Quantum Electronics* (New York, 1969).
32. L. D. Landau and E. M. Lifshitz, *Quantum Electrodynamics* ("Nauka", Moscow, 1989).
33. W. Van Roosbroeck and W. Shocklev, *Phys. Rev.* **94**, 1558 (1954).
34. G. G. Zegrya, *Doctor of Physical Sciences Thesis*, A. F. Ioffe Institute (1995).
35. G. G. Zegrya, *Semiconductors*, **31**, N11 (1997).
36. W. J. Zawadski, *J. Phys. C* **16**, 229 (1983).
37. A. Ya. Shik, *Fiz. Tekh. Poluprovodn.* **22**, 1843 (1988).
38. G. P. Agrawal and N. K. Dutta, *Long-Wavelength Semiconductor Lasers* (Van Nostrand Reinhold, New York, 1986).
39. *Quantum Well Lasers*, Ed. by P.S. Zory, Jr. (Academic, New York, 1993).
40. E. P. O'Reilly and A. R. Adams, *IEEE J. Quantum Electron.* **30**, 366 (1994).
41. E. P. O'Reilly and M. Silver, *Appl. Phys. Lett.* **63**, 3318 (1993).
42. J. Wang, P. Von Allmen, J.-P. Leburton, and K. J. Linden, *IEEE J. Quantum Electron.* **31**, 864 (1995).
43. G. G. Zegrya and A. D. Andreev, *Appl. Phys. Lett.* **67**, 2681 (1995).
44. A. D. Andreev and G. G. Zegrya, *Physics and Technology, International Symposium* (St.-Petersburg, Russia), 99 1996.
45. M. I. Dyakonov and V. Yu. Kachorovskii, *Phys. Rev. B* **49**, 17130 (1994).
46. M. P. Mikhailova and A. N. Titkov, *Semicond. Sci. Technol.* **9**, 1279 (1994).
47. M. P. Mikhailova, G. G. Zegrya, K. D. Moiseev, et al., *Proc. Soc. Photo-Opt. Instrum. Eng.* **2397**, 166 (1995).
48. J. R. Meyer, C. A. Hoffman, F. J. Bartoli, and L. R. Ram-Mohan, *Appl. Phys. Lett.* **67**, 757 (1995).
49. P. J. P. Tang, M. J. Pullin, S. J. Chung, et al., *Proc. Soc. Photo-Opt. Instrum. Eng.* **2397**, 389 (1995).
50. A. R. Adams, *Electron. Lett.* **22**, 249 (1986).
51. G. G. Zegrya and A. D. Andreev, *Proc. SPIE*, **2682**, 224 (1996).
52. M. P. Mikhailova, G. G. Zegrya, K. D. Moiseev, et al., *Proc. SPIE* **2397**, 166 (1995).
53. R. F. Nabiev, C. J. Chang-Hasnain, and H. K. Choi, *Semiconductor Lasers Advanced Devices and Applications*, Keystone, Colorado, USA, Technical Digest Series, **20**, 31 (1995).
54. H. K. Choi, G. W. Turner, and Z. L. Liau, *Appl. Phys. Lett.* **65**, 2251 (1994).
55. H. K. Choi, S. L. Eglash, and G. W. Turner, *Appl. Phys. Lett.* **64**, 2474 (1994).
56. H. Lee, P. K. York, R. J. Menna, R. U. Martinelli, D. Z. Garbuzov, S. Y. Nazayan, and J. C. Connolly, *Appl. Phys. Lett.* **66**, 1942 (1995).
57. S. R. Kurtz, R. M. Biefeld, L. R. Dawson, K. C. Baucom, and A. J. Howard, *Appl. Phys. Lett.* **64**, 812 (1994).
58. H. K. Choi, G. W. Turner, and Z. L. Liau, *Appl. Phys. Lett.* **65**, 2251 (1994).
59. G. G. Zegrya and S. G. Yastrebov, *Patent of Russian 2025010, Bulletin of inventions* (in russian) **15**, 12 (1994).
60. G. G. Zegrya, *Semiconductors*, **31**, N12 (1997).
61. G. G. Zegrya, *Appl. Phys. Lett.* to be published (1997).
62. V. L. Gurevich, *Sov. Phys. Solid State* **4**, 1252 (1962).

CHAPTER 8

Antimonide-Based Mid-Infrared Quantum-Well Diode Lasers

G. W. TURNER and H. K. CHOI

Lincoln Laboratory, Massachusetts Institute of Technology, Lexington, MA 02173-9108, USA

8.1.	Introduction and Brief History of Antimonide-Based Lasers	370
8.2.	Properties of Antimonide-Based Materials	373
	8.2.1. GaInAsSb	373
	8.2.2. AlGaAsSb	374
	8.2.3. InAlAsSb	378
	8.2.4. Band offset	379
8.3.	Effect of Strain on Mid-IR Laser Performance	381
	8.3.1. Effect of strain on band structure	382
	8.3.2. Threshold current density	385
8.4.	MBE Growth and Characterization of Strained QW Structures	387
	8.4.1. Introduction and background	387
	8.4.2. MBE equipment	388
	8.4.3. Growth procedures	389
	8.4.4. Growth of AlAsSb and AlGaAsSb	390
	8.4.5. Doping	392
	8.4.6. Growth of GaInAsSb/AlGaAsSb QW structures	395
	8.4.7. Growth of InAsSb/InAlAsSb QW structures	400
8.5.	Laser Performance	404
	8.5.1. GaInAsSb/AlGaAsSb QW lasers emitting in the 2-μm region	404
	8.5.2. QW lasers emitting beyond 3 μm	414
	8.5.2.1. InAsSb/InAlAsSb QW lasers on InAs	415
	8.5.2.2. InAsSb/InAlAsSb QW lasers on GaSb	422
8.6.	Conclusions and Future Projections	425
	Acknowledgments	427
	References	427

8.1 INTRODUCTION AND BRIEF HISTORY OF ANTIMONIDE-BASED LASERS

Semiconductor diode lasers emitting in the mid-infrared (mid-IR) spectrum of 2–5 µm would be very useful for a variety of applications exploiting the characteristics of this spectral range, which include strong molecular absorption lines and important atmospheric transmission windows. Although solid state lasers or gas lasers emitting in the mid-IR are available, diode lasers are far preferable because they are compact, efficient, reliable, and potentially inexpensive. Tunable single-frequency mid-IR diode lasers are ideally suited for highly sensitive detection of trace gases for such applications as pollution or toxic gas monitoring and industrial process control. The sensitivity is typically higher by two orders of magnitude at the fundamental absorption lines than at overtones that appear in the near infrared. For example, the detection sensitivity of methane is 1.7 parts per billion at 3.26 µm for a 1-m path length, while it is 600 parts per billion at 1.65 µm [1]. In addition, high-power mid-IR lasers emitting in the atmospheric transmission windows can be very useful for laser radars and for target designation and illumination.

Diode lasers fabricated from antimony-containing III–V compounds are being developed for mid-IR sources because other III–V diode lasers with high performance and reliability are commercially available at wavelengths between 0.6 and 2.0 µm. As shown in Fig. 1, which is a plot of the room-temperature (RT) bandgap vs lattice constant for a number of antimonide-based alloys, GaInAsSb lattice matched to GaSb or InAs can be used for the active layer because it has a direct bandgap between 1.7 and 4.4 µm. In addition, AlGaAsSb, which has a larger bandgap and lower refractive index than GaInAsSb, can be used for the cladding.

Although the III–V mid-IR diode lasers, based on InAs homojunctions with low-temperature emission at 3.1 µm, were first reported [2] in 1963, their development has proceeded very slowly because of technical difficulties and lack of large-scale applications. In addition, development of lead-salt lasers [3] in 1964 attracted much more attention because they can cover a wider wavelength range (3–30 µm) by varying alloy compositions, and also have a large temperature tuning range. Indeed, lead-salt lasers have been the workhorse for diode-laser-based high-resolution spectroscopy research in the past. Other applications of lead-salt lasers, however, have been limited because of their low-temperature operation, low power, and lack of reliability as compared to conventional III–V diode lasers.

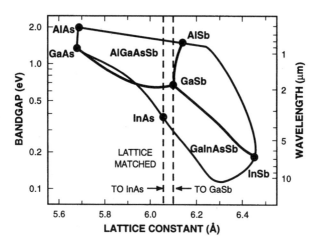

Fig. 1 Room-temperature bandgap vs lattice constant diagram for GaInAsSb, AlGaAsSb, and GaInAlSb.

With development of the liquid phase epitaxy (LPE) technique, which is capable of growing lattice-matched quaternary alloys and heterostructures, research in III–V mid-IR diode lasers was revived. Dolginov et al. [4] in 1976 reported double-heterostructure (DH) lasers consisting of GaSb active and AlGaAsSb cladding layers. They obtained RT pulsed operation at 1.78 µm with threshold current density J_{th} of 6.2 kA/cm^2. Dolginov et al. [5] subsequently reported DH lasers consisting of GaInAsSb active and GaSb cladding layers, with emission wavelengths up to 1.9 µm. However, the performance was rather poor, probably because GaSb has a larger refractive index than GaInAsSb, resulting in insufficient optical confinement. In 1980, Kobayashi et al. [6] recognized this problem and fabricated GaInAsSb/AlGaAsSb DH lasers. These lasers exhibited RT pulsed operation at 1.8 µm with J_{th} as low as 5 kA/cm^2. The value of J_{th} for GaInAsSb/AlGaAsSb DH lasers was gradually reduced by improving the quality of the epilayer and increasing the Al content in the cladding [7–11]. The lowest J_{th} obtained from LPE-grown lasers was 1.5 kA/cm^2, which contained 55% of Al in the cladding layers [10]. The emission wavelength was also increased to 2.48 µm by increasing the In and As content in GaInAsSb [9]. RT cw operation at 2.3 µm was achieved in similar structures in 1988 [10,11].

For emission beyond 3 µm, alloys with compositions close to InAs were utilized for the active layer to avoid the difficulty of growing

GaInAsSb alloys close to the miscibility gap [12]. In 1978, Dolginov et al. [13] obtained lasing between 3.1 and 3.7 μm by e-beam pumping of InAsSbP grown on InAs by LPE. In 1985, van der Ziel et al. [14,15] reported optically pumped laser action of InAsSb/InAsSbP DHs on InAs and InAsSb/AlGaAsSb DHs on GaSb. The emission wavelength was between 3.1 and 3.9 μm, and the maximum operating temperature was 140 K. Subsequently, InAsSb/InAsSbP DH diode lasers were reported [16,17]. Values of J_{th} as low as 39 A/cm^2 at 77 K were obtained for these lasers [17]. However, because of a low refractive index step and small band offsets between the active and cladding layers, the laser performance has been limited, with pulsed operation up to 140 K and cw operation near 80 K.

Although significant progress was made by LPE, it is molecular beam epitaxy (MBE), a nonequilibrium growth technique, that has pushed the performance of mid-IR lasers beyond the limits imposed by LPE materials. The first GaInAsSb/AlGaAsSb DH lasers grown by MBE were reported by Chiu et al. [18] in 1986. RT J_{th} as low as 4.2 kA/cm^2 was obtained at 2.2 μm. By improving the growth conditions and increasing the Al content in the cladding to 75%, Choi and Eglash [19] in 1991 obtained J_{th} as low as 0.94 kA/cm^2 for GaInAsSb/ AlGaAsSb DH lasers emitting at 2.2 μm. MBE also enabled growth of GaInAsSb alloys in the miscibility gap, resulting in lasers at ~ 3 μm with pulsed operation up to 255 K and cw up to 170 K [20]. At ~ 4 μm, InAsSb/AlAsSb DH lasers grown by MBE exhibited cw operation up to 105 K and pulsed up to 170 K [21].

However, real breakthroughs were achieved by employing strained quantum-well (QW) structures. The first GaInAsSb/AlGaAsSb strained QW lasers were reported by Choi and Eglash [22] in 1992, with J_{th} as low as 260 A/cm^2 and emission wavelength at ~ 2.1 μm. Since then, various strained QW structures have been demonstrated for wavelengths between 1.9 and 4.5 μm [23–28]. These include the InAsSb/InAlAsSb structure and type-II structures utilizing GaInSb/ InAs or InAsSb/InAs superlattices, in which the conduction- and valence-band minima are physically separated.

In this chapter, we present an overview of the current status of antimonide-based mid-IR QW lasers. We will focus on strained type-I structures, which have been grown by MBE. (Type-II approaches are discussed in Chapters 9 and 10.) In Section 8.2, properties of antimonide-based materials relevant to laser performance will be described. In Section 8.3, effects of strain on the performance of mid-IR lasers will be discussed. Growth of antimonide-based materials and QW structures by MBE and their characterization will be discussed in Section 8.4. Performance of antimonide-based QW diode lasers will be described in Section 8.5. Finally, future directions to improve the performance of the mid-IR lasers will be given in Section 8.6.

8.2 PROPERTIES OF ANTIMONIDE-BASED MATERIALS

8.2.1 GaInAsSb

As the active layer material for mid-IR lasers, GaInAsSb has been extensively studied. It has a direct bandgap for all the alloy compositions, and can be grown lattice matched to either InAs or GaSb. The GaSb substrate has a larger bandgap energy E_g than all of the GaInAsSb alloys, while InAs has a smaller E_g than most of the alloys. Growth of GaInAsSb has been performed mostly on GaSb substrates because the region of stable growth and the wavelength coverage are larger.

Fig. 2 plots the calculated binodal and spinodal stability curves for GaInAsSb at 615°C [12]. The binodal curve is the boundary between stable and metastable alloys, while the spinodal curve is the boundary between the metastable and unstable alloys. The two lines correspond to alloys lattice matched to GaSb or InAs. No alloys inside the binodal curve have been grown by LPE because it is close to an equilibrium growth technique. By using nonequilibrium techniques such as MBE or organometallic vapor phase epitaxy (OMVPE), alloys inside the miscibility gap have been grown [21,29]. However, the growth becomes increasingly more difficult as the composition moves further into the miscibility gap.

For $Ga_{1-x}In_xAs_ySb_{1-y}$ lattice matched to GaSb $[y = 0.866x/(1 - 0.048x)]$, E_g at 300 and 77 K as well as the split-off energy Δ are plotted in Fig. 3 [30]. These parameters can be fit using the following formulae:

$$E_g(300\,K) = 0.725(1 - x) + 0.290x - 0.6x(1 - x) \quad (1)$$

$$E_g(77\,K) = 0.801(1 - x) + 0.354x - 0.6x(1 - x) \quad (2)$$

$$\Delta = 0.761(1 - x) + 0.325x + 0.26x(1 - x). \quad (3)$$

The RT bandgap decreases steeply from 0.725 to 0.30 eV (with a corresponding increase in wavelength from 1.7 to 4.2 μm) as x increases from 0 to 0.6. However, E_g stays nearly constant for $x > 0.6$,

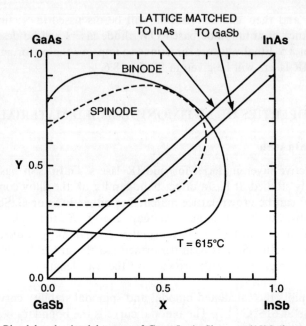

Fig. 2 Binodal and spinodal curves of $Ga_{1-x}In_xAs_ySb_{1-y}$ at 615°C (from Ref. [12]). The binodal curve is the boundary between stable and metastable alloys, while the spinodal curve is the boundary between the metastable and unstable alloys.

and has a minimum value of 0.28 eV (4.4 μm) at $x \sim 0.8$. The value of Δ does not depend appreciably on temperature.

Fig. 4 plots the calculated refractive index of GaInAsSb lattice matched to GaSb as a function of Ga content and energy [31]. As the Ga content is lowered, the refractive index is decreased. Because E_g is also reduced with decreasing Ga content, this behavior is opposite to those for most of the other III–V alloys. This peculiar behavior results from the fact that InAs has a lower refractive index than GaSb. Comparison of experimental values of GaSb with those of Fig. 4 indicates that the calculated value for GaSb underestimates the refractive index by ~ 0.05.

8.2.2 AlGaAsSb

The lattice constant of AlSb is about 0.6% larger than that of GaSb, which is much larger than the difference of 0.13% for GaAs and AlAs. As a result, AlGaAsSb with a small amount of As is required for lattice

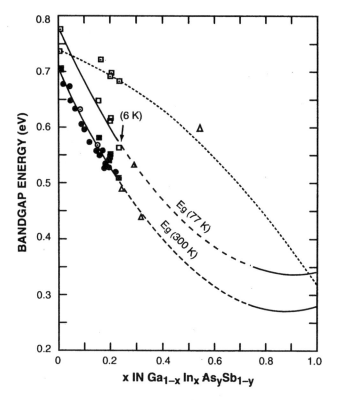

Fig. 3 Bandgap energy at 300 and 77 K and split-off energy Δ of GaInAsSb lattice matched to GaSb (from Ref. [30]). The portions drawn by dotted curves correspond to alloys in the miscibility gap. The experimental points are from many references listed in Ref. [30]. The value of Δ does not depend appreciably on temperature.

matching to GaSb. For an Al content of 1, an As content of 0.08 is needed. The calculated spinodal curves for AlGaAsSb are shown in Fig. 5 [32]. For alloys lattice matched to GaSb, all compositions are in the stable regime. However, for alloys lattice matched to InAs, there is a considerable region of unstable growth.

Fig. 6 shows bandgap energies of $Al_xGa_{1-x}Sb$ grown on GaSb as a function of x [33]. No band structure data on AlGaAsSb are available, but for alloys lattice matched to GaSb they should be very close to the AlGaSb values because only a small amount of As is present. For GaSb, the difference between the Γ and L band is only ∼ 80 meV. The direct–indirect transition occurs at $x = 0.2$, where $E_g = 0.98$ eV. For $x > 0.5$, the lowest lying conduction band is the X band.

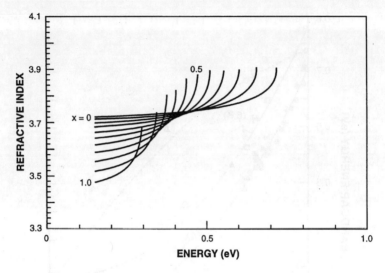

Fig. 4 Calculated refractive index of $Ga_{1-x}In_xAs_ySb_{1-y}$ lattice matched to GaSb as a function of x and energy (from Ref. [31]).

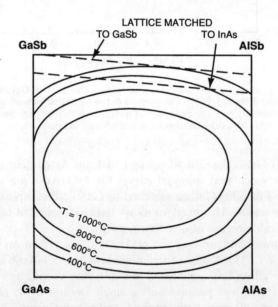

Fig. 5 Calculated spinodal curves for AlGaAsSb (from Ref. [32]). Alloys lattice matched to GaSb are in the stable region, while some alloys lattice matched to InAs are inside the miscibility gap.

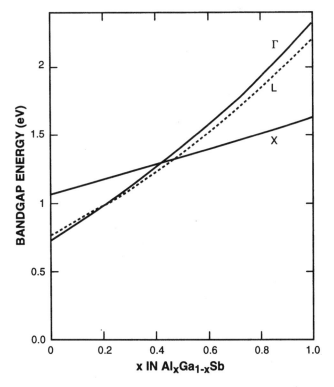

Fig. 6 Bandgap energies of $Al_xGa_{1-x}Sb$ grown on GaSb as a function of x (from Ref. [33]). The direct–indirect transition occurs at $x \sim 0.2$, where $E_g \sim 0.98$ eV. For $x > 0.5$, the lowest-lying conduction band is the X band.

The refractive index of AlGaAsSb lattice matched to GaSb was determined by measuring the reflectance as a function of incident angle [34]. The measured refractive index in the transparent wavelength region can be fairly accurately expressed as a function of energy by the single-effective-oscillator model [35]:

$$n^2 - 1 = E_0\, E_d/(E_0^2 - E^2), \qquad (4)$$

where

$$E_0 = 1.89(1-x) + 3.2x - 0.36x(1-x), \qquad (4a)$$

and

$$E_d = 24.5(1-x) + 28x - 4.4x(1-x). \qquad (4b)$$

The expressions for E_0 and E_d were derived by fitting the experimental values to Eq. (4) [34]. The calculated refractive index of AlGaAsSb from Eq. (4) is plotted in Fig. 7.

The thermal resistance of the laser structure is mainly determined by that of the upper cladding layer if the laser is mounted junction-side down. Thermal resistivity of AlGaAsSb lattice matched to GaSb was measured at 300 K [36], and is plotted in Fig. 8. As expected, the largest thermal resistivity is at $x \sim 0.5$, where it is 13 K-cm/W. For comparison, the thermal resistivity of $Al_{0.5}Ga_{0.5}As$ is 8 K-cm/W [37]. Although it has not been measured, the thermal resistivity of AlGaAsSb with x close to 1 should be much smaller. The thermal resistivity of AlSb is 1.8 K-cm/W [38].

8.2.3 InAlAsSb

There has been very little work on the material properties of InAlAsSb. Using LPE, Charykov et al. [39] were able to grow lattice-matched InAlAsSb on InAs with Al content up to 8%. The value of E_g at RT increases linearly from 0.36 to 0.48 eV as the Al content is

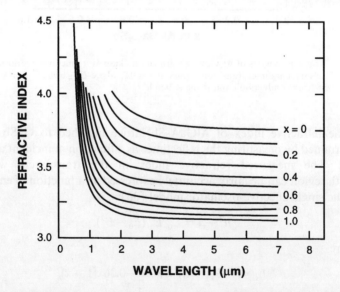

Fig. 7 Calculated refractive index (Eq. (4)) of AlGaAsSb lattice matched to GaSb as a function of energy (from Ref. [34]). Eq. (4) was obtained by fitting the experimental data to the single-effective-oscillator model.

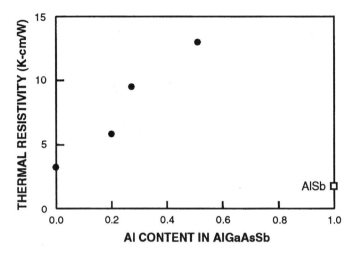

Fig. 8 300 K thermal resistivity of AlGaAsSb lattice matched to GaSb. The thermal resistivity of AlSb is also shown for comparison.

increased from 0 to 8%. Fig. 9 shows the calculated spinodal curves for InAlAsSb [32]. It is noted that the region of stable growth for InAlAsSb is limited to Al content less than 6% on GaSb and up to 12% on InAs. No data on the refractive index are available.

8.2.4 Band Offset

For the design of the laser structures, one of the most important parameters is the band offset. The larger the band offset is, the smaller the leakage current will be. Even though there are many reports of both theoretical and experimental investigation on the band offsets, there are still uncertainties in the band offset values of binary compounds, let alone ternary or quaternary compounds. Table I lists the relative valence-band positions of the binary compounds that are deemed most reliable. These values are from Ref. [40], except for InSb, which is from Ref. [41]. For valence-band positions of the quaternary compounds, linear interpolation of the binary values is assumed. As a result, any effect of bowing in the bandgap vs composition relationship appears in the conduction band.

Fig. 10 plots the relative conduction- and valence-band positions of $Ga_{1-x}In_xAs_ySb_{1-y}$ lattice matched to GaSb as a function of x. The band positions of $AlAs_{0.08}Sb_{0.92}$ are also shown for comparison. For

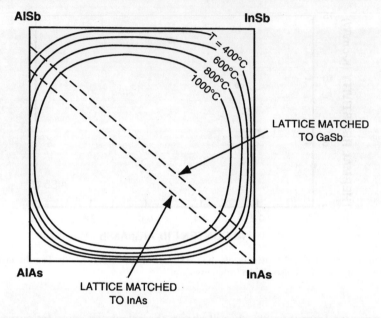

Fig. 9 Calculated spinodal curves for InAlAsSb (from Ref. [32]). For alloy compositions near InAs, the region of stable growth on GaSb is much smaller than on InAs.

TABLE I
Relative valence-band positions of the binary compounds. These values are from Ref. [40], except for InSb, which is from Ref. [41].

Compound	Valence-Band Position(eV)
AlAs	0.00
GaAs	0.48
InAs	0.65
AlSb	0.76
GaSb	1.12
InSb	1.25

wavelengths near 2 μm, both the conduction- and valence-band offsets are large, making it easy to design good laser structures. As x increases, both the conduction- and valence-band positions of $Ga_{1-x}In_xAs_ySb_{1-y}$ become lower. This lowering results in a large increase in the conduction-band offset, while decreasing the valence-band offset. For $x = 1$, the valence-band offset is slightly negative. The

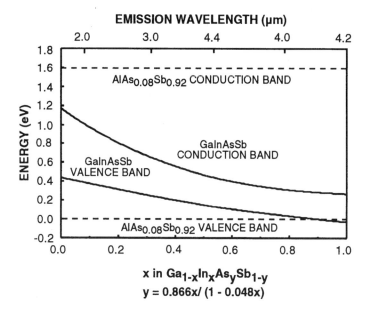

Fig. 10 Relative conduction- and valence-band positions of $AlAs_{0.08}Sb_{0.92}$ and $Ga_{1-x}In_xAs_ySb_{1-y}$ lattice matched to GaSb. As x increases, the valence-band offset decreases and the conduction-band offset increases. For $x \sim 1$, the valence-band offset is negative.

small valence-band offset is a major issue for the design of GaInAsSb/AlGaAsSb diode lasers emitting beyond 3 μm.

For emission between 3 and 4 μm, InAsSb has been used as the active layer because it is easier to grow than GaInAsSb alloys which have comparable bandgaps but lie closer to the miscibility gap. The most promising barrier material with respect to the band offset is InAlAsSb. The calculated valence-band position of InAlAsSb lattice matched to GaSb is plotted as a function of Al content in Fig. 11, with the position of $InAs_{0.91}Sb_{0.09}$ as the reference. For most of the Al content, InAlAsSb has a lower valence-band position than $InAs_{0.91}Sb_{0.09}$, although the difference is small. Because of the severe miscibility gap, however, the Al content in the barrier is limited to $\sim 15\%$.

8.3 EFFECT OF STRAIN ON MID-IR LASER PERFORMANCE

The performance of mid-IR lasers degrades rapidly with increasing wavelength. The main reason for this degradation is nonradiative

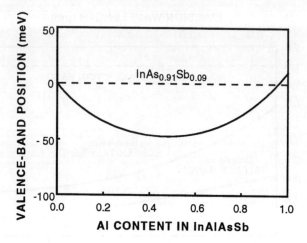

Fig. 11 Calculated valence-band position of InAlAsSb lattice matched to GaSb as a function of Al content with respect to the valence-band position of $InAs_{0.91}Sb_{0.09}$.

Auger recombination, which increases exponentially with wavelength. Another reason is increased internal loss due to higher free-carrier and intervalence-band absorption, which raises the threshold gain g_{th} and lowers the differential quantum efficiency η_d.

The use of compressively strained QW structures to improve the laser performance was proposed by Yablonovich and Kane [42], and independently by Adams [43]. They recognized that the valence band of all the III–V compounds has a much larger density of states than the conduction band, which results in a significant penalty in threshold current density, Auger recombination, and other nonradiative losses. The application of compressive strain changes the valence-band structure to alleviate these problems. It is now well documented that the compressively strained QW lasers indeed exhibit much superior performance to the DH or unstrained QW lasers. Although tensile strain has also been shown to significantly improve the performance of InP-based lasers [44] and GaInP red lasers [45], it will not be considered for the antimonide-based lasers because the valence-band offset for the tensile-strained active layer is not suitable for type-I QW lasers.

8.3.1 Effect of Strain on Band Structure

Fig. 12 compares the valence-band structure near the Brillouin zone center for unstrained and compressively strained bulk materials. There

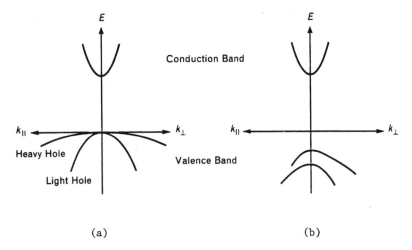

Fig. 12 Schematic valence-band structure near Brillouin zone center in the direction parallel ($k_\parallel$) and perpendicular ($k_\perp$) to the epilayer for (a) unstrained and (b) compressively strained bulk material. The compressive strain increases the bandgap energy, removes the degeneracy at the zone center, and makes the band structure anisotropic.

are three effects of the compressive strain: (1) E_g is increased, (2) the degeneracy at the zone center is removed, and (3) the band structure becomes anisotropic. In particular, the uppermost band in the direction parallel to the epilayer has a smaller effective mass. Since the density of states in 2-dimensional structures is proportional to the effective mass in the plane, the compressive strain reduces the density of states, and thus the threshold carrier density n_{th}. The smaller density of states can also reduce the intervalence-band absorption because of a reduced transition probability.

Because the strain splits the valence band and changes E_g, the band alignment is altered. Under the perturbation approximation, changes in the band positions for the conduction (δE_c), heavy-hole (δE_{hh}), light-hole (δE_{lh}), and split-off bands (δE_{so}) are [46]

$$\delta E_c = A a_c \varepsilon \tag{5a}$$

$$\delta E_{hh} = A a_v \varepsilon + B b \varepsilon / 3 + \Delta / 3 \tag{5b}$$

$$\delta E_{lh} = (A a_v + Bb/6) - \Delta/6 + 0.5 \, [B^2 b^2 \varepsilon^2 / 3 - 2 B b \varepsilon \Delta / 3 + \Delta^2]^{0.5} \tag{5c}$$

$$\delta E_{so} = (A a_v + Bb/6) - \Delta/6 - 0.5 \, [B^2 b^2 \varepsilon^2 / 3 - 2 B b \varepsilon \Delta / 3 + \Delta^2]^{0.5}, \tag{5d}$$

where ε is the strain, $A = 2(C_{11} - C_{12})/C_{11}$, $B = 3(C_{11} + 2C_{12})/C_{11}$ with C_{11} and C_{12} the elastic constants, a_c, a_v are conduction- and valence-band hydrostatic deformation potentials, respectively, and b is the shear deformation potential. Table II lists these parameters for the binary compounds. For quaternaries, linearly interpolated values are used.

The band structure of QW structures is often calculated by using the Luttinger–Kohn **k·p** perturbation theory [47]. Because of the complexity of the problem, many approximations are usually made. We follow the procedure described in Ref. [48]. To simplify the calculation, the interaction between the conduction and valence bands is ignored, while coupling between the valence and split-off bands is taken into account. Decoupling of the conduction band reduces the 8×8 **k·p** matrix to a 2×2 conduction matrix and a 6×6 valence matrix. With the axial approximation, the valence matrix is further simplified by applying a unitary transformation, which results in two 3×3 blocks. A finite-element method is then used to solve one of the two matrix equations. Luttinger parameters for the binary compounds are listed in Table III [49]. For quaternary compounds, linearly interpolated values are used. Results of band-structure calculations for the laser structures will be given in Section 8.5.

TABLE II
Values for elastic constants C_{11} and C_{12}, hydrostatic deformation potentials for the conduction and valence band a_c, a_v, and the shear deformation potential b (from Ref. [46]). Values for C_{11} and C_{12} are in units of 10^{11} dyne/cm^2.

Compound	C_{11}	C_{12}	a_c(eV)	a_v(eV)	b(eV)
AlAs	12.02	5.7	−5.3	2.7	−1.5
GaAs	11.88	5.38	−5.4	2.7	−1.7
InAs	8.33	4.53	−3.3	2.5	−1.8
AlSb	8.94	4.43	−3.8	2.2	−1.35
GaSb	8.84	4.03	−6.3	2.2	−3.3
InSb	6.67	3.65	−5.1	2.1	−2.0

TABLE III
Luttinger parameters for the binary compounds (from Ref. [49]).

	AlAs	GaAs	InAs	AlSb	GaSb	InSb
γ_1	3.45	7.65	19.67	4.15	11.80	35.08
γ_2	0.68	2.41	8.37	1.01	4.03	15.64
γ_3	1.29	3.28	9.29	1.75	5.26	16.91

The band-structure information is used to calculate the gain of the QW lasers [50–52]. The agreement between the theoretical and experimental results has been very good for the case of GaAs- and InP-based QW lasers. For antimonide-based QW lasers, there have been three reports of gain calculations; Ghiti and O'Reilly [53] on GaInAsSb/AlGaAsSb QW lasers for 2.5-μm emission, Nabiev et al. [54] on InAlAsSb QW lasers for emission at 3.2 and 4.0 μm, and Choi et al. [55] on InAsSb/InAlAsSb QW lasers for 3.5-μm emission. Results of gain calculations will be discussed in Section 8.5.

8.3.2 Threshold Current Density

There are four components in J_{th}:

$$J_{th} = J_R + J_L + J_I + J_A, \tag{6}$$

where J_R is radiative, J_L is leakage, J_I is interfacial recombination, and J_A is Auger recombination current density. The intrinsic component J_R is directly linked to g_{th} by a fundamental formula which is analogous to the Einstein A and B relationship [56]

$$J_R = 8\pi e \, g_{th} \, d\Delta v / \lambda^2, \tag{7}$$

where e is the electron charge, d is the total thickness of active layers, λ is the wavelength inside the semiconductor, Δv is the spontaneous emission bandwidth, and

$$g_{th} = [\alpha_i + \ln(1/R)/L]/\Gamma. \tag{8}$$

Here, α_i is the internal loss coefficient, R is the facet reflectivity, L is the cavity length, and Γ is the confinement factor. Because Δv is independent of λ and d is roughly proportional to λ, J_R would be inversely proportional to λ.

The leakage current is determined by the magnitude of conduction- and valence-band offsets. The interfacial recombination depends on the quality and number of the interfaces. In many of the antimonide-based laser structures, the active and barrier layers have very different group-V (As or Sb) compositions, which makes it more difficult to grow QW structures with good interface quality. More detailed discussion of the interface quality is given in Section 8.4.

The Auger recombination current is proportional to the cube of the threshold carrier density n_{th}:

$$J_A = r_A e \, n_{th}^3 d, \tag{9}$$

where r_A is the Auger coefficient. To reduce J_A, we should decrease r_A, n_{th}, or d. All of these parameters can be smaller in strained QW structures. First, d is much smaller than in DH structures. Second, the strain reduces n_{th}. In addition, a smaller g_{th} is expected in strained QWs because of reduced intervalence-band absorption, resulting in a smaller n_{th}. Another way to reduce g_{th}, according to Eq. (8), is by increasing L and Γ. Because of the cubic dependence of J_{th} on n_{th} in Auger-dominated lasers, the lowest J_{th} may be obtained in QW structures with a large number of wells.

The Auger coefficient can also be smaller in strained QW structures. There are two important Auger recombination processes for the antimonide lasers, as shown schematically in Fig. 13. The first is CHSH in which an electron in the conduction band recombines with a heavy hole, while an electron in the split-off band is excited to the heavy-hole band. Another is CHCC in which the energy released by the recombination of an electron–hole pair excites another electron in the conduction band to a higher energy level. By neglecting nonparabolicity of the energy bands, the Auger coefficient for the CHCC

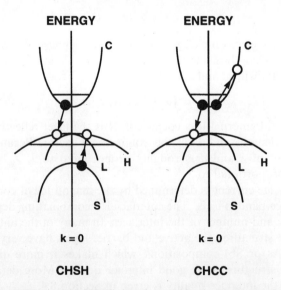

Fig. 13 (a) CHSH and (b) CHCC Auger recombination processes. In the CHSH process, an electron in the conduction (C) band recombines with a heavy hole (H), while an electron in the split-off (S) band is excited to the heavy-hole (H) band. In the CHCC process, energy released by recombination of an electron–hole pair excites another electron in the conduction band to a higher energy level.

is approximately represented by [57]

$$r_A \sim \exp(-\mu E_g/kT), \tag{10}$$

where $\mu = m_e/(m_e + m_h)$, with m_e and m_h the electron and heavy-hole effective mass, respectively. The CHCC Auger rate increases exponentially with wavelength, but it can be substantially reduced by increasing μ, which is possible in the strained QW structures. For the CHSH, the value of r_A is strongly dependent on the difference between E_g and Δ. Again by neglecting nonparabolicity of the energy bands, the CHSH Auger coefficient is approximately [57]

$$r_A \sim \exp[-(\Delta - E_g)/kT] \tag{11}$$

for $E_g < \Delta$, and for $E_g > \Delta$

$$r_A \sim \exp[-\mu'(E_g - \Delta)kT], \tag{12}$$

where $\mu' = m_s/(2m_h + m_e - m_s)$ with m_s the split-off-band effective mass. It is clear that $E_g \cong \Delta$ is not desirable. For the same magnitude of $E_g - \Delta$, the condition $E_g < \Delta$ is better than $E_g > \Delta$ because μ' is always smaller than 1. For $E_g > \Delta$, however, the CHSH Auger coefficient can be reduced by employing compressively strained QW active layers.

Accurate determination of the Auger coefficients is very difficult because of extensive calculation requirements and nonparabolicity of the band structures. In addition to the band-to-band Auger processes, there are phonon-assisted Auger processes which do not require momentum conservation from carriers. The phonon-assisted Auger process can become more important for QW structures [58], especially with thin layers, because interactions at heteroboundaries can be significant. The phonon-assisted Auger processes, however, have a smaller temperature dependence than the band-to-band processes.

8.4 MBE GROWTH AND CHARACTERIZATION OF STRAINED QW STRUCTURES

8.4.1 Introduction and Background

MBE is a nonequilibrium growth process, and offers the advantages of precise control of interfacial chemistry at heterojunctions and essentially no cross contamination between sources. Early work from 1977 to 1982 demonstrated that the conventional solid-source MBE was well suited to the growth of antimonide-based compounds including InSb [59], AlSb [60], GaSb [61], GaAsSb [62], and InGaSb [63]. In 1986, Chiu et al. [18] were able to grow the first GaInAsSb/AlGaAsSb DH lasers by MBE. In 1992, Choi and Eglash [22] first reported the growth of GaInAsSb/AlGaAsSb strained QW lasers. In addition to the growth of diode lasers, MBE has been used to grow antimonide-based compounds for a variety of electronic devices such as heterostructure transistors and resonant tunneling diodes [64–67]. These devices can make use of the large conduction-band offset available for certain heterojunction combinations such as InAs/AlSb. While a discussion of the work on electronic devices is beyond the scope of this chapter, it should be noted that there is significant overlap between the development of such electronic devices and the diode lasers in many aspects of the MBE growth process, especially the control of interfacial chemistry at heterojunctions. For a more extensive discussion of the basic MBE technology and its application to the growth of antimonide compounds, the reader is referred to review texts [68,69].

8.4.2 MBE Equipment

The antimonide-based materials were grown at Lincoln Laboratory in a solid-source 3-inch MBE system equipped with a valved As cracking source and conventional ion pumps. Liquid nitrogen was used as the coolant in the main cryopanel during growth, and isopropyl alcohol, cooled to $-40°C$ by a separate refrigerated chiller, was used for continuous cooling of the source shroud. The valved As cracking source allowed the choice of As_4 or cracked As_2 as the As species as well as rapid variation in the As flux by remote operation of the motorized valve control. Because no proven valved Sb cracking source is yet commercially available, two Sb_4 sources were employed to provide various combinations of Sb_4 fluxes. For the growth of antimonide-based QW laser structures, rapid flux changes in group-V elements are essential for the minimum interrupt time because the wells and barriers have very different alloy compositions.

The dopant sources were Be and Te (provided from GaTe) for p- and n-type conductivity, respectively. The more common n-type dopant, Si, has been found to be unsuitable for the antimonide compounds. Although other group-VI elements such as Se and S were tried as donor species for the antimonide-based materials, Te has been selected because it is capable of producing controllable n-type doping over a wide concentration range. Additional discussion on doping issues is presented later.

8.4.3 Growth Procedures

The (100) n-GaSb, n-InAs, and semi-insulating GaAs substrates were prepared by solvent cleaning and wet chemical etching procedures, mounted on substrate holders using In, and loaded into the MBE system. After outgassing and oxide desorption in the appropriate As or Sb fluxes, the substrates were ready for the subsequent epitaxial layer growth. The growth rates of the various binary, ternary, and quaternary alloys were between 0.6 and 1.0 μm/h, and the growth temperatures were in the range 400–550°C. Because of the importance of accurate substrate temperature control, especially in the growth of low-bandgap alloys for active regions, careful measurements of sample temperature were made during growth by optical pyrometry (calibrated by separate melting point measurements), and substrate thermocouple readings were used as secondary standards [70].

The growth parameters such as V/III flux ratios and temperature were empirically optimized for the best material quality by evaluating both *in situ* reflection high-energy electron diffraction (RHEED) images from the epilayer surface and *ex situ* characterization such as double-crystal X-ray diffraction (DCXRD) and photoluminescence (PL) spectroscopy. For accurate DCXRD measurements, care must be taken to chemically remove the In-alloyed region from the back side of the substrate, which can be formed during a long growth time by interaction between the substrate and In used for mounting samples to the substrate holder. Otherwise, the DCXRD results can be misleading because of substantial wafer warpage. In PL measurements at the mid-IR wavelengths, issues such as atmospheric absorption, background blackbody effects, and opacity of conventional glass lenses beyond 3 μm should be taken into consideration. For the growth of alloys containing both As and Sb, optimization of the V/III ratio requires numerous iterations because the sticking coefficients for As and Sb are different. Adjustment of the V/III ratio by changing only one of the species results in a larger than expected compositional change because of competing surface reactions between the two group-V species [71].

A general rule of thumb for the growth of mid-IR laser structures is that the total thickness of the epitaxial material scales with wavelength; a laser operating at 4 μm requires about twice the epitaxial layer thickness as a laser operating at 2 μm. This scaling of thickness with operating wavelength places a substantial burden on the stability and control of the MBE growth process, especially during the thick ternary or quaternary cladding layer growth. These cladding layers must be closely lattice matched to the substrate to prevent unwanted dislocation formation. In addition, the compositions of the active region materials should also be precisely controlled for the desired strain. As a result of all of these materials requirements, mid-IR diode lasers are one of the most challenging devices to be grown by the state-of-the-art epitaxial technology.

A number of techniques were used for the calibration of the necessary alloy compositions. Source fluxes were obtained from ionization-gauge readings before and after each run, and numerous growth rate and composition measurements were made by the analysis of RHEED oscillation frequencies [72] on test samples as well as during noncritical layer growth of the actual laser structures. The data from these *in situ* measurements were combined with *ex situ* data from measurements of lattice parameter by DCXRD, of bandgap by low-temperature PL spectroscopy, and of approximate chemical composition by Auger microprobe analysis. This combination of data provided the necessary feedback to establish correction curves, which were used to adjust source temperatures and fluxes to achieve the desired composition and lattice parameter. Estimates also had to be established for thermal expansion effects, since for thick structures lattice matched at RT, small differences in thermal expansion coefficients could lead to a significant strain at the growth temperature, causing dislocation formation [73]. The procedure at Lincoln Laboratory is to grow the entire structure lattice matched at the growth temperature, which results in a small mismatch at RT. An alternative approach is to lattice match at an intermediate temperature.

8.4.4 Growth of AlAsSb and AlGaAsSb

The best material quality for AlAsSb and AlGaAsSb was obtained by minimizing the V/III ratio and growing at relatively high temperatures, while still maintaining a 1×3 surface reconstruction as measured by RHEED, an indication of an Sb-stabilized growth condition. Such an approach of a minimal V/III ratio, combined with high-temperature growth, has also been used in the growth of high-quality AlGaAs on GaAs [74]. For the growth of the complete laser structures, however, it was often necessary to reduce the growth temperature for the upper cladding layer from the optimum value in order to prevent degradation of the underlying active region, which is grown at an even lower temperature.

To first approximation, the bandgap of AlGaAsSb is most sensitive to the Al/Ga flux ratio and the lattice matching is most sensitive to the As/Sb flux ratio. DCXRD analysis is a good test of compositional control and MBE stability required for the growth of thick AlGaAsSb cladding layers. Fig. 14 shows a DCXRD scan of a 2-μm-thick AlAsSb layer grown on GaSb using a (004) symmetric reflection from a GaSb first crystal. The AlAsSb layer has a diffraction peak at -150 arc sec, indicating that it is under a small compressive strain at RT. However, it would be very close to lattice matched at the growth temperature of $\sim 500°C$, which is a confirmation of good compositional control by using the simultaneous calibration procedures described above. The narrowness of the DCXRD peak also shows that the composition was maintained during the thick layer growth.

Compared to the growth on GaSb substrates, it is more challenging to grow lattice-matched AlGaAsSb and AlAsSb on InAs substrates

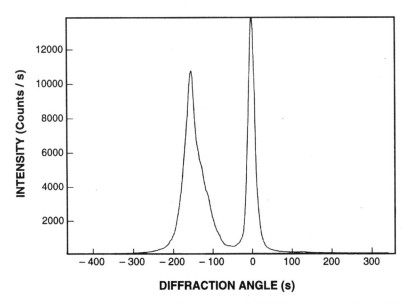

Fig. 14 Double-crystal X-ray diffractometry (DCXRD) scan of 2-μm-thick $AlAs_{0.08}Sb_{0.92}$ grown on GaSb using a (004) reflection from a GaSb first crystal. The AlAsSb layer has a diffraction peak at -150 arc sec, indicating that it is under a small compressive strain at room temperature, but very close to lattice matched at the growth temperature of $\sim 500°C$.

because they are near the region of the miscibility gap [32]. Lattice-matched $AlAs_{0.16}Sb_{0.84}$ was grown on InAs substrates by conventional MBE. The surface morphology of AlAsSb grown on InAs has slight texture and is somewhat inferior to that obtained on GaSb, which is mirror smooth. As will be discussed later, however, there is no indication that the inferior surface morphology of the cladding layer has any serious effect on the performance of InAs-based lasers. It has been reported that modulated MBE growth of AlAs and AlSb resulted in improved AlAsSb material quality [75]. More details of the modulated growth are discussed in Chapter 10.

8.4.5 Doping

Because no semi-insulating GaSb or InAs substrates are readily available, samples for Hall measurements to determine doping levels and mobilities must be grown either on semi-insulating GaAs substrates or on epitaxially grown insulating buffer layers. The large lattice mismatch (7–8%) between GaSb or InAs and semi-insulating GaAs substrates results in a large density of dislocations in the heteroepitaxial layers. However, the majority-carrier transport properties of the heteroepitaxial layers, as determined by Hall measurements, do not seem to be significantly affected by the presence of dislocations. The carrier concentration values determined by Hall measurements of GaSb layers grown on semi-insulating GaAs were in good agreement with the values obtained by capacitance–voltage (C–V) measurements on companion samples with p–n junctions grown on GaSb [76].

There is a relatively large ($> 1 \times 10^{16} \, cm^{-3}$) background acceptor concentration in high-quality unintentionally doped GaSb-based alloys. Conversely, there is a high background donor level ($\sim 1 \times 10^{16} \, cm^{-3}$) in high-quality InAs-based materials (such as InAsSb) grown by MBE. These high background levels are attributed to antisite defects or defect complexes [77]. The background concentration has been found, at least in GaSb, to be dependent on both the growth technique and growth conditions [78]. In spite of this high background acceptor concentration, high-quality n-GaSb has been demonstrated by Te doping [76], with RT mobilities as high as $7600 \, cm^2/V$-s at a carrier concentration of $1.2 \times 10^{16} \, cm^{-3}$.

The results of Be doping for p-type GaSb and AlGaAsSb are straightforward and comparable to p doping of GaAs. Table IV shows a summary of electrical measurements for heavily doped p-type AlGaAsSb grown by MBE under a variety of conditions. Values of p

TABLE IV

Electrical properties of p-Al$_x$Ga$_{1-x}$As$_y$Sb$_{1-y}$ grown under various conditions by MBE. T_{Be} is the Be cell temperature, and p_{300}, p_{77}, μ_{300}, μ_{77} are hole concentrations and mobilities at 300 and 77 K, respectively.

As Source	x	y	T_{Be}(°C)	p_{300}(cm^{-3})	μ_{300} (cm^{-2}/V-s)	p_{77}(cm^{-3})	μ_{77} (cm^{-2}/V-s)
As$_2$	0.9	0.07	1050	6.6×10^{16}	56	3.6×10^{16}	95
As$_2$	1.0	0.08	1150	1.2×10^{18}	130	8.0×10^{16}	1433
As$_2$	1.0	0.16	1050	6.6×10^{16}	227	1.5×10^{16}	342
As$_2$	1.0	0.16	1150	1.7×10^{19}	97	1.4×10^{19}	160
As$_4$	1.0	0.08	940	3.6×10^{17}	140	9.8×10^{16}	380

up to 1.7×10^{19} cm^{-3}, with mobilities of ~ 100 cm^2/V-s, have been obtained at RT.

Obtaining highly conducting n-AlGaAsSb, however, is much more complicated than the case for p-AlGaAsSb. The n-type conductivity of AlGaAsSb is very sensitive to the Al content as well as growth conditions such as growth temperature, V/III ratio, and the use of As$_4$ or As$_2$. Table V lists the results of electrical measurements for n-AlGaAsSb samples grown under various conditions. Wide variations in electron concentration and mobility illustrate the complexity of the problem. The mobilities at high electron concentrations are quite low. This problem may be partly due to DX centers, the deep-level traps which have been widely investigated in AlGaAs [79]. The presence of DX centers renders the doping efficiency of Te in n-AlGaAsSb much lower than in GaSb. As a result, higher levels of Te flux are required for AlGaAsSb than for GaSb to achieve the same

TABLE V

Electrical properties of n-Al$_x$Ga$_{1-x}$As$_y$Sb$_{1-y}$ grown under various conditions by MBE. T_{GaTe} is the GaTe cell temperature, and n_{300}, n_{77}, μ_{300}, μ_{77} are electron concentrations and mobilities at 300 and 77 K, respectively.

As Source	x	y	T_{GaTe}(°C)	n_{300}(cm^{-3})	μ_{300} (cm^{-2}/V-s)	n_{77}(cm^{-3})	μ_{77} (cm^{-2}/V-s)
As$_2$	0.75	0.06	435	7.0×10^{16}	30	3.4×10^{15}	6
As$_2$	0.9	0.07	475	3.0×10^{16}	90	4.3×10^{16}	10
As$_2$	1.0	0.08	440	5.2×10^{16}	910	3.2×10^{15}	9500
As$_2$	1.0	0.08	450	1.7×10^{17}	445	3.2×10^{16}	1975
As$_2$	1.0	0.16	425	3.2×10^{17}	144	2.4×10^{16}	248
As$_2$	1.0	0.16	450	6.9×10^{18}	9	1.8×10^{16}	210
As$_4$	1.0	0.08	435	2.6×10^{17}	75	1.9×10^{15}	380

doping concentration, increasing the impurity scattering. Fig. 15 plots the RT electron concentration measured by the Hall technique for $Al_xGa_{1-x}As_{0.03}Sb_{0.97}$ grown on semi-insulating GaAs with the same Te flux [80]. The electron concentration exhibits a W shape, with two minima at $x \sim 0.2$ and 0.6 that coincide with the Γ–L and L–X crossover points, respectively. Increasing the Al content from 0.6 to 0.8 increases the electron concentration significantly, indicating that the DX concentration is reduced as the alloy approaches the ternary composition. Such reduction in DX-trap density has also been observed in AlGaAs as the alloy approaches the binary [79]. While the maximum electron concentration in AlGaAsSb lattice matched to GaSb is $\sim 1 \times 10^{17}\,cm^{-3}$, it is much higher for alloys lattice matched to InAs. Electron concentrations up to $\sim 10^{19}\,cm^{-3}$ have been obtained for $AlAs_{0.16}Sb_{0.84}$. Compared to MBE, obtaining n-AlGaAsSb by OMVPE is even more difficult because conventional precursors such as trimethylantimony and trimethylaluminum result in too high p-doping concentrations due to C incorporation. Only very recently,

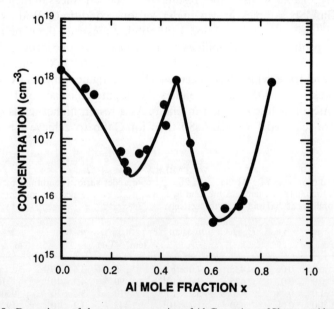

Fig. 15 Dependence of electron concentration of $Al_xGa_{1-x}As_{0.03}Sb_{0.97}$ on Al content (from Ref. [80]). The layers were grown by molecular beam epitaxy using the same Te flux. The concentration exhibits a W shape, with two minima at $x \sim 0.2$ and 0.6 that coincide with the Γ–L and L–X crossover points, respectively.

n-AlGaAsSb has been obtained by OMVPE using alternative precursors such as tritertiarybutylaluminum and triethylgallium [81,82].

The carrier concentration of AlGaAsSb decreases at low temperatures due to freezeout, especially for samples with a large density of DX centers. Fig. 16 shows the dependence of carrier concentration on temperature [83], as determined by Hall measurements, for $Al_xGa_{1-x}As_ySb_{1-y}$ grown lattice matched to GaSb but on semi-insulating GaAs using As_4. The freezeout is especially severe for $x = 0.75$, but it is much smaller for $x = 1$. This carrier freezeout can significantly increase the series resistance of mid-IR diode lasers operating at low temperatures by reducing the n-cladding conductivity. Such reduced operating temperatures are still necessary for diode lasers emitting at wavelengths longer than 3 µm.

8.4.6 Growth of GaInAsSb/AlGaAsSb QW Structures

In developing MBE materials for strained QW devices, consideration must be given to the concept of critical layer thickness [84] of the various antimonide-based alloys. There are a number of theories that predict the relationship between lattice mismatch and critical layer

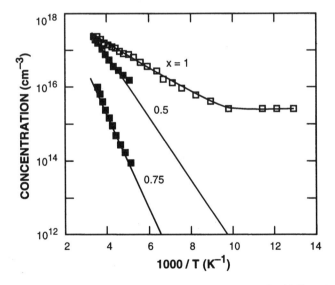

Fig. 16 Dependence of electron concentration on temperature for $Al_xGa_{1-x}As_ySb_{1-y}$ lattice matched to GaSb (from Ref. [83]). The freezeout is especially severe for $x = 0.75$, but it is much smaller for $x = 1$.

thickness, based on various assumptions on the energetics of relaxation in the materials involved. The onset of dislocation formation in lattice-mismatched semiconductors, however, is not adequately described by a sudden transition from no relaxation to complete relaxation once the critical layer thickness is exceeded. It has been found in other III–V materials that complete (100%) relaxation of lattice-mismatched epilayers does not occur until the epilayer thickness greatly exceeds the estimated critical thickness values [85].

The approach at Lincoln Laboratory is to use the theoretical predictions of critical thickness as rough estimates for the range of lattice mismatch and thicknesses that can be safely incorporated in QW active regions. The compositions of the appropriate well and barrier materials are determined by characterizing thick test layers ($> 1\,\mu$m) using DCXRD and PL with the assumption that these thick layers are fully relaxed. Such assumptions of almost complete relaxation in thick layers have been verified by careful evaluation of X-ray diffraction using asymmetric reflections, which are sensitive to both the in-plane and perpendicular lattice parameter, as well as energy shifts determined from PL data of both strained and unstrained epilayers of the same nominal composition. With the calibration of the intended compositions from the test layers, QW structures are grown. The DCXRD and PL data from the QW structure are then compared to the expected values, and further corrections are made to the growth conditions. With this iterative procedure, conditions for the optimum growth of the actual laser structure are determined.

As discussed in Section 8.3, strain in the active layer can have beneficial effects on the operating characteristics of the mid-IR diode laser, and large compressive strains are preferred in general. A balance has to be achieved, however, between the values of strain in the QWs, the number and thickness of wells and barriers, and the practical limit for incorporating such multiple strained layers into a complete structure without causing strain relaxation and subsequent dislocation formation. Our design approach has been to attempt to grow a strain-balanced active region, where the strain-thickness product in the wells is of the same value but opposite sign as the strain-thickness product of the barriers. We have grown compressively strained wells and tensile-strained barriers, with the maximum strain value of $\sim 1\%$. The number of wells and barriers has been chosen to balance optical gain, confinement factor, and threshold current requirements.

To illustrate the effect of strain relaxation on the analysis of QW laser structures, Fig. 17 plots the simulated X-ray diffraction patterns

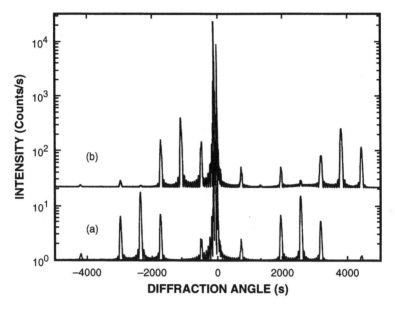

Fig. 17 Simulated X-ray diffraction patterns for two hypothetical multiple quantum-well (QW) structures. In structure (a), both the wells and barriers are strained, and in structure (b) the barriers are under strain, but the wells are relaxed. For structure (a), the maximum of the envelope associated with the satellite peaks for the wells falls at approximately twice the value for the fully relaxed layer of the same composition. For structure (b), the maximum of the envelope associated with the wells moves closer to the substrate peak, while the envelope maximum for the barriers shifts further away from the substrate peak because the effective "substrate" lattice parameter has changed due to the relaxation of the wells.

of two hypothetical structures. In structure (a), both the wells and barriers are fully strained (0% relaxation). In this case, the maximum of the envelope associated with the satellite peaks of the wells falls at approximately twice the value for the fully relaxed layer of the same composition. In structure (b), only the wells are allowed to relax. Then the maximum of the envelope associated with the wells moves to a value closer to that of the substrate, while the maximum of the envelope for the barriers shifts further away from the substrate peak, implying that as the wells relax the barriers are further strained because the effective "substrate" lattice parameter has changed. Understanding such interactions is essential in interpreting DCXRD data from the QW laser structures.

While the growth of the AlGaAsSb barrier materials was found to be straightforward, the exact shutter sequence used for the growth had a major impact on the quality of the QW structures [69]. Because the $Al_{0.25}Ga_{0.75}As_{0.02}Sb_{0.98}$ barrier contained only a small amount of As, it was initially believed that it could be left under an Sb_4 flux only (the As shutter closed) while the As valve position was changed to an appropriate setting for the growth of the GaInAsSb well. However, it was found that both the structural quality of the QW and the measured laser performance were significantly improved if the As shutter was left open throughout the growth of the entire QW region. A representative diagram of the altered shutter sequence is illustrated in Fig. 18. There is a delay time t_D at each interface while the As valve position is adjusted. Two five-well QW laser structures with the active layers under a compressive strain of $\sim 5 \times 10^{-3}$ were grown using the original (the As shutter closed) and altered (the As shutter open) sequences, respectively. Fig. 19 compares DCXRD scans of the laser structures grown with the two shutter sequences. While the satellite peaks are substantially broadened for the structure grown with the original shutter sequence, well-resolved satellite peaks can be observed in the structure grown with the altered sequence, indicating higher structural perfection. As will be discussed in Section 8.5, diode lasers fabricated from a wafer grown using the new shutter sequences showed a dramatic performance improvement.

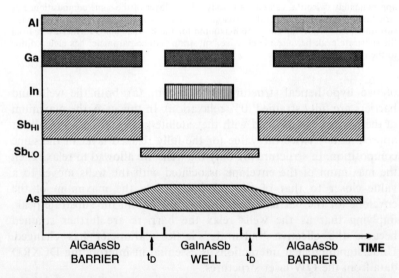

Fig. 18 Schematic diagram of the improved shutter sequence. There is a delay time t_D at each interface during which the As valve position is adjusted.

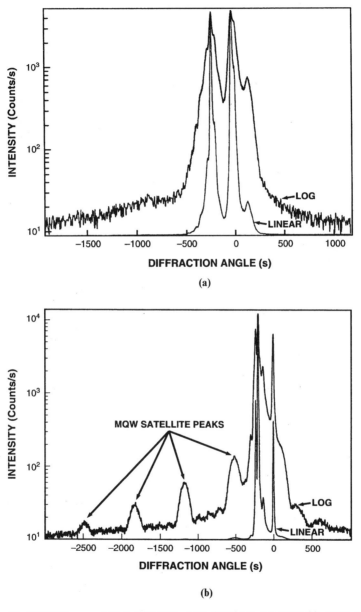

Fig. 19 DCXRD scan of GaInAsSb/AlGaAsSb QW laser structures (a) grown with original shutter sequence, and (b) with improved shutter sequence. The improved shutter sequence resulted in a much better structural quality as shown by the sharp satellite peaks.

8.4.7 Growth of InAsSb/InAlAsSb QW Structures

To extend the operating wavelength of GaInAsSb-based QW lasers beyond 3 μm required changes not only in the alloy compositions of GaInAsSb active layers, but also in the barrier material because of the valence-band offset problem described earlier. A major concern for the choice of the active layer material was the stability of the alloy at the growth temperature. Although the metastable GaInAsSb alloy grown by MBE has been used for DH lasers [20], it showed signs of alloy decomposition while the upper cladding layer was grown. As a result, InAsSb, which is farthest from the miscibility gap region, has been chosen for the active layer. For the barrier, InAlAsSb has been used because it can provide adequate offsets for both the conduction and valence bands.

Although InAsSb is not close to a region of miscibility, there have been reports of the observation of CuPt-type ordering in this alloy grown by OMVPE [86] and MBE [87]. For $InAs_{1-x}Sb_x$ ($0 \leqslant x \leqslant 0.2$) that we have grown on GaSb by MBE in the temperature range 430–470°C, careful low-temperature PL measurements of the compositional dependence of E_g rule out the possibility of CuPt-type ordering [88], although our measured E_g is slightly smaller than previously reported values. Details of this bandgap discrepancy need further investigation.

In growing the InAsSb/InAlAsSb QW structures, the most difficult problem is the stability of the InAlAsSb barrier material. As was shown in Fig. 9, the region of stable growth for InAlAsSb is limited to Al content less than 6% on GaSb, while it is up to 12% on InAs. As a result, the growth of InAsSb/InAlAsSb QW structures was somewhat easier on InAs than on GaSb. However, the longest emission wavelength is shorter for lasers grown on InAs than on GaSb.

Very high quality InAsSb/InAlAsSb QWs could be grown on InAs at 430°C. Fig. 20 shows a DCXRD scan of a ten-well/eleven-barrier test structure, along with the simulated curve of the expected X-ray diffraction spectra. The close agreement between the observed data and the theoretical curve indicates excellent structural perfection and abrupt interfaces for this QW structure. Low-temperature PL of the sample showed very strong emission with an FWHM of 6 meV.

For the QW laser structures on InAs, an $AlAs_{0.16}Sb_{0.84}$ cladding layer was selected instead of AlGaAsSb because $AlAs_{0.16}Sb_{0.84}$ provides the least carrier freezeout at low temperatures, and the largest refractive index step and the maximum valence-band offset relative to

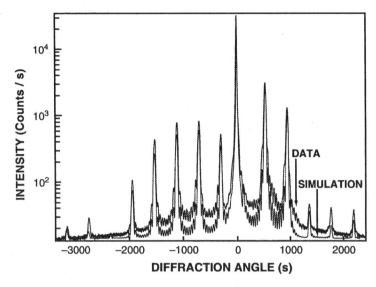

Fig. 20 DCXRD scan and simulation of InAsSb/InAlAsSb QW structure grown on InAs. The sharpness of the satellite peaks indicates excellent structural quality with abrupt interfaces.

the active region. However, AlAsSb is much more susceptible to oxidation in atmosphere during device processing. For good optical quality, the AlAsSb upper cladding layer needs to be grown at temperatures higher than 430°C used for the growth of the active region. The growth temperature of the upper cladding layer was found to have a large effect on the underlying InAsSb/InAlAsSb QWs. Fig. 21 shows the DCXRD scans of samples from a QW wafer after they were subjected to different furnace annealing temperatures for 3 h under an excess As environment. For annealing at temperatures above $\sim 510°C$, degradation in the DCXRD pattern was observed, indicating some intermixing or roughening of interfaces. Similar annealing behavior has been observed in AlSb/GaSb superlattices [89].

As in the growth of the 2-µm QW lasers, the growth of InAsSb/InAlAsSb QW lasers on InAs also requires optimization of the shutter sequence and delay times. Compared with the growth of GaInAsSb/AlGaAsSb QW lasers under continuous Sb-dominated flux conditions, however, the growth of InAsSb/InAlAsSb QW lasers is even more complicated because they require additional growth transitions from Sb-dominated cladding growth to As-dominated well/barrier growth, and then back to Sb-dominated upper cladding growth. The exact sequence

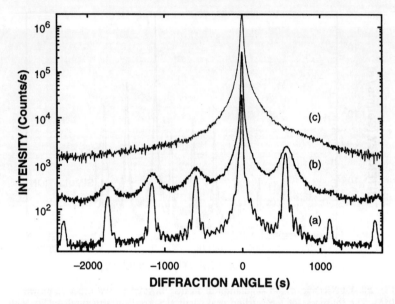

Fig. 21 DCXRD scans of the QW structure of Fig. 21. (a) the control sample, (b) after annealing at 510°C for 3 h, and (c) after annealing at 530°C for 3 h. The satellite peaks broadened after 510°C annealing, and they are completely smeared out after 530°C annealing. The scans (b) and (c) are shifted up by one and two units, respectively.

of these group-V growth transitions, the shutter sequence, the delay times used for the As-dominated well/barrier growth, and the temperature of the upper cladding growth were all found to have a significant impact on the structural quality of the QW lasers. Fig. 22 shows a comparison of DCXRD scans for three different QW laser structures grown with various combinations of shutter sequences and temperatures. The performance of these lasers will be described in Section 8.5.

Growth of InAsSb/InAlAsSb QW lasers on GaSb substrates was much more difficult because of the proximity of InAlAsSb to the miscibility gap. With $In_{0.85}Al_{0.15}As_{0.9}Sb_{0.1}$ barriers, which have given our best laser results to date, we have observed some problems with alloy stability after upper cladding growth. Fig. 23 plots the DCXRD scan of the laser structure with ten $InAs_{0.85}Sb_{0.15}$ wells and eleven $In_{0.85}Al_{0.15}As_{0.9}Sb_{0.1}$ barriers. The broadening of the satellite peaks indicates that there was roughening or intermixing at the interface. Attempts to increase the Al content beyond 0.15 for better confinement led to partial relaxation in the wells and barriers, resulting in formation of misfit dislocations.

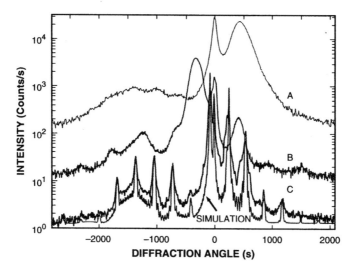

Fig. 22 DCXRD comparison of QW laser structures grown on InAs with different growth conditions. Wafer A has broadened satellite peaks in the X-ray pattern, wafer B shows an intermediate-quality pattern, and wafer C exhibits the best pattern. Scans B and A are shifted up by one and two units, respectively.

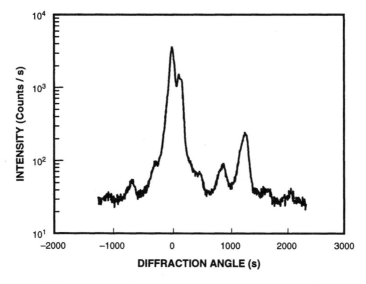

Fig. 23 DCXRD scan of a QW structure with ten $InAs_{0.85}Sb_{0.15}$ wells and eleven $In_{0.15}Al_{0.85}As_{0.9}Sb_{0.1}$ barriers grown on GaSb. The broadening of the satellite peaks indicates that there was roughening or intermixing at the interface.

8.5 LASER PERFORMANCE

8.5.1 GaInAsSb/AlGaAsSb QW Lasers Emitting in the 2-μm Region

The first GaInAsSb/AlGaAsSb QW lasers were reported in 1992 with an emission wavelength of $\sim 2.1\,\mu m$ [22]. The laser structure consisted of the following layers: n^+-GaSb buffer, 2-μm n-$Al_{0.9}Ga_{0.9}As_{0.07}Sb_{0.93}$ cladding, active region consisting of five 10-nm $Ga_{1-x}In_xAs_ySb_{1-y}$ wells and six 20-nm $Al_{0.2}Ga_{0.8}As_{0.02}Sb_{0.98}$ barriers, 2-μm p-$Al_{0.9}Ga_{0.1}As_{0.07}Sb_{0.93}$ cladding, and 0.05-μm p^+-GaSb contacting. All the layers were grown lattice matched to the GaSb substrate, except for the active wells, which were under slight compressive strain. Broad-stripe lasers exhibited RT pulsed J_{th} as low as $260\,A/cm^2$ and the characteristic temperature T_0 up to 113 K. Values of η_d as high as 70% and internal quantum efficiency η_i of 87% were obtained. The maximum pulsed operating temperature was 155°C, which was limited by melting of In used for packaging. Under cw operation, output power of 190 mW/facet was obtained from a 100-μm aperture. Subsequently, the series resistance was reduced by introducing linearly graded AlGaAsSb layers between the substrate and n-AlGaAsSb cladding and between p-cladding and p^+-cap layers. With high- and low-reflection coatings at the back and front facets, respectively, single-ended cw output power of 600 mW was obtained from a 300-μm aperture. The maximum power conversion efficiency was 9%. The performance did not change appreciably between 1.9 and 2.2 μm.

Much higher performance was obtained [90] by optimizing the shutter sequence to get smoother interfaces as described earlier. The laser structure was similar to the one described above, except that the alloy compositions in the well/barrier/cladding were slightly different. Based on the growth parameters and the amount of strain, the best estimate for the active layer composition is $x = 0.86$ and $y = 0.05$. The valence-band offset between the active and barrier layers for the above laser structure was calculated to be $\sim 90\,meV$, while the conduction-band offset was $\sim 350\,meV$. Fig. 24 shows the RT pulsed J_{th} vs inverse cavity length L^{-1} of the improved lasers emitting at 1.9 μm as well as the previous lasers emitting at 2.1 μm. For $L = 2\,mm$, $J_{th} = 143\,A/cm^2$. The value of J_{th} per well is only $29\,A/cm^2$, which is about half the lowest value obtained for InGaAs/AlGaAs lasers emitting at $\sim 1\,\mu m$. As L becomes smaller, the value of J_{th} increases

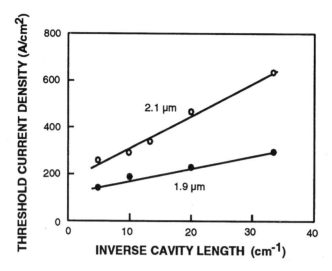

Fig. 24 Room-temperature pulsed threshold current density J_{th} vs inverse cavity length L^{-1} of the improved GaInAsSb/AlGaAsSb QW lasers emitting at 1.9 μm as well as the previous QW lasers emitting at 2.1 μm.

gradually, to a value of 280 A/cm² for 300 μm. For comparison, J_{th} for the earlier QW lasers increased much more rapidly as L was decreased, indicating that their gain coefficient is much smaller.

The band structure and gain of the GaInAsSb QW lasers were calculated using a procedure described in Ref. [52]. The valence subband structure for the laser described above is shown in Fig. 25. There are three confined heavy-hole bands and one confined light-hole band. The splitting between the light-hole band and the highest heavy-hole band is ~ 60 meV. The effective mass of the uppermost band near the zone center is $0.08 \, m_0$. The conduction band has two confined states because of a large potential barrier. Fig. 26 shows the gain vs energy at several carrier densities. The transparency carrier density is $1.1 \times 10^{18} \, cm^{-3}$. Fig. 27 plots the gain as a function of current density as well as a few experimental data points assuming $\alpha_i = 10 \, cm^{-1}$. The calculated J_{th} is much smaller than the experimental values. There are substantial differences in both the transparency current density and the gain coefficient. The experimental data can be fit reasonably well by assuming $r_A = 7 \times 10^{-29} \, cm^6 s^{-1}$, which is close to the value for GaSb [91]. The actual value of r_A would be smaller because the carrier injection efficiency is less than 1.

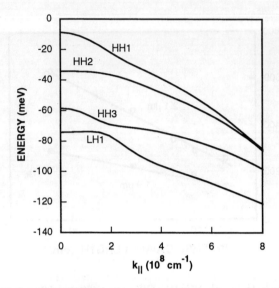

Fig. 25 Calculated valence-subband structure of $Ga_{0.86}In_{0.14}As_{0.05}Sb_{0.95}/Al_{0.25}Ga_{0.75}As_{0.02}Sb_{0.98}$ QW laser. Thicknesses of the active and barrier layers are 10 and 20 nm, respectively.

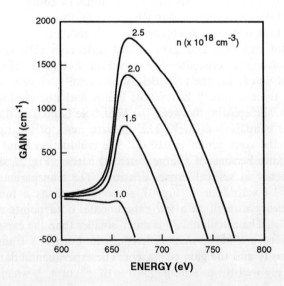

Fig. 26 Calculated gain vs energy at several carrier densities as a function of energy for the laser structure of Fig. 25.

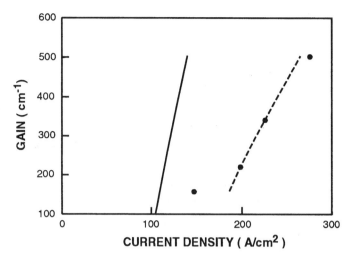

Fig. 27 Calculated gain as a function of current density for $Ga_{0.86}In_{0.14}As_{0.05}Sb_{0.95}/Al_{0.25}Ga_{0.75}As_{0.02}Sb_{0.98}$ QW laser is plotted in solid curve. A few experimental data points assuming the internal loss coefficient $\alpha_i = 10\,\text{cm}^{-1}$ are also plotted. The experimental data can be fit reasonably well (dashed line) by assuming the Auger coefficient $r_A = 7 \times 10^{-29}\,\text{cm}^6\,\text{s}^{-1}$.

Fig. 28 shows the cw output power and voltage vs current curves for a 300-μm-wide by 1000-μm-long device at a heatsink temperature of 12°C. The front and back facets were coated to have reflectivities of 4 and 95%, respectively. The threshold current is $\sim 650\,\text{mA}$ and the initial slope efficiency is $\sim 0.3\,\text{W/A}$, corresponding to η_d of 47%. The maximum output power is 1.3 W. The turn-on voltage is $\sim 0.6\,\text{V}$, which is almost the same as E_g of the active layer. The series resistance is initially $\sim 0.3\,\Omega$, and decreases to $\sim 0.1\,\Omega$ at high current levels. Even though the interfaces at the cladding/substrate and cladding/cap were graded, the superlinear current vs voltage curve indicates that there still is a potential barrier, most likely at the n-cladding/barrier interface because of a large conduction-band offset.

The power conversion efficiency η, defined as the ratio of optical output power and electrical input power, was as high as 15.5%. Although this is a very respectable number, it is still much smaller than $> 50\%$ obtained for GaAs-based lasers. The power efficiency can be approximately expressed by the following equation:

$$\eta = \eta_d\,(E/e)(I - I_{th})/I(V_0 + IR_s), \tag{13}$$

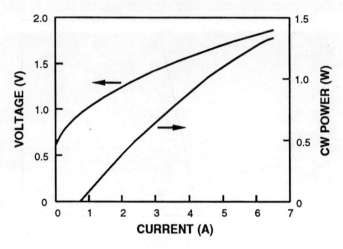

Fig. 28 CW output power and voltage vs current curve for a 1000-μm-long by 300-μm-wide GaInAsSb/AlGaAsSb QW laser at a heatsink temperature of 12°C.

where E is the photon energy, I_{th} is threshold current, V_0 is turn-on voltage and R_s is series resistance. It is clear from Eq. (13) that η becomes smaller as E decreases unless V_0 and R_s scale with E. Typically V_0 scales with E if there is no big potential barrier. However, R_s normally increases with wavelength because thicker cladding layers are required. In addition, the sheet resistance of AlGaAsSb is higher than that of GaAlAs or InP because of its inferior carrier mobility. As a result, η would decrease at longer wavelengths.

Reliability of GaInAsSb/AlGaAsSb lasers is expected to be good because of the small photon energy and the use of quaternary active and cladding layers with very different atomic sizes, which is believed to slow dislocation propagation [92]. A preliminary reliability test is very encouraging. One device operating at 700 mW did not show any degradation for 500 h, the maximum duration of the test. All other devices that operated for shorter periods exhibited stable operation. Because of the very high Al content in the cladding, the facets have to be coated immediately after cleaving to prevent oxidation and degradation. Near RT, the output power of the coated devices is limited by the junction-temperature rise rather than the catastrophic facet damage.

The emission wavelength at RT was extended to 2.78 μm by Lee et al. [93] by increasing the In and Sb content in the GaInAsSb active layers. The laser structure incorporated a QW active region consisting of four 10-nm $Ga_{0.76}In_{0.24}As_{0.16}Sb_{0.84}$ wells and five 18-nm

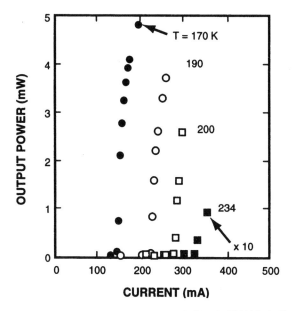

Fig. 29 CW output power vs current curves of GaInAsSb/AlGaAsSb QW laser emitting at ~2.7 μm at several temperatures (from Ref. [94]).

$Al_{0.25}Ga_{0.75}As_{0.02}Sb_{0.98}$ barriers. Devices with 22-μm stripes were fabricated, and 90% and 30% reflection coatings were deposited on the back and front facets, respectively. Under pulsed operation, J_{th} and T_0 at RT were ~10 kA/cm² and 58 K, respectively. At 15°C, the maximum output power was 90 mW, with η_d of 9%. At the maximum operating temperature of 60°C, the emission wavelength was 2.9 μm. Fig. 29 shows the cw output power vs current curves at several temperatures. The maximum cw operating temperature was 234 K (−39°C), where the emission wavelength was 2.7 μm [94]. A maximum power of 5 mW was obtained at 170 K. The laser operated predominantly in a single longitudinal mode under certain operating conditions, and it was used for very sensitive detection of water vapor at ~2.6 μm [95], where its fundamental absorption lines are located.

Both the spontaneous and differential efficiency of 2.7-μm lasers decreased rapidly with temperature. Garbuzov et al. [94] studied this decrease by observing the spontaneous emission above threshold at 190 K. As the drive current was increased from 220 mA (I_{th} = 215 mA) to 270 mA, the intensity and peak position of the spontaneous emission remained constant, even though the laser output power increased by 1000 fold. This implies that above threshold the quasi-Fermi level was pinned, and all the carriers injected into the lasing state underwent stimulated emission. Therefore, the decrease in differential efficiency was explained by the inefficient injection of carriers into the lasing state.

For low threshold and good beam quality, ridge-waveguide (RW) lasers were fabricated by Choi et al. [96]. Ridges were formed by reactive ion etching in a BCl_3/Ar plasma. The 8-μm-wide RW lasers emitting at $\sim 2.1\,\mu m$ exhibited pulsed threshold current as low as 29 mA, and the maximum cw power was 28 mW at RT [96]. These lasers exhibited predominantly a single longitudinal mode at $\sim 2.13\,\mu m$, which could be continuously tuned without mode hopping over 1.2 nm by changing the heatsink temperature from 18 to 24°C or over 0.8 nm by changing the current. Such extended tuning without mode hopping has also been observed for GaAs/AlGaAs lasers and was attributed to spectral hole burning [97].

Much higher performance was obtained from 5-μm-wide RW lasers fabricated in the improved wafer with emission at $\sim 1.9\,\mu m$ [98]. For uncoated devices with $L = 500\,\mu m$, the RT threshold current was ~ 20 mA. One 1000-μm-long device with high-reflection and antireflection coatings on the back and front facets, respectively, exhibited maximum cw power of 100 mW at RT. The threshold current was 40 mA, and the initial slope efficiency was 0.25 W/A corresponding to η_d of 39%. The lateral far-field pattern maintained a single mode up to the maximum output power. Fig. 30 shows cw power vs current curves at several heatsink temperatures of a 1000-μm-long device with high-reflection and passivation coatings on the back and front facets, respectively. The maximum cw operating temperature is 130°C, which is 100°C higher than that for GaInAsSb/AlGaAsSb DH lasers [18]. For temperatures between 20 and 80°C, T_0 is 85 K. The slope efficiency does not change appreciably between 20 and 60°C, but becomes smaller at higher temperatures.

For single-frequency operation, York et al. [99] fabricated distributed Bragg-reflector (DBR) lasers. The structure of the RW DBR laser is shown in Fig. 31. For the gain section, 5-μm-wide ridges were defined by reactive ion etching. For the grating section, first-order gratings ($\Lambda \sim 0.3\,\mu m$) as well as the ridges were formed by reactive ion beam etching. The lasers operated pulsed up to 37°C, with T_0 of 96 K. The light output vs current at 15°C is shown in Fig. 32. The threshold current is ~ 250 mA, and η_d is $\sim 8\%$. The high threshold current is partly due to a relatively high J_{th} (1.5 kA/cm^2) of the wafer. The sudden

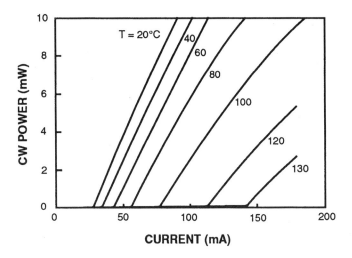

Fig. 30 CW output power vs current curves at several heatsink temperatures of a GaInAsSb/AlGaAsSb ridge-waveguide (RW) laser with high-reflection and passivation coatings on the back and front facets, respectively.

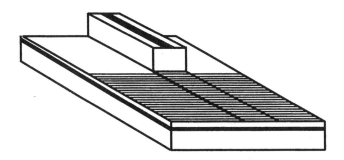

Fig. 31 Schematic structure of the RW distributed Bragg-reflector (DBR) laser (from Ref. [99]). The grating section is unpumped.

jump in the output power at threshold is caused by the loss in the unpumped grating section. The lasers operated in a single longitudinal mode, which could be continuously tuned between 1.966 and 1.972 μm at a rate of 0.16 nm/K.

For high output power in a diffraction-limited beam, tapered lasers were fabricated in GaInAsSb/AlGaAsSb QW structures [100]. The schematic of the tapered structure is shown in Fig. 33. The gain region is defined by a linearly tapered metallized contact area whose angular

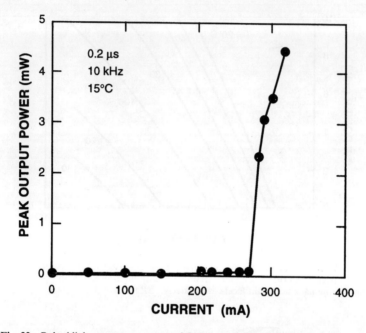

Fig. 32 Pulsed light output vs current of GaInAsSb/AlGaAsSb RW DBR laser at 15°C (from Ref. [99]). The sudden jump in the output power at threshold is caused by the loss in the unpumped grating section.

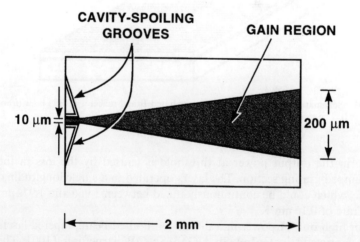

Fig. 33 Schematic structure of the tapered laser. The gain region is defined by a linearly tapered metallized contact area whose angular width is chosen to match the diffraction angle of the beam from the narrow section.

width is chosen to match the diffraction angle of the beam from the narrow section. The low reflectivity of the output facet and cavity-spoiling grooves at the narrow end suppress the Fabry-Perot oscillation. For InGaAs/AlGaAs tapered lasers emitting at $\sim 1\,\mu m$, cw output power more than 1 W has been obtained in a near-diffraction-limited beam [101].

The tapered lasers were fabricated from a wafer with emission wavelength of $2.03\,\mu m$ and pulsed J_{th} of $330\,A/cm^2$. The facet near the etched grooves was coated to $\sim 95\%$ reflectivity, and the output facet to $\sim 2\%$ reflectivity. Fig. 34 shows the cw output power vs current at a heatsink temperature of 12°C. The threshold current is $\sim 0.7\,A$ and the initial η_d is 23%. The maximum output power is 750 mW. The far-field pattern along the junction plane is shown in Fig. 34 at three current levels. A near-diffraction-limited pattern is observed for current up to 1.5 A, where the output power is 120 mW. The FWHM at 1.5 A is 0.55°, which is ~ 1.05 times diffraction limited assuming a uniform 200-μm aperture. As the current is increased further, the side-lobes begin to grow more rapidly than the central lobe. The central-lobe power saturates at ~ 200 mW, as shown by the dots in Fig. 35. The degradation of the far-field pattern at high power levels may be attributed to filament formation [102].

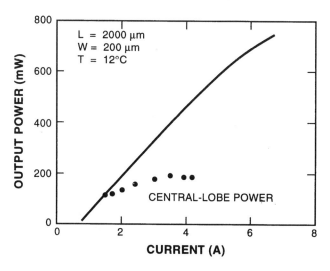

Fig. 34 CW output power vs current of a GaInAsSb/AlGaAsSb tapered laser at a heatsink temperature of 12°C. The dots represent the central-lobe power.

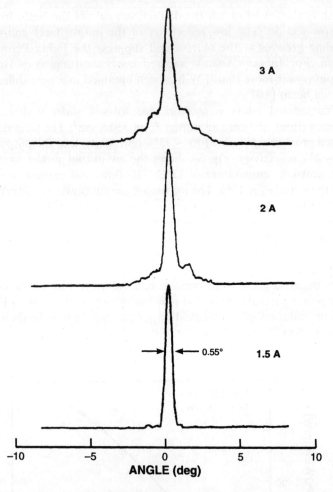

Fig. 35 Far-field pattern of the tapered laser along the junction plane at three current levels. The degradation of the far-field pattern at high power levels may be attributed to filament formation.

8.5.2 QW Lasers Emitting Beyond 3 μm

For QW lasers with emission beyond 3 μm, InAsSb has been used for the active layer because it is stable at the growth temperature and is less complicated to grow than GaInAsSb that lies closer to the miscibility gap. For the barrier, either InGaAs, InAlAs, or InAlAsSb has been utilized. Lasers with the InGaAs barriers were reported by Kurtz *et al.*

[23]. The structure, grown on InAs substrates by OMVPE, had a 2-μm $InP_{0.65}Sb_{0.35}$ cladding, a 1.5-μm active region consisting of 95-Å $InAs_{0.9}Sb_{0.1}$ wells and 133-Å $In_{0.93}Ga_{0.07}As$ barriers, followed by a 100-Å $InP_{0.65}Sb_{0.35}$ cap layer. Under optical pumping, the lasers operated pulsed up to ~ 100 K with emission at ~ 3.9 μm. Although the InGaAs barrier provides an adequate valence-band offset, the conduction-band offset is too small to be effective [103].

The InAlAs barrier can provide a larger conduction-band offset than InGaAs. The conduction-band offset increases rapidly with Al content, but the strain becomes excessive for high Al contents. QW lasers with the InAlAs barriers were reported by Choi et al. [24]. The structure, grown on a GaSb substrate by MBE, had an active region consisting of fifteen 10-nm $InAs_{0.8}Sb_{0.2}$ wells and sixteen 20-nm $In_{0.95}Al_{0.05}As$ barriers, 3-μm cladding layers of $AlAs_{0.08}Sb_{0.92}$, and a GaSb cap. Pulsed operation up to 85 K was obtained at 4.5 μm. The poor performance was attributed to nonuniform carrier distribution in the QW region, which results from a relatively large valence-band offset (85 meV) and a small conduction-band offset (~ 25 meV) due to a small Al content in the barrier layers. Because the thermal energy is only 7 meV at 80 K, holes tend to be populated preferentially in the first few wells. Electrons, with small barriers, are then easily attracted to the wells where holes are populated. This nonuniform distribution results in gains from only a few QWs and losses from the other wells. By optical pumping, however, much more uniform carrier distribution is obtained, yielding a substantially higher performance from the same structure. Pulsed operation was observed up to 144 K, with a maximum peak power of 0.27 W/facet at 80 K.

The barrier material that has provided the best laser performance is InAlAsSb [25]. Fig. 36 shows the calculated band offsets as a function of Al content in InAlAsSb with respect to $InAs_{0.9}Sb_{0.1}$ on InAs. The effects of strain are also taken into account. For InAlAsSb, the tensile strain is fixed at 3.5×10^{-3}. The valence-band offset increases very gradually with the Al content, while the conduction-band offset increases rapidly. As discussed earlier, the region of stable growth for InAlAsSb is larger on InAs than on GaSb. However, QW lasers grown on InAs substrates do not reach as long wavelengths as those grown on GaSb.

8.5.2.1 InAsSb/InAlAsSb QW lasers on InAs

Three laser structures summarized in Table VI were grown on InAs. No graded layers were incorporated in these laser structures. Because

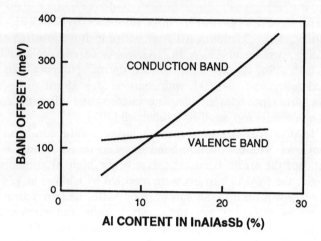

Fig. 36 Calculated band offsets as a function of Al content in InAlAsSb with respect to InAs$_{0.9}$Sb$_{0.1}$ on InAs. The tensile strain in InAlAsSb is fixed at 3.5×10^{-3}.

TABLE VI
Detailed structures of three InAsSb/InAlAsSb/AlAsSb QW lasers grown on InAs substrates.

	InAs$_{0.935}$Sb$_{0.065}$ Well		In$_{0.85}$Al$_{0.15}$As$_{0.9}$Sb$_{0.1}$ Barrier		AlAs$_{0.16}$Sb$_{0.84}$ Cladding
	Thickness (nm)	Number	Thickness (nm)	Number	Thickness, n or p (μm)
Wafer A	15	10	30	11	2.0
Wafer B	10	10	20	11	2.5
Wafer C	20	8	40	9	2.0

different growth conditions were used for the three wafers, their interfacial qualities were not the same. As shown in Fig. 22, wafer A had broadened satellite peaks in the X-ray pattern, wafer B had an intermediate-quality pattern, and wafer C had the best pattern.

Fig. 37 plots pulsed J_{th} as a function of temperature for 1000-μm-long devices fabricated from the three wafers. At 80 K, the values of J_{th} are 60, 50, and 30 A/cm^2 for lasers A, B, and C, respectively. At this temperature, the contribution of Auger recombination to J_{th} may be small. The main difference in J_{th} for the three structures may be due to interfacial recombination current. A very high T_0 (> 50 K) for laser A below 100 K may be attributed to the slowly increasing interfacial

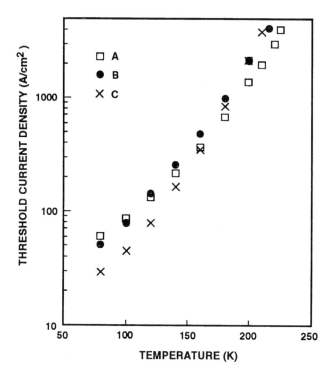

Fig. 37 Pulsed J_{th} vs temperature of three 1000-μm-long InAsSb/InAlAsSb QW lasers on InAs. See Table III for details of the laser structures.

recombination current. Between 100 and 200 K, T_0 is 30–40 K for laser A, which is typical for Auger-dominated devices. For lasers B and C, T_0 is slightly smaller than that for laser A at the same temperature. Even though laser A has the highest J_{th} at 80 K, its maximum operating temperatures is 225 K, which is higher than 215 and 210 K for lasers B and C, respectively. It is possible that the Auger rates for the 15-nm wells are smaller than those for the 10- and 20-nm wells.

Calculations were made for structure A using the procedure of Ref. [52]. The conduction- and valence-band offsets for this structure were 195 and 65 meV, respectively. Fig. 38 shows the valence-subband structure. The compressive strain splits the heavy- and light-hole bands, resulting in the highest subband with a small in-plane effective mass ($m = 0.063 m_0$) over a range of ~15 meV. However, the top heavy-hole band is separated from the light-hole band by only ~35 meV because of a relatively small strain (−0.45%) and a large thickness (15 nm) of

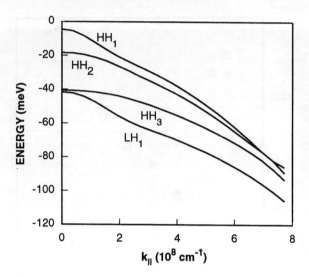

Fig. 38 Valence-subband structure of laser A.

the well. The density of states for the conduction band is much smaller than that of the valence band. As the carrier density is increased, the electron quasi-Fermi level rises rapidly above the quantized level, while the hole quasi-Fermi level remains very close to the band edge.

The gain vs energy at 80 K for several carrier densities is shown in Fig. 39. As a result of the very different density of states for the conduction and valence bands, there is a strong blue shift of the gain peak as the carrier density is increased. The gain-peak energy at transparency is about 25 meV higher than the value calculated from the bandgap and the quantized energy levels. Fig. 40 shows the gain per well as a function of carrier density and radiative current density at 80, 150, and 200 K. For structure A, the confinement factor is $\sim 15\%$. By assuming $\alpha_i = 10\,\text{cm}^{-1}$, g_{th} per well is $\sim 150\,\text{cm}^{-1}$ for $L = 1000\,\mu\text{m}$. The values of n_{th} at 80, 150, and 200 K are 1.15, 2.3, and $3.4 \times 10^{17}\,\text{cm}^{-3}$, respectively. These values are comparable to the ones calculated for GaInSb/InAs broken-gap superlattice structures [104], which have 0.91 and $2.1 \times 10^{17}\,\text{cm}^{-3}$ for a gain of $50\,\text{cm}^{-1}$ at 80 and 150 K, respectively.

The radiative J_{th} at 80, 150, and 200 K are 25, 30, and $35\,\text{A/cm}^2$, respectively. It is clear that the radiative component is a very small fraction of J_{th} at high temperatures. Assuming that the difference between the experimental and the radiative J_{th} is the Auger recombination current density, we can estimate an upper limit for r_A from Eq. (9).

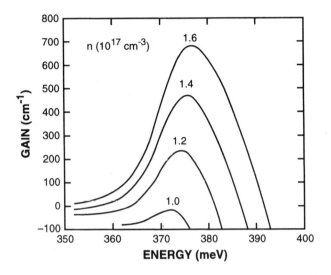

Fig. 39 Gain vs energy of laser A at 80 K for several carrier densities.

The values of r_A at 150 and 200 K are calculated to be 0.8 and 1.3×10^{-26} cm^6 s^{-1}, respectively. These values are in line with values reported for InAs bulk [105] or InAsSb/InAlAsSb QW structures [106], but the actual values may be substantially smaller because the injection efficiency can be much less than 1.

The emission wavelengths at 80 K for lasers A, B, and C were 3.32, 3.21, and 3.38 μm, respectively. This difference results from the difference in quantum confinement. At low temperatures, the wavelength shifted to longer values at a rate of ~ 1 nm/K, but the rate gradually increased at higher temperatures and reached ~ 2 nm/K at 220 K. This increased rate follows the change in E_g, which decreases more rapidly at higher temperatures. For devices with small values of T_0, however, the rate decreased with temperatures because of strong bandfilling [21].

RW lasers 8 μm wide were fabricated from wafer A. Fig. 41 shows the cw power vs current curves for a 1000-μm-long device at several temperatures. At 100 K, the cw threshold current is 12 mA. The initial η_d is ~ 30% from both facets, which is slightly smaller than ~ 35% obtained for the broad-stripe lasers. The maximum cw operating temperature is 175 K, where the threshold current is 102 mA. In general, the emission spectrum showed multilongitudinal modes. The emission wavelength at 175 K was 3.48 μm. The lateral far-field pattern showed a single lobe with a FWHM of 24°.

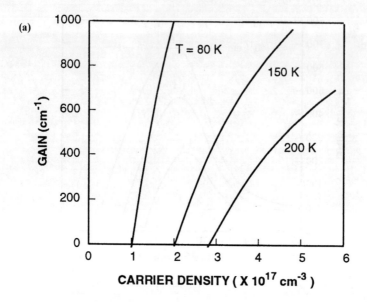

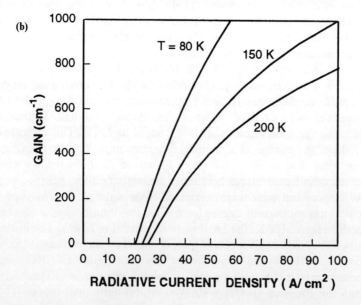

Fig. 40 Gain per well of laser A at 80, 150, and 200 K as a function of (a) carrier density and (b) radiative current density.

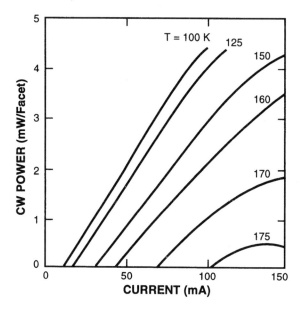

Fig. 41 CW power vs current curves at several temperatures for a 1000-μm-long RW laser fabricated from wafer A.

Broad-stripe lasers 250-μm wide were fabricated using SiO_2 patterning from another wafer with a similar structure as wafer C. Values of initial η_d at 80 K for $L = 500$, 1000, and 1500 μm were 45, 35, and 28%, respectively. From these values, η_i and α_i were estimated to be 63% and 9 cm^{-1}, respectively. The differential efficiency remained more or less constant up to 140 K, and then decreased gradually. Even at 200 K, however, the efficiency was $\sim 1/3$ of the low-temperature values. In contrast, the efficiency of the DH lasers decreased much more rapidly with temperature [107]. Fig. 42 shows the cw output power vs current of a 1500-μm-long device at several temperatures. At 80 K, the threshold current is 150 mA, and the maximum cw power is 215 mW/facet limited by junction heating. The maximum output power decreases with temperature, and at 150 K it is ~ 35 mW/facet. The operating voltage for the maximum power was ~ 4 V at 80 K, which is more than 10 times the photon energy. Substantially higher output power is expected if the operating voltage can be reduced.

The high operating voltage is caused by large barriers at the substrate/n-cladding as well as n-cladding/barrier interfaces. The schematic band diagram of InAs/AlAsSb/InAlAsSb under the flat-band

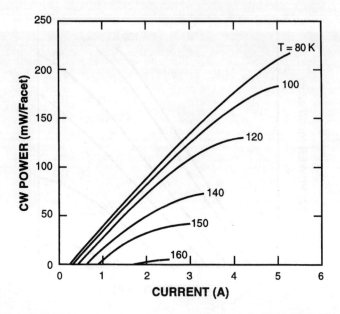

Fig. 42 CW power vs current curves for a 250 × 1500-μm InAsSb/InAlAsSb/InAs QW laser at several temperatures.

condition is illustrated in Fig. 43. The conduction-band offsets for InAs/AlAsSb and AlAsSb/InAlAsSb are 1.2 and 1.0 V, respectively. The actual barrier heights, which depend on doping concentrations in the layers, are smaller than these values, but still very large. Such large barriers substantially increase the turn-on voltage and series resistance. Fig. 44 shows the current vs voltage for the laser structure at 300, 200, and 100 K. The turn-on voltage increases as the temperature is lowered. However, the series resistance does not change with temperature, indicating that the carrier freezeout in AlAsSb is not significant at 100 K. It should be possible to reduce the turn-on voltage and series resistance by grading the interfaces. Incorporating compositionally graded AlGaAsSb on InAs, however, is not as straightforward as on GaSb because some AlGaAsSb lattice matched to InAs is inside the miscibility gap region.

8.5.2.2 *InAsSb/InAlAsSb QW lasers on GaSb*

For wavelengths longer than obtained by InAs-based lasers, the following laser structure was grown on GaSb by MBE [24]: 1-μm

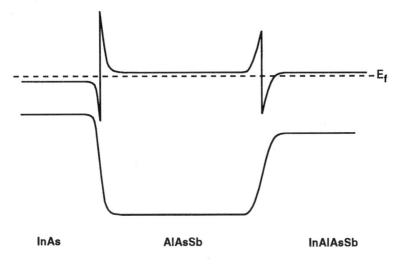

Fig. 43 Schematic band diagram for InAs/AlAsSb/InAlAsSb under flat-band condition.

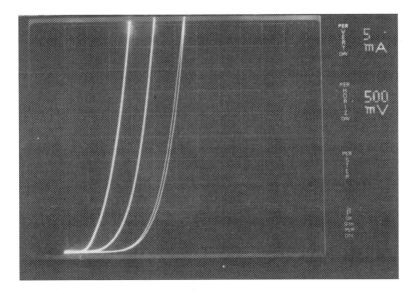

Fig. 44 Current vs voltage curves for a 100 × 1000-μm InAsSb/InAlAsSb/InAs QW laser at 300, 200, and 100 K.

n-GaSb buffer, linearly graded n-AlGaAsSb layers, 3-μm n-AlAs$_{0.08}$Sb$_{0.92}$ cladding, active region consisting of ten 15-nm InAs$_{0.85}$Sb$_{0.15}$ active wells and eleven 30-nm In$_{0.85}$Al$_{0.15}$As$_{0.86}$Sb$_{0.14}$ barriers, 3-μm p-AlAs$_{0.08}$Sb$_{0.92}$ cladding, linearly graded p-AlGaAsSb layers, and a 50-nm p$^+$-GaSb cap. As shown in Fig. 23, DCXRD of the laser structure exhibits somewhat broadened satellite peaks, indicating that the interfaces are not very smooth. The growth of InAsSb/InAlAsSb QW structures on GaSb is more difficult than on InAs because the region of stable growth for InAlAsSb is smaller.

Broad-stripe lasers 100-μm wide were fabricated by using SiO$_2$ patterning. Pulsed J_{th} of a 1000-μm-long device for temperatures between 80 and 165 K is plotted in Fig. 45. For comparison, data for the best DH lasers emitting at 3.9 μm are also plotted in the Fig. 45. At 80 K, J_{th} of the QW device is 78 A/cm^2. The value of T_0 up to 120 K is 30 K, which is larger than the 20 K observed for the DH lasers. Both the lower J_{th} and higher T_0 for the QW devices indicate that the Auger recombination has been decreased by the employment of the strained QW structure. As the temperature is increased above 120 K, T_0 becomes smaller. At the maximum operating temperature of 165 K, J_{th} is 3.5 kA/cm^2. Besides the Auger recombination, large leakage current due to the relatively small conduction- and valence-band offsets might have limited the operating temperature.

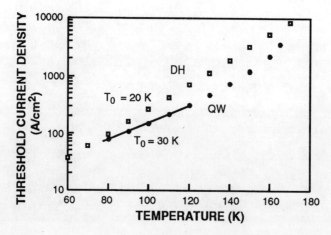

Fig. 45 Pulsed J_{th} of a 1000-μm-long InAsSb/InAlAsSb/GaSb QW laser at several temperatures.

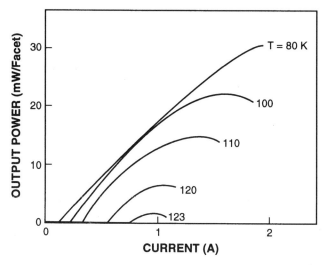

Fig. 46 CW power vs current curves at several temperatures of a 100 × 1000-μm InAsSb/InAlAsSb/GaSb QW laser.

Fig. 46 shows the cw power vs current for the same device. The maximum power at 80 K is 30 mW/facet, which is much higher than 24 mW obtained for the DH lasers with high-reflection/antireflection coatings and junction-down mounting [21]. The highest cw operating temperature is 123 K, which is higher than 105 K obtained for the 60-μm-wide DH devices mounted junction-side down. The voltage at the maximum power was ~ 3 V, somewhat smaller than the 4 V for lasers on the InAs substrates because of the incorporation of graded AlGaAsSb layers. RW lasers 15 μm wide exhibited cw threshold current of 35 mA at 80 K, and they operated cw up to 128 K. The far-field pattern in the transverse direction had an FWHM of $\sim 60°$. In the lateral direction, the far-field pattern showed a single lobe with an FWHM of $\sim 15°$.

8.6 CONCLUSIONS AND FUTURE PROJECTIONS

As a result of better material quality obtained by MBE and reduced Auger recombination rates in strained QW structures, substantial improvements have been made in the mid-IR laser performance. Fig. 47 plots the maximum cw operating temperature of III–V diode lasers as a function of wavelength. At ~ 2 μm, cw operation at significantly higher than RT has been achieved. However, the maximum temperature decreases rather steeply at longer wavelengths primarily due to Auger recombination. For practical purposes, it is desirable to have the cw operation temperature higher than 200 K, which is attainable by thermoelectric cooling.

In order to reduce the Auger rates, the following approaches can be tried; higher strain for reduced n_{th}, use of GaInAsSb active wells rather than InAsSb for increased Δ, and optimizing the doping profile near the active region for reduced free-carrier absorption and g_{th}. As is discussed in Chapter 10, the use of carefully designed type-II structures may also reduce the Auger rates. However, interfacial recombination current from many interfaces involving both antimony-dominated and arsenic-dominated layers must be substantially reduced in order not to mask the intrinsic advantage.

For high-power operation, the operating voltage should be reduced. Grading the interfaces with high potential barriers would substantially reduce the turn-on voltage and series resistance. As the photon energy is reduced, however, the operating voltage can still be several times the photon energy at high drive current. Optical pumping with near-infrared diode lasers is a good way of obtaining high power because the equivalent voltage is fixed at the pump photon energy regardless of the input power. Optical pumping of InAsSb/AlAsSb DHs by

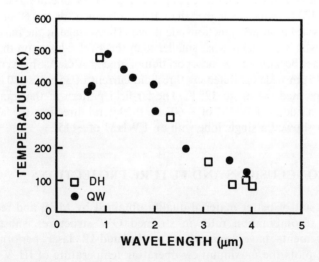

Fig. 47 Maximum cw operating temperature vs wavelength of III–V diode lasers.

0.98-μm diode laser arrays yielded pulsed power up to 2.7 W at 4 μm, with average power of 350 mW at 55 K [108]. QW structures with appropriate designs to exploit advantages of optical pumping may yield much higher power. However, the optical pumping requires the additional cost of pump lasers and focusing optics.

Another approach for high power is to use a cascaded active region, in which one electron can generate many photons by cascading through each gain stage. In this case, the importance of series resistance is much reduced. Quantum cascade lasers with emission between 4 and 11 μm have been demonstrated utilizing intersubband transitions inside InGaAs/InAlAs QWs on InP substrates. Pulsed operation up to 300 K and cw operation up to 110 K have been obtained at ~ 5 μm [109]. However, these lasers are not expected to generate high cw power because J_{th} is very high due to intrinsically fast phonon scattering times. Recently, quantum cascade lasers utilizing type-I or type-II interband transitions have been proposed [110]. It will be interesting to see the properties of this type of laser.

Acknowledgments

We would like to acknowledge the pioneering contributions of S. J. Eglash to the understanding of MBE growth of antimonide-based materials. We would also like to acknowledge technical discussions with N. G. Anderson, A. Baliga, M. Falcon, T. Y. Fan, S. Forouhar, D. K. Johnstone, H. Q. Le, Z.-L. Liau, M. A. Marciniak, D. McDaniel, I. Melngailis, C. D. Nabors, M. W. Prairie, A. Sanchez-Rubio, D. L. Spears, A. J. Strauss, B.-Y. Tsaur, J. N. Walpole, and C. A. Wang as well as the technical assistance of D. R. Calawa, J. W. Chludzinski, M. K. Connors, M. Finn, D. F. Kolesar, L. Krohn, M. J. Manfra, W. L. McGilvary, P. M. Nitishin, J. R. Ochoa, and J. V. Pantano. We would like to thank K. J. Challberg for expert editing of the manuscript. This work was sponsored by the Air Force Phillips Laboratory and the Jet Propulsion Laboratory under the NASA Office of Space Access and Technology.

References

1. D. S. Bomse, D. H. Hovde, D. B. Oh, J. A. Silver, and A. C. Stanton, *SPIE Proc.* **1681**, 138 (1992); assuming a minimum detectable absorbance of 10^{-5} in gas at 1 atm and 300 K.

2. I. Melngailis, *Appl. Phys. Lett.* **2**, 176 (1963).
3. J. F. Butler, A. R. Calawa, R. J. Phelan, T. C. Harman, A. J. Strauss, and R. H. Rediker, *Appl. Phys. Lett.* **5**, 75 (1964).
4. L. M. Dolginov, L. V. Druzhinina, P. G. Eliseev, M. G. Mil'vidskii, and B. N. Sverdlov, *Sov. J. Quantum Electron.* **6**, 257 (1976).
5. L. M. Dolginov, L. V. Druzhinina, P. G. Eliseev, A. N. Lapshin, M. G. Mil'vidskii, and B. N. Sverdlov, *Sov. J. Quantum Electron.* **8**, 416 (1978).
6. N. Kobayashi, Y. Hiroshi, and C. Uemura, *Jpn. J. Appl. Phys.* **19**, L30 (1980).
7. A. E. Drakin, P. G. Eliseev, B. N. Sverdlov, A. E. Bochkarev, L. M. Dolginov, and L. V. Druzhinina, *IEEE J. Quantum Electron* **23**, 1089 (1987).
8. C. Caneu, J. I. Zyskind, J. W. Sulhoff, T. E. Glover, J. Centanni, C. A. Burrus, A. G. Dentai, and M. A. Pollack, *Appl. Phys. Lett.* **51**, 764 (1987).
9. A. N. Baranov, E. A. Gusseinov, B. E. Dzhurtanov, T. N. Danilova, A. N. Imenkov, and Yu. P. Yakovlev, *Sov. Tech. Phys. Lett.* **14**, 798 (1988).
10. A. E. Bochkarev, L. M. Dolginov, A. E. Drakin, P. G. Eliseev, and B. N. Sverdlov, *Sov. J. Quantum Electron.* **18**, 1362 (1988).
11. A. N. Baranov, T. N. Danilova, B. E. Dzhurtanov, A. N. Imenkov, S. G. Konnikov, A. M. Litvak, V. E. Usmanskii, and Yu. P. Yakovlev, *Sov. Tech. Phys. Lett.* **14**, 727 (1988).
12. E. Tournie, J. L. Lazzari, H. Mani, F. Pitard, C. Alibert, and A. Joullie, *SPIE Proc.* **1361**, 641 (1990).
13. L. M. Dolginov, Yu. N. Kochergin, I. V. Kryukova, V. I. Leskovich, E. V. Matveenko, M. G. Mil'vidskii, and B. M. Stephanov, *Sov. Tech. Phys. Lett.* **4**, 580 (1978).
14. J. P. van der Ziel, R. A. Logan, R. M. Mikulyak, and A. A. Ballman, *IEEE J. Quantum Electron.* **21**, 1827 (1985).
15. J. P. van der Ziel, T. H. Chiu, and W. T. Tsang, *Appl. Phys. Lett.* **47**, 1139 (1985).
16. N. V. Zotoba, S. A. Karandashev, B. A. Matveev, N. M. Stus', and G. N. Talalakin, *Sov. Tech. Phys. Lett.* **12**, 599 (1986).
17. M. Aidaraliev, N. V. Zotova, S. A. Karandashev, B. A. Matveev, N. V. Stus', and G. N. Talalakin, *Sov. Tech. Phys. Lett.* **15**, 600 (1989).
18. T. H. Chiu, W. T. Tsang, J. A. Ditzenberger, J. P. van der Ziel, *Appl. Phys. Lett.* **49**, 1051 (1986).
19. H. K. Choi and S. J. Eglash, *Appl. Phys. Lett.* **59**, 1165 (1991).
20. H. K. Choi, S. J. Eglash, and G. W. Turner, *Appl. Phys. Lett.* **64**, 2474 (1994).
21. H. K. Choi, G. W. Turner, and Z. L. Liau, *Appl. Phys. Lett.* **65**, 2251 (1994).
22. H. K. Choi and S. J. Eglash, *Appl. Phys. Lett.* **61**, 1154 (1992).
23. S. R. Kurtz, R. M. Biefeld, L. R. Dawson, K. C. Baucom, and A. J. Howard, *Appl. Phys. Lett.* **64**, 812 (1994).
24. H. K. Choi, G. W. Turner, and H. Q. Le, *Appl. Phys. Lett.* **66**, 3543 (1995).
25. H. K. Choi and G. W. Turner, *Appl. Phys. Lett.* **67**, 332 (1995).
26. Y. H. Zhang, *Appl. Phys. Lett.* **66**, 118 (1995).
27. T. C. Hasenberg, D. H. Chow, A. R. Kost, R. H. Miles, and L. West, *Electron. Lett.* **31**, 275 (1995).
28. K. D. Moiseev, M. P. Mikhailova, O. G. Ershov, and Yu. P. Yakovlev, *Tech. Phys. Lett.* **21**, 482 (1995).
29. M. J. Cherng, H. R. Jen, C. A. Larsen, G. B. Stringfellow, H. Lundt, and P. C. Taylor, *J. Cryst. Growth* **77**, 408 (1986).
30. F. Korouta, H. Mani, J. Bahn, F. J. Hua, and A. Joullie, *Rev. Phys. Appl.* **22**, 1459 (1987).
31. S. Adachi, *J. Appl. Phys.* **61**, 4869 (1987).
32. K. Onabe, *NEC Res. Dev.* **72**, 1 (1984).
33. C. Alibert, A. Joullie, and A. M. Joullie, *Phys. Rev. B* **27**, 4946 (1983).
34. C. Alibert, M. Skouri, A. Joullie, M. Benouna, and S. Sadiq, *J. Appl. Phys.* **69**, 3208 (1991).

35. M. A. Afromowitz, *Solid State Commun.* **15**, 59 (1974).
36. A. Bochkarev, A. Drakin, and B. Sverdlov, *Electron. Lett.* **26**, 418 (1990).
37. M. A. Afromowitz, *J. Appl. Phys.* **44**, 1292 (1973).
38. P. D. Maycock, *Solid-State Electron.* **10**, 161 (1967).
39. N. A. Charykov, A. M. Litvak, K. D. Moiseev, and Y. P. Yakovlev, *SPIE Proc.* **1512**, 198 (1991).
40. Y. Tsou, A. Ichii, and E. Garmire, *IEEE J. Quantum Electron.* **28**, 1261 (1992).
41. R. H. Miles, D. H. Chow, J. N. Shulman, and T. C. McGill, *Appl. Phys. Lett.* **57**, 801 (1990).
42. E. Yablonovich and E. O. Kane, *IEEE J. Lightwave Technol.* **4**, 504 (1986).
43. A. R. Adams, *Electron. Lett.* **22**, 249 (1986).
44. C. E. Zah, R. Bhat, B. Pathak, C. Caneu, F. J. Favire, N. C. Andreadakis, D. M. Hwang, M. A. Koza, C. Y. Chen, and T. P. Lee, *Electron. Lett.* **27**, 1414 (1991).
45. D. P. Bour, R. S. Geels, D. W. Treat, T. L. Paoli, F. Ponce, R. L. Thornton, B. S. Krusor, R. D. Bringans, and D. F. Welch, *IEEE J. Quantum Electron.* **30**, 593 (1994).
46. N. G. Anderson, Ph.D. thesis, North Carolina State University, Raleigh, North Carolina, 1988.
47. J. M. Luttinger and W. Kohn, *Phys. Rev.* **97**, 869 (1955).
48. A. Baliga, D. Trivedi, and N. G. Anderson, *Phys. Rev. B* **49**, 10402 (1994).
49. *Landolt-Bornstein: Numerical Data and Functional Relationships*, Vol. 17a, O. Madelung, Ed. (Springer-Verlag, Berlin, 1982).
50. D. Ahn and S.-L. Chuang, *IEEE J. Quantum Electron.* **26**, 13 (1990).
51. S. W. Corzine, R.-H. Yan, and L. A. Coldren, in *Quantum Well Lasers*, P. S. Zory, Jr. Ed. (Academic, Orlando, FL, 1993).
52. A. Baliga, F. Agahi, N. G. Anderson, K. M. Lau, and S. Cadambi, *IEEE J. Quantum Electron.* **32**, 29 (1996).
53. A. Ghiti and E. P. O'Reilly, *Semicond. Sci. Technol.* **8**, 1655 (1993).
54. R. F. Nabiev, C. J. Chang-Hasnain, and H. K. Choi, presented at *Semiconductor Lasers: Advanced Devices and Applications*, August 21–23, 1995, Keystone, Colorado, Paper MB6.
55. H. K. Choi, G. W. Turner, M. J. Manfra, M. K. Connors, F. P. Herrmann, A. Baliga, and N. G. Anderson, *SPIE Proc.* **2682**, (1996).
56. C. H. Henry, R. A. Logan, H. Temkin, and F. R. Merritt, *IEEE J. Quantum Electron.* **19**, 941 (1983).
57. R. I. Taylor, A. R. Abram, M. G. Burt, and C. Smith, *IEE Proc. J. Optoelectron.* **132**, 364 (1985).
58. G. G. Zegrya and V. A. Kharchenko, *Sov. Phys. JETP* **74**, 173 (1992).
59. A. J. Norieka, M. H. Francombe, and C. E. C. Wood, *J. Appl. Phys.* **52**, 7416 (1981).
60. G. A. Chang, H. Takaoka, L. L. Chang, and L. Esaki, *Appl. Phys. Lett.* **40**, 938 (1982).
61. M. Yano, Y. Suzuki, T. Ishii, Y. Matsushima, and M. Kimata, *Jpn. J. Appl. Phys.* **17**, 2091 (1978).
62. T. Waho, S. Ogawa, and S. Maruyama, *Jpn. J. Appl. Phys.* **16**, 1875 (1977).
63. M. Yano, T. Takase, and M. Kimata, *Jpn. J. Appl. Phys.* **18**, 387 (1977).
64. C. R. Bolognesi, E. J. Caine, and H. J. Kroemer, *IEEE Electron Device Lett.* **15**, 16 (1994).
65. J. N. Schulman, D. H. Chow, and T. C. Hasenberg, *Solid-State Electron.* **37**, 981 (1994).
66. K. F. Longenbach, L. F. Luo, and W. I. Wang, in *Resonant Tunneling in Semiconductors*, L. L. Chang, Ed. (Plenum, New York, 1991).
67. B. R. Bennett, B. V. Shanabrook, R. J. Wagner, J. L. Davis, J. R. Waterman, and M. E. Twigg, *Solid-State Electron.* **37**, 733 (1994).
68. E. H. C. Parker, Ed. *The Technology and Physics of Molecular Beam Epitaxy*, (Plenum, New York, 1985).
69. M. A. Herman and H. Sitter, Eds. *Molecular Beam Epitaxy-Fundamentals and Current Status*, (Springer-Verlag, Berlin, 1989).

70. G. W. Turner, H. K. Choi, D. R. Calawa, J. V. Pantano, and J. W. Chludzinski, *J. Vac. Sci. Technol. B* **12**, 1266 (1994).
71. S. J. Eglash, H. K. Choi, and G. W. Turner, *J. Cryst. Growth* **111**, 696 (1991).
72. G. W. Turner, B. A. Nechay, and S. J. Eglash, *J. Vac. Sci. Technol. B* **8**, 283 (1992).
73. H. Kressel and J. K. Butler, *Semiconductor Lasers and Heterojunction LEDs* (Academic, Orlando, FL, 1977), p. 437.
74. T. Hayakawa, M. Kondo, T. Suyama, K. Takahashi, S. Yamamoto, S. Yano, and T. Hijikata, *Appl. Phys. Lett.* **49**, 788 (1986).
75. Y.-H. Zhang, *J. Cryst. Growth* **150**, 838 (1995).
76. G. W. Turner, S. J. Eglash, and A. J. Strauss, *J. Vac. Sci. Technol. B* **11**, 864 (1993).
77. A. Y. Polyakov, M. Stam, A. G. Milnes, R. G. Wilson, Z. Q. Fang, P. Rai-Choudhury, and R. J. Hillard, *J. Appl. Phys.* **72**, 1316 (1992).
78. M. E. Lee, I. Poole, W. S. Truscott, I. R. Cleverly, K. E. Singer, and D. M. Rohlfing, *J. Appl. Phys.* **68**, 131 (1990).
79. P. M. Mooney, *J. Appl. Phys.* **67**, R1 (1990).
80. A. Z. Li, J. X. Wang, Y. L. Zheng, G. P. Ru, W. G. Bi, Z. X. Chen, and N. C. Zhu, *J. Cryst. Growth* **127**, 566 (1993).
81. C. A. Wang, K. F. Jensen, A. C. Jones, and H. K. Choi, *Appl. Phys. Lett.* **68**, 400 (1996).
82. R. M. Biefeld, A. A. Allerman, and M. W. Pelczynski, *Appl. Phys. Lett.* **68**, 932 (1996).
83. A. Y. Polyakov, S. J. Eglash, A. G. Milnes, M. Ye, S. J. Pearton, and R. G. Wilson, *J. Cryst. Growth* **127**, 728 (1993).
84. R. Hull and J. C. Bean, in *Strained-Layer Superlattices: Materials Science and Technology*, Vol. 33, T. P. Pearsal Ed. (Academic, San Diego, 1991).
85. D. I. Westwood and D. A. Wolf, *J. Appl. Phys.* **73**, 1187 (1993).
86. S. R. Kurtz, L. R. Dawson, R. M. Biefield, D. M. Follstaedt, and B. L. Doyle, *Phys. Rev. B* **46**, 1909 (1992).
87. T.-Y. Seong, G. R. Booker, A. G. Norman, and I. T. Ferguson, *Appl. Phys. Lett.* **64**, 3593 (1994).
88. M. Marciniak, Ph.D. Thesis, Air Force Institute of Technology, Dayton, Ohio, 1995.
89. N. Iwata, Y. Nakahara, and I. Hirosawa, *Inst. Phys. Conf. Proc.* **106**, 459 (1990).
90. H. K. Choi, G. W. Turner, and S. J. Eglash, *IEEE Photon. Technol. Lett.* **6**, 7 (1994).
91. M. Takeshima, *Jpn. J. Appl. Phys.* **22**, 491 (1983).
92. J. Matsui, *Mater. Res. Soc. Symp. Proc.* **14**, 477 (1983).
93. H. Lee, P. K. York, R. J. Menna, R. U. Martinelli, D. Z. Garbuzov, S. Y. Naryan, and J. C. Connolly, *Appl. Phys. Lett.* **66**, 1942 (1995).
94. D. Z. Garbuzov, R. U. Martinelli, R. J. Menna, P. K. York, H. Lee, S. Y. Naryan, and J. C. Connolly, *Appl. Phys. Lett.* **67**, 1346 (1995).
95. R. U. Martinelli, D. Z. Garbuzov, H. Lee, P. K. York, R. J. Menna, J. C. Connolly, and S. Y. Naryan, *SPIE Proc.* **2382**, 250 (1995).
96. H. K. Choi, S. J. Eglash, and M. K. Connors, *Appl. Phys. Lett.* **63**, 3271 (1993).
97. M. Nakamura, K. Aiki, N. Chinone, R. Ito, and J. Umeda, *J. Appl. Phys.* **49**, 4644 (1978).
98. H. K. Choi, G. W. Turner, M. K. Connors, S. Fox, C. Dauga, and M. Dagenais, *IEEE Photon. Technol. Lett.* **7**, 281 (1995).
99. P. K. York, R. J. Menna, D. Z. Garbuzov, H. Lee, R. U. Martinelli, and S. Y. Naryan, presented at CLEO'95, May 21–26, 1995, Baltimore, Maryland, Paper CPD31.
100. H. K. Choi, J. N. Walpole, G. W. Turner, S. J. Eglash, L. J. Missaggia, and M. K. Connors, *IEEE Photon. Technol. Lett.* **5**, 1117 (1993).
101. J. N. Walpole, E. S. Kintzer, S. R. Chinn, C. A. Wang, and L. J. Missaggia, *SPIE Proc.* **1850**, 51 (1993).

102. L. Goldberg, M. R. Surette, and D. Mehuys, *Appl. Phys. Lett.* **62**, 2304 (1993).
103. Z. L. Liau and H. K. Choi, *Appl. Phys. Lett.* **65**, 3219 (1995).
104. M. E. Flatte, C. H. Grein, H. Ehrenreich, R. H. Miles, and H. Cruz, *J. Appl. Phys.* **78**, 4552 (1995).
105. V. L. Dalal, W. A. Hicinbothem, Jr., and H. Kressel, *Appl. Phys. Lett.* **24**, 184 (1974).
106. J. R. Lindle, J. R. Meyer, C. A. Hoffman, F. J. Bartoli, G. W. Turner, and H. K. Choi, *Appl. Phys. Lett.* **67**, 3153 (1995).
107. H. Q. Le, G. W. Turner, S. J. Eglash, H. K. Choi, and D. A. Coppeta, *Appl. Phys. Lett.* **64**, 152 (1994).
108. H. Q. Le, G. W. Turner, H. K. Choi, J. R. Ochoa, A. Sanchez, J. M. Arias, M. Zandian, R. R. Zucca, and Y.-Z. Liu, *SPIE Proc.* **2382**, 262 (1995).
109. J. Faist, F. Capasso, C. Sirtori, D. L. Sivco, A. L. Hutchinson, and A. Y. Cho, *Electron. Lett.* **32**, 560 (1996).
110. J. R. Meyer, I. Vurgaftman, R. Q. Yang, and L. R. Ram-Mohan, *Electron. Lett.* **32**, 45 (1996).

CHAPTER 9

Mid-Wave Infrared Sources Based on GaInSb/InAs Superlattice Active Layers

R. H. MILES and T. C. HASENBERG

Hughes Research Laboratories, 3011 Malibu Canyon Road, Malibu, CA 90265, USA

9.1.	Introduction	433
9.2.	Benefits of GaInSb/InAs Superlattices	434
	9.2.1. Intrinsic properties	435
	9.2.2. Practical advantages	439
9.3.	Growth and Materials Properties	439
	9.3.1. GaInSb/InAs superlattice active layers	440
	9.3.2. AlSb/InAs cladding layers	442
9.4.	Diode Laser Design and Characteristics	444
	9.4.1. Device design	444
	9.4.2. Device fabrication	449
	9.4.3. Laser results and discussion	452
9.5.	Conclusions	458
	Acknowledgments	459
	References	459

9.1 INTRODUCTION

Lasers emitting in the 2–5 μm wavelength range have numerous potential applications, both commercial and military [1–4]. Recent years have seen significant progress towards the goal of room temperature operation for diode lasers in this spectral range, promising compact and inexpensive sources more widely applicable than conventional technology such as optical parametric oscillators (OPOs). Several distinct classes of devices have been fabricated and a considerable variety of materials employed. These include IV–VI [5] and II–VI [6] diodes and III–V devices based on both interband [7–16] and intersubband [17] transitions.

Here we address diodes based on III–V GaInSb/InAs superlattice active layers, employing interband optical transitions. This class of superlattices holds several performance records and promises to eclipse others [21]. However, it should be mentioned that GaInAsSb/AlGaAsSb diodes [7–13] grown by molecular beam epitaxy (MBE) are currently the more mature III–V devices, having exhibited room temperature operation at 2.78 μm [7] and maximum operating temperatures of 255 K, 165 K, and 85 K for emission at 3.0 μm, 3.9 μm, and 4.5 μm, respectively [8–10]. Noteworthy successes have also been realized for laser diodes grown by both liquid phase epitaxy (e.g. InAsSbP [18]) and metalorganic chemical vapor deposition (e.g. In(Ga)AsSb with InPSb clads [19,20]). Nevertheless, intrinsic properties favor GaInSb/InAs superlattices over the competition; devices based on these structures are projected to have the best mid-wave infrared (MWIR) performance if brought to maturity [21].

In this chapter we briefly review the growth of GaInSb/InAs superlattices and AlSb/InAs superlattices employed as clads, and summarize the intrinsic and extrinsic properties salient to diode lasers. This is followed by a more comprehensive presentation of laser results derived to date. However, we first motivate the choice of active layer by summarizing theoretical and practical advantages relative to competing injection devices.

9.2 BENEFITS OF GaInSb/InAs SUPERLATTICES

Interest in GaInSb/InAs-based devices stems both from favorable intrinsic properties and from practical advantages associated with MBE growth of superlattices. Threshold current densities should be reduced by a comparatively low valence band density of states and by large in-plane oscillator strengths. Losses associated with Auger recombination have been predicted to be lower than those in competing III–V and II–VI systems. Predicted reduction of these processes has been confirmed at long wavelengths [22]. Unlike the cascade lasers, which are characterized by rapid phonon relaxation processes, radiative recombination should dominate other intrinsic processes except at very high excitation levels. Problems associated with the more conventional quaternary III–V devices should also be mitigated. Specifically, the energy gap of the GaInSb/InAs class of superlattices can be reduced to zero, in principle allowing lasers of arbitrarily long emission wavelength to be fabricated. Further, the valence band edge energy of these structures is appreciably higher than that of candidate cladding layer materials such as AlGaAsSb alloys and AlSb/InAs superlattices, allowing confinement of both holes and electrons in GaInSb/InAs active layers for any emission wavelength. From a practical standpoint, MBE growth of these superlattices appears to be easier and more reproducible than for alloys containing more than one group-V species.

9.2.1 Intrinsic Properties

An unusual broken gap band alignment between GaInSb and InAs distinguishes this system from other candidate MWIR emitters, and underlies much of the interest in GaInSb/InAs superlattices. The conduction band edge of InAs lies *below* the valence band edge of GaInSb. This yields a semimetallic band structure in the absence of quantum confinement effects, and imparts considerable freedom to tailor the electronic band structure of the conduction and valence bands in both semiconducting and semimetallic superlattices. Of particular importance is the ability to tailor structures to enhance radiative processes and to suppress Auger recombination and inter-subband absorption.

As illustrated in Fig. 1, the extreme broken-gap band alignment in this system yields a valence band with in-plane dispersion well matched to that of the conduction band for a considerable distance from zone-center. In addition, degeneracy in the light and heavy hole bands at the Γ-point is removed. The result is that holes and electrons are distributed similarly in k-space near $k_\perp = 0$, appreciably increasing the probability of radiative electron–hole recombinations and hence reducing the threshold carrier density. This point is illustrated in Fig. 2, in which threshold carrier densities for GaInSb/InAs superlattices are compared to those of several competing systems. It should be noted that the calculation compares thick superlattices, for which growth-direction dispersion is ill-matched. Meyer *et al.* have suggested that sandwiching very thin GaInSb/InAs wells between AlSb barriers lends itself to realizing a 2-dimensional density of states, addressing this shortcoming [23]. While growth of such structures is not straightforward, preliminary results suggest merit for this approach if the density of extrinsic recombination centers can be reduced [24].

Auger processes allow electron–hole pairs to recombine without emitting photons, instead conserving energy and momentum by promoting a third carrier to a higher-lying state. As such, these processes

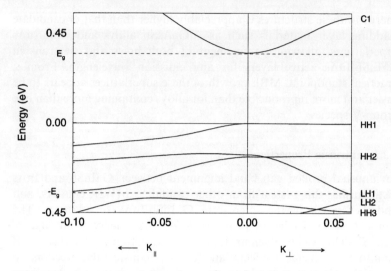

Fig. 1 Electronic band structure of a 35 Å/16.7 Å $Ga_{0.75}In_{0.25}Sb/InAs$ superlattice, illustrating the low density of states at an energy one fundamental gap below the valence band edge (indicated by $-E_g$). This "minigap" reduces Auger recombination rates by reducing the joint density of states for regions of high thermal occupation probability. From Ref. [21].

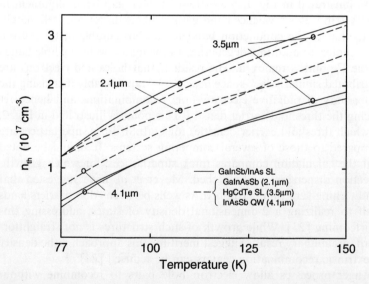

Fig. 2 Comparison of threshold carrier concentration for GaInSb/InAs superlattices and competing MWIR systems, for a gain of $25\,cm^{-1}$. Adapted from Ref. [21].

can be viewed purely as loss mechanisms in a laser. Reduction of Auger recombination and intersubband absorption losses is achieved by reducing the joint densities of states for these processes, particularly near zone-center, where both electrons and holes are likely to reside. To this end, it is advantageous to introduce minigaps in the growth-direction dispersion relation at energies one fundamental gap above the conduction band edge and below the valence band edge [21]. It is of further benefit to distance higher lying band extrema from the Γ-point, as this requires that carriers involved in these loss processes reside in states with large wavevectors, to satisfy requirements of momentum conservation. The considerable dispersion of the ground state valence and conduction bands means that such states are only significantly occupied at exceptionally high temperatures or excitation levels.

In general, Auger recombination can be reduced in this system by increasing the InSb fraction. Constraining the $Ga_{1-x}In_xSb$ composition to a value unlikely to result in misfit dislocations (typically $x < 0.4$), the Auger rate is found to be reduced for several discrete ranges of superlattice periods, for which superlattice minigaps fall an energy gap above or below the conduction or valence band edges, respectively. The choice between these ranges and the particular GaInSb:InAs thickness ratio are dictated by the desired emission energy and need for high optical efficiency. The heavy hole mass is sufficiently great that quantum confinement effects are small for the hole states. Consequently the fundamental energy gap is comparatively insensitive to the thickness of the GaInSb layer in which holes are localized, and is instead much more readily controlled through the InAs layer thickness. In practice, the InAs layer thickness is largely determined by the emission wavelength, while Auger processes are suppressed by choosing the GaInSb thickness and composition to place minigaps at the appropriate energies. We note that it is further desirable to increase the ratio of GaInSb:InAs layer thicknesses where possible, as the InAs layers are essentially optically dead due to the effective confinement of holes in the GaInSb layers.

The benefits of reduced Auger recombination are illustrated in Figs. 3 and 4, which show threshold current densities J_{th} for the structures of Fig. 2 and for a single structure as a function of gain, respectively. Fig. 3 illustrates that benefits of reduced Auger recombination are particularly great at longer wavelengths; reductions in J_{th} at the shortest wavelength are negligible. That Auger processes are not often significant in 3.5 μm GaInSb/InAs superlattices is well illustrated

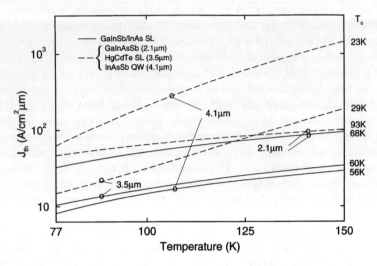

Fig. 3 Threshold current densities for the structures of Fig. 2. Reductions in Auger recombination rates and threshold carrier concentrations result in increased benefits for the GaInSb/InAs superlattices at longer wavelengths. Adapted from Ref. [21].

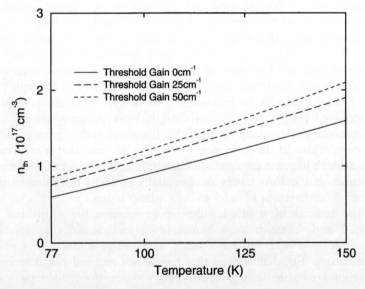

Fig. 4 Weak dependence of threshold carrier density on gain for the 35 Å/16.7 Å $Ga_{0.75}In_{0.25}Sb/InAs$ superlattice of Fig. 1. From Ref. [21].

by the comparatively small rise in threshold current for increased gain illustrated in Fig. 4.

A further limitation of some III–V systems which is circumvented with GaInSb/InAs superlattices is that of carrier confinement, particularly at high temperatures and/or long wavelengths. As is evident from Fig. 8 in Section 9.4.1, the conduction and valence band edges of these structures are comparatively high in energy, making it possible to effectively confine both electrons and holes in GaInSb/InAs active layers for arbitrarily long emission wavelengths.

Last, optical confinement factors are similar to those of competing ternary and quaternary approaches, allowing effective waveguiding employing a variety of Al-containing cladding layers. As will be discussed in Section 9.3.2, we have chosen to employ InAs/AlSb superlattice clads rather than AlGaAsSb alloys, due to ease of lattice match with these structures. However, both yield good optical confinement.

9.2.2 Practical Advantages

MBE growth of lasers based on superlattices appears to be easier and more reproducible than for devices based on alloys containing more than one group-V species. Changing the emission wavelength of a GaInSb/InAs-based diode is achieved simply by changing the constituent layer thicknesses, which is readily, predictably, and precisely controlled employing current MBE technology, obviating wavelength-specific calibrations. By contrast, incorporation of group-V elements depends sensitively upon both fluxes and growth temperature, typically demanding prior calibrations and considerable attention to growth conditions to achieve specific emission wavelengths in lasers based upon GaInAsSb active layers. These calibrations become particularly onerous when group-V alloys are employed as cladding layers, as requirements of lattice match place stringent tolerances on the compositions of these thick layers.

9.3 GROWTH AND MATERIALS PROPERTIES

Recent strides in MBE growth techniques have greatly reduced the density of nonradiative recombination centers in GaInSb/InAs superlattices, allowing lasers to be based on structures which had negligible luminescence as grown two years ago. Key to these improvements have been use of a monomeric Sb source and subsequent use of an *in situ* anneal, in combination with growth conditions previously shown to yield dislocation-free structures with crisp compositional profiles. In this section we review the growth and properties of these superlattices, as well as those of AlSb/InAs superlattices employed as cladding layers.

9.3.1 GaInSb/InAs Superlattice Active Layers

Growth of these superlattices has been described in detail elsewhere [25], and will only be briefly reviewed here. Samples were grown in VG V80 and Perkin Elmer 430P MBE machines equipped with valved, cracked As sources and EPI 175 cracked Sb sources. Only (100) oriented GaSb substrates were used; high dislocation densities have been shown to appreciably degrade the optical quality of structures grown on GaAs. All $Ga_{0.75}In_{0.25}Sb$/InAs layers were grown at substrate temperatures of approximately 380°C, which has been found to yield superlattices of nearly ideal structural quality (i.e. dislocation densities too low to be detectable by transmission electron microscopy and sharp, square-wave compositional modulation) [25]. Five second Sb soaks were employed at each InAs-GaInSb interface, yielding predominantly InSb-like interfacial bonds, which have been shown to result in oscillator strengths higher than those for GaInAs-like interfaces for band-to-band transitions [26]. Growth of each superlattice was followed by a 30 min, 500°C anneal, performed *in situ* in an Sb flux. This has been shown to improve the optical efficiencies of these structures [27]. Although the microstructural details remain unexplored, the propensity of antimonides to defect dope suggests that this anneal changes the density and/or optical activity of native defects in these layers or at the interfaces.

Laser diodes were grown employing an atomic Sb source, achieved by employing a cracking zone temperature of 900°C. The intensity of luminescence from these superlattices has been shown to degrade rapidly as the dominant impinging species is changed to dimers or tetramers [27]. We note in passing that luminescent efficiencies currently degrade upon heating our EPI 175 cracking cell much above 900°C, despite the continued prevalence of monomeric Sb at these temperatures. We attribute this result to contaminants introduced in the films at the higher temperatures, and regard this as both a cell-dependent and tractable problem given the successful operation of more mature As cracking cells at much higher temperatures. We have also found improvements in the radiative efficiencies of MWIR $Ga_{0.75}In_{0.25}Sb$/InAs superlattices upon reducing the As overpressure to a point just sufficient to maintain a group-V stabilized surface during growth of the InAs layers, and on closing the As valve for growth of all but the InAs layers. The nature of the nonradiative processes resulting when low levels of As are incorporated in the antimonide layers has not yet been studied.

The primary emission wavelength of these superlattices has been shown to be continuously variable throughout the infrared, being a function of the constituent layer thicknesses and GaInSb composition chosen at the time of growth. Experimental values are in good agreement with expectations based on 8-band $k \cdot p$ theory. Also in agreement with calculation are optical absorption coefficients, which are found to be comparable to those in bulk zinc blende materials near the energy gap.

The effectiveness of Auger suppression has been demonstrated even in non-optimized, long wavelength structures, as illustrated in Fig. 5 [22].

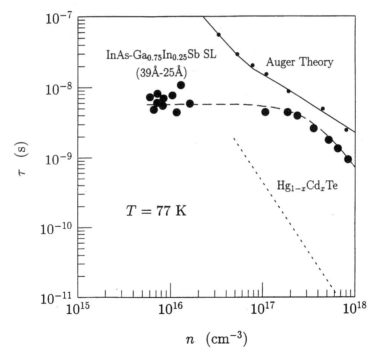

Fig. 5 Auger and Shockley-Read limited lifetimes of a GaInSb/InAs superlattice with an 11 μm gap. The S-R lifetimes of 5 ns have recently been significantly improved upon. The reduction of Auger recombination at higher carrier concentrations is evident even in this non-optimized structure. From Ref. [22].

Given the complexity of the Auger calculation, agreement between experiment and theory is excellent, as is also the case for effective masses [28]. All measurements to date support the calculations on which performance benefits for this class of laser are predicated. As such, theory has been invaluable in guiding design of the laser active layers, especially in view of the considerable degrees of freedom afforded by use of this superlattice.

9.3.2 AlSb/InAs Cladding Layers

Rather than undertake the careful growth control and time-consuming calibrations necessary to fabricate lasers with AlGaAsSb cladding layers, we have chosen to employ AlSb/InAs superlattice clads [29]. Precise lattice match to a GaSb substrate, which is necessary for the 2+μm thick clads required for MWIR emitters, is readily achieved with this superlattice owing to the small and opposite strains in the two layers of the superlattice. Lattice mismatches relative to GaSb are approximately +0.5% and −0.5% for AlSb and InAs, respectively, hence superlattice layer thicknesses should be roughly equal to eliminate the driving force for the generation of dislocations. The superlattices benefit further from a Si modulation doping scheme which allows n-type layers to be fabricated without the use of Te, which has in some cases been found to have long-term detrimental effects on device performance.

The energy gap of an AlSb/InAs superlattice, and thus carrier confinement and injection efficiency in the active layer, is set by the period of the superlattice, and is adjustable over a range easily sufficient to cover the MWIR spectral band. While the index of refraction of the superlattice is not as low as for some AlGaAsSb alloys, it is sufficiently low to achieve good optical confinement factors, as will be shown in Section 9.4.1.

AlSb/InAs superlattices were grown under conditions similar to those employed for the GaInSb/InAs active layers, with the exceptions that the substrate temperature was roughly 450°C and the growth rate somewhat higher (≈ 2 Å/s). Sb soaks were employed at the interfaces. The As flux was not modulated during growth, but was maintained at a value sufficient to preserve a group-V stabilized surface. n-type doping was achieved by depositing InAs:Si/AlSb superlattices, while p-layers were formed with InAs/AlSb:Be structures. Doping densities of 10^{18} cm^{-3} were readily achieved, but 10^{17} cm^{-3} was found to be adequate for the lasers fabricated here. Diode characteristics display low forward bias drops, compatible with the injection levels necessary for diode laser operation.

X-ray diffraction such as that in Fig. 6 shows most devices to be of excellent structural quality, with no evidence of misfit or threading dislocations. However, while nearly ideal structures can be obtained for appreciable lattice mismatches between the cladding layers and substrate, as illustrated by sizable splittings between the substrate and zeroth-order superlattice diffraction peaks, we believe these devices to be ill-suited to application as lasers. Specifically, cleaving such wafers often yields facets of poor quality, displaying cracks or even delamination of the epilayers. We attribute this to differences in thermal expansion coefficients between the substrates and epilayers. While structures can be grown virtually defect free by matching lattice constants of the layers at the elevated ($\approx 450°C$) growth temperature, differential thermal contraction on cooling to room temperature results in significant stresses in the epilayers at 300 K. Relief of these stresses is catalyzed by laser processing steps such as cleaving, and results in inoperable or inferior devices. This problem is readily avoided by growing structures closely lattice matched to the substrates at room temperature.

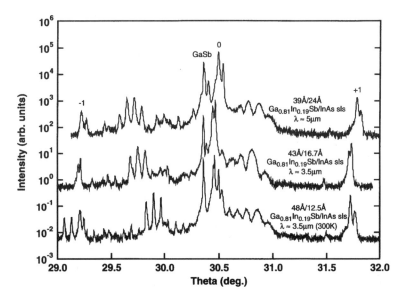

Fig. 6 High resolution X-ray diffraction from three GaInSb/InAs-based lasers.

9.4. DIODE LASER DESIGN AND CHARACTERISTICS

9.4.1 Device Design

Beyond the considerations of reduced non-radiative recombination rates and high oscillator strengths for the optically active GaInSb/InAs superlattices, addressed in Section 9.2.1, laser design relies upon efficient injection of carriers into these regions and effective optical waveguiding within these media. The multi-quantum well (MQW) and separate confinement heterostructure (SCH) designs employed here achieve each of these goals, and allow the constraint of lattice mismatch to the GaSb substrate to be relieved. Consequently the GaInSb/InAs superlattice quantum wells can be designed primarily to suppress loss mechanisms such as Auger recombination and intersubband absorption, and to achieve a particular emission wavelength. Mismatch stresses that typically result when the optically active regions are designed in this manner are then counterbalanced by opposite stresses in GaInAsSb or AlGaInAsSb barriers. Such an approach is successful until strain energies rise to such an extent that individual wells relax. For the study described here, well thicknesses were maintained below about 300 Å, and no evidence of stress relaxation was observed.

A schematic summarizing typical GaInSb/InAs-based laser structures studied to date is shown in Fig. 7 and a simplified band edge diagram in Fig. 8. Typical active layers are 5 to 9 period multi-quantum wells, bounded on both sides by GaInAsSb barriers. Wells studied to date have been comprised of thin superlattices with three to five 30–48 Å thick $Ga_{1-x}In_xSb$ layers interleaved between InAs layers 12.5–24 Å thick. InSb fractions x are usually 0.25. Wells terminated with GaInSb layers are most common (i.e. a barrier/GaInSb/InAs ••• InAs/GaInSb/barrier growth sequence), but properties of lasers with wells bounded by InAs layers have also been examined. Parameters varied have largely been directed towards examining optical confinement and effects of increased numbers of wells. Emission at a particular wavelength is usually best achieved for a particular set of superlattice parameters. While this wavelength is a function of temperature of operation, with the consequence that different superlattices are best suited to 300 K or low temperature operation at a specific wavelength, few complete laser structures have been based on GaInSb/InAs superlattices other than those believed to be optimal for a given set of conditions.

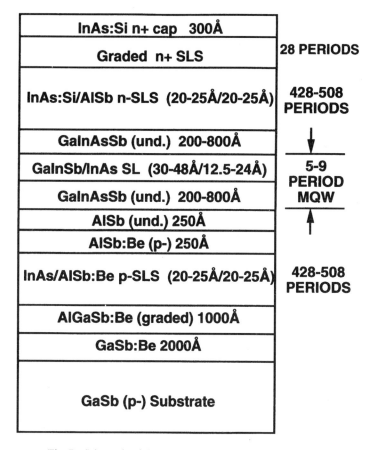

Fig. 7 Schematic of the MQW laser structures described here.

The energies at which electrons and holes are injected into the active region and the magnitude of the barrier blocking their escape is set by the energy gap of the AlSb/InAs superlattice cladding layers, which is in turn determined by the period of this superlattice. In addition, as shown in the figure, we employ an additional 500 Å thick AlSb layer to further block electron escape from the active region. Use of such a layer is facilitated by the choice of the GaInSb/InAs active layer; while the AlSb valence band energy is higher than that of InAs ($\Delta E_v \approx -110$ meV), the valence band edge of the GaInSb/InAs superlattice is considerably higher than either. We note that this would *not* be the case for active layers such as GaInAsSb.

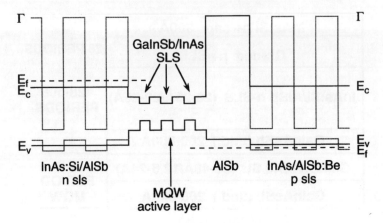

Fig. 8 Simplified band edge diagram for the MQW lasers of Fig. 7. Solid lines reflect conduction and valence band edges. Heavy dotted lines denote Fermi energies and light dotted lines Γ-point band edges of the AlSb and InAs layers comprising the cladding superlattices.

Laser design has also taken account of optical confinement factors, quantifying the waveguiding properties of particular active and cladding layer combinations to analyze single mode losses for epilayer thicknesses within the range readily grown by MBE ($\approx 5\,\mu m$). As is shown in Fig. 9, achieving edge emitting lasers requires that the MQW active layers act as slab waveguides, with the n-type and p-type superlattices functioning as optically confining cladding layers. The

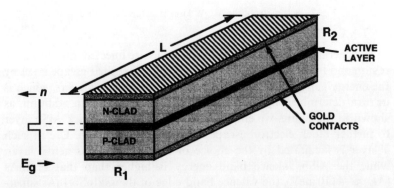

Fig. 9 A slab waveguide showing the waveguide layer (MQW laser active region) and the n-type and p-type layers which function as the cladding layers.

efficiency of the waveguide can be quantified by the optical confinement factor Γ, defined as the fraction of the electric field intensity confined in the optically active, waveguide region.

Because exact refractive indices of many of the constituent materials are unknown, a simplified model has been adopted in calculating optical confinement factors. The index of a $Ga_{1-x}In_xSb$ alloy is assumed to be a linear interpolation of the GaSb and InSb values. The index of $Ga_{1-x}In_xAs_ySb_{1-y}$ barriers is likewise assumed to be a linear interpolation between the values for InAs and GaSb, since lattice match to a GaSb substrate requires that $x \approx y$. Indices of both superlattices and multi-quantum wells are assumed to be thickness-weighted averages of the constituents. Values used here are enumerated in Table I. Γ is calculated using a standard three layer, slab waveguide theory [31], employing the refractive indices of the MQW active layer and two cladding layers.

Thick active regions and cladding layers and large differences in the refractive indices of active and cladding layers are essential to achieving the high values of Γ necessary for low lasing thresholds. Since Γ scales inversely with wavelength, particularly thick active and cladding layers are necessary for MWIR lasers. Cladding layers at least 2 µm thick are typically found to be necessary for good optical confinement, while increasing the thickness of the active layers likewise provides a better waveguide. Typical results of optical confinement calculations are shown in Fig. 10, which illustrates refractive index and electric field intensity profiles for sample D in Table I. A value of 0.446 is obtained for Γ for this sample. Values for the lasers we have grown so far range between 0.2 and 0.5, which is roughly equivalent to those of commercial AlGaAs based diodes.

In practice, we note that an optimum MQW active layer thickness exists for achieving minimum laser threshold current densities, owing to the trade-off between higher optical confinement and increased carrier densities required to achieve population inversion for thick optically active regions. However, an unusual benefit of this system is the high refractive index of GaInAsSb barriers, typically exceeding that of the GaInSb/InAs superlattice wells. This situation is unlike that in AlGaAs lasers, and allows us to employ active layers with thick barriers and thin wells, compromising nothing in Γ while benefitting in principle from a lower threshold due to a reduced optically active volume.

By extension, a separate confinement heterostructure [32] (SCH) design is particularly attractive for this class of MWIR laser. This

TABLE I
Refractive indices of selected III–V binary semiconductors and compounds employed in GaInSb/InAs-based MQW lasers.

Wavelength	AlAs	AlSb	InAs	GaSb	InSb	$Ga_{0.75}In_{0.25}As_{0.33}Sb_{0.67}$	$Ga_{0.85}In_{0.15}As_{0.13}Sb_{0.87}$
3.0 μm	2.870	3.241	3.523	3.813	4.026	3.743	3.772
3.5 μm	2.868	3.211	3.523	3.819	4.020	3.748	3.778
4.0 μm	2.867	3.182	3.523	3.833	4.015	3.759	3.790
4.5 μm	2.865	3.174	3.510	3.834	4.021	3.756	3.789

Wavelength	$Ga_{0.75}In_{0.25}Sb/InAs$ 39 Å/12.5 Å	$Ga_{0.81}In_{0.19}Sb/InAs$ 43 Å/16.7 Å	$Ga_{0.75}In_{0.25}Sb/InAs$ 35 Å/24 Å	InAs/AlSb 20 Å/20 Å
3.0 μm	3.783	3.761	—	3.382
3.5 μm	3.785	3.764	—	3.367
4.0 μm	—	3.771	3.734	3.353
4.5 μm	—	—	3.730	3.342

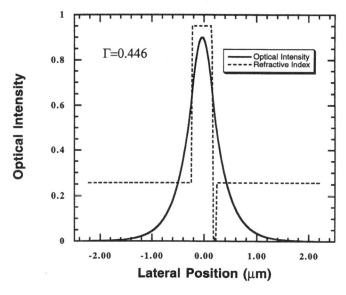

Fig. 10 Electric field intensity plotted as a function of lateral position for a baseline MQW laser (sample D).

comprises a thin MQW region bracketed by thicker waveguide regions, yielding a small active volume for injected carriers while maintaining efficient waveguiding. Fig. 11 shows an SCH MWIR laser with $Ga_{0.67}In_{0.33}As_{0.30}Sb_{0.70}$ barriers and $Ga_{0.85}In_{0.15}As_{0.13}Sb_{0.87}$ waveguide layers. Although the high gallium mole fraction in the waveguide layers results in a lower refractive index and larger bandgap than in the barriers, the index remains higher and the bandgap smaller than those of the InAs/AlSb cladding layers, preserving efficient waveguiding and efficient carrier injection.

The merits of the SCH structure are illustrated by direct comparisons to conventional MQW lasers with equal waveguide thicknesses. Fig. 12 shows the optical modes and Γ values of two such structures to be nearly identical. However, the active volume in the SCH structure is 3 times smaller than that in the MQW, resulting in about a factor of two reduction in threshold current density.

9.4.2 Device Fabrication

A schematic of a typical GaInSb/InAs-based laser is depicted in Fig. 7. Most devices have been deposited upon p+ GaSb substrates. Epitaxial growth begins with a Be-doped GaSb buffer layer followed by

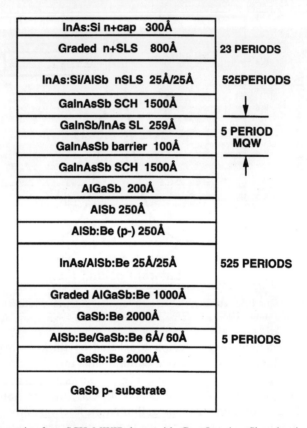

Fig. 11 Schematic of an SCH MWIR laser with $Ga_{0.67}In_{0.33}As_{0.30}Sb_{0.70}$ barriers and $Ga_{0.85}In_{0.15}As_{0.13}Sb_{0.87}$ separate confinement waveguide layers.

an $Al_xGa_{1-x}Sb$ layer compositionally graded from $x = 0.2$ to $x = 1.0$ to promote hole transport through the device. This is followed by growth of an AlSb:Be/InAs superlattice cladding layer, doped p-type to a density of approximately $10^{17}\,cm^{-3}$. The typical superlattice period is 40–50 Å, with InAs layers only slightly thicker than AlSb, to achieve a strain-balanced structure. The 2+ μm thick (≈ 500 period) clad is readily grown lattice matched to a GaSb substrate; ease of control of the thicknesses of the InAs and AlSb layers routinely yields lattice mismatches between the superlattice and the substrate of less than 0.1%. The excellent resultant structural quality yields material which is relatively free of misfit dislocations, serving as a good laser cladding layer [14,15] free of nonradiative recombination centers.

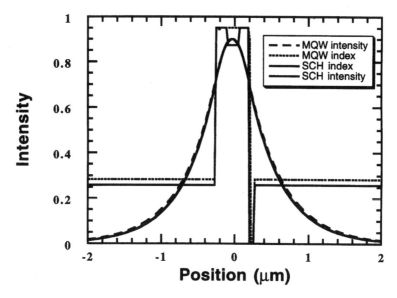

Fig. 12 Comparison of the optical mode profiles in a conventional MQW laser and a SCH MQW laser.

Although AlGaAsSb cladding layers have been investigated as alternatives to InAs/AlSb superlattices, we have observed greater reproducibility and have achieved consistently smaller lattice mismatches for the superlattice clads. The composition of AlGaAsSb is extremely sensitive to temperature (the As incorporation rate is strongly temperature dependent) and therefore difficult to control within required tolerances to thicknesses of 2 μm or greater required for the cladding layers.

After the p-type clad, an AlSb electron blocking layer is grown. This layer does not hinder hole injection from the p-type clad, but serves as an electron barrier, preventing electrons injected from the n-type clad from leaking out of the active region.

Most active layers are MQWs, consisting of GaInSb/InAs superlattice quantum wells and GaInAsSb barriers. $Ga_{0.75}In_{0.25}Sb$ layers are typically employed. The $Ga_{1-x}In_xAs_ySb_{1-y}$ composition is chosen to balance stresses in the superlattice wells and to confine both electrons and holes in the wells, without imposing a barrier to injection from the cladding layers.

Referring again to Fig. 7, growth of the active layer is followed by an n-type cladding layer. This consists of a strain-balanced AlSb/InAs:Si superlattice, doped to a density of approximately $10^{17}\,cm^{-3}$.

As mentioned in Section 9.3.2, use of this modulation doped superlattice is particularly desirable as it obviates use of Te, which would be required with a conventional cladding layer such as AlGaAsSb. The n-type clad is followed by a digitally-graded AlSb/InAs:Si superlattice in which the AlSb layer thicknesses are reduced in each successive period. This serves to grade the band edges to those of InAs. The structure is capped with a 2000 Å Si-doped InAs layer. This and the substrate are good choices for contacting the devices; unannealed, evaporated gold forms excellent ohmic contacts to both p-type GaSb and n-type InAs.

Several different GaInSb/InAs superlattices have been employed to achieve lasing in the 3–5 μm spectral region. The thicknesses of the GaInSb and InAs layers and the $Ga_{1-x}In_xSb$ composition are varied to achieve wells tailored to each desired wavelength. As explained in Section 9.2.1, the parameters are carefully chosen to achieve band structures which minimize loss mechanisms such as Auger recombination. Much of the experimental effort to date has been devoted to 3.5 μm lasers, employing 39 Å/12.5 Å and 35 Å/16.7 Å $Ga_{0.75}In_{0.25}Sb$/InAs superlattices predicted to have favorable room temperature and cryogenic characteristics, respectively [21]. However, we note that the present theory is strictly applicable only to infinitely thick GaInSb/InAs superlattices, ignoring effects due to the finite widths of the superlattices employed in the MQW active layers. Small changes in the calculated optimum parameters can be expected as these quantum effects are added to the model.

9.4.3 Laser Results and Discussion

While not yet meeting their projected performance levels, initial MWIR diode lasers based on GaInSb/InAs broken-gap superlattices already hold several records for operating temperature [14–16]. As illustrated in Fig. 13, devices demonstrated to date span the 2.8–4.3 μm wavelength range. Rapid progress promises significant further improvements in temperatures and output powers, and is likely to significantly extend the spectral range over which these lasers are competitive.

Table II lists key characteristics of selected laser diodes. Although attempts have been made to investigate some critical device design issues, current studies and data are far from complete. Sample D is representative of "baseline" MQW laser samples. The laser employs an MQW active region consisting of eight superlattice wells bracketed by

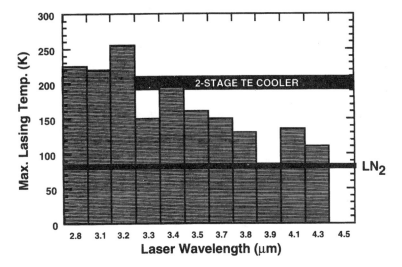

Fig. 13 Maximum operating temperature of GaInSb/InAs superlattices demonstrated to date, as a function of emission wavelength. Operation was pulsed with a 0.1% duty cycle. CW operation has been observed at roughly the same temperatures in those samples prepared in junction-side-up and -down mounts.

nine, 200 Å thick $Ga_{0.75}In_{0.25}As_{0.23}Sb_{0.77}$ barriers. Each well is comprised of five, 39 Å thick $Ga_{0.75}In_{0.25}Sb$ layers bracketing four, 12.5 Å thick InAs layers. This combination of a high number of relatively thick wells yielded the highest maximum lasing temperature (255 K) under 0.1% duty cycle, pulsed operation. Emission was at 3.19 μm at this temperature. Fig. 14 depicts a family of light output vs. current input $(L-I)$ characteristics for this device at various temperatures. Greater power output at 160 K than at 140 K is likely due to carrier freezeout in the cladding layers. The lowest threshold current density for this structure was found to be 760 A/cm^2, and was observed at 140 K for a broad stripe device with uncoated facets. A plot of the threshold current as a function of temperature is shown in Fig. 15. Fitting the expression $I = I_0 \exp(T/T_0)$ to the data yields a T_0 value of 86 K up to a temperature of 200 K. However, the significance of this number is unclear due to complications such as carrier freezeout. At higher temperatures T_0 drops to 33 K, a typical value for 3–4 μm lasers [18,33,34]. However, we stress that the particular GaInSb/InAs structure employed here is likely not ideal for reduced Auger recombination as the calculations on which the choice was

TABLE II
Broken-gap superlattice laser parameters.

Sample	Active layer	Well	Barrier	Wells	Wavelength
A	$3.5 \times \{Ga_{0.75}In_{0.25}Sb(39\ \text{Å})/InAs(12.5\ \text{Å})\}$	193 Å	200 Å	8	2.80 μm (225 K)
B	$3.5 \times \{Ga_{0.75}In_{0.25}Sb(39\ \text{Å})/InAs(12.5\ \text{Å})\}$	167 Å	200 Å	8	3.08 μm (190 K)
C	$4.5 \times \{Ga_{0.75}In_{0.25}Sb(39\ \text{Å})/InAs(12.5\ \text{Å})\}$	245 Å	400 Å	5	3.09 μm (220 K)
D	$4.5 \times \{Ga_{0.75}In_{0.25}Sb(39\ \text{Å})/InAs(12.5\ \text{Å})\}$	245 Å	200 Å	8	3.19 μm (255 K)
E	$4.5 \times \{Ga_{0.81}In_{0.19}Sb(43\ \text{Å})/InAs(16.7\ \text{Å})\}$	282 Å	1050 Å	5	3.40 μm (195 K)
F	$4.5 \times \{Ga_{0.75}In_{0.25}Sb(35\ \text{Å})/InAs(24\ \text{Å})\}$	271 Å	400 Å	5	4.13 μm (135 K)
G	$4.5 \times \{Ga_{0.75}In_{0.25}Sb(35\ \text{Å})/InAs(24\ \text{Å})\}$	271 Å	800 Å	5	4.32 μm (110 K)

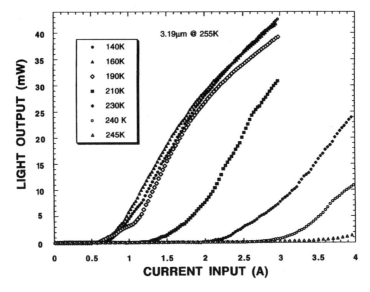

Fig. 14 Family of light output vs. current input ($L-I$) curves at various temperatures for sample D.

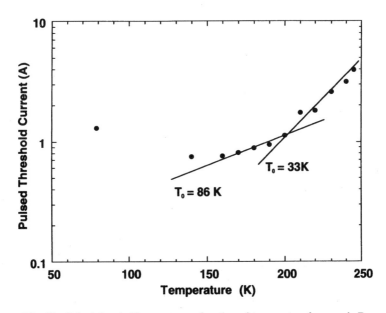

Fig. 15 Pulsed threshold current as a function of temperature for sample D.

based ignore effects due to the finite thickness of the wells. Examination of the threshold current density as a function of stripe width indicates that there is significant lateral current spreading in these devices, suggesting that currents can be reduced in ridge waveguide structures. Peak pulsed output powers of over 6 mW have been demonstrated.

Sample C is similar to sample D but has only five quantum wells and 400 Å thick barriers. Although sample D has a significantly larger optically active, quantum well volume, the two structures have similar total waveguide thicknesses and consequently similar optical confinement factors. The reduced number of wells yields a lower threshold current density (400 A/cm^2 at 140 K), but the maximum operating temperature is reduced to 220 K. Because they have identical quantum wells, the carrier concentration to reach population inversion and the gain per well should be the same in the two samples. Laser threshold is reached when the gain (which becomes positive when population inversion is attained) equals the external losses. Losses due to free carrier absorption will likely be greater in sample D than in C for wells populated to equivalent levels, but the greater number of wells yields higher overall gain in sample D. That the sample with more wells lases at higher temperatures suggests that the latter is the greater effect; over-coming external losses evidently requires populating the wells to an appreciable extent, so that losses which are nonlinear in carrier concentration (such as nonradiative Auger recombination) are significant, favoring structures for which a fixed gain can be achieved for lower well occupation. Assuming the dominant mechanism to be Auger recombination, the effect can be expected to become more important at higher temperatures. As is now widely recognized for a variety of MWIR laser systems, it is imperative to maintain low carrier concentrations to minimize losses due to Auger recombination, thereby reducing threshold currents and maximizing operating temperatures.

Sample A is again similar to sample D, but employs only 3.5 period GaInSb/InAs superlattice wells. Greater quantum confinement effects for these thin wells raise the emission wavelength to 2.79 μm at 225 K, which is the maximum lasing temperature for this device. Increased Auger recombination in the thinner wells is likely responsible for the poorer high temperature performance of this structure. In addition to a reduced active volume due to the thinner wells, Auger processes are likely promoted by a new electronic band structure for a system with such thin wells. Specifically, the greater energy gap of this structure brings into resonance different conduction and valence band excited states relevant to the Auger process, and quantum confinement itself changes the dispersion of the system.

The termination of the superlattice wells is changed in sample B from GaInSb to InAs. The sample is in other respects like sample C, utilizing 3.5 period, 39 Å/12.5 Å $Ga_{0.75}In_{0.25}Sb/InAs$ superlattice wells. The change of well termination alters the band structure of the active region and hence the Auger rate in the laser. This change is illustrated by the shift in lasing wavelength to 3.08 μm, at a maximum operating temperature of 190 K. The shift is in accord with simple theoretical considerations, as is the reduced T_{max}, in view of the altered intrinsic loss mechanisms.

Different samples have been grown to achieve other emission wavelengths and to confirm the utility of improved optical confinement factors. Sample E utilizes a 43 Å/16.7 Å $Ga_{0.81}In_{0.19}Sb/InAs$ superlattice to achieve lasing at 3.4 μm, up to 195 K. Five quantum wells are bracketed by six, 1050 Å $Ga_{0.81}In_{0.19}As_{0.17}Sb_{0.83}$ barriers, yielding a good optical confinement factor despite the small number of wells. Facet coated devices with pulsed threshold current densities of 2.5 kA/cm^2 at 140 K have been demonstrated. Peak powers of 5.3 mW are typical for this sample. We speculate that higher temperature operation can likely be achieved by employing more quantum wells.

A 4.17 μm diode (sample F) has lased pulsed up to 135 K with a 4.4 mW peak power. A minimum threshold current density of 2.1 kA/cm^2 was measured at 120 K. The structure consists of five 35 Å/24 Å $Ga_{0.75}In_{0.25}Sb/InAs$ superlattice quantum wells bracketed by six 400 Å thick $Ga_{0.75}In_{0.25}As_{0.23}In_{0.77}$ barriers. Multiple longitudinal modes are evident in the laser spectrum for a broad area device (Fig. 16).

The longest laser of this type reported to date (sample G) is a 4.32 μm device nominally identical in all respects to the 4.17 μm diode described above, except that the barriers are 800 Å thick, rather than 400 Å. We would expect the thicker barriers to increase the quantum confinement slightly, resulting in a shorter emission wavelength. However, probable depletion of the indium supply in the MBE machine likely resulted in slightly thinner InAs layers in the superlattices of sample F than in sample G, resulting in the observed decrease in wavelength. Sample G lased up to 110 K, with a threshold current density of 4.9 kA/cm^2. The peak pulsed power at 110 K is 0.54 mW.

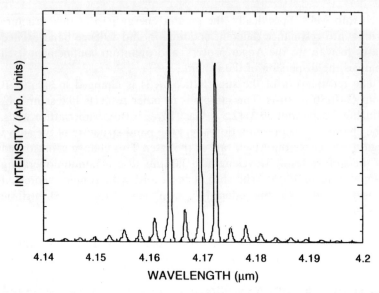

Fig. 16 Emission spectrum for a 4.17 μm broad area laser operating pulsed at 110 K.

9.5 CONCLUSIONS

Calculations of reduced Auger and intersubband absorption losses suggest that GaInSb/InAs superlattice-based lasers are the most promising of the candidates currently being considered for higher temperature MWIR operation [21]. The observation of slightly higher values of T_0 supports the contention that these losses are reduced in the superlattice-based lasers relative to those based on GaInAsSb, although measured values do not yet match predictions. This discrepancy between observed and ideal performance is likely due in part to nonoptimal choices of superlattice active regions, as the GaInSb/InAs wells employed to date are sufficiently thin to have electronic properties deviating appreciably from those of thick superlattices. This does not represent a fundamental limitation, but requires that barriers also be taken into account in calculating band structures conducive to low Auger and absorption losses. Such work is presently underway.

While results do not yet match expectations, progress in these lasers has been exceptionally rapid, and has already yielded state-of-the-art devices. Devices employing thicker (4.5 period) superlattices appear to display higher maximum operating temperatures than those with thinner (3.5 period) wells, although it remains unclear whether InAs-terminated wells are better than those terminated with GaInSb. Increasing the number of quantum wells from five to eight appears to raise maximum operating temperatures, at the expense of raising low temperature threshold currents.

It seems probable that considerable further improvements will be realized by employing recent enhancements in growth techniques and by refining device design. As with extended band structure models accounting for the finite thicknesses of the GaInSb/InAs wells, transport calculations will likely contribute significantly to the design of better lasers, e.g. improving the homogeneity of carrier distributions from one optically active well to another. Fabrication processes can also be improved, and high reflection coatings employed. Higher maximum lasing temperatures, lower thresholds, and higher powers should result.

Acknowledgments

We gratefully acknowledge the guidance of H. Ehrenreich of Harvard University, C. H. Grein of the University of Illinois at Chicago, and M. E. Flatté of the University of Iowa on issues of superlattice design. We are indebted to our collaborators A. R. Kost, L. West, and D. H. Chow at HRL, and thank H. Dunlap, C. Haeussler, and L. Warren for expert technical assistance. This work was supported in part by the Air Force/Phillips Laboratory under Contract No. F29601-93-C-0037.

References

1. C. R. Webster, R. D. May, C. A. Trimble, R. G. Chave, and J. Kendall, *Applied Optics* **33**, 454 (1994).
2. P. S. Lee, R. F. Majkowski, and R. M. Schreck, "Method for determining fuel and engine oil consumption using tunable diode laser spectroscopy," US Patent No. 4,990,780, issued February 5, 1991.
3. P. S. Lee, R. F. Majkowski, and T. A. Perry, *IEEE Transactions on Biomedical Engineering* **38**, 966 (1991).
4. G. Papen, private communication.
5. Z. Shi, M. Tacke, A. Lambrecht, and H. Böttner, *Appl. Phys. Lett.* **66**, 2537 (1995).
6. M. Zandian, J. M. Arias, R. Zucca, R. V. Gil, and S. H. Shin, *Appl. Phys. Lett.* **59**, 1022 (1991).
7. H. Lee, P. K. York, R. J. Menna, R. U. Martinelli, D. Z. Garbuzov, S. Y. Narayan, and J. C. Connolly, *IEEE International Laser Conference Digest* **14**, PD15 (1994).
8. H. K. Choi, S. J. Eglash, and G. W. Turner, *Appl. Phys. Lett.* **64**, 2474 (1994).
9. H. K. Choi and G. W. Turner, *Appl. Phys. Lett.* **67**, 332 (1995).
10. H. K. Choi, G. W. Turner, and H. Q. Le, *Appl. Phys. Lett.* **66**, 3543 (1995).
11. S. J. Eglash and H. K. Choi, *Appl. Phys. Lett.* **64**, 833 (1994).

12. T. H. Chiu, W. T. Tsang, J. A. Ditzenberger, and J. P. van der Ziel, *Appl. Phys. Lett.* **49**, 1051 (1986).
13. H. K. Choi and S. J. Eglash, *Appl. Phys. Lett.* **59**, 1165 (1991).
14. T. C. Hasenberg, D. H. Chow, R. H. Miles, A. R. Kost, and L. West, *Electr. Lett.* **31**, 275 (1995).
15. R. H. Miles, D. H. Chow, T. C. Hasenberg, A. R. Kost, and Y.-H. Zhang, to appear in *Proceedings of the 7th International Narrow Gap Semiconductor Conference* (Santa Fe, 1995).
16. D. H. Chow, R. H. Miles, T. C. Hasenberg, A. R. Kost, Y.-H. Zhang, H. L. Dunlap, and L. West, *Appl. Phys. Lett.* **67**, 3700 (1995).
17. Carlo Sirtori, Jerome Faist, Federico Capasso, Deborah L. Sivco, Albert L. Hutchinson, and Alfred Y. Cho, *Appl. Phys. Lett.* **66**, 3242 (1995).
18. A. N. Baranov, A. N. Imenkov, V. V. Sherstnev, and Yu. P. Yakovlev, *Appl. Phys. Lett.* **64**, 2480 (1994).
19. R. J. Menna, D. R. Capewell, Ramon U. Martinelli, P. K. York, and R. E. Enstrom, *Appl. Phys. Lett.* **59**, 2127 (1991).
20. S. R. Kurtz, R. M. Biefeld, A. A. Allerman, A. J. Howard, M. H. Crawford, and M. W. Pelczynski, *Appl. Phys. Lett.* **68**, 1332 (1996).
21. M. E. Flatté, C. H. Grein, H. Ehrenreich, R. H. Miles, and H. Cruz, *J. Appl. Phys.* **78**, 4552 (1995).
22. E. R. Youngdale, J. R. Meyer, C. A. Hoffman, F. J. Bartoli, C. H. Grein, P. M. Young, H. Ehrenreich, R. H. Miles, and D. H. Chow, *Appl. Phys. Lett.* **64**, 3160 (1994).
23. J. R. Meyer, C. A. Hoffman, F. J. Bartoli, and L. R. Ram-Mohan, *Appl. Phys. Lett.* **67**, 757 (1995).
24. J. I. Malin, C. L. Felix, J. R. Meyer, C. A. Hoffman, J. F. Pinto, C.-H. Lin, P. C. Chang, S. J. Murray, and S.-S. Pei, *Electron. Lett.* **32**, 1593 (1996).
25. R. H. Miles, D. H. Chow, and W. J. Hamilton, *J. Appl. Phys.* **71**, 211 (1992).
26. D. H. Chow, T. C. McGill, and R. H. Miles, *J. Cryst. Growth* **111**, 683 (1991).
27. R. H. Miles, D. H. Chow, Y.-H. Zhang, P. D. Brewer, and R. G. Wilson, *Appl. Phys. Lett.* **66**, 1921 (1995).
28. J. P. Omaggio, J. R. Meyer, R. J. Wagner, C. A. Hoffman, M. J. Yang, D. H. Chow, and R. H. Miles, *Semicond. Sci. and Technol.* **8**, S112 (1993).
29. D. H. Chow, Y.-H. Zhang, R. H. Miles, and H. L. Dunlap, *J. Cryst. Growth* **150**, 879 (1995).
30. Landolt-Börnstein, "*Numerical Data and Functional Relationships in Science and Technology*", New Series III/17a, Eds. O. Madelung, M. Schulz, and H. Weiss (Springer, Heidelberg, 1982).
31. "Integrated Optics," edited by T. Tamir, "Chapter 2 – Theory of dielectric wave guides," H. Kogelnik (Springer-Verlag, New York, 1975).
32. H. C. Casey, Jr. and M. B. Panish, "*Heterostructure Lasers, Part B*", pp. 197–207 (Academic Press, Orlando, 1978).
33. H. K. Choi and G. W. Turner, *Appl. Phys. Lett.* **67**, 332 (1995).
34. S. R. Kurtz, R. M. Biefeld, L. R. Dawson, K. C. Baucom, and A. J. Howard, *Appl. Phys. Lett.* **64**, 812 (1994).

CHAPTER 10

InAs/InAs$_x$Sb$_{1-x}$ Type-II Superlattice Midwave Infrared Lasers

YONG-HANG ZHANG*

Hughes Research Laboratories, 3011 Malibu Canyon Road, Malibu, CA 90265, USA

10.1.	Introduction	462
10.2.	Theoretical Modeling	465
	10.2.1. Strain effects in InAs/InAs$_x$Sb$_{1-x}$ materials system	467
	10.2.2. Wavefunction overlaps	470
	10.2.3. Possible LWIR detectors applications	472
	10.2.4. Optical confinement	474
10.3.	Laser Structures, MBE Growth, and Materials Characterization	475
	10.3.1. Laser structures	475
	10.3.2. Modulated molecular beam epitaxial growth of group-V alloys	476
	10.3.3. Materials characterization	484
	10.3.3.1. X-ray diffraction measurements	484
	10.3.3.2. STM and TEM study of atomic scale structures	485
	10.3.3.3. Optical spectroscopy	488
10.4.	Optically Pumped High-Power MWIR Lasers	492
	10.4.1. Advantages of optically pumped MWIR lasers	492
	10.4.2. CW optically pumped MWIR lasers	493
	10.4.3. Quasi-CW optically pumped high-power MWIR lasers	495
10.5.	Conclusions	496
	Acknowledgments	497
	References	497

*Present address: Electrical Engineering Department, Arizona State University, Tempe, AZ 85287, USA.

10.1 INTRODUCTION

Midwave infrared (MWIR) (2–5 μm) light sources possessing very narrow spectral linewidth and high power output have great potential in applications such as trace-gas monitoring (remote pollution monitoring, medical diagnostics, explosive detection, etc.) and military countermeasures. Semiconductor lasers are ideal candidates for these light sources and possess advantages such as broad coverage of wavelength range, straightforward modulation, simplicity of operation, and compact size. At present, three semiconductor material systems are used to make these laser diodes: Pb salts, II/VI, and III/V compound semiconductors. Commercially, only Pb-salt laser diodes are available. Due to the poor thermal conductance of these materials, the maximum power output of Pb-salt double-heterostructure (DH) and multiple quantum well (MQW) lasers is smaller than 1 mW with pulsed operation near room temperature and with cw operation below 203 K [1–3]. In the case of narrow bandgap II/VI compound semiconductors, such as HgCdTe alloys, there are few publications available for lasers, although the material properties have been extensively studied for IR-detector applications. Only recently, HgCdTe lasers are reported to operate at 2.86 μm up to 90 K under pulsed electric injection [4] and at 3.2 μm up to 154 K under pulsed optical pumping [5]. Beside these two material systems, the quest for MWIR semiconductor lasers is mainly focused on III/V compound semiconductors. Not very long ago the most developed III/V semiconductor MWIR lasers were based on GaInAsSb/AlGaAsSb quaternary alloys grown on GaSb [6,7]. Room temperature or near room temperature operation of laser diodes at wavelengths shorter than 3.0 μm has been realized [7,8]. However, GaInAsSb quaternary alloys with bandgaps narrower than 0.37 eV (> 3.4 μm) are metastable because of a miscibility gap [9]. The limited availability of substrates and the miscibility gap in GaInAsSb quaternary alloys strongly restrict the possible choices of material compositions for MWIR laser structures. Due to these restrictions, DH and quaternary GaInAsSb MQW lasers can barely operate beyond 3.0 μm.

In order to increase operating temperature and to extend the operating wavelength range beyond 3.0 μm for MWIR semiconductor lasers, several other III/V material systems such as strained InAs/InGaAs [10] and InGaAs/InP [11] on InP substrates, InAsSb/AlAsSb on GaSb [12,13], InAsSb/InGaAs [14] and InAs(Sb,P)/InAsSbP on InAs [15–17] are explored. Since active regions of these lasers consist of type-I MQW or superlattices (SL), i.e. their bandedge diagram is direct in real space, the longest operating wavelength is then limited by the natural availability of the narrower bandgap alloys that can be practically grown on available substrates. The "quantum confinement" effect and compressive strains in the active layers result in further increase of effective bandgaps, such as the cases for $In_xAs_{1-x}Sb (x > 0.1)$ alloys on GaSb or InAs substrates. The increase of the bandgap is not favorable to broader coverage of longer operating wavelengths. In addition to all these efforts on traditional heterojunction lasers there is a "non-traditional" approach, i.e. the unipolar "quantum cascade laser" reported [18]. Tremendous progresses in providing broader wavelength coverage and larger power output have been made in the past a few years. Room temperature operation of 5 μm quantum cascade lasers has been demonstrated recently [19].

In contrast to type-I MQWs or SLs, the interband optical transition in type-II SLs is spatially indirect. Therefore, its effective bandgap can be narrower than any of the constituent materials. In general, there are two categories of type-II alignments for heterostructures, broken-gap alignment and staggered alignment (see Fig. 1). In the late 70's and

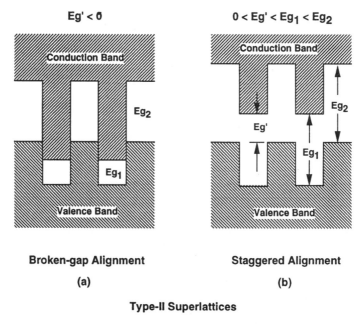

Fig. 1 Two kinds of bandedge alignment of type-II SLs: (a) Broken-gap alignment; (b) Staggered alignment. E'_g indicates the effective bandgap.

early 80's, Esaki and his group carried out detailed study of optical and transport properties of InAs/GaSb SLs. Their research has revealed that the bandedge alignment in these InAs/GaSb heterostructures and SLs is type-II (broken-gap alignment). Due to large potential barriers for both electrons and holes, the overlaps between electron and hole wavefunctions in InAs/GaSb SLs is very small. Such small wavefunction overlap restricts the potential applications of InAs/GaSb SL to optical devices such as detectors and lasers. Motivated by the search for new III/V compound semiconductor materials for IR-detector applications, InAs/InAs$_x$Sb$_{1-x}$ SL ($x < 0.6$, Sb rich) based on InSb substrate [20] was proposed. This SL has a type-II bandedge alignment (staggered alignment). Due to the large lattice mismatch to InSb substrate, thick device structures made of this material system have cracks. In order to improve the lattice-match condition while still remain the optical advantages of type-II SL for IR detectors, Smith and Mailhiot proposed InAs/Ga$_x$In$_{1-x}$Sb SLs which can be grown lattice matched (strain-balanced) to GaSb [21]. Theoretical calculations have shown that the wavefunction overlaps in these structures are much larger than those in InAs/GaSb SL cases, indicating very promising potential for IR-detector applications. Detailed theoretical modeling of band structures and Auger coefficients for InAs/Ga$_x$In$_{1-x}$Sb SLs has also shown that one can properly design SL structures to reduce Auger recombinations. In addition to the theoretical studies, successful material growth [22,23] and demonstration of prototype detectors have been reported [24].

Although various types of type-II SLs have been studied for IR-detector applications, it was not very clear whether they can also be utilized for lasers until recently. Kroemer proposed the use of type-II (staggered) heterostructures for tunable below-bandgap light emission [25]. In order to avoid additional uncertainties in the band alignments due to strain effects and to take advantage of the best interfaces that could be attained with existing technologies, he restricted the choices to lattice-matched systems, Al$_{0.48}$In$_{0.52}$As/InP, AlAs$_{0.56}$Sb$_{0.44}$/InP, and Ga$_{0.5}$In$_{0.5}$P/AlAs. Stimulated light emission from a type-II SL was first observed in one of these three material systems, Al$_{0.48}$In$_{0.52}$As/InP, at 2 K under pulsed high-power optical pumping [26]. The operation wavelengths are longer than that of the bandgap wavelengths of either InP or Al$_{0.48}$In$_{0.52}$As. Compared with type-I SLs or MQWs the overlap of the electron and hole wavefunctions in these Al$_{0.48}$In$_{0.52}$As/InP type-II SLs is extremely small, yielding a relative oscillator strength of only 0.04 compared with ~ 1 for a conventional type-I SL or MQW.

Due to such a small relative oscillator strength, there was some doubt about practical applications of these structures for lasers. However, for the case of mid- and far-infrared wavelength regions, theoretical [21,27] and experimental [22,23] studies on the optical properties of InAs/In$_x$Ga$_{1-x}$Sb type-II (broken-gap alignment) SLs have shown a reasonably large oscillator strength. Recently, MWIR lasers covering a broad wavelength range have been reported [28–30]. Details of the development of the MWIR laser based on this material system have been discussed in one of the previous chapters.

In this chapter we will focus on another material system, InAs/InAs$_x$Sb$_{1-x}$ type-II (staggered alignment) SLs, and their application to MWIR lasers [31]. Different from the Al$_{0.48}$In$_{0.52}$As/InP type-II SL, the present InAs/InAs$_x$Sb$_{1-x}$ type-II SL has a much extended electron state in conduction band, which results in a much larger electron- and hole-wavefunction overlap. Owing to a large valence band offset between InSb and InAs, this material system has the following advantages for MWIR laser applications [32]: (i) Only a small amount of InSb is needed in the InAs$_x$Sb$_{1-x}$ layers to reach 5 µm; (ii) The compressive strain introduced in the InAs$_x$Sb$_{1-x}$ layers and their layer thicknesses can be properly designed to reduce nonradiative Auger recombinations. In addition, this structure has an average lattice constant close to that of the InAs substrate, upon which a lattice-matched group-V alloy AlAs$_{0.16}$Sb$_{0.84}$ can be used as cladding layers, providing both optical and hole confinement. Our experimental results demonstrate that use of InAs/InAs$_x$Sb$_{1-x}$ type-II SL for high-power MWIR lasers is practical and very promising. The present chapter is organized as follows. In the Section 10.2, theoretical consideration and modeling of type-II InAs/InAs$_x$Sb$_{1-x}$ heterostructures, SL, and laser structures will be presented. In the Section 10.3, detailed growth technique and material characterization will be discussed. The results of cw and quasi-cw optically pumped MWIR lasers are given in the Section 10.4 followed by the conclusions in the Section 10.5.

10.2 THEORETICAL MODELING

Before we discuss strain effects and other electronic properties of InAs/InAs$_x$Sb$_{1-x}$ SLs, it is helpful to examine the bandedge alignments of the constituent semiconductors. Bandedge alignments are very important parameters in the design of type-II SL for lasers. In

Fig. 2 the relative bandedge alignments of most frequently used free-standing III/V semiconductors are drawn to the scale. The valence band offset values are listed in Table I [23,33–37]. It is worth noting that the valence band offset between free standing InAs/InSb is not well established. The number listed in the table for InAs/InSb is derived from the optical measurements reported in Ref. [23]. Another independent optical study of InSb/InAs$_x$Sb$_{1-x}$ ($x < 0.2$, Sb rich) on InSb substrates revealed a value of 410 ± 100 meV [38]. Recently, Imperial College group has reported a detailed study of the valence offset in InAs/InAs$_x$Sb$_{1-x}$ MQWs and SL grown on GaAs substrates

TABLE I
Band offsets between several III/V semiconductors.

Material Systems	GaSb/AlSb	GaSb/InAs	GaAs/AlAs	InAs/GaAs	InSb/GaSb	InSb/InAs
Valence Band Offset (eV)	0.40 [33]	0.51 [34]	0.55 [35]	0.17 [36]	0.1 [23]	0.61 [*]

*This value is derived from the optical measurements reported in Ref. [23].

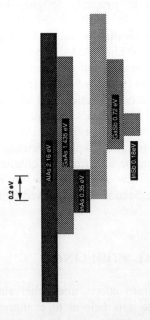

Fig. 2 Bandedge alignments of some unstrained III/V semiconductors. The figure is drawn to scale and the band offsets used are listed in Table I.

by using photoluminescence spectroscopy (PL), photoluminescence excitation spectroscopy (PLE), and theoretical modeling [39–41]. They have reported that a valence band offset between free standing InAs/InSb is between 710 meV to 880 meV. A theoretical value of 500 meV is predicted based on the first-principle band-structure calculations [42].

10.2.1 Strain Effects in InAs/InAs$_x$Sb$_{1-x}$ Materials System

Due to a large difference between lattice constants of InAs and InSb, InAs and InAs$_x$Sb$_{1-x}$ ($x > 0$) alloys are lattice mismatched. For sufficiently thin layers the lattice-mismatch is accommodated by coherent strain in the individual layers. This strain modifies the bandgap and splits the heavy-hole and light-hole bands. As an example, let us consider a heterojunction consisting of unstrained InAs and coherently strained InAs$_x$Sb$_{1-x}$ grown on an InAs substrate. As shown in Fig. 3, the bandedge alignment is type-II (staggered alignment). The optical

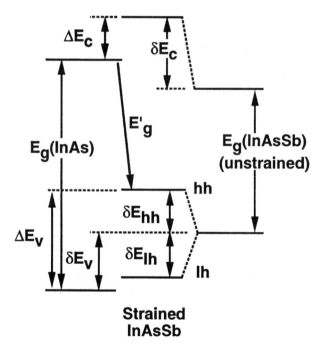

Fig. 3 Schematic bandedge alignment between unstrained InAs and compressively strained InAs$_x$Sb$_{1-x}$.

transition across the heterojunction is spatially indirect. Neglecting the quantum confinement energies, the optical transition across the heterojunction gives the lower limit of the effective bandgap of InAs/InAs$_x$Sb$_{1-x}$ MQWs or SLs. By assuming that the hydrostatic strain only causes a shift of conduction bandedge and biaxial strain splits the valence band heavy-hole and light-hole maxima, the shifts of heavy-hole bandedge (δE_{hh}), light-hole bandedge (δE_{lh}), and conduction bandedge (δE_c) in a coherently strained layer can be described by [38]

$$\varepsilon_{\parallel} = (a_{sub} - a_{epi})/a_{sub}$$

$$\delta E_c = 2[-5.7 + 0.13(1-x)](1 - C_{12}/C_{11})\varepsilon_{\parallel}$$

$$\delta E_{lh} = 1.8(1 + 2C_{11}/C_{12})\varepsilon_{\parallel}$$

$$\delta E_{hh} = -1.8(1 + 2C_{11}/C_{12})\varepsilon_{\parallel}$$

where $\varepsilon_{\parallel}$ is the in-plane (perpendicular to the growth direction) compressive strain, a_{sub} the substrate lattice constant, a_{epi} the epilayer lattice constant, and C_{11} and C_{12} the elastic constants. The bandgap of unstrained InAs$_x$Sb$_{1-x}$ as function of InAs mole fraction (x) and temperature (T) is given in eV by [43]

$$E_g(\text{InAs}_x\text{Sb}_{1-x}) = 0.411 - \frac{3.4 \times 10^{-4} T^2}{210 + T} - 0.876(1-x)$$
$$+ 0.70(1-x^2) + 3.4 \times 10^{-4}(1-x)xT^2.$$

Thus, as shown schematically in Fig. 4, the effective bandgap (E'_g) of the spatially indirect optical transition across the InAs/InAs$_x$Sb$_{1-x}$ heterojunction is

$$E'_g = E_g(\text{InAs}) - \delta E_{hh} - (1-x)\Delta E_{gv}(\text{InAs/InSb}),$$

where $\Delta E_{gv}(\text{InSb/InAs})$ is the valence band offset between free standing InSb and InAs. Based on the material parameters [14,44] given in

TABLE II

Materials properties of InSb and InAs. C_{11}, C_{12}, a_0, m_e^*, m_{hh}^*, and m_{lh}^* are elastic constants, lattice constant, and effective masses for electron, heavy hole, and light hole, respectively [14,44]. The corresponding parameters for alloy are obtained by using linear interpolation.

Materials	m_e^*/m_0	m_{hh}^*/m_0	m_{lh}^*/m_0	C_{11} (10^{12} dynes/cm^2)	C_{12} (10^{12} dynes/cm^2)	a_0 (Å)
InSb	0.014	0.34	0.015	0.6652	0.3351	6.497
InAs	0.024	0.35	0.025	0.8465	0.5001	6.058

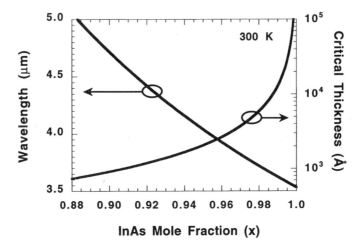

Fig. 4 Optical transition energy across InAs/InAs$_x$Sb$_{1-x}$ heterojunction and critical thickness of coherently strained In$_x$As$_{1-x}$Sb grown on InAs are plotted as function of InAs mole fraction (x) in the InAs$_x$Sb$_{1-x}$ layer.

Table II and assuming $\Delta E_{gv}(\text{InSb/InAs}) = 0.61$ eV, the calculated optical transition wavelength, $\lambda = 1.24/E'_g$, at 300 K is plotted in Fig. 4 as a function of InAs mole fraction (x) in InAs$_x$Sb$_{1-x}$. Only 12% InSb is needed to reach 5 µm at 300 K. Such a small InSb mole fraction facilitates the growth of strained InAs$_x$Sb$_{1-x}$ on InAs substrates. As to be discussed below the actual $\Delta E_{gv}(\text{InSb/InAs})$ is probably larger than what is used for the calculations. If that is the case, even less Sb is needed to cover the same wavelength range.

In order to properly design the InAs/InAs$_x$Sb$_{1-x}$ type-II SL active region for lasers without causing any misfit dislocations, it is necessary to know the critical thickness of coherently strained InAs$_x$Sb$_{1-x}$ grown on InAs. Based on the Matthews–Blakeslee model [45], the critical thickness (h_c) for InAs$_x$Sb$_{1-x}$ capped by InAs is given by

$$h_c = \frac{b}{\pi \varepsilon_\parallel} \frac{1 - 0.25 v}{(1+v)} \left[\ln\left(\frac{h_c}{b}\right) + 1 \right],$$

where $b = 4.3$ Å is the Burgers vector length for InAs while v the Poison's ratio. The calculated h_c as function of InAs mole fraction (x) in InAs$_x$Sb$_{1-x}$ are plotted in Fig. 4. For capped InAs$_{0.88}$Sb$_{0.12}$ the h_c on InAs substrate is close to 1000 Å. Such a thick h_c makes it practical to use strained InAs/InAs$_x$Sb$_{1-x}$ SL or MQWs for laser structures.

10.2.2 Wavefunction Overlaps

Electron and hole quantum-confinement energies need to be taken into account to determine the effective bandgap of a SL. Based on the envelope function approximation [46] and the Kronig-Penney model, the ground state confinement energy E_n can be obtained by numerically solving the following eigenvalue equation

$$\frac{(\beta(E_n)/m_b^*)^2 - (\alpha(E_n)/m_w^*)^2}{2[\alpha(E_n)/m_w^*][\beta(E_n)/m_b^*]} \sin[\alpha(E_n)d_w]\sinh[\beta(E_n)d_b]$$
$$+ \cos[\alpha(E_n)d_w]\cosh[\beta(E_n)d_b] = \cos[k_i(d_w + d_b)],$$

where

$$\alpha(E_n) = \frac{\sqrt{2m_w^* E_n}}{\hbar},$$

$$\beta(E_n) = \frac{\sqrt{2m_b^*(V_b - E_n)}}{\hbar},$$

$k_i = 0$, and m_w^*, m_b^*, d_w, and d_b are the carrier effective masses and layer thicknesses for wells and barriers, respectively. V_b is the barrier potential height. As shown in Fig. 3, for the case of conduction (valence) band, V_b is equal to the corresponding band offset ΔE_c (ΔE_v) between InAs and compressively strained $InAs_xSb_{1-x}$ layers, where

$$\Delta E_v = \delta E_v + \delta E_{hh},$$
$$\Delta E_c = E_g(\text{unstrained } InAs_xSb_{1-x}) + \delta E_v + \delta E_c - E_g(\text{InAs}).$$

We mentioned earlier some concerns about $AlAs_{0.56}Sb_{0.44}/InP$ type-II SLs used for the laser active region because of the small overlap between electron and hole wavefunctions. However, it is not the case for $InAs/InAs_xSb_{1-x}$ type-II SLs. A schematic bandedge diagram is shown in Fig. 5 by dash lines for a typical 77 Å-InAs/23 Å-$InAs_{0.926}Sb_{0.074}$ SL structure used in this study. By using the formula given above, the barrier heights for electrons and heavy-holes are obtained as 11 meV and 65 meV, respectively. The calculated electron ground state has a confinement energy of 2.5 meV while that for hole ground state is 38 meV. The nonparabolicity of the InAs conduction band has not been taken into account because of the small electron confinement energy. A comparison of these theoretical calculations with experimental results will be given in Section 10.4.

Due to the small electron effective masses in InAs and InSb and low potential barriers in conduction band, the amplitude of the electron

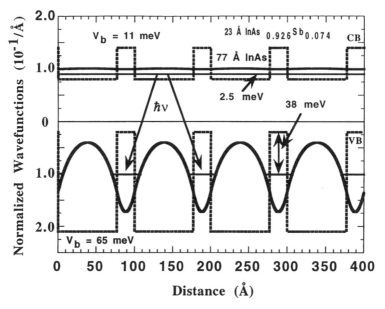

Fig. 5 Thick solid lines give the normalized wavefunctions for electrons and holes. The optical transition in this structure is spatially indirect. The overlap of electron and hole wavefunctions is 0.89, yielding a relative oscillator strength of 0.79 along the growth direction. The dash lines present the bandedge diagram while thin solid lines show the ground states of the conduction band and the valence band, respectively. The electron confinement energy is 2.5 meV while that for heavy holes is 38 meV. The corresponding barrier heights (V_b) are 11 meV and 65 meV for electrons and heavy holes, respectively.

wavefunction in the $InAs_xSb_{1-x}$ barriers is almost the same as in the InAs wells. In order to quantitatively evaluate the ground-state wavefunction overlap, the electron and heavy-hole wavefunctions along the growth direction normalized to the unit cell of the SL are calculated and plotted as solid lines in Fig. 5. The electron wavefunction extends like a plane wave. In contrast, the heavy-hole wavefunction in the valence band is localized because of the heavy effective masses of the constituent materials. The overlap between these two wavefunctions is calculated to be 0.89. Since the oscillator strength of optical transitions in QW or SL structures is proportional to the square of the wavefunction overlap, the 77 Å-InAs/23 Å-$InAs_{0.926}Sb_{0.074}$ type-II SL structure thus has a relative oscillator strength of 0.79 along the growth direction, as compared to ~ 1 for a conventional type-I quantum well or SL. Such a large relative oscillator strength, which is desired for large radiative recombination probability, is one of the most important properties making the InAs/InAs$_x$Sb$_{1-x}$ type-II SL practical for laser active regions. It is worth noting that no Coulomb interaction has been included in the previous calculation. In real laser structures, since the injection level is relatively high (on the order of 10^{18} cm^{-3}) the Coulomb interaction between carriers should not be neglected. If this effect is taken into account, the potential profile of the InAs/InAs$_x$Sb$_{1-x}$ SL will be changed and both electron and hole wavefunctions will penetrate deeper into their corresponding barrier layers, yielding an even larger overlap of electron and hole wavefunctions and narrower effective bandgap.

In addition to the typical sample structure, theoretical modeling of other structures with different compositions and layer thicknesses shows that the overlap between the electron and heavy-hole wavefunctions is a slow function of well width and InAs$_x$Sb$_{1-x}$ composition within the MWIR region (3–5 μm) and beyond. In contrast to the case of type-I SL such as GaAs/AlGaAs SL, where the effective bandgap is a strong function of well width, this finding provide the foundation that why optical measurements can determine the band offset for type-II SLs accurately.

10.2.3 Possible LWIR Detectors Applications

InAs/InAs$_x$Sb$_{1-x}$ SLs are promising candidates not only for IR lasers but also for IR detectors. A 71 Å-InAs/21 Å-InAs$_{0.61}$Sb$_{0.39}$ SL strain-balanced structure lattice matched to GaSb substrate for 10 μm (longwave IR, LWIR) detector application has been proposed [47]. The strains in this structure are balanced by carefully choosing the thickness ratio between InAs and InAs$_{0.61}$Sb$_{0.39}$. Fig. 6 shows the calculated band structure of the 71 ÅInAs/21 Å-InAs$_{0.61}$Sb$_{0.39}$ SL ($E_g \sim 10$ μm) [48]. Biaxial strains lift the degeneracy of the heavy-hole and light-hole bandedges. The corresponding SL heavy-hole and light-hole bands are therefore split. When the splitting exceeds the energy gap, the Auger recombinations of the carriers are suppressed in p-type materials. The values of the theoretical detectivity (D*) for the 10 μm 71-Å-InAs/21 Å-InAs$_{0.61}$Sb$_{0.39}$ SL, 11 μm 49.8 Å-InAs/15 Å-In$_{0.4}$Ga$_{0.6}$Sb SL, and 11 μm bulk HgCdTe are plotted in Fig. 7 [48]. Nonparabolic bands and momentum-dependent matrix elements are employed for the calculations of Auger and radiative recombination lifetimes. Due to suppression of Auger recombination pathways in both SLs, the minority carrier lifetime (τ_n) in the p-type materials of an n-on-p photodiode is dominated by radiative recombination below 100 K in both SL

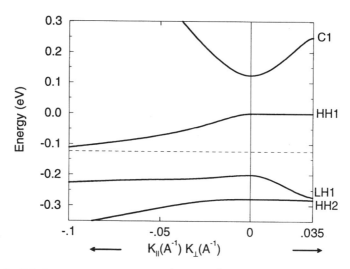

Fig. 6 Calculated band structure of a 71 Å-InAs/21 Å-InAs$_x$Sb$_{1-x}$ type-II SL structure for long-wavelength IR detector applications at 10 μm. Bands are plotted in the in-plane (∥) and growth-axis (⊥) directions [48].

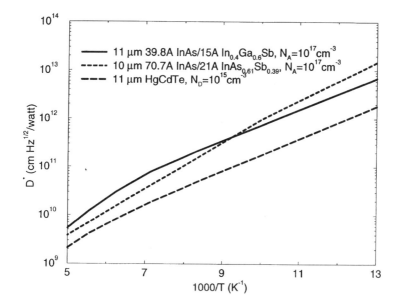

Fig. 7 Comparison of the theoretical specific detectivities for long-wavelength IR detectors with 71 Å-InAs/21 Å-InAs$_x$Sb$_{1-x}$ type-II SL (10 μm), 39.8 Å-InAs/15 Å-InGa$_{0.4}$Sb$_{0.6}$ type-II SL (11 μm), and bulk HgCdTe (11 μm) active layers [48].

systems. The larger bandgap of 70.9 Å-InAs/21 Å-InAs$_{0.61}$Sb$_{0.39}$ SL structure results in lower minority carrier concentrations, hence somewhat higher detectivity than that of 49.8 Å-InAs/15 Å-In$_{0.4}$Ga$_{0.6}$Sb SL below 100 K at which most high performance LWIR detectors operate. In comparison, the minority carrier lifetimes of bulk Hg$_x$Cd$_{1-x}$Te are dominated by Auger recombination over the plotted temperature range. Therefore, both LWIR SLs are predicted to possess greater detectivities than bulk Hg$_x$Cd$_{1-x}$Te. Detailed discussion about the theoretical modeling can be found in Ref. [48]. PL study of a 172 Å-InAs/52 Å-InAs$_{0.76}$Sb$_{0.24}$ SL reveals a bandgap of 7.2 μm at 10 K, which is in agreement with the model described in the previous sections [49].

10.2.4 Optical Confinement

In the previous sections we have discussed mainly the structure of active region for MWIR lasers. In addition to the active region, the optical confinement layers (cladding layers) play also important roles in practical laser structures. AlAs$_{0.16}$Sb$_{0.84}$ lattice matched to InAs substrate is a very attractive candidate for cladding layers. It has large

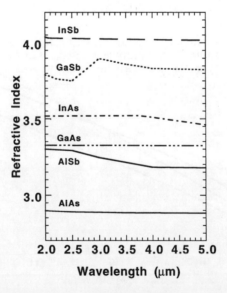

Fig. 8 Refractive indices for AlAs, AlSb, GaAs, GaSb, InAs, and InSb as function of wavelength within the wavelength range of 2–5 μm. The curve for AlAs is obtained based on the model given in Ref. [50] and the rest are experimental data [51].

bandgap and small refractive index. As shown in Fig. 8, refractive indexes of InAs, InSb, AlAs, and AlSb are plotted for MWIR wavelength range (2–5 μm) [50]. Since there are no detailed experimental data available for refractive indexes of AlAs in the wavelength range of 2–5 μm, a theoretical curve obtained by using the formalism in the Ref. [51] is plotted. By linear interpolation, we can derive the refractive index difference of 0.36 between InAs and $AlAs_{0.16}Sb_{0.84}$. Such a large refractive index difference make it possible for thin cladding layers to provide sufficient optical confinement, facilitating the MBE growth of these layers.

10.3 LASER STRUCTURES, MBE GROWTH, AND MATERIALS CHARACTERIZATION

10.3.1 Laser Structures

The layer structure and schematic bandedge diagram of the investigated laser structures are given in Fig. 9. A 0.2 μm InAs buffer is first grown on unintentionally doped InAs substrates and followed by a 1.0 to 1.5 μm thick $AlAs_{0.16}Sb_{0.84}$ ordered-alloy cladding layer. The active region, an $InAs/InAs_xSb_{1-x}$ type-II SL, is sandwiched by two 0.2 to 0.4 μm thick InAs waveguide layers. These two waveguide layers play also another very important role in absorbing the optical pumping light. For different samples, the $InAs/InAs_xSb_{1-x}$ SL active regions have different compositions in $InAs_xSb_{1-x}$, layer thicknesses, and numbers of periods. On top of the active and waveguide regions, another 1.0 to 1.5 μm thick $AlAs_{0.16}Sb_{0.84}$ ordered-alloy cladding layer and a 500 Å to 1000 Å thick InAs cap layer are grown. This relatively thick InAs cap layer is used to prevent the oxidization of $AlAs_{0.16}Sb_{0.84}$ in air. As already mentioned in the Section 10.2.4, due to large refractive index difference between InAs and $AlAs_{0.16}Sb_{0.84}$, only 1-μm cladding layers are sufficient thick to provide strong optical confinement. It is also worth mentioning that from Fig. 2 one can see that the valence bandedge of AlAs lies very low in energy. As a result, $AlAs_{0.16}Sb_{0.84}$ lattice-matched to InAs has a valence bandedge of 25 meV lower than that of InAs. Such a low lying valence bandedge is extremely important for MWIR lasers to provide hole confinement.

Since there is no n-type dopant (such as Te) for antimonides in the MBE chamber used for the growths, the structures in present study are unintentionally doped and specially designed for optical pumping. However, the structures discussed in this chapter are ready for

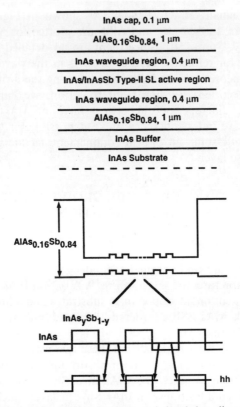

Fig. 9 Schematic sample layer structure and bandedge diagram of the InAs/InAs$_x$Sb$_{1-x}$ type-II SL laser structures.

electrically injected laser diodes if Te doping is available. As a matter of fact, Choi and Turner from MIT Lincoln Lab have just demonstrated state-of-the-art high-power electrically injected MWIR laser by using Te doped AlAs$_{0.16}$Sb$_{0.84}$ cladding layers grown on InAs substrates [52]. In addition to laser structures, InAs/InAs$_x$Sb$_{1-x}$ type-II SLs lattice matched to GaSb have also been studied. The sample structures used in present study are summarized in Table III.

10.3.2 Modulated Molecular Beam Epitaxial Growth of Group-V Alloys

Molecular Beam Epitaxy (MBE) is the most frequently used growth technique for antimonides and arsenic/antimonide alloys, although

TABLE III
Sample structures.

Sample No.	Substrate	Cladding Layers	Waveguide Layers	Active Layer	Cap Layer
#1	InAs	$1.5\,\mu m$ $AlAs_{0.16}Sb_{0.84}$	—	$1\,\mu m$ InAs	200 Å InAs
#2	InAs	$1\,\mu m$ $AlAs_{0.16}Sb_{0.84}$	$0.2\,\mu m$ InAs	30 period 77 Å-InAs/23 Å-InAs$_{0.92}$Sb$_{0.08}$	200 Å InAs
#3	InAs	$1\,\mu m$ $AlAs_{0.16}Sb_{0.84}$	$0.5\,\mu m$ InAs 22 Å-InAs/22 Å $AlAs_{0.16}Sb_{0.84}$	100 period 37 Å-InAs/15 Å-InAs$_{0.92}$Sb$_{0.08}$	1500 Å InAs
#4	GaSb	—	—	30 period 172 Å-InAs/52 Å-InAs$_{0.76}$Sb$_{0.24}$	0
#5	InAs	$1\,\mu m$ $AlAs_{0.16}Sb_{0.84}$	$0.2\,\mu m$ InAs	10 period 54 Å-InAs/54 Å-InAs$_{0.91}$Sb$_{0.09}$	200 Å InAs
#6	InAs	$1\,\mu m$ $AlAs_{0.16}Sb_{0.84}$	$20.2\,\mu m$ InAs	6 period 70 Å-InAs/97.4 Å-InAs$_{0.92}$Sb$_{0.08}$	200 Å InAs
#7	InAs	$1\,\mu m$ $AlAs_{0.16}Sb_{0.84}$	$0.24\,\mu m$ InAs	6 period 72.5 Å-InAs/101.5 Å-InAs$_{0.74}$Sb$_{0.26}$	200 Å InAs

Metal–Organic Chemical Vapor Deposition (MOCVD) has demonstrated the capability of growing high quality GaSb and InAsSb. In the present study a conventional solid-source MBE system equipped with Si, Be, Ga, In, Al, and cracking As and Sb cells is used to grow the samples. Unintentionally doped InAs wafers are used as substrates. The two cracking cells are used for providing As_2 and Sb beams [53]. The growth temperatures are measured by both a calibrated pyrometer and a thermocouple. The growth rate for all the epilayers is kept around 0.50 µm/hr. The As_2 and Sb beam equivalent pressures are measured with a flux gauge possessing two-digit accuracy. *In situ* reflection high energy electron diffraction (RHEED) is used to probe surface reconstruction patterns.

As mentioned above, $AlAs_{0.16}Sb_{0.84}$ and $InAs_xSb_{1-x}$, two group-V alloys, are used for our laser structures. For group-III alloys, such as InGaAs, GaAlAs etc., due to the near-unit sticking coefficients of most group-III elements, the control of compositions is relative straightforward, simply by changing the group-III flux ratios during growth. However, this is not true for group-V alloys such as GaInAsP, AlGaAsSb, InAsSb etc. The control of compositions of group-V alloys becomes problematic for MBE because of large differences in the incorporation rates of different group-V species [54–57]. Difficulties in achieving accurate control of group-V ratios in $GaAs_ySb_{1-y}$ and $AlAs_xSb_{1-x}$ layers lattice matched to InAs or GaSb are particularly acute because of: (i) The very high incorporation coefficient of As relative to that of Sb; (ii) The low As mole fractions in these ternary alloys [56–59]. Precise control of these alloy compositions is extremely important for the growth of optoelectronic devices consisting of several-micron thick epilayers. For conventional MBE, due to the long growth time of these thick layers, both fast fluctuations and slow drifts in group-V cell temperatures can result in composition fluctuations and gradient. Even very small deviations from the exact lattice-matched condition for thick cladding layers can cause relaxation of the strain, resulting in the formation of misfit dislocations. When these dislocations propagate through the active region in a laser structure, the device performance will then be degraded.

Different from conventional MBE, we use a straightforward and accurate means of controlling the incorporation of As and Sb by rapidly modulating As_2 and Sb beams during MBE [60,61]. This method will be referred to MMBE later on. By exploiting this method the MBE growth yields ordered $AlAs_xSb_{1-x}$ alloys (or AlAs/AlSb short-period superlattices) rather than random alloys. The procedures used in the present study for the MMBE growth of the $AlAs_xSb_{1-x}$ ordered alloys are as follows. During the growth the Al shutter is kept open all the time, while the As and Sb shutter are alternately opened and closed, respectively. The average AlAs mole fraction is then controlled by the As-shutter duty-cycle [$As_{shutter-time}/(As_{shutter-time} + Sb_{shutter-time})$]. A schematic shutter-sequence diagram is shown in Fig. 10(a). Typical As and Sb shutter-time are 0.4 s and 8.1 s, respectively. For an average growth rate of 1.4 Å/s, this modulation of the group-V beams results in a spatial periodic composition variation with a nominal period of about 12 Å (~ 4 atomic monolayers). The As_2 flux is kept around 2.0×10^{-6} Torr while Sb flux around 1.0×10^{-6} Torr. During the growth of these $AlAs_xSb_{1-x}$ ordered alloys the RHEED reveals a clear and streaky 1×3 pattern when the Sb shutter is open, and a mixed 1×3 and 2×4 pattern when the As shutter is open. It is known that the composition of $AlAs_xSb_{1-x}$ grown by MBE depends on group-III fluxes, As_2/Sb flux ratios, and growth temperatures [58–59]. A systematic comparison between the samples grown by conventional MBE and MMBE reveals that the latter has several advantages over the former [60,61]: (i) MMBE takes advantage of the precise and extremely reproducible control of shutters to obtain the desired alloy composition straightforwardly; (ii) The composition of the $AlAs_xSb_{1-x}$ ordered alloys grown by MMBE has a much weaker dependence on As_2 flux; (iii) MMBE permits the use of much higher As_2 fluxes. MMBE grown samples show dramatic improvements in their surface morphology and crystalline quality. In the following paragraphs, detailed discussion of these advantages will be provided.

In Fig. 10(b) the AlAs mole fraction of a set of $AlAs_xSb_{1-x}$ samples is depicted as a function of the As-shutter duty-cycle. These experimental data reveal clearly that the AlAs mole fraction is a linear function of the shutter time within the experimental range. A linear least square fit of the data is given by the tilted straight line. Control of the AlAs mole fraction is therefore straightforward, as it can be achieved by setting the shutter time in a growth program rather than tedious optimizing the As_2 or Sb flux through changing the corresponding cell temperature. It is necessary to mention that the AlAs mole fraction is much larger (a factor of more than 3 larger) than the As-shutter duty-cycle. This finding indicates incorporation of the As physically absorbed on the surface and the residual As in growth chamber after the closure of the As shutter. This effect is actually in favor of improving the crystalline quality in the sense that it smears out interfaces and reduces local strains. It is important to point out that although the

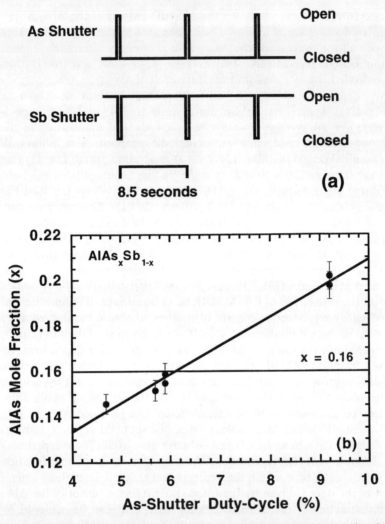

Fig. 10 (a) Schematic time sequence of the As and Sb shutter positions during the MMBE growth of $AlAs_xSb_{1-x}$ lattice-matched to InAs; (b) AlAs mole fraction is plotted as a function of the As-shutter duty-cycle for a set of $AlAs_xSb_{1-x}$ samples.

shutter time for the MBE system used in present study is about 0.1 s, which is comparable to the As shutter time, 0.4 s, the accuracy of the control of the AlAs mole fraction is not jeopardized because only the stability of the shutter time is more important here. The random fluctuations of the shutter time are expected to be very small.

In order to compare the As-flux dependence of the AlAs mole fractions in $AlAs_xSb_{1-x}$ layers grown with different methods, a set of $AlAs_xSb_{1-x}$ reference samples is grown by conventional MBE, i.e. both As and Sb shutters are kept open during the growth. These $AlAs_xSb_{1-x}$ layers will be called random alloys below. During these growths, the Sb flux is kept at 1.6×10^{-6} Torr. The As_2 flux is fine tuned around 2.0×10^{-7} Torr by changing the As cell temperature to reach the lattice-match condition. This beam flux is so low, almost an order of magnitude lower than that of Sb and the As_2 flux usually needed for the growth of GaAs at a rate of 1 μm/h, that it cannot even maintain an As-stable surface during the growth of InAs at a rate of 0.17 μm/h at a substrate temperature of 400°C. The dependence of the AlAs mole fraction in these samples is plotted as function of As_2 flux in Fig. 11(a). A linear least square fit of the experimental data reveals a slope $dx/dP_{As_2} = 0.13$. For direct comparison, another set of data obtained from the ordered alloys, i.e. ordered $AlAs_xSb_{1-x}$ layers grown by MMBE, are plotted in Fig. 11(b). A linear least square fit of these data yields a slope $dx/dP_{As_2} = 0.014$, which is one order of magnitude smaller than that of random alloys grown by conventional MBE. This finding demonstrates that MMBE facilitates the control of the average alloy composition. It is not difficult to understand this finding because the incorporation of As_2 (or Sb) is almost independent of group-V overpressure during the MBE growth of III/V binaries like AlAs and AlSb. Therefore, the average AlAs mole fraction is much less sensitive to the actual As_2 (or Sb) flux for MMBE and can be controlled accurately. The remaining weak dependence of the AlAs mole fraction on the As-flux is expected to be due to the incorporation of the residual As_2 in growth chamber after the closure of the As shutter.

Typical images of surface morphology of $AlAs_{0.16}Sb_{0.84}$ random alloy samples grown at 520°C and 450°C are shown in Fig. 12(a) and (b), respectively. At the higher growth temperature, the surface appears shiny to the naked eye under room light. Under a Nomarski microscope, a large number of cluster-like surface defects are clearly visible on the sample. These defects do not have any preferential orientation. However, the surface morphology of the sample grown at lower temperature looks slightly different [see Fig. 12(b)]. Although the surface morphology of samples grown at lower temperatures is slightly improved, high-density surface defects remain visible. The shape of these defects becomes elongated along $\langle 110 \rangle$ azimuth from random-cluster-like ones for the sample grown at the higher temperature. A sputtering Auger

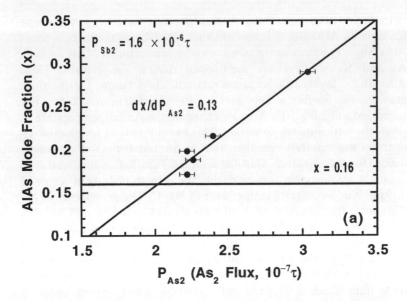

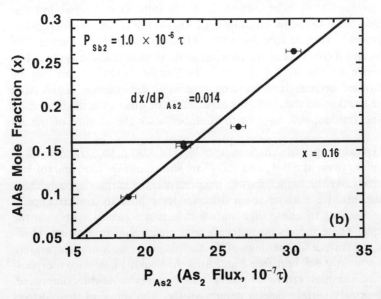

Fig. 11 AlAs mole fraction (x) in AlAs$_x$Sb$_{1-x}$ alloys as a function of As$_2$ flux: (a) Results obtained from samples grown by conventional MBE; (b) Results obtained from samples grown by MMBE.

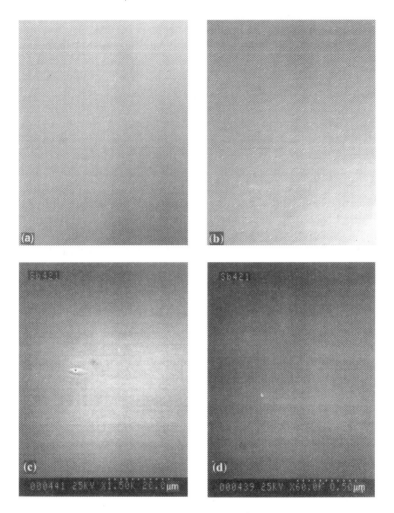

Fig. 12 Typical surface morphology of $AlAs_{0.16}Sb_{0.84}$ grown by different methods under different conditions. Images (a) and (b) show the surface morphology of $AlAs_{0.16}Sb_{0.84}$ random alloy samples grown by conventional MBE at 480°C and 520°C, respectively, as observed under a Nomarski microscope. Figures (c) and (d) are scanning electron microscope images with different magnifications of an $AlAs_{0.16}Sb_{0.84}$ ordered alloy grown by MMBE. The oval defect in (c) is utilized to optimize the focus.

electron spectroscopy analysis of these defects reveals no obvious spatial composition variation, neither lateral on the surface nor vertical along the growth direction. The vertical resolution is on the order of 100 Å and the electron beam spot size is on the order of 1 μm.

The cause of these defects is speculated to be the low As_2 flux and a possible miscibility gap that is found to be a serious problem for liquid phase epitaxy (LPE) [62].

In a striking contrast to the reference random-alloy samples, $AlAs_{0.16}Sb_{0.84}$ ordered alloys grown by MMBE possess a great improvement in the surface morphology. Fig. 12(c) and (d) show scanning electron microscope images of these samples' morphology with different magnifications. The images reveal featureless mirror-like surfaces. Typically, only a very small number of surface defects at a density of a few hundred per cm^2 are visible under a Nomarski microscope. The oval defect shown in Fig. 12(c) is used to obtain optimum focus. Under higher magnification the surface still looks smooth [Fig. 12(d)]. Such mirror-like morphology can be obtained at growth temperatures between 450°C to 540°C. For the growths of $InAs_xSb_{1-x}$ ordered-alloys, similar MMBE growth technique is applied. Very smooth surface is also obtained.

10.3.3 MATERIALS CHARACTERIZATION

10.3.3.1 X-ray diffraction measurements
In addition to the excellent surface morphology, the $AlAs_xSb_{1-x}$ ordered alloys possess very high crystalline quality. As an example, the X-ray diffraction spectrum of an $InAs/AlAs_{0.16}Sb_{0.84}$ DH laser (Sample #1) consisting of two 1.5 µm $AlAs_{0.16}Sb_{0.84}$ ordered-alloy cladding layers and a 1-µm InAs active region is shown in Fig. 13. Distinct satellite peaks corresponding to a short-period superlattices with a period of 11 Å is observed, which agrees well with the nominal period of 12 Å. The narrow linewidth indicated by the well resolved $CuK_{\alpha 1}$ and $CuK_{\alpha 2}$ splitting for a satellite peak and the 0-order peak implies that the spatial modulation of the composition is homogeneous along the growth direction. In addition, a high resolution rocking curve taken at the InAs (400) reflection by using a four-crystal X-ray diffractometer reveals that the lattice mismatch of the $AlAs_xSb_{1-x}$ layers to the substrate is less than 6.5×10^{-4}. The 0-order peak has a FWHM of 58 arc seconds. As mentioned earlier, the growth rate used for this study is at 0.5 µm/h. The growth time for this specific laser structure is more than 8 h. By taking these facts into account, these findings are consistant with highly stable beam fluxes and shutter times during the growth. The best FWHM observed for high

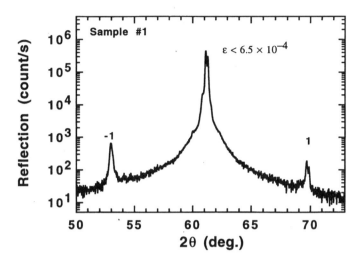

Fig. 13 An X-ray diffraction curve from a double heterojunction laser structure consisting of two 1.5 μm $AlAs_xSb_{1-x}$ cladding layers and a 1 μm InAs active region grown on an InAs substrate by MMBE. The diffraction curve is recorded around the InAs (400) reflection. The distinct satellite peaks correspond to a superlattice with a period of 11 Å. The lattice mismatch between the thick $AlAs_xSb_{1-x}$ and the substrate is less than 6.5×10^{-4}. The FWHM of the $AlAs_xSb_{1-x}$ peak is 58 arc seconds.

resolution four-crystal X-ray diffraction peaks of thick $AlAs_xSb_{1-x}$ ordered-alloy layers is 35 arc seconds.

High resolution X-ray diffraction rocking curve in the vicinity of InAs (400) reflection of an $InAs/InAs_xSb_{1-x}$ laser structure (Sample #2) is depicted in Fig. 14. Clear satellite peaks corresponding to a SL with a 100 Å period are distinctively visible. The lack of apparent broadening and splitting of the satellite peaks indicates a homogeneous periodicity of the SL along the growth direction. Taking into account the ratio of shutter times for each individual layer, the InAs layer thickness is then determined as 77 Å while that for $InAs_xSb_{1-x}$ ordered-alloy layer 23 Å. An average InAs mole fraction in the $InAs/InAs_xSb_{1-x}$ SL is determined to be 0.983 from the 0-order peak position. Consequently, the mole fraction in the 23-Å $InAs_xSb_{1-x}$ alloys layers is derived to be 0.926.

10.3.3.2 STM and TEM study of atomic scale structures

The group-V interfaces, such as arsenide/antimonide interface, have been the topic of extensive study. It has been reported that the

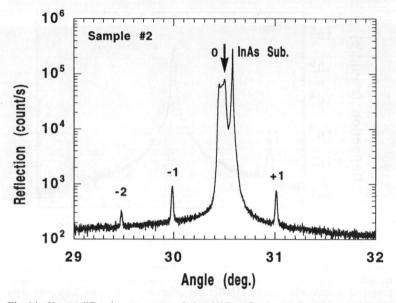

Fig. 14 X-ray diffraction curve at InAs (400) reflection taken from an InAs/InAs$_{0.926}$Sb$_{0.074}$/AlAs$_x$Sb$_{1-x}$ laser structure grown on an InAs substrate by MMBE. The main InAs/InAs$_{0.926}$Sb$_{0.074}$ SL peak gives the average InSb composition while the corresponding satellite peaks provide the periodicity of the SL. By taking into account the shutter times, the InAs layer thickness is determined to be 77 Å while that for InAs$_{0.926}$Sb$_{0.074}$ is 23 Å.

interface quality has strong impact on the material properties. As mentioned in the previous paragraphs, the alloy composition in the group-V alloys are controlled by the shutter duty cycle. Therefore, it is extremely helpful to understand the detailed atomic structural properties of MMBE-grown samples. In this section, we will discuss the results of cross-section Scanning Tunneling Microscopy (STM) and Transmission Electron Microscopy (TEM) studies of InAs/InAs$_x$Sb$_{1-x}$ SLs.

For STM study, we focus on a 172 Å-InAs/52 Å-InAs$_{0.76}$Sb$_{0.24}$ SL (Sample #4). Each of the 53 Å InAs$_{0.76}$Sb$_{0.24}$ layers consists of 4 periods of nominal 7.8 Å-InAs/5.2 Å-InSb superlattice structures. Fig. 15(a) shows a high-resolution constant-current cross-section STM image of a single InAs$_{0.76}$Sb$_{0.24}$ ordered-alloy layer surrounded by InAs layers [63]. Contrast between the InAs layers and the InAs$_{0.76}$Sb$_{0.24}$ ordered-alloy layer can be seen clearly in the image. The growth direction is indicated in the figure. Fig. 15(b) shows an averaged topographic line scan of the InAs$_{0.76}$Sb$_{0.24}$ layer and the surrounding InAs layers, taken

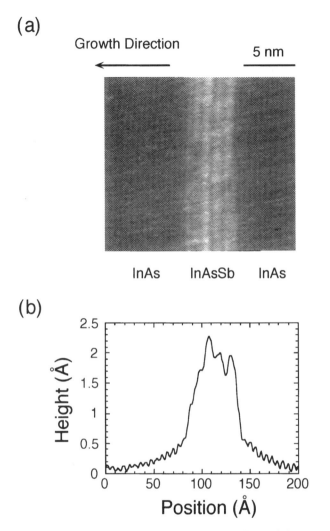

Fig. 15 (a) Constant-current image of a single $InAs_{0.76}Sb_{0.24}$ alloy and the surrounding InAs layers. (b) Topographic line scan of the $InAs_{0.76}Sb_{0.24}$ alloy layer and the surrounding InAs layers, averaged over an area approximately 225 Å × 170 Å shown in the image.

over 200 individual scan lines covering an area approximately 200 Å (170 Å shown in the image). Also visible in the image is contrast caused by the ordered structure within the $InAs_{0.76}Sb_{0.24}$ layer produced by the MMBE growth technique. These features are observed consistently with different cleaved samples and tips. The averaged line scan makes more apparent the asymmetric appearance of the $InAs_{0.76}Sb_{0.24}$ layers in the constant-current STM images. Differences in topographic height, caused by differences in electronic structure, between the short-period superlattices and the darker InAs layers become larger as more $InAs_{0.76}Sb_{0.24}$ is grown. The maximum contrast in height (corresponding to maximum content) is consistently observed at the next-to-last (i.e. the third) grown InSb-like layer in the $InAs_{0.76}Sb_{0.24}$ alloy region. This non-uniformity in the electronic structure in the $InAs_{0.76}Sb_{0.24}$ layers suggests the presence of atomic cross-incorporation within the $InAs_{0.76}Sb_{0.24}$ layers during the MMBE growth, which is in consistent with the observation discussed in Section 10.3.1.

TEM study of $InAs/InAs_{0.924}Sb_{0.076}$ SL in a laser structure (Sample #2) reveals additional detailed information about the structure properties of the SL. The sample is examined in a $\langle 110 \rangle$ direction perpendicular to the [001] growth direction. TEM diffraction pattern of the $InAs/InAs_{0.924}Sb_{0.076}$ SL active region is shown in Fig. 16. Only zinc-blende reflections appear on the photographs. Special attention is paid to the possible reflections caused by CuPt ordering. But no reflection spots at half the distance between (000) and $\{111\}$ reflection spots are visible, indicating that there is no CuPt ordering in the studied $InAs_{0.924}Sb_{0.076}$ ordered alloys. This finding is further confirmed by photoluminescence study of MMBE grown thick $InAs_{0.9}Sb_{0.1}$ ordered-alloy layers lattice matched to GaSb substrates. The bandgaps of these samples show no reduction and agree very well with the data published in the literature. In contrast, CuPt orderings in $InAs_xSb_{1-x}$ random alloys grown by MOCVD and MBE are evidenced by detailed TEM and optical spectroscopy study [64–66]. Due to the spontaneous nature, this kind of ordering is difficult to control and not desirable for device applications. In view of the present results, the ordered alloys grown by MMBE technique provides an alternative way to bypass the ordering issue.

10.3.3.3 *Optical spectroscopy*
Theoretical modeling discussed in 10.2 has predicted that $InAs/InAs_xSb_{1-x}$ SLs can cover a broad wavelength range with a small change of the InSb mole fraction in the $InAs_xSb_{1-x}$ layers. In this section, we will discuss the experimental results of optical spectroscopy on a set of $InAs/InAs_xSb_{1-x}$ SL samples.

Low temperature PL and Fourie Transformation Photoluminescence Excitation spectroscopy (FTPLE) [67] provide detailed information of the effective bandgaps and shape of joint density of states. In Fig. 17 the

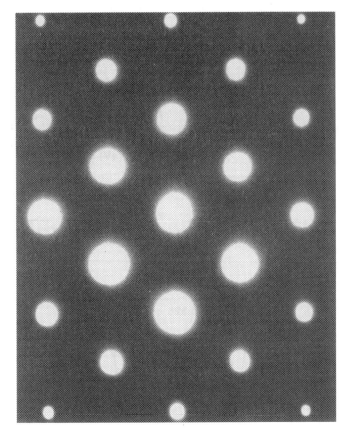

Fig. 16 TEM diffraction image of the InAs/InAs$_{0.926}$Sb$_{0.074}$ active region in a laser structure (Sample #2). Only zinc-blende reflections are visible.

PL and FTPLE spectra from the set of samples (Sample #2, #5, and #7) at 10 K are plotted. For Sample #2, the FWHM of PL peak is only 15 meV. Different from PL that may result from below-gap impurity- or defect-related transitions, FTPLE spectra reveal the exact bandgap of SLs and the qualitative shape of the joint density of states. The onset of the FTPLE spectra, i.e. the onset of absorption edge, is used here as the measure of bandgap for the type-II SLs. Stokes shifts between the PL peaks and the FTPLE onsets are listed in Table IV. It is known that the Stokes shift between PL and FTPLE spectra from QW or SL structures results from the composition variations and well-width fluctuations. The small Stokes shifts are clear evidences of high

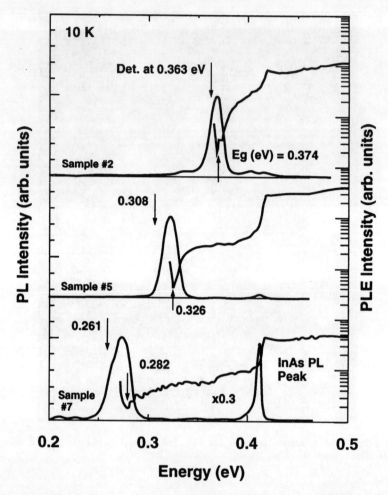

Fig. 17 PL and FTPLE spectra of a set of MWIR InAs/InAs$_x$Sb$_{1-x}$ type-II SL laser structures samples at 10 K. The sample structures and compositions are listed in Table IV. The FTPLE spectra are taken by monitoring the low energy side of the PL peak.

crystalline quality of the samples. Inspection of the FTPLE spectra in Fig. 17 reveals no obvious subband structures. As an expected feature for type-II SLs, the onset is soft in contrast to the typical feature of type-I GaAs/AlGaAs SL where strong exciton absorption peak appears at the bandedge and a step-like plateau follows on the high energy side [68]. Detailed discussion about the FTPLE results is given in Ref. [69].

Fig. 18 shows room temperature photoluminescence spectra from a set of samples having similar structures as the one discussed above. The major differences in these samples are the compositions and layer thicknesses of the constituent layers in the InAs/InAs$_x$Sb$_{1-x}$ SLs.

TABLE IV
Sample parameters and experimental results are listed in comparison with theoretical modeling results. The calculated bandgaps [E_g(theo.)] are obtained by assuming a valence band discontinuity of 610 meV between free standing InSb and InAs. No fitting parameter is used in the calculation. The stokes shift observed from PL and FTPLE of the InAs waveguides is 6 meV.

Sample	PL Peak (eV)	E_g (FTPLE) (eV)	E_g (theo.) (eV)	Stokes shift (meV)
#2	0.369	0.374	0.385	5
#5	0.323	0.330	0.354	7
#7	0.276	0.280	0.294	4

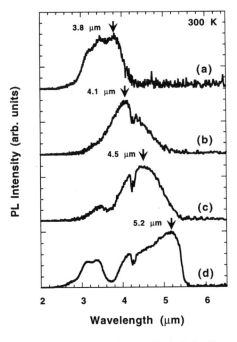

Fig. 18 Room temperature PL spectra of a set of InAs/InAs$_x$Sb$_{1-x}$ type-II SL samples with different layer thicknesses and InAs mole fractions in the layers. The peak at 3.4 μm for spectra (c) and (d) is the PL from InAs layers while the dips at 4.2 μm in these spectra are due to CO_2 absorption. The photoexcitation density is approximately 200 W/cm^2.

Strong room temperature PL spectra from 3.8 μm up to 5.2 μm have been observed. Careful inspection of these spectra reveals that the FWHMs range from 66 to 80 meV. The longer the PL wavelength, the broader the FWHM is. This finding is attributed to the stronger band filling effect due to decreasing of wavefunction overlap as the Sb composition is increased in the $InAs_xSb_{1-x}$ layers. Compared with a reference InAs bulk sample, which gives very strong PL at room temperature, the integrated PL intensities of these $InAs/InAs_xSb_{1-x}$ type-II SL samples are comparable or even stronger. For spectra (c) and (d) there is a peak at 3.4 μm. This peak is attributed to the PL from InAs substrate or InAs waveguide regions. The sharp dips at 4.2 μm in the spectra are due to the absorption of CO_2 in the atmosphere. All these PL results shown above demonstrate that it is practical to use $InAs/InAs_xSb_{1-x}$ type-II SL to cover a broad MWIR wavelength range.

10.4 OPTICALLY PUMPED HIGH-POWER MWIR LASERS

10.4.1 Advantages of Optically Pumped MWIR Lasers

The heat generated in the active region of high-power MWIR lasers limits the highest operating temperature and degrades quantum efficiency. For electrically injected lasers, contact resistance, series resistance in the cladding and buffer layers, and the relaxation of hot injected electrons in the active region all contribute to the heat generation. Even for an ideal MWIR laser, which does not have any contact and series resistance, the thermal relaxation of injected hot electrons in the active region still produces tremendous heat. By use the typical laser structure mentioned earlier as an example, which consists of $AlAs_{0.16}Sb_{0.84}$ barriers with a bandgap of 1.6 eV and an active region with a bandgap of 0.365 eV (3.4 μm), the maximum power efficiency for such an ideal electrically injected DH laser diode cannot be higher than $0.365/1.6 = 23\%$. 77% of the input power will be dissipated as heat in the active region. Such a low power efficient is refered to "quantum defect". However, if one uses a below-cladding-layer-bandgap pumping source and a proper design of active region to optimize the absorption of pumping power, then the maximum power efficiency of an ideal optically pumped laser can be drastically increased. Such a high power efficiency reduces the heat and results in high operating temperature and quantum efficiency. Compact optical pumped MWIR laser systems have been successfully demonstrated for practical applications [70]. Detailed discussion of optically pumped high power semiconductor lasers can be found in Ref. [71].

10.4.2 CW Optically Pumped MWIR Lasers

Fig. 19 shows the cw light output of a 3.8 μm InAs/InAs$_x$Sb$_{1-x}$ SL laser (Sample #6) as function of absorbed excitation power at different temperatures. The pumping light is from a cw Ti:sapphire laser at 800 nm. AT 54 K, lasing is observed at an equivalent threshold current density (J_{th}) of 14 A/cm^2. Linear output behavior remains under excitations up to 3 times of the threshold. As temperature is increased, the lasing threshold increases accordingly. At 126 K the equivalent J_{th} is 430 A/cm^2. The measured maximum operating temperature is higher than 130 K. Higher operating temperature is expected if higher pump power was available. Use the empirical equation for the temperature dependence of $J_{th} = J_{th0} \cdot e^{T/T_0}$, where T is the laser operating temperature, J_{th0} the threshold current density at 0 K, and T_0 is the characteristic temperature. For this specific laser, T_0 is determined to be 20 K.

Fig. 20 shows the temperature dependent optical spectra of the edge emitted light from Sample #6 under a cw optical pumping of 500 mW.

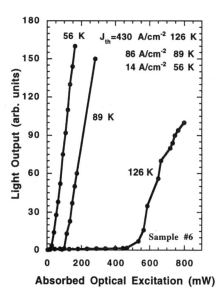

Fig. 19 Light-output as a function of pumping power for a typical InAs/InAs$_x$Sb$_{1-x}$ type-II SL laser structure (Sample #6) at different temperatures. The calculated equivalent injected current densities are given in the figure.

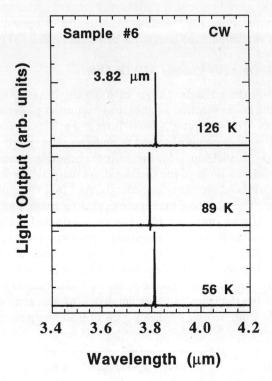

Fig. 20 Lasing spectra of the typical InAs/InAs$_x$Sb$_{1-x}$ type-II SL laser structure (Sample #6) at various temperatures. The peak positions are arround 3.8 μm.

As the sample temperature is increased, the lasing wavelength does not change much with temperature. This observation is explained in terms of stronger band filling as the J_{th} increases with temperature.

Fig. 21 shows a typical high-resolution lasing spectrum at 40 K for Sample #3. Only one set of longitudinal modes is observed. No obvious splitting of the longitudinal modes is visible, which is usually indicative of a coexisting higher lateral mode(s) along the growth direction. The separation ($\Delta\lambda$) of the longitudinal modes is measured to be 9.0 Å with a cavity length (L) of 1.5 mm. The equivalent index (n_e) is determined to be 4.28 by using the simple formula

$$n_e = n - \frac{\lambda}{n}\frac{dn}{d\lambda} = \frac{\lambda^2}{2\Delta\lambda L}$$

where n is the averaged refractive index of the active and waveguide regions. The center wavelength is 3.4 μm at 40 K.

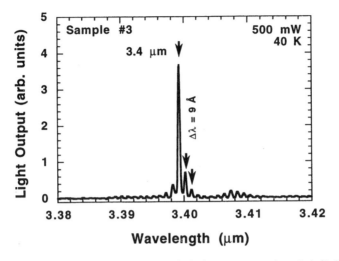

Fig. 21 High-resolution longitudinal-mode lasing spectrum of an InAs/InAs$_x$Sb$_{1-x}$ type-II SL laser structure (Sample #3) at 40 K under a pumping power of 500 mW. The cavity length is 1.5 mm. The measured longitudinal modes separation is 9.0 Å.

10.4.3 Quasi-CW Optically Pumped High-Power MWIR Lasers

In order to obtain high power output, a broad area device with ~ 0.3-mm $\times$ 3.1-mm is used for quasi-cw optical pumping experiments [72]. The quasi-cw optical pumping light source is a laser-diode-array stack with a 250 W peak power at 940 nm. Only single lens is used to focus the excitation light. Such a compact set-up is very practical for various applications. Fig. 22 gives the light output power per facet of an InAs/InAs$_{0.926}$Sb$_{0.074}$ type-II SL laser vs. absorbed pumping power at 90 K. The pumping pulse length is 10 μs and the duty-cycle is 1%. A least square fitting of the experimental data in Fig. 22 reveals that the external differential quantum efficiency is 2.5%/facet for optical pumping power less than 20 W. A kink is observed around a pumping power of 25 W. Above the kink, the linear behavior of the output power vs. pumping power remains up to the maximum power measured, but the quantum efficiency drops to 2%/facet. The peak power measured is 1.3 W/facet under the maximum absorbed pumping power of 60 W. At maximum pumping power, neither deviation from the linear behavior of the power output vs. pumping power nor catastrophic degradation of the device is observed. Therefore, it is reasonable to believe that a higher maximum peak power could be achieved by using even higher pumping power. By increasing the optical

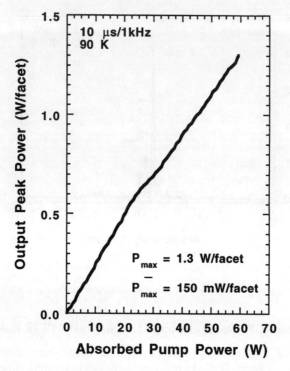

Fig. 22 Peak power vs. absorbed pumping power for an InAs/InAsSb type-II SL lasers under pulse optical pumping at 90 K.

pumping duty cycle, a 150 mW/facet average power is achieved at 95 K. For this case, the external differential quantum efficiency is also increased to 8%/facet. With the 60 W absorbed pump power, the maximum pulse operation temperature is 195 K.

10.5 CONCLUSIONS

In present chapter we reviewed the theoretical considerations, device design, material growth and characterization, and device characteristics of $InAs/InAs_xSb_{1-x}$ type-II SL MWIR lasers. As predicted by theory and demonstrated by experiments, the use of $InAs/InAs_xSb_{1-x}$ type-II SL relaxed the restrictions of available III/V material alloys that can be possibly grown on existing substrates with device quality, providing a wide wavelength coverage range (MWIR to far IR) below the bandgap of the constituent materials of the SL. Detailed discussion was also given to the MMBE growth technique which facilitated the growth of high quality As/Sb group-V alloys. X-ray diffraction measurements, STM, TEM and optical spectroscopy study of the InAs/InAsSb SLs demonstrated very high crystalline and interface qualities. Optically pumped MWIR lasers consisting of these InAs/InAs$_x$Sb$_{1-x}$ SLs showed record high power output of 1.3 W/facet at 3.4 µm for semiconductor lasers. The maximum cw and quasi-cw operating temperature are higher than 130 K and 195 K for a 3.8 µm laser and a 3.4 µm laser, respectively. It is also expected that electrically injected lasers can be readily made by using the identical laser structures after introducing proper doping.

Acknowledgments

The author would like to thank H. Q. Le from MIT Lincoln Lab for high-power optical pumping experiments, A. Lew and E. Yu from University of California at San Diego for cross-section STM study, Y. Chen from Lawrence Berkeley Lab for TEM study, F. Fuchs from Faunhofer Institute for Applied Solid State Physics for FTPLE measurements, C. Grein from University of Illinois at Chicago for the theoretical modeling of the band structure and D* for InAs/InAs$_x$Sb$_{1-x}$ SLs.

References

1. B. Spanger, U. Schiessl, A. Lambrecht, H. Böttner, and M. Tacke, *Appl. Phys. Lett.* **53**, 2582–2584 (1988).
2. Z. Feit, D. Kostyk, R. J. Woods, and P. Mak, *Appl. Phys. Lett.* **58**, 343–345 (1991).
3. Z. Shi, M. Tacke, A. Lambrecht, and H. Böttner, *Appl. Phys. Lett.* **66**, 2537–2540 (1995).
4. M. Zandian, J. M. Arias, R. Zucca, R. V. Gil, and H. Shin, *Appl. Phys. Lett.* **59**, 1022–1024 (1991).
5. H. Q. Le, J. M. Arias, M. Zandian, R. Zucca, and Y.-Z. Liu, *Appl. Phys. Lett.* **65**, 810–812 (1994).
6. H. K. Choi and S. J. English, *Appl. Phys. Lett.* **61**, 1154 (1992).
7. H. K. Choi, S. J. English, and G. W. Turner, *Appl. Phys. Lett.* **64**, 2474 (1994).
8. H. Lee, P. K. York, R. J. Menna, R. U. Martinelli, D. Z. Garbuzov, S. Y. Narayan, and J. C. Connolly, "Room-temperature 2.78 µm AlGaAsSb/GaInAsSb quantum-well lasers", *Appl. Phys. Lett.* **66**, 1942–1944 (1995).
9. M. J. Cheng, G. B. Stringfellow, D. W. Kisker, A. K. Srivastava, and J. L. Zyskind, *Appl. Phys. Lett.* **48**, 419–421 (1986).
10. E. Tournie, P. Grunberg, C. Fouillant, S. Kadret, G. Boissier, A. Baranov, A. Joullie, E. Gaumont-Goarin, and K. H. Ploog, *Electron. Lett.* **29**, 1255–1256 (1993).
11. S. Forouhar, A. Ksendzov, A. Lasson, and H. Temkin, *Electron. Lett.* **28**, 1431–1432 (1992).

12. S. J. Eglash and H. K. Choi, *Appl. Phys. Lett.* **64**, 833–835 (1994).
13. H. K. Choi, G. W. Turner, and H. Q. Le, Narrow Gap Semiconductors 1995, *Proceedings of the International Conference on Narrow Gap Semiconductors*, Ed. J. L. Reno, Institute of Physics Publishing, London, 1995, p. 1 and references therein.
14. S. R. Kurtz, R. M. Biefeld, L. R. Dawson, K. C. Baucom, and A. J. Howard, *Appl. Phys. Lett.* **64**, 812–814 (1994).
15. A. N. Baranov, T. N. Danilova, O. G. Ershov, A. N. Imenkov, V. V. Sherstnev, and Yu. P. Yakovlov, *Sov. Tech. Phys. Lett.* **18**, 725–726 (1992).
16. A. N. Baranov, A. N. Imenkov, V. V. Sherstnev, and Yu. P. Yakovlov, *Appl. Phys. Lett.* **64**, 2480–2482 (1994).
17. M. Razeghi *et al.*, private communication. InAsSb/InAsSbP MQW lasers grown by MOCVD on InAs substrates demonstrate a threshold current density of $40\,\text{A/cm}^2$ with output power of 260 mW and 60 mW for pulse and CW operations around 80 to 100 K, respectively. The lasers have a cavity length of 300 μm and 100 μm stripe width. T_0 is 42 K.
18. J. Faist, F. Capasso, D. L. Sivco, C. Sirtori, A. L. Hutchinson, and A. Y. Cho, *Science* **264**, 553–556 (1994).
19. J. Faist, F. Capasso, C. Sirtori, D. L. Sivco, A. L. Huntingson, and A. L. Cho, to be published.
20. G. C. Osbourn, *J. Vsc. Sci. Technol. B* **2**, 176 (1984).
21. D. L. Smith and C. Mailhiot, *J. Appl. Phys.* **62**, 2545–2548 (1987).
22. D. H. Chow, R. H. Miles, J. R. Sonerstrom, and T. C. McGill, *Appl. Phys. Lett.* **56**, 1418–1420 (1990).
23. R. H. Miles, D. H. Chow, J. N. Schulman, and T. C. McGill, *Appl. Phys. Lett.* **57**, 801–803 (1990).
24. R. H. Miles and J. A. Wilson, to appear in *Proceedings of the 3rd Electrochemical Society Meeting on Long Wavelength Infrared Detectors*.
25. H. Kroemer, *IEEE Electron Device Lett.* **EDL-4**, 20–22 (1983);
    E. J. Caine, S. Subbanna, H. Kroemer, and J. L. Merz, *Appl. Phys. Lett.* **45**, 1123–1125 (1984).
26. E. Lugagne-Delpon, P. Voisin, M. Voos, and J. P. Andre, *Appl. Phys. Lett.* **60**, 3087–3089 (1992).
27. C. H. Grein, P. M. Young, and H. Ehrenreich, *J. Appl. Phys.* **76**, 1940–1942 (1994).
28. R. H. Miles, D. H. Chow, Y.-H. Zhang, P. D. Brewer, and R. G. Wilson, *Appl. Phys. Lett.* **66**, 1921 (1995).
29. T. C. Hasenberg, D. H. Chow, A. R. Kost, R. H. Miles, and L. West, *Electron. Lett.* **31**, pp. 275–276 (1995).
30. D. H. Chow, R. H. Miles, T. C. Hasenberg, A. R. Kost, Y.-H. Zhang, H. L. Dunlap, and L. West, *Appl. Phys. Lett.* **67**, 3700 (1995).
31. Y.-H. Zhang, *Appl. Phys. Lett.* **66**, 118 (1995).
32. Y.-H. Zhang, R. H. Miles, and D. H. Chow, *IEEE J. of Selected Topics in Quantum Electronics* **1**, 749 (1995).
33. G. J. Gualtieri, G. P. Schwartz, R. G. Nuzzo, and W. A. Sunder, *Appl. Phys. Lett.* **49**, 1037–1039 (1986).
34. G. J. Gualtieri, G. P. Schwartz, R. G. Nuzzo, R. J. Malik, and J. F. Walker, *J. Appl. Phys.* **61**, 5337–5341 (1987).
35. J. Batey and S. L. Wright, *J. Appl. Phys.* **59**, 200–209 (1985).
36. J. Batey and S. L. Wright, *Surf. Sci.* **174**, 320–323 (1986).
37. S. P. Kowalczyk, W. J. Schaffer, E. A. Kraut, and R. W. Grant, *J. Vac. Sci. Technol.* **20**, 705–708 (1982).
38. S. R. Kurtz, G. C. Osbourn, R. M. Biefeld, and R. S. Lee, *Appl. Phys. Lett.* **53**, 216–218 (1988).
39. Y. B. Li, R. A. Stradling, A. G. Norman, P. J. P. Tang, S. J. Chung, and C. C. Phillips, *Proceedings of the 22nd Int. Conf. on the Phys. of Semicond.*, Vancouver, 1994.

40. P. J. P. Tang, M. J. Pullin, S. J. Vhung, C. C. Phillips, R. A. Stradling, A. G. Norman, Y. B. Li, and L. Hart, *Semicond. Sci. Technol.* **10**, 1177 (1995).
41. M. J. Pulling, P. J. P. Tang, S. J. Chung, C. C. Phillips, R. A. Stradling, A. G. Norman, Y. B. Li, and L. Hart, Narrow Gap Semiconductors 1995, *Proceedings of the International Conference on Narrow Gap Semiconductors*, Ed. J. L. Reno, Institute of Physics Publishing, London, 1995, p. 8.
42. S. Wei and A. Zunger, *Phys. Rev. B* **52**, 12039 (1995).
43. H. H. Wieder and A. R. Clawson, *Thin Solid Film* **15**, 217–221 (1973).
44. Landolt-Börnstein, *Numerical Data and Functional Relationships in Science and Technology*, New Series III/17a, Eds. by O. Madelung, M. Schulz, and H. Weiss (Springer, Heidelberg, 1982).
45. J. W. Matthews and A. E. Blakeslee, *J. Crystal Growth* **27**, 118–121 (1974).
46. G. Bastard, "Wave mechanics applied to semiconductor heterostructures", Les Edition de Physique, France.
47. Y.-H. Zhang and F. Fuchs, unpublished.
48. C. H. Grein, M. E. Flatte, and H. Ehrenreich, to be published.
49. Y.-H. Zhang, unpublished.
50. B. O. Seraphin and H. E. Bennett, *Semiconductor and Semimetals* Vol. 3, Eds. R. K. Willardson and A. C. Beer, Academic Press, N. Y., p. 499.
51. S. Adachi, *J. Appl. Phys.* **58**, R1 (1985).
52. H.K. Choi and G. W. Turner, private communication.
53. It is discovered by using another identical Sb cracking cell with similar cracking temperatrure that 80% of the flux is antimony monomar. For details please see P. Brewer and K. P. Killeen, *Mat. Res. Soc. Symp. Proc.*, **406**, 109–114 (1996) and P. D. Brewer, D. H. Chow, and R. M. Miles, *J. Vac. Sci. Technol. B*, in press.
54. Y. Matsushima and S. Gonda, *Jap. J. Appl. Phys.* **15**, 2093 (1976); K. Woodbridge, J. P. Gowers, and B. A. Joyce, *J. Crystal Growth* **49**, 132 (1980).
55. C. A. Chang, R. Ludeke, L. L. Chang, and L. Esaki, *Appl. Phys. Lett.* **31**, 759 (1977).
56. C. T. Foxon, B. A. Joyce, and M. T. Norris, *J. Crystal Growth* **78**, 342 (1980).
57. T. H. Chiu, J. L. Zyskind, and W. T. Tsang, *J. Electron. Mater.* **16**, 57 (1987).
58. J. A. Lott, L. R. Dawson, E. D. Jones, and J. F. Klem, *Appl. Phys. Lett.* **56**, 1242 (1990).
59. J. A. Lott, L. R. Dawson, E. D. Jones, I. J. Fritz, J. S. Nelson, and S. R. Kurtz, *J. Electron. Mater.* **19**, 989 (1990).
60. Y.-H. Zhang and D. H. Chow, *Appl. Phys. Lett.* **65**, 3239 (1994).
61. Y.-H. Zhang, "Accurate Control of As and Sb Incorporation Ratio during Solid-Source Molecular-Beam Epitaxy" *J. of Crystal Growth* **150**, 838 (1995).
62. H. C. Casey, Jr. and M. B. Panish. *Heterostructure Lasers* (Academic, NY, 1978), Part A and B.
63. A. Y. Lew, E. T. Yu, and Y.-H. Zhang, *J. Vac. Sci. Technol., B* **14**, 2940 (1996).
64. H. R. Yen, K. Y. Ma, and G. B. Stringfellow, *Appl. Phys. Lett.* **54**, 1154 (1989).
65. S. R. Kurtz, L. R. Dawson, R. M. Biefeld, D. M. Follstaedt, and B. L. Doyle, *Phys. Rev. B* **46**, 1909 (1992).
66. D. M. Follstaedt, R. M. Biefeld, S. R. Kurtz, L. R. Dawson, and K. C. Baucom, "Ordering and bandgap reduction in $InAs_xSb_{1-x}$ alloys", Narrow Gap Semiconductors 1995, *Proceedings of the International Conference on Narrow Gap Semiconductors*, Ed. J. L. Reno, Institute of Physics Publishing, London, 1995, p. 225.
67. B. Hamilton and G. Clark, *Mater. Sci. Forum*, 38–41, 1337 (1989).
68. R. Dingle, Festkörperproblem/*Advances in Solid State Physics*, Vol. 15, Ed. H. J. Queisser, Pergamon Press, Oxford, 1975, p. 21.
69. F. Fuchs and Y.-H. Zhang, unpublished.
70. H. Q. Le, G. W. Turner, S. J. Eglash, H. K. Choi, D. A. Coppeta, *Appl. Phys. Lett.* **64**, 152 (1994) and references therein.

71. H. Q. Le, G. W. Turner, J. R. Ochoa, H. K. Choi, and J. M. Arias, M. Zandian, R. R. Zucca, and Y. -Z. Liu, *Proceeding of SPIE*, Vol. 2382: "Laser diodes and applications" Ed. K. J. Linden, p. 262 (1995).
72. Y. H. Zhang, H. Q. Le, D. H. Chow, R. H. Miles, "Mid infrared lasers grown on InAs by modulated-molecular-beam epitaxy," Narrow Gap Semiconductors 1995, *Proceedings of the International Conference on Narrow Gap Semiconductors*, Ed. J. L. Reno, Institute of Physics Publishing, London, 1995, p. 36.

Index

ab initio self-consistent
  method   20
AlAs mole fraction
  in $AlAs_xSb_{1-x}$ alloys   482
  in $AlAs_xSb_{1-x}$ layers   480
$AlAs_xSb_{1-x}$ layers
  lattice matched to InAs or
    GaSb   478
$AlAs_{0.16}Sb_{0.84}$
  surface morphology   481
AlAsSb
  growth   390
Al(Ga)As-like interface
  structure   116
AlGaAsSb   374
  growth   390
(AlGa)Sb/InAs-based epilayers
  and heterostructures
  undoped
    MBE growth and
      characterization   99
(AlGa)Sb/InAs interface
  formation   115
(AlGa)Sb/InAs type-II
  heterostructure
  electronic properties   125
AlSb
  lattice constant   374
  layer   123
  nucleation layer
    growth on GaAs
      substrate   106
AlSb/GaSb short period strained-
  layer superlattice (see superlattice)

AlSb/InAs
  cladding layers   442
  quantum well
    band structure   137
  superlattice (see superlattice)
AlSb/InAs/AlSb single
  quantum well
  unintentionally doped
    carrier mobility and
      concentration   133
anisotropy
  optical   210
    in InAs/AlSb
      heterostructures   211
  valence band   220
antimonide-family
  asymmetric double quantum
    well (see double quantum
    well)
  electro-optical modulator
    (see modulator)
antimony tetramer dissociation
  thermodynamic analysis   101
atomic scale
  interface structure   115
  structure
    STM study   485
    TEM study   485
Auger coefficient   326, 386
  for CHCC   386
Auger current   335
  dependence on quantum well
    parameters   346
  dependence on strain   339, 346

threshold
  strain dependence 351
  vs. conduction band
    offset 350
  vs. carrier density 338
Auger electron 281
Auger events
  CHCC 259
Auger process 435
  CHCC threshold 359
  non-threshold 292
Auger recombination 281, 416
  CHCC 336
  CHHH 259
  current 344
    mechanism of
      suppression 340
  current density 418
  in quantum wells 292
  mechanism
    in type-I heterostructures 299
    in type-II heterostructures 308
  non-threshold mechanism 325
  process 360
  rate 296, 302, 316
    effective suppression of 321
    logarithm of 321
  reduction 437

band alignment
  broken-gap 435
band gap
  effective 56, 468
  energies of $Al_xGa_{1-x}Sb$
  grown on GaSb 375
  energy 25
    III-V semiconductor 173
  in [111] and [001]
    ordered superlattices 38

InAs/AlSb SPSLs with different
  types of interfaces 131
  room temperature
    vs. lattice constant, for antimonide-based alloys 370
band offset 379
  conduction 9
  between GaSb and AlSb 249
  valence 9, 129
  $In_{0.18}Ga_{0.82}Sb/GaSb$
    system 13
bandedge alignment 465
bandwidth
  spontaneous emission 385
beam equivalent pressure (BEP)
  ratio 100
binding energies 61
bond length 32
bonding energies 60
Born-Oppenheimer
  approximation 14
Bragg angle 65
Brillouin zone 21
  integration 21
  zinc blende 26
bulk modulus 25
Burstein-Moss shift 209

calorimetric absorbance 208
Car-Parrinello simulations 20
carrier
  confinement 439
  density
    threshold 332
  free
    absorption by 286
  leakage 329
cladding layer 400
conventional $\Gamma-\Gamma$ asymmetric
  designs 248

cross-sectional scanning
 tunneling microscopy
 (XSTM) 67
current
 Auger (see Auger current)
 density
  threshold 385
 injection 261
 leakage 344
 recombination
  interfacial 416
  threshold 331
  pulsed 455
  RT 410
  temperature
   dependence 353
CW output power
 vs. current 413, 421

deep-level transient
 spectroscopy (DLTS) 12
defect
 native
  optical activity 440
 nonuniform
  distribution in
   epilayer 108
deformation potential
 conduction band maximum
  (CBM) 25
 constants 277
 valence band maximum
  (VBM) 25
density functional
 formalism 15
 implementation 18
 theory 14
detector
 long wavelength infrared
  applications 472

diode (see also photodiode)
 mid-infrared quantum well
  high-power
   operation 426
direct-indirect transition 375
dislocation formation 396
doping 392
 Be, for p-type GaSb and
  AlGaAsSb 392
 n-type
  of GaSb 113
  of (AlGa)Sb and InAs in
   molecular beam
    epitaxy 112
 p-type
  of (AlGa)Sb and InAs
   in molecular beam
    epitaxy 112
double crystal X-ray diffraction
 (DCXRD)
double quantum well (see also
 quantum well)
 asymmetric 241
  antimonide-family 253
 InAs/AlSb
  reflectance spectra 194

efficiency
 differential 409
 spontaneous 409
elasticity
 macroscopic theory 32
electron plasma 200
ellipsometric spectroscopy
 of higher lying band
  gaps 222
emission rate
 spontaneous 285
energy eigenvalue 26
equation of state

INDEX

Murnahan, semi-empirical 25
equilibrium constant 25
equivalent index 494
exchange correlation energy 17

far-infrared spectroscopy 194
fast Fourier transform (FFT)
  two-dimensional 77
Fermi-Dirac distribution
  function 288, 293
Fermi distribution function 243
Fermi quasi-levels of electrons
  and holes 285
finite element method (FEM)
  formalism 238
Fröhlich mechanisms 217
full potential augmented plane
  wave method 20
full potential linearized augmented
  plane wave method 10

gain 260, 283
  maximum, vs. radiative
    contribution 262
GaAs-like bond 59
$GaAs_ySb_{1-y}$ layer lattice matched
  to InAs or GaSb 478
GaInAsSb
  binodal and spinodal
    stability curve 373
GaInAsSb/AlGaAsSb quaternary
  alloy grown on GaSb 462
$Ga_{1-x}In_xAs_ySb_{1-y}$ lattice
  matched to GaSb 373
GaSb
  buffer layer 62, 82
  substrates 100
GaSb/GaAs
  growth
    lattice mismatched 103

MBE 102
interface
  threading dislocations
    107
GaSb/InSb
  interface 42
  superlattice (see superlattice)
GaSb/InSb (InAs/InSb)
  system 31
GaSb/GaSb
  unstrained
    MBE growth 102
Γ–L and L–X crossover
  points 394
Γ and L band edge profile 241
Γ–L electro-optic modulator
  (see modulator)
Γ–L intervalley-transfer
  mechanism 250
Γ–L intersubband
  interaction 250
Ge_xSi_{1-x} system 81
group-V alloy
  modulated molecular beam
    epitaxial growth 476
growth
  AlAsSb (see AlAsSb)
  AlGaAsSb (see AlGaAsSb)
  AlSb (see AlSb)
  GaSb/GaAs
    (see GaSb/GaAs)
  GaSb/GaSb
    unstrained
      (see GaSb/GaSb)
  group-V alloy (see group-V
    alloy)
  parameters 389
  procedures 389
  strained quantum well
    structure (see quantum well
    structure)

temperature 391
  low 121

Hall technique 394
Hartree-Fock approach 17
Hedin-Lundqvist
  parametrization 18
heterojunction
  type-I (AlSb/GaSb) 96
  type-II staggered (AlSb/
    InAS) 96
  type-II "broken-gap" (GaSb/
    InAs) 96
heterostructure
  InAs/AlSb/GaSb (see InAs/
    AlSb/GaSb heterostructure)
  InAs/AlSb/GaSb (see InAs/
    AlSb/GaSb heterostructure)
    171
  type-II
    active region 146
    (AlGa)Sb/InAs (see
      (AlGa)Sb/InAs type-II
      heterostructure)
    antimonide-based 260
    InAs/(AlGa)Sb (see InAs/
      (AlGa)Sb type-II
      heterostructure)
    InAs/AlSb (see InAs/AlSb
      heterostructure)
high-resolution X-ray diffraction
  (HRXRD) 181
hole-like states 127

imaging theory
  non-linear 72
InAlAsSb 378
InAs
  buffer layer 82
  layer 123

InAs/(AlGa)Sb type-II
  heterostructure 99
InAs/AlSb
  double quantum well
    (see double quantum well)
  HFET 162
  heterostructure
    infrared photoluminescence 204
    type-II
      strong optical in-plane anisotropy in
      PL spectrum 212
  interface 42
InAs/AlSb/GaSb
  heterostructure 171
  interband-tunneling structure
    (see interband tunneling
    structures)
InAs/GaSb (InAs/AlSb)
  heterointerface 178
InSb layer
  Sb-terminated 87
InSb-like bonds 59
interaction
  carrier-interface 275
  electron-electron 293
  intersubband (see Γ–L
    intersubband interaction)
interband transition 203
  shift 223
interband-tunneling
  structure 203
    InAs/AlSb/GaSb 225
interface (see also local interface
  abruptness)
  bonds
    InSb-like 178
    GaAs-like 178
  epitaxial 85
  GaAs-like 178

INDEX

Sb-terminated 87
InAs/InSb (see InAs/InSb)
InAs-on-GaSb 69
microscopic symmetry 213
modes
  GaAs-like 172
  InSb-like 172
  Raman
    pure AlAs 121
interface-bond-related
 behavior 124
interfacial
  bonding
    control 62
    disorder 59
     assessment via high
      resolution TEM 79
    morphology 84
interpolation functions
 locally-defined 238
intersubband plasmon energies
 (see plasmon energies)
intersubband second-harmonic
 generator (see second-harmonic
 generator)
intersubband spectroscopy by
 inelastic light scattering 216
intersubband transition
 absorption coefficient 244
intraband absorption
  at heteroboundaries 286
  by electrons 287
  by holes 291
  mechanism of
    suppression 362
intraband transition
  matrix element of 288

$\mathbf{k} \cdot \mathbf{p}$ model 238
Kane matrix element 286

Kane multiband model 275
Kohn-Sham single-particle
 wave function
  LDA 18
Kohn-Sham energy
 functional 20
Kohn-Sham equations 14, 16
Kohn-Sham-Gàspàr exchange
 potential 18
Kramers-Kronig relations 245
Kronig-Penney model 470

laser
  diode
    design 444
    characteristics 444
  GaInAsSb quantum
   well 404
  GaInAsSb/AlGaAsSb 381
  infrared, middle-wavelength 141
    CW optically
     pumped 492, 493
    type-II 256
  interband cascade
    type-II 264
  non-injection 148
  quantum well
    emitting beyond
     $3\mu m$ 414
    GaInAsSb
     emitting in the $2\mu m$
      region 404
    InAsSb/InAlAsSb
     growth
      on GaSb
       substrate 402, 422
      on InAs
       substrate 415
    InGaAs/InGaAsP 113
    mid-infrared

threshold
  characteristics   331
type-I $InxGa_{1-x}As_ySb_{1-y}/Al_xGa_{1-x}As_ySb_{1-y}$
  heterostructure   142
type-II heterostructure   146
laser structures   475
  GaInSb/InAs-based   444
lattice
  harmonics   22
  mismatch   82, 124
  parameter   32
    III-V semiconductor
    173
light output vs. current
  output   455
liquid phase epitaxy (LPE)   11
local density approximation
  (LDA)   10, 17
local interface abruptness   63

Mach-Zender
  configurations   250
magnetoresistance
    low temperature   134
magnetotransport measurements   127
matrix element
  dipole   245
  Kane (see Kane matrix element)
  of absorption   289
  of Auger transition   295
  of electron-electron
    interaction   294
metal organic chemical vapor
  phase deposition (MOCVD)   8
metal organic vapor phase
  epitaxy (MOVPE)   12, 173, 266
microscopy (see cross-sectional scanning tunneling microscopy)

migration enhanced epitaxy
  (MEE)   63
modulator
  electro-optical
    antimonide-family   240
  Γ–L   244
  infrared electro-optical   239
molecular beam epitaxy
  (MBE)   8
momentum-space reservoir   252
Monkhorst-Pack scheme   24
Monte-Carlo methods
  quantum   18
muffin tin region   21
multi-slice
  approach   71
  theory   71

nodal variational principle   239
normal-incidence operator   249

optical anisotropy
  (see anisotropy)
optical confinement   446, 474, 504
optical transmission   178

perturbation
  approximation   383
perturbation theory
  first-order   293
  Luttinger-Kohn $\mathbf{k} \cdot \mathbf{p}$ 384
  many-body   18
phase shift   288
phonon doublets
  zone-folded LA   187
phonon spectra
  zone-folded LA   189
photoconductivity
  persistent
    low-temperature   140

negative 140
photocurrent spectroscopy
  (see superlattice)
photodiode (see also diode)
  p-i-n superlattice 225
photoluminescence emission
  free exciton 111
photoluminescence excitation
 spectroscopy
  Fourier transformation 488
photoluminescence spectra
  anomalous temperature
   behavior 151
photoluminescence spectroscopy
  low temperature 488
plasma contribution 248
plasmon
  energies
    intersubband 220
  modes
    coupled LO-
     phonon-intersubband 218
Poisson equation 22
Poisson ratio 104, 277
power efficiency 407
propagation length 248
pseudo-wave function 19
pseudopotential
  local 19
  non-local 19
  norm-conserving 19

quantum-confinement energies
  electron 470
  hole 470
quantum efficiency
  internal 342
quantum well heterostructure
  strained
    wave functions 275

carrier spectra 275
quantum well laser structure
  Auger engineering 356
quantum well structure
  GaInAsSb/AlGaAsSb
    growth 395
  InAsSb/InAlAsSb
    growth 400
  strained
    characterization 387
    MBE growth 387
quasi-particle approach 18

Raman
  scattering
    interface vibrational
     mode 117
  spectra 66
    room-temperature 193
  spectroscopy 63, 175, 187
recombination
  Auger (see Auger
   recombination)
  electron hole
    radiative 435
  rate 281
reflection high-energy electron
 diffraction (RHEED) 177, 389
reflectivity
  facet 385
refractive index 247
  AlGaAsSb lattice matched
   to GaSb 377
  averaged 494
  GaInAsSb lattice matched to
   GaSb 374
scanning tunneling microscopy
 (see cross-sectional scanning
 tunneling microscopy)
second-harmonic generator 250

coefficient 251
conversion efficiency 251
intersubband 251
L-valley 250
self-consistency cycle 24
semiconductor
  binary
    conduction band 26
    valence band 26
  III-V 173
single-effective-oscillator
  model 377
single-frequency operation 410
spectral structure
  derivative-like 200
spectroscopy (see deep-level transient spectroscopy; ellipsometric spectroscopy; far-infrared spectroscopy; intersubband spectroscopy by inelastic light scattering; optical spectroscopy; photocurrent spectroscopy; photoluminescence excitation spectroscopy; photoluminescence spectroscopy; Raman spectroscopy; X-ray photoelectron spectroscopy)
spin-orbit
  interaction 276, 301
  splitting 26
split-off energy Δ 373
stochastic processes 60
strain
  compressive 417
  distribution
    inhomogeneous 184
  effect
    on band structure 382
    on mid-infrared laser
      performance 381
  hydrostatic component 201
  inhomogeneous
    distribution 184
  relaxation 396
substrate
  GaSb (see GaSb)
  n-InAs 389
  terrace 89
  (100) n-GaSb 389
superlattice
  active layer
    GaInSb/InAs 440
  AlSb/GaSb short period
    strained-layer 110
  AlSb/InAs
    energy gap 442
  broken-gap
    GaInSb/InAs 452
  GaInSb/InAs
    intrinsic properties 435
  GaSb/InSb
    bulk binary
      constituents 23
  InAs/Ga$_{1-x}$In$_x$Sb 259
  InAs/GaSb
    effect of n-type doping
      on FMR spectra 200
    HRTEM contrast 71
    photocurrent spectroscopy 214
  InAs/InAs$_x$Sb$_{1-x}$ type-II
    (staggered alignment) 465
  InAs/InSb
    bulk binary
      constituents 23
  semiconductor
    small-period 58
  1×1 29
  Charge density distribution 40
  electronic levels 33
    ternary 30

structural properties 31
  3 × 3 44
    InAs/GaSb interface 48
    valence band offset 41
      GaSb/InSb 46
      InAs/InSb 47

transistor
  (AlGa)Sb/In(AsSb)
  based 158
transport characteristic
  room temperature 96
tunneling transport
  resonant 155

valence band offset (see band
  offset)
valence band anisotropy (see
  anistropy)

Vegard's rule 32

wavefunction engineering 238
Wigner-Seitz radius 18

X-ray diffraction (see also double
  crystal X-ray diffraction; high-
  resolution X-ray diffraction) 63,
  118, 443, 485
X-ray photoelectron
  spectroscopy 10

III-V compounds
  energy gaps vs. lattice
    parameters 98
8×8 **k·p** matrix 384